Regression with Social Data

Regression with Social Data

Modeling Continuous and Limited Response Variables

ALFRED DEMARIS

Bowling Green State University
Department of Sociology
Bowling Green, Ohio

WILEY-INTERSCIENCE

A JOHN WILEY & SONS, INC., PUBLICATION

Learning Resources
Centre

1269763X

Published by John Wiley & Sons, Inc., Hoboken, New Jersey.
Published simultaneously in Canada.

For general information on our other products and services please contact our Customer Care
Department within the U.S. at 877-762-2974, outside the U.S. at 317-572-3993 or fax 317-572-4002.

Wiley also publishes its books in a variety of electronic formats. Some content that appears in print,
however, may not be available in electronic format.

Library of Congress Cataloging-in-Publication Data:

DeMaris, Alfred, 1946–
    Regression with social data : modeling continuous and limited response variables / Alfred DeMaris.
       p. cm. — (Wiley series in probability and statistics)
    Includes bibliographical references and index.
    ISBN 0-471-22337-9 (cloth)
    1. Regression analysis. 2. Social sciences—Statistics—Methodology. 3.
Statistics—Methodology. I. Title. II. Series.
HA31.3.D46 2004
519.5'36—dc22   2004041183

Printed in the United States of America

10   9   8   7   6   5   4   3   2   1

To Gabrielle

Contents

Preface

Here is all the invisible world, caught, defined and calculated.
In these books the Devil stands stripped of all his brute disguises.
Here are all your familiar spirits—your incubi and succubi;
your witches that go by land, by air, and by sea;
your wizards of the night and of the day.

—Arthur Miller, The Crucible

My students often seem to regard statistics as only slightly removed from sorcery and witchcraft. Hence I begin with the words uttered by Reverend Hale in Arthur Miller's (1954) classic play. Like Hale's books, this one also promises to demystify the arcane—in this case, regression analysis.

Regression models, in some form or another, are ubiquitous in social data analysis. Although classic linear regression assumes a continuous dependent variable, later incarnations of the technique allowed the response to take on a variety of more limited forms: binary, multinomial, truncated, censored, strictly integer, and others. Increasingly, regression texts are incorporating some limited-dependent-variable techniques—typically, binary response models—along with classic linear regression in their coverage. However, other than in econometrics texts, it is rare to find regression models for the full spectrum of continuous and limited response variables treated in one volume. This monograph aims to provide just such a treatment.

In particular, the first six chapters of the book parallel the coverage of the typical monograph on linear regression: an introduction to regression modeling (Chapter 1), simple linear regression (Chapter 2), multiple linear regression (Chapter 3), regression with categorical predictors (Chapter 4), regression with nonlinear effects (Chapter 5), and finally, a consideration of advanced topics such as generalized least squares, omitted-variable bias, influence diagnostics, collinearity diagnostics, and alternatives to ordinary least squares for heavily collinear data (Chapter 6). The second half, however, considers models for dependent variables that are limited in one way or another. Examples of such data are event counts, categorical responses, truncated responses, or censored responses. The topic coverage in the second half of the book is therefore: binary response models (Chapter 7), multinomial response models (Chapter 8), censored and truncated regression (Chapter 9), regression models for count data (Chapter 10), an introduction to survival analysis (Chapter 11), and multistate, multiepisode, and interval-censored survival models (Chapter 12).

The book is intended both as a reference for data analysts working primarily with social data and as a graduate-level text for students in the social and behavioral sciences. As a text it is most suited to a two-course sequence in regression. As an example, I normally employ the material in Chapters 1 through 7 for a doctoral-level course on regression analysis. This course focuses primarily on linear regression but includes an introduction to binary response models. In a more advanced course on regression with limited dependent variables, I use Chapters 2 through 4 to review the multiple linear regression model, and then use Chapters 7 through 12 for the heart of the course. On the other hand, a survey of regressionlike models using the generalized linear model as the guiding framework might conceivably employ Chapters 1 through 5, and then 7 through 10. Other chapter combinations are also possible.

This book is not intended to be one's first exposure to regression. It is assumed that the reader has had a thorough introduction to probability theory, statistical inference, and applied bivariate statistics, along with an introduction to correlation and regression. Having covered the material in, say, Agresti and Finlay (1997) or Knoke et al. (2002), for example, would be good preparation for the current monograph. The basics of probability and statistical inference are nevertheless reviewed in the appendix to Chapter 1 in case the reader needs to refresh his or her understanding of these topics. It is also assumed that the reader has a solid grasp of college-level algebra. Beyond these requirements, no specialized mathematical or statistical skills are required. Some differential calculus is employed here and there in the exposition, and a smattering of matrix algebra appears—primarily in Chapter 6. Those unfamiliar with these topics will find a fairly thorough discussion of them in Appendix A. This collection of math tutorials also discusses basic algebra, summation notation, functions, and covariance algebra. These tutorials are self-contained sections that can be referred to as necessary during the course of reading through the book.

The book's emphasis tends to be on the estimation, interpretation, and evaluation of theoretically driven models in the social sciences. Due to the variety of regression models considered, coverage of specific techniques (e.g., linear regression) is necessarily more selective than found in books devoted entirely to one type of model. In particular, I have avoided discussion of exploratory model-building techniques, such as stepwise regression, along with the extensive examination of model residuals. Readers interested in these topics can find ample coverage in other works. Instead, the focus is on the substantive and statistical plausibility of models, the correct interpretation of model parameters, the global evaluation of model adequacy, and a variety of inferential procedures of interest to those working with social data. As maximum likelihood estimation is central to the models considered in Chapters 7 through 12, in the second half of the book considerable emphasis is placed on the expression for the likelihood function. This allows the reader to see how models are estimated, since once the function is written, algorithms for parameter estimation are readily available.

My writing style is the product of an attempt to marry rigor with accessibility. Rigor comes in the form of mathematical development in places where it is necessary for conveying a deeper level of understanding. Accessibility is achieved (hopefully) by providing enough steps so that the math is clear, and by explaining the steps "in English" whenever possible. It is also hoped that the reader with more modest math skills will invest a little time and energy in the math tutorials in Appendix A. These are designed to give the reader the tools needed to at least follow the mathematical expositions in the text. As someone who developed mathematics skills rather late in life, I appreciate the trepidation with which some readers approach mathematical explication. Nonetheless, a complete understanding of this material is not possible without some math. Ideally, the returns to the reader in terms of statistical comprehension will be worth the effort.

A number of resources are available to help readers assimilate the material in the book. First, there are approximately 275 end-of-chapter exercises in Chapters 2 through 12, plus another 63 in Appendix A. The *Instructor's Manual* that accompanies the book contains complete solutions to all the exercises. Additionally, 10 datasets are available so that readers can practice the techniques taught herein using their favorite regression software. The datasets are incorporated into several of the end-of-chapter exercises. The datasets can be downloaded through the Wiley Web site, as discussed in Chapter 1 (see the section "Datasets Used in This Volume" in Chapter 1 for further information).

Acknowledgments

Many people have contributed in one way or another to the production of this work. First, I would like to thank the following statisticians for reading and commenting on preliminary book chapters: Alan Agresti, Kenneth A. Bollen, Nancy Boudreau, William H. Greene, David W. Hosmer, and James A. Sullivan. Most of these professors are people with whom I have had little or no prior connection, but who were simply gracious and collegial enough to take the time to help me produce a better product. If there are flaws in this work, it is undoubtedly due to my failure to take their sage advice. I also wish to thank Bowling Green State University for providing the resources—in particular, time and computing support—that have made it possible to complete this project in a timely fashion. Thanks are also due to my colleagues in the Department of Sociology at Bowling Green State University for being supportive of this project and for politely excusing (or at least not complaining about) my absence at department colloquia and other functions while at work on this project. Additionally, I wish to express my appreciation to Steve Quigley and the production staff at John Wiley & Sons for their professionalism as well as their encouragement of this monograph. Finally, I wish to give sincere thanks to my lovely wife, Gabrielle, for her unfailing love and support throughout the writing of this work.

CHAPTER 1

Introduction to Regression Modeling

The last several decades in the social sciences have been characterized by the increasing use of mathematical models of social behavior. The ready availability of quantitative data on social phenomena, generated by large-scale social surveys, is certainly a contributing factor in this development. Although models for social data vary widely in complexity and sophistication, most can be considered to be variants of the technique known as *linear regression*. Classic linear regression, however, was predicated on the notion that the outcome variable being modeled was continuous in nature. Many outcomes of interest, on the other hand, are limited in their measurement in some way or another. In this monograph, I define a *limited response variable* to be any outcome that is not continuous—or approximately continuous—throughout its logical range. Such measures include a continuous response that is truncated or censored, one that is categorical, and one that represents a count of some phenomenon. Also included under this definition are measures of survival time in a given state, as this type of response is also typically characterized by restrictions imposed by censoring and/or truncation. Linear regression has been extended over the years to the modeling of limited dependent variables, via the *generalized linear model*, discussed below. The purpose of this book, therefore, is to present an integrated treatment of regression modeling that weaves seamlessly through the various metrics that the response variable can take. By collecting a variety of seemingly disparate techniques under the regression umbrella, this book will hopefully render these methods easier to assimilate.

CHAPTER OVERVIEW

In this chapter I introduce the concept of a statistical model: in particular, a linear regression model. It turns out that linear regression models are special cases of what

Regression with Social Data: Modeling Continuous and Limited Response Variables,
By Alfred DeMaris
ISBN 0-471-22337-9   Copyright © 2004 John Wiley & Sons, Inc.

is referred to as the *generalized linear model* (Gill, 2001; McCullagh and Nelder, 1989; Nelder and Wedderburn, 1972), which subsumes all the models discussed in this book. The important components of such a model are therefore sketched out in this chapter to foreshadow what is to follow in subsequent chapters. I then outline three major components of model evaluation, which are considered throughout the book for assessing model adequacy. Next, I consider the role of regression models in causal inference. Whether or not acknowledged explicitly, regression modeling in the social and behavioral sciences is frequently designed to illustrate causal dynamics. I therefore devote some space to a discussion of recent developments in, and controversies pertaining to, the use of regression models for causal inference. The chapter concludes with a description of the data sets used for this volume, some of which the reader may download to practice the techniques taught herein. Finally, the chapter appendix contains a review of important statistical principles relied on throughout the volume.

MATHEMATICAL AND STATISTICAL MODELS

In the social and behavioral sciences, a model is often a *set of one or more equations describing the processes that generated the observations on one or more response variables.* I use the term *generated* here in a causal sense, since that is what is typically implied in researchers' models, as well as the language used to describe them. (I shall have more to say about causal language shortly.) When coupled with a set of assumptions about the manner in which observations were *sampled* from a larger *population*, it becomes a *statistical model.* Like many "models" of real-world phenomena, such models are not to be taken too literally. As others have observed, "All models are wrong. Some are useful" [attributed to George Box in Gill (2001, p. 3)]. Nonetheless, to the extent that a model provides a broad outline of the dynamics underlying behavioral phenomena, it can be useful for advancing knowledge.

Linear Regression Models

A linear regression model is an *equation* in which a random response, or outcome, variable Y, is posited to be a *linear function* of a set of input, or explanatory variables, denoted $X_1, X_2, \ldots$. (These labels are, of course, purely arbitrary. The outcome could just as well be denoted W, U, or η, and the explanatory variables—also called *regressors*—could be labeled V, Z, or ξ.) To give this discussion substantive flesh from the start, suppose that the "population" of interest is the population of all persons over 18 years of age in the United States in 1998. Suppose further that Y is a continuous measure of *attitude toward abortion*, with a higher score indicating a more liberal, or unrestrictive, attitude. And let's say that X_1 is *marital status*, where "married" is coded 1, and "any other status" is coded 0. (Called *dummy variables* these types of variables are explored in detail in Chapter 4.) Additionally, say that X_2 is *education*, coded from 0 for "no formal schooling" to 20 for "four or more years of graduate study." A regression model for *attitude toward abortion* for the ith observation sampled from the population based on these two regressors takes the form

$$Y_i = \beta_0 + \beta_1 X_{i1} + \beta_2 X_{i2} + \varepsilon_i. \tag{1.1}$$

This is a linear equation, in the sense that Y is defined to be a weighted sum of constants times explanatory variables (see Sections I and II of Appendix A for definitions of functions, linear functions, and weighted sums). But—you might object—there's no variable multiplied by β_0 and no constant multiplying ε_i. Well, both are actually present. The "variable" corresponding to β_0 is X_0, which equals 1 for all cases. This factor is, therefore, easily omitted from the equation. The constant multiplier of ε_i is simply assumed to be 1. Hence, this multiplier can also be omitted. The β's—β_0, β_1, β_2—are the *parameters* of the equation: They are assumed to take on constant values for each person in the population. The last term, ε, is an equation *disturbance*, or error term. It is a random variable that represents all factors affecting Y other than X_1 and X_2. Both the parameters and ε are unobserved in any given sample. That is, even though we can observe the values of Y and the X's for any sample of n cases from the population, we cannot observe either the parameters or the error term. These factors, however, can be estimated with the sample data. In fact, the major purpose of regression modeling is to estimate the β's and to use these to describe the relationship between Y and the X's in the population, as well as to make predictions about the value of Y for cases with particular combinations of values of the X's.

Model (1.1) is for individual observations. The model for the *expected value*, or mean, or arithmetic average, of Y in the population, conditional on the X's, is instead simply

$$E(Y_i | X_{i1}, X_{i2}) = \mu_i = \beta_0 + \beta_1 X_{i1} + \beta_2 X_{i2}. \tag{1.2}$$

The β's quantify the manner in which the mean of Y is related to the explanatory variables in the model. In particular, β_1 indicates the expected, or average, difference in Y in the population for those who are 1 unit apart in marital status—that is, for marrieds versus everybody else in our substantive example. β_2 indicates the expected difference in Y in the population for those who have a year's difference in formal schooling. So, for example, in the prediction of one's attitude toward abortion, if β_1 is -1.5 and β_2 is 2.3, these would be interpreted as follows: Married persons' attitude toward abortion, on average, is 1.5 units lower than others', holding education constant. (The precise meaning of "holding other variables constant" will be taken up in subsequent chapters.) Those with a year's more formal schooling, on average, are 2.3 units higher on attitude toward abortion than others, holding marital status constant. Furthermore, if β_0 is 7.5, a married person with a college degree is estimated to have mean abortion attitude equal to $7.5 - 1.5(1) + 2.3(16) = 42.8$.

This "model" of attitude toward abortion is certainly an oversimplification of the set of factors associated with such attitudes. But it is parsimonious, and its adequacy in accounting for variation in attitude toward abortion can be evaluated (more about this later). To estimate the β's with sample data employing the most common technique—ordinary least squares (OLS)—we make some additional assumptions about the equation errors. First, we assume that they are uncorrelated with one another. That is, there is no tendency for a large error for the first observation, say, to presage a larger or smaller error for the second observation than would occur by chance. If sampling is random and the data are cross-sectional rather than longitudinal, this assumption is usually pretty safe. Second, we assume that they have a mean of zero at each *covariate pattern*, or combination of predictor values. As an example, being married and having 16 years of education is one covariate pattern; being other-than-married with 12 years of education is another covariate pattern; and so on. Hence, this assumption is that the mean of the errors at any covariate pattern is zero. Finally, we assume that the variance of the error terms is the same at each covariate pattern. Given a random sample of n persons from the population, along with their measures on Y, X_1, and X_2, we can proceed with an estimation of this equation and employ it to further our understanding of abortion attitudes.

Generalized Linear Model

A linear regression model is a special case of the *generalized linear model* (GLM). A generalized linear model is a *linear model for a transformed mean of a response variable whose probability distribution is a member of the exponential family* (Agresti, 2002). What does this mean? Well, for starters, let's apply this definition to the regression model delineated in equation (1.2) and corresponding assumptions above. The quantity μ_i in equation (1.2) is referred to as the *conditional mean* of the response variable. It is the mean of the Y_i conditional on a particular covariate pattern. (The ε_i are, moreover, more properly called the *conditional errors*—the errors, at each covariate pattern, in predicting the individual Y_i using the conditional mean.) The model is therefore a model for the *mean* of the response variable. It is also for the *transformed mean* of Y, although the transformation employed here is the *identity* transformation, which is "transparent" to us. That is, if $g(\mu_i)$ indicates a transformation of the mean using the function $g(\cdot)$, then $g(\mu_i)$ in the classic regression model is just μ_i. Also, in the classic regression model, it is assumed that the errors are *normally* distributed. (This assumption is not essential if n is large, however.) Because Y is a linear combination of the regressors plus the error term, and assuming that the regressor values are fixed, or held constant, over repeated sampling, Y is also normally distributed. The normal distribution is a member of the *exponential family* of probability distributions.

   Essentially, there are three components that specify a generalized linear model. First, the *random* component identifies the response variable, Y, its mean, μ, and its probability distribution. Second, the *systematic* component specifies a set of explanatory variables used in a linear function to predict the transformed mean of the response variable. The systematic component, referred to as the *linear predictor* (Agresti, 2002), has the form $\sum_{k=0}^{K} \beta_k X_{ik}$ for the ith case, where the X's are the explanatory variables and the β's are the parameters representing the variables' "effects" on the mean of the response. In the example of attitude toward abortion, $\sum_{k=0}^{K} \beta_k X_{ik}$ is just $\beta_0 + \beta_1 X_{i1} + \beta_2 X_{i2}$. Third, the *link function*, $g(\mu)$, specifies the transformation function for the mean of Y, which the model equates to the systematic component.

The linear regression model is especially simple because the response variable is continuous—at least theoretically—and the link function is the *identity link*. That is, $g(\mu) = \mu$, and hence the regression model is $\mu_i = E(Y_i) = \sum_{k=0}^{K} \beta_k X_{ik}$, as we saw in equation (1.2). An important characteristic about this equation is that the left- and right-hand sides are equally unrestricted. That is, if Y is continuous, its theoretical range is from minus to plus infinity, which implies a similar range for μ. The right-hand side is also free to take on any values in that range, since there are no restrictions on either the parameters or the values of the predictors. However, later in this book we consider other regressionlike models in which the response variable is either binary, nonnegative discrete, or otherwise limited in its range. The link function is therefore designed to ensure that the response is converted into an unrestricted form, to match the unrestricted nature of the linear predictor. Let's consider how the GLM framework extends to those situations.

First, we need to describe the exponential family of density functions. (Readers unfamiliar with the concept of a density function may want to review that material in the chapter appendix.) A density is a member of the exponential family if it can be written in the form

$$f(y|\mu) = a(\mu)\,b(y)e^{yg(\mu)}, \tag{1.3}$$

where, as before, μ is the mean of Y, $a(\mu)$ is a function involving only μ, and perhaps constants, and $b(y)$ is a function involving only Y, and perhaps constants (Agresti, 2002). Once the density is written in this form, the link function that equates the mean of Y to the linear combination of explanatory variables is $g(\mu)$. As an example, suppose that the response variable, Y, is binary, taking on values 1 if a person has had sexual intercourse any time in the preceding month, and 0 otherwise. Suppose further that we are interested in modeling having had sexual intercourse in the preceding month as a function of several predictors, such as *marital status, education, age, religiosity*, and so on. Such a response variable is said to have the *Bernoulli distribution* with parameter π, and its density function (see the chapter appendix) is

$$f(y|\pi) = \pi^y (1-\pi)^{1-y}.$$

For binary Y, $E(Y) = \pi$, so π is the mean of the response in this case. Now, since

$$\pi^y(1-\pi)^{1-y} = \pi^y(1-\pi)(1-\pi)^{-y}$$

$$= \pi^y \frac{1-\pi}{(1-\pi)^y}$$

$$= (1-\pi)\left(\frac{\pi}{1-\pi}\right)^y$$

$$= (1-\pi)e^{y\ln[\pi/(1-\pi)]}$$

we see that the Bernoulli density is a member of the exponential family, with $a(\mu) = (1 - \pi)$, $b(y) = 1$, and $g(\mu) = \ln[\pi/(1 - \pi)]$. Thus, $\ln[\pi/(1 - \pi)]$ is the link function for this model, and the model for the transformed mean becomes

$$\ln \frac{\pi_i}{1 - \pi_i} = \sum_{k=0}^{K} \beta_k X_{ik}.$$

This type of model is called a *logistic regression* model. Notice that since π ranges from 0 to 1, $\pi/(1 - \pi)$ ranges from 0 to infinity, and therefore $\ln[\pi_i/(1 - \pi_i)]$ ranges from minus to plus infinity. The left-hand side of this model is thus an unrestricted response, just as in the case of linear regression.

As a second example, suppose that the response on *sexual frequency* really is recorded in terms of the number of separate acts of sexual intercourse that the person has engaged in during the preceding month. This type of outcome is referred to as a *count variable*, since it represents a count of events. It is a discrete variable whose distribution is likely to be very right-skewed. We may want to utilize this information to inform the regression. One appropriate density for this type of variable is the *Poisson density*. Hence, if Y takes on values 0, 1, 2, . . . and $\mu > 0$, the Poisson density is

$$f(y \mid \mu) = \frac{e^{-\mu} \mu^y}{y!}.$$

To see that this is a member of the exponential family, we rewrite this density as

$$\frac{e^{-\mu} \mu^y}{y!} = e^{-\mu} \frac{1}{y!} e^{y \ln \mu},$$

where $a(\mu) = e^{-\mu}$, $b(y) = 1/y!$, and $g(\mu) = \ln \mu$. Therefore, $\ln \mu$ is the link function, and the model for the transformed mean becomes

$$\ln \mu = \sum_{k=0}^{K} \beta_k X_{ik}.$$

This model is referred to as a *Poisson regression model*. Here, in that μ ranges from 0 to infinity, $\ln \mu$ ranges from minus to plus infinity. Once again, the left-hand side of the model is an unrestricted response.

The advantage to the GLM approach is that the link function connects the linear predictor, $\sum_{k=0}^{K} \beta_k X_{ik}$, to the mean of the response variable rather than to the response variable itself, so that the outcome can now take on a variety of nonnormal forms. As Gill (2001, p. 31) states: "The link function connects the stochastic [i.e., random] component which describes some response variable from a wide variety of forms to all of the standard normal theory supporting the systematic component through the mean function, $g(\mu)$ . . . ." Once we assume a particular density function for Y, we can then employ maximum likelihood estimation (see the chapter appendix for an explanation of the maximum likelihood technique) to estimate the parameters of the model. For the classic linear regression model with normally distributed errors (and thus a normally distributed response), maximum likelihood (ML) and ordinary least squares (OLS) estimation are equivalent (OLS estimation is covered in Chapter 2).

Model Evaluation

Models in the social sciences are useful only to the extent that they effectively encapsulate real-world processes. In this section we therefore consider ways of evaluating model adequacy. The assessment of a model encompasses three major evaluative dimensions. The first dimension is *empirical consistency*, or as many call it, *goodness of fit*. *A model is empirically consistent if the response variable behaves the way the model says that it should.* In other words, a model is empirically consistent to the extent that the response variable behaves in accordance with model assumptions and follows the pattern dictated by the model's structure. Moreover, if the model's predictions for Y match the actual Y values quite closely, the model is empirically consistent. The second dimension is *discriminatory power*, which is *the extent to which the structural part of the model is able to separate, or discriminate, different cases' scores on the response from one another.* Since separation, or dispersion, constitutes variability in the response, discriminatory power is typically assessed by examining how much of the variability in the response is due to the structural part of the model. The third dimension is *authenticity*, also called *model-reality consistency* by Bollen (1989). *A model is authentic to the extent that it mirrors the true processes that generated the response.*

To illustrate the differences in these dimensions, I draw on a particular variant of regression modeling called a *path model*, essentially a model for a causal system in which one or more response variables is a function of a set of predictors. A path model is an example of what is referred to as a *covariance structure model* or *structural equation model* [see DeMaris (2002a) or Long (1983) for an introduction to such models]. In this type of model, the goal is to account for the correlations (or covariances) among the variables in the system, using the structural coefficients of the model. For example, suppose that we have three continuous, standardized variables measured for a random sample of married adult respondents: Z_1 is the the degree of physical aggression in the respondent's marriage in the past year, Z_2 is the frequency of verbal disagreements in the respondent's marriage in the past year, and Z_3 is the frequency of verbal disagreements in the respondent's parents' marriage when the respondent was a teenager. The sample correlations among these variables are $\mathrm{corr}(Z_1,Z_2) = .45$, $\mathrm{corr}(Z_1,Z_3) = .6125$, and $\mathrm{corr}(Z_2,Z_3) = .2756$. In path analysis, these correlations are the observations that are to be accounted for by the model.

A path model can be specified using either a diagram or a series of equations. Using the latter approach, suppose that a researcher arrives at the following OLS sample estimates for a simple path model for Z_1, Z_2, and Z_3:

$$Z_2 = .45(Z_1) + e_2,$$
$$Z_3 = .5(Z_1) + .25(Z_2) + e_3. \tag{1.4}$$

The model suggests that the frequency of verbal disagreements in the respondent's marriage in the past year is a function of the degree of physical aggression in the respondent's marriage in the past year, plus a random error term (e_2). It also maintains that the frequency of verbal disagreements in the respondent's parents' marriage when the respondent was a teenager is a function of the degree of physical aggression in the respondent's marriage in the past year and the frequency of verbal disagreements in the respondent's marriage in the past year, plus a random error term (e_3). (Okay, this doesn't make much substantive sense, but *that* will be the point, as the reader can see below.) It can (and, in fact, will) be shown that the sample correlations among Z_1, Z_2, and Z_3 are functions of the model's estimated parameters. The total number of "observations" in path analysis consists of the number of nonredundant correlations among the variables in the system. In the present example, that number is three. There are also three parameters in the system: the three coefficients. Whenever the number of correlations is the same as the number of parameters in the system of equations, the model is *saturated*, or *just-identified*. In this case, the structural parameters will reproduce perfectly the correlations among the variables. When there are fewer parameters than correlations to explain, the model is *overidentified*. In that case, the model is a more parsimonious description of the correlations. The model will no longer perfectly reproduce the correlations. But we can assess how *closely* the model's parameters will reproduce the correlations in order to gauge its performance in "fitting" the data.

Let's see how the correlations can be shown to be functions of the structural parameters of the model. (Those unfamiliar with covariance algebra may want to read Section III of Appendix A before continuing.) First, note that since the variables are standardized, their covariances are also their correlations. Thus, $\text{corr}(Z_1,Z_2) = \text{cov}(Z_1,Z_2) = \text{cov}(Z_1, .45Z_1 + e_2) = .45\,\text{Cov}(Z_1,Z_1) + \text{cov}(Z_1,e_2) = .45$ (since the covariance of a variable with itself is its variance, which for standardized variables equals 1, and the covariance between OLS residuals and regressors in the same equation is zero). Moreover, $\text{corr}(Z_1,Z_3) = \text{cov}(Z_1, .5Z_1 + .25Z_2 + e_3) = .5v(Z_1) + .25\,\text{cov}(Z_1,Z_2) = .6125$; and $\text{corr}(Z_2,Z_3) = \text{cov}(.45Z_1 + e_2, .5Z_1 + .25Z_2 + e_3) = .45(.5)\,v(Z_1) + .45(.25)\,\text{cov}(Z_1,Z_2) = .2756$. (Note that OLS residuals in different equations are uncorrelated with each other.) We see that the correlations are reproduced exactly from the model parameters, because the model is saturated.

The structural coefficients also allow us to determine how much the model accounts for variation in the response variables. The part of the variance of a response variable that is accounted for by the model can be determined by considering the overall variance of each response. Recalling that the variance of a standardized variable is 1, the variance in Z_2 can be decomposed into the proportion due to the structural part of the model and the proportion due to error. Thus, we have $1 = v(Z_2) = \text{cov}(Z_2,Z_2) = \text{cov}(.45Z_1 + e_2, .45Z_1 + e_2) = .45^2\,v(Z_1) + v(e_2) = .2025 + v(e_2)$. That is, 20.25% of the variation in Z_2 is due to the structural (as opposed to the random) part of the model. Similarly, $1 = v(Z_3) = \text{cov}(.5Z_1 + .25Z_2 + e_3, .5Z_1 + .25Z_2 + e_3) = (.5)(.5)\,v(Z_1) + (.5)(.25)\,\text{cov}(Z_1,Z_2) + (.5)(.25)\,\text{cov}(Z_1,Z_2) + (.25)(.25)\,v(Z_2) + v(e_3) = .5^2 + (2)(.5)(.25)(.45) + .25^2 + v(e_3) = .425 + v(e_3)$. Here we see that 42.5% of the variation in Z_3 is due to the model.

At this point, let's consider the three aspects of model evaluation. First, notice that the model is *perfectly* empirically consistent, since the data—the correlations—"behave" exactly the way the model says they should; they are predicted perfectly by the model. Discriminatory power, on the other hand, is only moderate; at most, 42.5% of the variation in any response variable is accounted for by the model. Another way of saying this is that we experience, at most, only a 42.5% improvement in the discrimination of scores on the response variable when using—as opposed to ignoring—the model, in predicting the responses. Finally, however, the model is completely *inauthentic*, in a causal sense. To begin, the frequency of verbal disagreements in the respondent's parents' marriage when respondents were teenagers cannot possibly be caused by the subsequent tenor of respondents' marriages. Additionally, physical aggression tends to be preceded by verbal conflict rather than the converse. It is therefore unreasonable to suggest that it is physical aggression that leads to verbal conflict. If anything, the occurrence of physical aggression should suppress the frequency of subsequent verbal altercations, since partners would be fearful of a reoccurrence of violence. From the foregoing it should be clear that empirical consistency, discriminatory power, and authenticity are three separate although related criteria by which models can be evaluated.

REGRESSION MODELS AND CAUSAL INFERENCE

Regression modeling of nonexperimental data for the purpose of making causal inferences is ubiquitous in the social sciences. Sample regression coefficients are typically thought of as estimates of the causal impacts of explanatory variables on the outcome. Even though researchers may not acknowledge this explicitly, their use of such language as *impact* or *effect* to describe a coefficient value often suggests a causal interpretation. This practice is fraught with controversy [see, e.g., McKim and Turner (1997) as well as the November 1998 and August 2001 issues of *Sociological Methods & Research* for recent debates on this topic in sociology]. In this section of the chapter I explore the controversy and provide some recommendations.

What Is a Cause?

Philosophers and others have debated the definition of *cause* for centuries without ever coming to complete agreement on it. However, current common use of the term implies that the application of a cause to some element changes its state or trajectory, compared to what that would be without application of the cause. Beyond this basic idea, however, there appear to be two primary "models" of causality in operation among social scientists. The *regression* or *structural equation modeling* perspective is that *a variable X is a cause of Y if, all else equal, a change in X is followed by a change in Y* (Bollen, 1989). The implicit assumption is that a cause is synonymous with an *intervention*, which, when applied, changes the nature of the outcome, on average. With nonexperimental data, the intervention has been executed by nature. Nonetheless, the implication is that if X is truly a cause of Y, changing its value should change Y for the cases involved, compared to what its value would be were X left unchanged. Should this reasoning be applied to equation (1.1), β_2 would be described as individuals' average change in attitude toward abortion were we to increase their schooling by one year.

A somewhat different perspective is encompassed by what is referred to as the *potential response model* of causality (Pearl, 1998), attributed to Rubin (1974), and therefore also referred to as the *Rubin model*. This viewpoint entails a counterfactual, or contrary-to-fact, requirement for causality: *X is a cause of Y if the value of Y is different in the presence of X from what it would have been in the absence of X (or under a different value for X)*. Although this sounds quite similar to the notion of intervention articulated above, there are some subtle differences. First, let's consider the potential response model more formally. Suppose that X represents a treatment with two values: t for the treatment itself and c for the absence of treatment. Define Y_t as the score on a response, Y, for the ith case if the case had been exposed to t, and Y_c as the response for the same case if that case had *instead* been exposed to c. Then the *true causal effect* of X on Y for the ith case is $Y_t - Y_c$. Notice that this definition of cause is counterfactual, since the ith case can be "freshly" exposed to either t or c but not to both. Repeated application of c followed by t is not considered equivalent. Similarly, the *average causal effect* for some population of cases is the average of all true causal effects for all cases. That is, the average causal effect is $E(Y_t - Y_c)$ over the population of cases. Neither the true causal effect nor the average causal effect can ever be observed, in practice. Notice the difference between this model and the intervention approach to causality discussed above. An intervention is an observable operation. What's more, it is indifferent to the case's prior history: We can *change* the case's value from c to t and observe what happens, on average, to Y. The potential response model, in contrast, defines causality in a way that is impossible to observe, since the values Y_t and Y_c presume that the case's history has been magically "erased" in each case before a particular level of X is applied.

Nonetheless, according to the potential response model, the average causal effect can be estimated in an unbiased fashion if there is random assignment to the cause. Unfortunately, this pretty much rules out making causal inferences from nonexperimental data. However, others acknowledge the possibility of making the assumption of "conditional random assignment" to the cause in observational data, provided that this assumption is theoretically tenable (Sobel, 1998). Still, hard-core adherents to the potential response framework would deny the causal status of most of the interesting variables in the social sciences because they are not capable of being assigned randomly. Holland and Rubin, for example, have made up a motto that expresses this quite succinctly: "No causation without manipulation" (Holland, 1986, p. 959). In other words, only "treatments" that can be assigned randomly to any case at will are considered candidates for exhibiting causal effects. All other attributes of cases, such as gender and race, cannot be causes from this perspective. I agree with others (e.g., Bollen, 1989) who take exception to this restrictive conception of causality, despite the intuitive appeal of counterfactual reasoning. Regardless of whether it can be randomly assigned, any attribute that exposes one to differential treatment by one's environment ought to be considered causal.

When Does a Regression Coefficient Have a Causal Interpretation?

Assuming that we could agree on the definition of a cause, perhaps a more pressing question is: When can a regression coefficient be given a causal interpretation? With nonexperimental data, of course, random assignment to the cause is not possible. In lieu of this, several scholars insist that a fundamental requirement for a causal interpretation to be given to the sample estimate of β in $Y = \beta X + \varepsilon$ is that $\text{Cov}(X,\varepsilon) = 0$, or that the equation disturbance, ε, is uncorrelated with the causal variable. This has been referred to variously as the *pseudoisolation assumption* (Bollen 1989), the *causal assumption* (Clogg and Haritou, 1997), or the *orthogonality condition* (Pearl, 1998). Let us see why this important condition is necessary to causal inferences. Suppose, indeed, that you wish to estimate the model $Y = \beta X + \varepsilon$ using sample data and you believe that the association of X with Y is causal, that is, X causes Y. Suppose, however, that, in truth, a latent variable, ξ, affects both X and Y. Hence, the true model is $X = \gamma_1 \xi + \upsilon$, with $\text{Cov}(\xi,\upsilon) = 0$, and $Y = \beta X + \gamma_2 \xi + \varepsilon'$, where $\text{Cov}(X,\varepsilon') = \text{Cov}(\xi,\varepsilon') = 0$. [We assume that all variables are centered (i.e., deviated from their means), obviating the need for intercept terms.] Notice, then, that ε is really equal to $\gamma_2 \xi + \varepsilon'$. Also, note that $\text{Cov}(X,\xi) = \text{Cov}(\xi, \gamma_1 \xi + \upsilon) = \gamma_1 V(\xi)$. Thus, $\text{Cov}(X,\varepsilon) = \text{Cov}(X, \gamma_2 \xi + \varepsilon') = \gamma_2 \text{Cov}(X,\xi) = \gamma_1 \gamma_2 V(\xi)$. So if $\text{Cov}(X,\varepsilon)$ is zero, this ensures that one or all of γ_1, γ_2, and $V(\xi)$ equal zero; and this means either that ξ is a constant for every case, in which case it has no real influence on X or Y, or that ξ has no influence on X, or that ξ has no influence on Y. In any of these cases, b from the sample regression is a consistent estimator of β (see the chapter appendix for a discussion of consistency). Otherwise, the sample estimator of β is

$$b = \frac{\text{cov}(X,Y)}{v(X)}$$

and the probability limit of b is

$$\text{plim } b = \frac{\text{plim cov}(X,Y)}{\text{plim } v(X)} \text{ (by the Slutsky theorem), which} = \frac{\text{Cov}(X,Y)}{\sigma_x^2}$$

(since sample estimators of variance and covariance—denoted by lowercase "cov" and "v"—are consistent for their population counterparts—denoted by uppercase "Cov" and "V"), where σ_x^2 denotes the population variance of X and

$$\frac{\text{Cov}(X,Y)}{\sigma_x^2} = \frac{\text{Cov}(X, \beta X + \gamma_2 \xi + \varepsilon')}{\sigma_x^2}$$

$$= \frac{\beta \sigma_x^2 + \gamma_2 \text{Cov}(X, \xi)}{\sigma_x^2}$$

$$= \beta + \frac{\gamma_2 \gamma_1 V(\xi)}{\sigma_x^2}.$$

Hence, b is consistent for $\beta + \gamma_2\gamma_1\, V(\xi)/\sigma_x^2$, which is, in general, not the same as β. In fact, if β in the true model is really zero, the value of b may mistakenly attribute the impact of ξ on X, represented by γ_1, and the impact of ξ on Y, represented by γ_2, to a causal effect of X on Y. For this reason, the orthogonality condition is necessary for attributing a causal interpretation to b.

Unfortunately, to assume that the orthogonality condition holds is a great leap of faith. Clogg and Haritou (1997) point out that there is no statistical technique, using the data under scrutiny, for determining whether or not the orthogonality condition obtains. So in practice, researchers often add one or more control variables to the model, inferring that the estimate of X's effect in the model with the "proper variables" controlled is unbiased for the "causal effect." In the words of Clogg and Haritou (1997, p. 84): "Partial regression coefficients or analogous quantities are assumed to be the same as causal effects when the right controls (additional predictors) are included in the model." However, adding variables that are *not* causes of Y to the equation can lead to a failure of the orthogonality condition in the expanded model. This can then result in what Clogg and Haritou (1997) call *included-variable bias*. That is, the estimate of X's effect in the expanded model is biased for the causal effect, due to inclusion of an extraneous variable.

Let's see how this works. Suppose that the true causal model for Y is $Y = \beta X + \varepsilon$ and that the orthogonality condition, $\mathrm{Cov}(X,\varepsilon) = 0$, holds. But you estimate $Y = \beta X + \gamma Z + \upsilon$, where Z is a "predictor" of Y but not a causal influence (e.g., as weight is a predictor of height). For this equation to be valid for causal inference, the necessary causal assumption is $\mathrm{Cov}(X,\upsilon) = \mathrm{Cov}(Z,\upsilon) = 0$. Now ε is actually $\gamma Z + \upsilon$ (the disturbance always contains all predictors of Y that are left out of the current equation). So, since $\mathrm{Cov}(X,\varepsilon) = 0$, we have that $\mathrm{Cov}(X,\,\gamma Z + \upsilon) = \gamma\mathrm{Cov}(X,Z) + \mathrm{Cov}(X,\upsilon) = 0$, or that $\mathrm{Cov}(X,\upsilon) = -\gamma\mathrm{Cov}(X,Z)$. Provided that neither γ nor $\mathrm{Cov}(X,Z)$ is zero, the orthogonality condition fails for the estimated model. Hence, the estimate of β from that model is biased for the true causal effect.

Recommendations

In light of the foregoing considerations, one might ask whether we should abandon causal language altogether when dealing with nonexperimental data, as has been suggested by some scholars (e.g., Sobel, 1998). Freedman (1997a,b) is especially critical of drawing causal inferences from observational data, since all that can be "discovered," regardless of the statistical candlepower used, is association. Causation has to be assumed into the structure from the beginning. Or, as Freedman (1997b, p. 182) says: "If you want to pull a [causal] rabbit out of the hat, you have to put a rabbit into the hat." In my view, this point is well taken; but it does not preclude using regression for causal inference. What it means, instead, is that *prior knowledge of the causal status of one's regressors* is a prerequisite for endowing regression coefficients with a causal interpretation, as acknowledged by Pearl (1998). That is, concluding that, say, $\beta \neq 0$ in the equation $Y = \beta X + \varepsilon$ doesn't *demonstrate* that X is a cause of Y. But if X *is* a cause of Y, we should find that β is nonzero in this equation, assuming that all relevant confounds have been controlled. That is, a nonzero β is at least *consistent* with a causal effect of X on Y. It remains for us to marshal theoretical and/or additional empirical—preferably experimental—grounds for attributing to X causal status in its association with Y. In other words, I think it is quite reasonable to talk of regression parameters as "effects" of explanatory variables on the response, provided that there is a flavor of tentativeness to such language.

Perhaps the proper attitude toward causal inference using regression is best expressed in the following quotes. Clogg and Haritou (1997) recommended that researchers routinely run several regressions that include the focus variable plus all possible combinations of potential confounds and assess the stability of the focus variable's effect across all regressions. They then say (p. 110): "The causal questions that social researchers ask are important ones that we ought to try to answer. If they can only be answered in the context of nonexperimental data, then a method that conveys the uncertainty inherent in the enterprise ought to be sought. We believe that the uncertainty in causal assumptions, not the uncertainty in statistical assumptions and certainly not sampling error, is the most important fact of this enterprise."

Sobel's (1998, p. 346) advice is in the same vein: "[s]ociologists might follow the example of epidemiologists. Here, when an association is found in an observational study that might plausibly suggest causation, the findings are treated as preliminary and tentative. The next step, when possible, is to conduct the randomized study that will more definitively answer the causal question of interest."

In sum, causal modeling via regression, using nonexperimental data, can be a useful enterprise provided we bear in mind that several strong assumptions are required to sustain it. First, regardless of the sophistication of our methods, statistical techniques only allow us to examine *associations* among variables. Thus, the most conservative approach to interpreting β in $Y = \beta X + \varepsilon$ is to say that β represents the expected *difference* in Y for those who are 1 unit apart in X. To say that β reflects the expected *change* in Y were we to *increase* X by 1 unit imparts a uniquely causal interpretation to the X–Y association revealed by the regression. Whether such an interpretation is justified requires additional information, in the form of theory and/or experimental work. At the least, we must assume that $Cov(X,\varepsilon)$ is zero. This means that no other variable, observed or unobserved, confounds the relationship between X and Y, as in the case of ξ above. As no empirical means exists for checking on this assumption, it is an act of faith. At most we will be able to argue that our findings are *consistent* with a causal effect of X on Y. But only the triangulation of various bits of evidence from many sources, over time, can establish this relation with any authority.

DATASETS USED IN THIS VOLUME

Several datasets are used for examples and exercises throughout the book. Ten of the datasets—those needed for the exercises—can be downloaded from the FTP site for this book at *http://www.wiley.com*. The datasets are in the form of raw data files, easily readable by statistical software programs such as SAS, SPSS, and STATA. Also included at the site are full codebooks in MS Word, listing all variable names and their descriptive labels as well as their order on the data records. Two of the datasets (*students* and *GSS98*, described below) contain missing values that must be imputed by the reader, as instructed in the exercises. All dataset names below in bold face type indicate data that are available for downloading. The following are brief descriptions of the datasets (names of all downloadable data files and associated codebooks are given in parentheses).

National Survey of Families and Households Datasets

The National Survey of Families and Households (NSFH) is a two-wave panel study of a national probability sample of households in the coterminous United States conducted between 1987 and 1994. Wave 1 of the NSFH, completed in 1988, interviewed 13,007 respondents aged 19 and over living in households in the United States. Certain targeted groups were oversampled: cohabitors, recently married couples, minorities, step-parent families, and one-parent families. For respondents who were cohabiting or married, a shorter, self-administered questionnaire was also given to the partner. The NSFH collected considerable demographic and family information as well as data on more sensitive couple topics such as the quality of the relationship and the manner of handling disagreements, including physical aggression. The survey is described in more detail in Sweet et al. (1988). In wave 2, completed in 1994, interviews were conducted with all 10,005 surviving members of the original sample and with the current spouse or cohabiting partner of the primary respondent. Question sets from the first wave were largely duplicated in the second. The six datasets described below are subsets of this survey.

Couples Dataset (couples.dat; couples.doc). This is a 6% random sample of all married and cohabiting couples from wave 1, with an *n* of 416 couples. The variables reflect various characteristics of the relationship from both partners' perspectives, as well as items tapping depressive symptomatology of the primary respondent.

Kids Dataset (kids.dat; kids.doc). This consists of a sample of 357 parents and their adult offspring from both waves of the NSFH. Information is contained on couples who were married or cohabiting, with a child between the ages of 12 and 18 in the household in 1987–1988, whose child was also interviewed in 1992–1994. Only cases in which the child had experienced sexual intercourse by 1992–1994 and in which the child had answered the items on sexual permissiveness and sexual behavior were included. Variables reflect attitudes, values, and other characteristics of the parents measured in wave 1, as well as sexual attitudes and behavior reported by their adult offspring in wave 2. Further detail is provided in DeMaris (2002a).

Union Disruption Dataset (disrupt.dat; disrupt.doc). These data consist of 1230 married and cohabiting couples in unions of no more than three years' duration at wave 1 who were followed up in wave 2. Primary interest was in the prediction of union disruption by wave 2, based on various characteristics of the relationship reported in wave 1, including intimate violence. This is a subset of the data used for the larger study reported in DeMaris (2000).

Cohabiting Transitions Dataset (cohabtx.dat; cohabtx.doc). This dataset consists of 411 cohabiting couples in wave 1, followed up in wave 2. It was used to examine the predictors of transition to separation or marriage, as opposed to remaining in the unmarried cohabiting state, by wave 2. Wave 1 characteristics of couples used as predictors of transitions were similar to those for the *union disruption dataset.* The full study is reported in DeMaris (2001).

Wave 1 Couples Dataset. These are the 7273 married and cohabiting couples in wave 1 who constitute the original pool of couples from which the longitudinal violence dataset (described below) was culled. Several characteristics of the relationship were measured in wave 1, with a focus on couple disagreements.

Violence Dataset. These data represent 4095 couples in wave 1 who were still intact in wave 2 and who provided information on patterns of intimate violence at both time periods. The response of interest is the *couple violence profile*, a three-category classification of violence patterns. Predictors are characteristics of the relationship as reported in wave 1. The full study is reported in DeMaris et al. (2003).

Datasets from the NVAWS

NVAWS is short for the national survey on Violence and Threats of Violence Against Women and Men in the United States, 1994–1996, collected by Tjaden and Thoennes (1999). The target population for the NVAWS included men and women from all 50 states and the District of Columbia, and includes 8000 men and 8000 women who were 18 years of age and older in 1994. Datasets employed in this book utilize only the women's data. Variables contain information about four types of victimization experienced over the life course: physical assault, sexual assault, stalking, and threats, as well as the mental health sequelae of such experiences. Three datasets are subsets of this survey.

Victims Dataset. This consists of the 1779 women who reported being victimized at least once by physical or sexual assault, stalking, or intimidation.

Current-Partner Victims Dataset. This is the subset of 331 women from the victims dataset who report being victimized by a current intimate partner.

Minority Women Dataset. These 1343 women are the minority subset of the original 8000 women in the NVAWS.

Other Datasets

Students Dataset (students.dat; students.doc). This is a sample of 235 students taking introductory statistics at Bowling Green State University (BGSU) from the author between the years 1990 and 1999. Variables include student characteristics collected in the first class session as well as the scores on the first two exams, given, respectively, in the sixth and tenth weeks of the course.

GSS98 Dataset (gensoc.dat; gensoc.doc). These data consist of the 2832 respondents from the 1998 General Social Survey. The GSS is conducted roughly biennially by the National Opinion Research Center. It is based on a multistage probability sample that is representative of all noninstitutionalized English-speaking persons 18 years of age and older living in the household population of the United States. Variables in the dataset represent selected demographic and attitudinal or opinion items deemed by the author to be of interest.

Faculty Salary Dataset (faculty.dat; faculty.doc). This consists of 725 faculty members employed at both the main and Firelands campuses of BGSU during the academic year 1993–1994. Data represent faculty salaries and factors deemed to predict variation in salaries, such as rank and years of seniority. The primary purpose of the study was to discover whether there was any evidence of gender inequity in salary allocation at the institution. Reports of the full studies utilizing these data can be found in Balzer et al. (1996) and Boudreau et al. (1997).

Introductory Sociology Dataset (introsoc.dat; introsoc.doc). These data were taken from all nine sections of introductory sociology offered at BGSU during the 1999 spring semester. The study involved four waves of data collection during the course of the semester. The total sample size is 751 students, but due to absenteeism at one or another data collection point, sample sizes vary in each wave. The focus of the study was an examination of the factors predicting academic performance, particularly self-esteem. Variables consist of measures such as prior and current academic performance, indexes of self-esteem and test anxiety, and related academic factors. Results of the study can be found in Bradley (2000).

Unemployment Transitions Dataset (jobs.dat; jobs.doc). These data are in the form of 620 unemployment spells for 283 Brazilian immigrants residing in the United States and Canada in 1990–1991. The purpose of the study was to test predictions from job search theory regarding the duration in, and rate of exit out of, unemployment for an immigrant population. The predictors consist largely of demographic, familial, and human capital variables. The full study is reported in Goza and DeMaris (2003).

Inmates Dataset (inmates.dat; inmates.doc). This dataset, collected by the Ohio Department of Rehabilitation, consists of information on 1485 male inmates admitted to the Ohio Department of Rehabilitation and Correction during September and October 1985. Variables reflect demographic and criminal history information for each inmate as well as individual lifestyle data and correctional-institution information regarding rule infractions during incarceration. The full study is reported in Clark (2001).

APPENDIX: STATISTICAL REVIEW

Overview

In this appendix I review basic statistical concepts and notation necessary to an understanding of the material in subsequent chapters. I assume that the reader has been exposed to most of this material at a previous time. However, those who are unfamiliar with probability and distribution theory, expectation, variance, covariance, correlation, sampling distributions, parameter estimation, and tests of hypotheses will probably want to read this appendix before proceeding with the rest of the book.

Variables and Their Measurement

The raw material of statistics consists of *data*. Data are essentially measurements for one or more variables, taken on one or more cases, from some population of cases of interest. Let's flesh this idea out a little more. We assume that there is a larger population of cases in which the researcher has an interest. The *population* is simply the collection of cases that the researcher is trying to make general statements about, or "generalize to." *Cases* in the social and behavioral sciences are typically people, but do not have to be. They are the individual units of observation in one's study. These can be individuals or organizations, just as they can be incidents or events. What we typically obtain in sampling cases from the population are attributes or characteristics of the cases, usually expressed as numerical values. These are our *measurements* on the cases. The attributes are called *variables*, and each variable typically exhibits some variability in realized *values* across the n cases in our sample. When the value of a variable for a given case cannot be predicted ahead of time, we refer to that variable as a *random variable*. For example, suppose that I randomly sample a person from the U.S. population and code his or her gender as 1 for male and 0 for female. Then the person's gender is a random variable—I don't know ahead of time what value it will take. If, on the other hand, I divide the population into males and females ahead of time and sample first from the males and second from the females, gender is no longer a random variable. In this case, we say that gender is *fixed*—its value is set ahead of time by the researcher prior to sampling, and there is no mystery about what each case's gender is. This distinction is important in regression modeling when we describe the regressors as random variables versus *fixed effects*.

Variables are distinguished by two major criteria in statistics, both having to do with the specificity of their measurement. The first distinction pertains to *level of measurement*. There are four commonly conceived levels: nominal, ordinal, interval, and ratio. *Nominal variables* are those whose values indicate only qualitative differences in the attribute of interest; they carry no information as to rank order on the attribute. For example, religious affiliation coded 1 for "Protestant," 2 for "Catholic," 3 for "Jewish," and 4 for "other denomination" is a nominal variable. All that can be said about cases with two different values on this attribute is that they are, well, different. Other than that, the numerical codes 1, 2, 3, and 4 convey no quantitative differences on the dimension of religious affiliation.

The values of *ordinal variables*, on the other hand, represent not only qualitative differences but also relative rank order on the attribute. *Religiosity*, for example, coded 1 for "not at all religious," 2 for "slightly religious," 3 for "moderately religious," and 4 for "very religious," is an ordinal variable. Given two people with different religiosity scores, say 3 versus 4, we can say that the second person is "more religious" than the first. How much more religious, however, cannot be specified precisely.

Interval variables represent an even more precise level of measurement. The values of interval variables are distinguished by the fact that they convey the exact amount of the attribute in question. Annual income in dollars, for example, is an interval variable. Further, given two people with different values of income, say $45,529.52 and $51,388.03, we can say not only that their incomes are qualitatively different and that the second person is higher in income but can also specify *precisely how much difference there is in their incomes*: $5858.51, to be exact. Notice, however, that if we collapse income categories into ranges, the variable loses its interval-level specificity and becomes ordinal. For example, suppose that we have income categories defined in $10,000 ranges and coded from 1 for [0–10,000) to 11 for [100,000 or more). Further, suppose that individual A is in category 5 [40,000–50,000) and individual B is in category 6 [50,000–60,000). Certainly, we can say that B has a higher income than A. But it is no longer possible to specify precisely how much higher B's income is.

Ratio variables are interval-level variables with a meaningful zero point. In this case, it makes sense to speak of the ratio of two values. Income is also an example of a ratio variable. If A makes $50,000 a year and B makes $100,000, B makes twice as much income as A.

The other major criterion for distinguishing variables is whether they are discrete or continuous. This distinction is central to the characterization of their probability distributions (see below). Technically, a *discrete variable* is one with a countable number of values. This is a technical concept which essentially means that the values have a one-to-one relationship with the collection of positive integers. Since there are an infinite number of positive integers, discrete variables could conceivably have an infinite number of values. In practice, discrete variables take on only a relatively few values. For example, the number of children ever borne by U.S. women is a discrete variable, taking on values 0, 1, 2, and so on, up to some maximum value delimited by biological possibility, say 25 or so. Nominal variables are always discrete, as are ordinal variables, since rank order can always be put in a one-to-one correspondence with positive integers.

Continuous variables are those with an uncountable number of values. These variables can, technically, take on any value in the real numbers, delimited only by their logical range. Realistically, measurement limitations prevent us from ever actually observing continuous variables in practice. For example, the weight of humans in pounds could conceivably take on any of an uncountably infinite number of values in the range [0–1000]. But limitations in instruments for weight measurement mean that we probably cannot discern weight differences smaller than, say, .001 pound between two people. No matter. We will find it expedient to *treat* variables as continuous if they are at least ordinal in nature, if they have a sufficient number of values, and if their probability distributions are not too skewed. Otherwise, they will be treated as discrete. For this book, therefore, the discrete–continuous distinction is the one that is most important.

Probability and Distribution Theory

In sampling cases from a population, we speak of the *probability* of observing a particular value for a given variable, for the ith individual, where i equals $1, 2, \ldots, n$. The technical definition of probability is quite arcane (see, e.g., Chung, 1974; Hoel et al., 1971). Intuitively, however, the probability of some outcome refers to the *relative frequency of its occurrence over an infinite repetition of the conditions that made its observation possible*. For example, if we toss an honest coin, the probability of observing a head is .5. This means that if we were to toss that coin an infinite number of times, 50% of the outcomes would be heads. Since we will never be able to conduct an infinite repetition of any experiment, probabilities are figured by a simple rule. For any event, E, the probability of event E, or $P(E)$, is defined as follows:

$$P(E) = \frac{\text{number of ways that } E \text{ can occur}}{\text{total number of observable outcomes}}$$

Hence, in the coin example, there is only one way to get a head, but there are two possible outcomes of a coin toss: a head or a tail. The probability of a head is therefore $\frac{1}{2} = .5$.

Although in this book we will not be concerned with probability problems per se, a few probability rules are important. First, for any event A, if $P(A)$ is the probability that A occurs, then $1 - P(A)$ is the probability that it doesn't occur (or that anything else occurs that isn't A). Further, consider any two events, A and B. Then the event (A and B), also denoted ($A \cap B$), refers to an event that is both A and B simultaneously, while the event (A or B), also denoted ($A \cup B$), refers to the event that at least one of A or B occurs. For example, if A is "being married" and B is "having a child," (A and B) is "being married with a child," while (A or B) is satisfied by any of these three events: being married but childless, having a child outside marriage, or being married with a child. The *conditional probability* of an event is the probability of an event under the restriction that some condition holds first. The conditional probability of some event B, given that event A holds, is denoted $P(B|A)$. For example, the conditional probability of B given A, from above, is the conditional probability of having a child given that the person is married. Two events are *independent* if $P(B) = P(B|A)$, and *dependent* otherwise. For example, the events "being married" and "having a child" are independent if the probability of having a child is unchanged by whether or not a subject is known to be married. In all likelihood, these events are not independent, since the probability of having a child when one is married is probably higher than the probability of having a child in general, called the *unconditional probability* of having a child. If A and B are independent events,

P(*A* and *B*) = P(*A*)P(*B*). This generalizes to: If events A_i are independent, for $i = 1$, $2, \ldots, n$, then P(A_1 and A_2 and $\cdot \cdot \cdot$ and A_n) = P(A_1)P(A_2) $\cdot \cdot \cdot$ P(A_n).

Probability Distributions. More important for the current work are probability distributions. (Readers with a limited math background may want to review Appendix A, Section I, before proceeding with this section.) A probability distribution for a random variable *X* is an enumeration of all possible values of *X*, along with the probability associated with each value, should one collect one observation on *X* from the population. Actually, this is too simple. In truth, we need to distinguish between the *distribution* and *density* functions for the variable *X*. The distribution function for *X*, denoted F(*x*), tells us P(*X* ≤ *x*) for any value *x* of *X*. That is, the distribution function tells us the probability of observing any value *up to and including x*, when we make a single observation on *X* from the population. (I follow the statistical convention here of using *X* to denote the variable generally and *x* to denote a specific value of the variable, e.g., 3.2, 5.93, etc.)

What the *density function* tells us, on the other hand, depends on whether *X* is discrete or continuous. If discrete, the density of *x*, denoted f(*x*), gives us the *probability* of getting the specific value *x* of *X* when we sample one value of *X* from the population. Figure 1.1 depicts a simple discrete density function for a variable *X*.

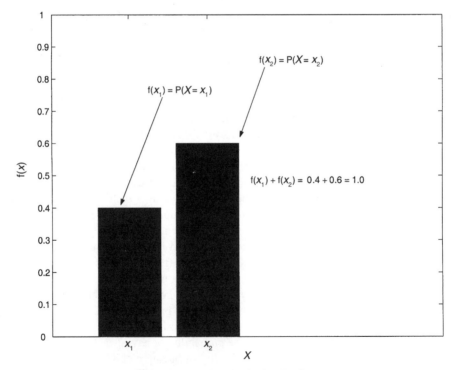

Figure 1.1   Discrete density function for *X*.

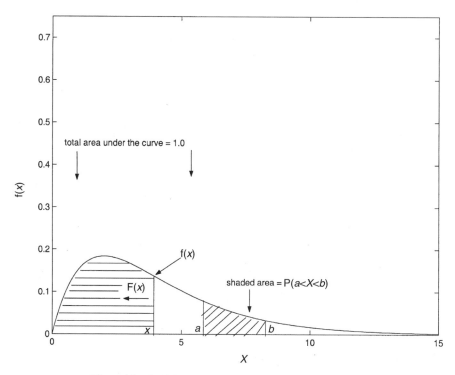

Figure 1.2 Continuous density and distribution functions for X.

There are two values of the variable: x_1 and x_2. The heights of the bars indicate the probabilities of observing each value, where $P(x_1) = .4$ and $P(x_2) = .6$. If X is continuous, on the other hand, then f(x) gives us the *density* associated with the specific value x of X when we sample one value of X from the population. The density is not a probability, although it is closely related to one. Rather, it is the function's value when the function is evaluated at the point x. In that the function describes a curve over the X-axis, the density is the point on the curve immediately above point x. Figure 1.2 depicts a typical continuous density function. If X is truly continuous, $P(X = x) = 0$. That is, there is zero probability of observing any particular value of X, although there is some nonzero probability of observing X to fall within some specific range of values. Thus, with continuous variables, we speak of $P(a < X < b)$ rather than $P(X = a)$, where a and b are specific values of X. The connection of the density to a probability lies in the fact that $P(a < X < b)$ is the area under the curve f(x) between the X-values of a and b. The higher the curve over a and b [i.e., the greater the density over the interval (a,b)], the greater the area under the curve and thus the greater the corresponding probability.

Discrete Density and Distribution Functions. Several discrete density functions will be important in this volume. One example is the *Bernoulli density*. Suppose that for any adult sampled from the population, we record whether he or she has ever been mugged. We call this variable X, with values 1 for "ever mugged" and 0 for "never mugged." Let us, further, denote the probability of being mugged, in general, as π. This means that the probability of not having been mugged is $1 - \pi$. Then the density function for X can be written $f(x) = \pi^x (1 - \pi)^{1-x}$. This gives us the probabilities associated with the two possible values of X, since $f(1) = \pi^1 (1 - \pi)^{1-1} = \pi$, and $f(0) = \pi^0 (1 - \pi)^{1-0} = 1 - \pi$. We refer to this function as the Bernoulli density with parameter π. Variables with a Bernoulli density have mean equal to π and variance equal to $\pi(1 - \pi)$. Once the parameter's value is known, the probability of any value of X is determined automatically. The Bernoulli distribution function, $F(x)$, is particularly simple, since $F(0) = P(X \leq 0) = P(X = 0) = 1 - \pi$ and $F(1) = P(X \leq 1) = 1.0$.

Another example of a discrete density is the *Poisson*. Its parameter will be denoted by θ (theta), where $\theta > 0$. For a discrete variable X taking on the values $0, 1, 2, \ldots$, the Poisson density is

$$f(x) = \frac{e^{-\theta}\theta^x}{x!}.$$

Here, again, once θ is known, the probability of any value of X is determined by the function. For example, if θ is 2.2, the probability of observing an X of 5 is

$$f(5) = \frac{e^{-2.2} \cdot 2.2^5}{5!} = .0476.$$

The Poisson distribution function that gives us $P(X \leq x)$ is just the sum of the probabilities for values $0, 1, 2, \ldots, x$. Hence the distribution function can be written

$$F(x) = \sum_{j=0}^{x} \frac{e^{-\theta}\theta^j}{j!}$$

A variable that is Poisson distributed has mean and variance both equal to θ. For discrete density functions, in general, the sum of the probabilities associated with all possible values of the variable is 1.0. This is easy to verify with the Bernoulli, since $\pi + (1 - \pi) = 1.0$.

Continuous Density and Distribution Functions. Continuous density functions give the densities associated with continuous variables. One of the simplest, for illustration, is the *exponential density*, with parameter λ (lambda). For $X \geq 0$ and $\lambda > 0$, the density is $f(x) = \lambda e^{-\lambda x}$ and its distribution function is $F(x) = 1 - e^{-\lambda x}$. Figure 1.3 depicts the exponential density with $\lambda = 2.2$. So if $\lambda = 2.2$, say, then $f(4) = 2.2e^{-2.2(4)} = .00033$. As mentioned above, .00033 is not the probability of observing a value of 4. Rather, it is the point on the curve $f(x) = \lambda e^{-\lambda x}$ directly above the value of 4 on the X-axis. On the other hand, $F(4) = 1 - e^{-2.2(4)} = .99985$ is the probability that X is less than 4. [For continuous variables, $P(X < 4)$ and $P(X \leq 4)$ are the same, since $P(X = 4) = 0$.] Here we see that 4 is an unusually high value for this distribution, since we are almost certain to observe values less than 4 if we sample from this distribution. Exponentially distributed variables have mean equal to $1/\lambda$ and variance equal to $1/\lambda^2$.

A couple of remarks are in order at this point. First, the total area under the curve of a continuous density function is scaled so that it always equals 1.0, as indicated

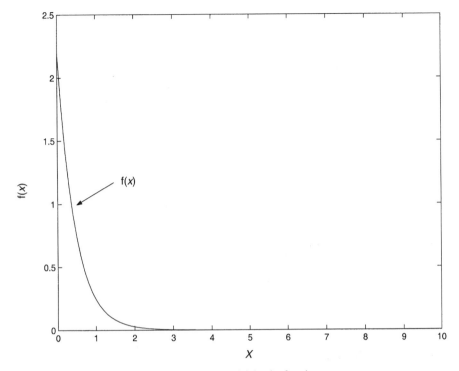

Figure 1.3 Exponential density function.

in Figure 1.2. This is the continuous-variable equivalent of the probabilities summing to 1.0 for a discrete variable. Second, to find the probability that X is between the values a and b in its range, we find the area under the curve between a and b.[1] (In Figure 1.2, this is shown as the second shaded area under the curve.) Given the distribution function, this is quite simple, since the area between a and b under any density function $f(x)$ is $F(b) - F(a)$. For example, what is the probability that X is between 2 and 4 for our exponentially distributed variable above? The answer is $P(2 \leq X \leq 4) = F(4) - F(2) = 1 - e^{-2.2(4)} - (1 - e^{-2.2(2)}) = .01213$.

Third, the exponential distribution function is called a *closed-form function*, since it can be evaluated by means of an algebraic formula. Not all distribution functions (or density functions, for that matter) are so easily evaluated, as we will see in the next example.

One of the most important densities in all of statistics is the *normal density*. Its graph, shown in Figure 1.4 for $X > 0$, is bell-shaped and is familiar to anyone who has ever taken a statistics class. Perhaps not as familiar is the density function itself. Its formula is

$$f(x) = \frac{1}{\sigma \sqrt{2\pi}} \exp\left[-\frac{1}{2} \left(\frac{x - \mu}{\sigma} \right)^2 \right],$$

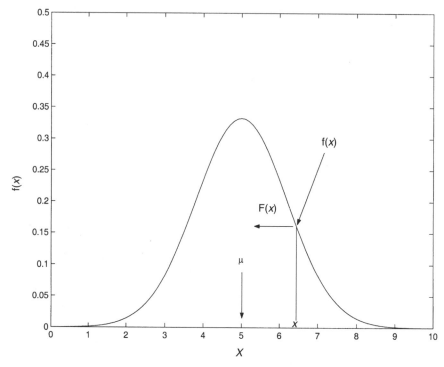

Figure 1.4 Normal density and distribution functions.

where μ and σ are the parameters of the distribution and π is the geometric constant, whose value is approximately 3.14159. Normally distributed variables have mean equal to μ and variance equal to σ^2. Many real-world continuous variables are approximately normally distributed. Its importance, however, arises from its theoretical significance. It turns out that the sampling distributions of effect estimators (i.e., regression coefficients) in regression models are usually asymptotically[2] normal, enabling statistical inference via t or z tests. To find the density for the value of a normally distributed variable, we must first specify μ and σ. For example, suppose that μ is 3 and σ is 1.5. Then the density associated with an X-value of 4.9 is

$$f(4.9) = \frac{1}{1.5\sqrt{2(3.14159)}} \; \exp\left[-\frac{1}{2}\left(\frac{4.9 - 3}{1.5}\right)^2\right] = .1192.$$

The distribution function for the normal distribution is not closed form. That is, although we can easily evaluate the density at any value x for a normally distributed variable, finding $F(x)$ requires finding the area to the left of x under the curve (as indicated in the figure). The distribution function, $F(x)$, is

$$F(x) = \int_{-\infty}^{x} \frac{1}{\sigma\sqrt{2\pi}} \; \exp\left[-\frac{1}{2}\left(\frac{t-\mu}{\sigma}\right)^2\right] dt$$

This expression, read

"the integral of $\dfrac{1}{\sigma\sqrt{2\pi}}\exp\left[-\dfrac{1}{2}\left(\dfrac{t-\mu}{\sigma}\right)^2\right]$ from negative infinity to x,"

indicates the area under the normal curve to the left of x. As there is no closed-form solution to this integral, it must be approximated using numerical techniques.

An especially important normal distribution is the *standard normal distribution*, which has a mean of zero and a standard deviation of 1. Given the parameters μ and σ, any normally distributed variable X can be made to have the standard normal distribution simply by converting X to standard-score form via the formula $Z = (X - \mu)/\sigma$. The density and distribution functions for the standard normal distribution have a special notation: $\phi(z)$ for the density function and $\Phi(z)$ for the distribution function. The functions are

$$\phi(z) = \frac{1}{\sqrt{2\pi}}\, e^{(-1/2)z^2}$$

$$\Phi(z) = \int_{\infty}^{z} \frac{1}{\sqrt{2\pi}}\, e^{(-1/2)t^2} dt.$$

Because areas under the standard normal density curve have been computed and tabled, it is a simple matter to find $P(a < X < b)$ for any normally distributed variable, X. We simply convert the problem into a comparable problem involving areas under the standard normal curve. For example, suppose that X is normally distributed with mean 3 and standard deviation 1.5. If we sample one observation from this distribution, what is the probability that its value will be between 4.9 and 5.5? Now if we were to standardize all of the X-values by subtracting 3 and dividing by 1.5, the resulting variable, Z, would have the standard normal distribution. The values 4.9 and 5.5, also converted to Z-scores, would be $(4.9 - 3)/1.5 = 1.27$ and $(5.5 - 3)/1.5 = 1.67$. Thus, the problem becomes: What is the probability of a standard normal variable being between the values of 1.27 and 1.67? As in the example of the exponential distribution above, $P(1.27 < X < 1.67) = F(1.67) - F(1.27)$, or, using the special notation for the standard normal distribution function, $\Phi(1.67) - \Phi(1.27)$. Using a standard normal table, we have $\Phi(1.67) = .9525$, while $\Phi(1.27) = .898$. The answer is, therefore, $.9525 - .898 = .0545$.

Expectation, Variance, Covariance, and Correlation. Several characteristics of the distributions of variables in the population are important in statistical analyses. I discuss four in this section: expectation, variance, covariance, and correlation.

Expectation. The expected value of X, denoted $E(X)$, is the mean of X in the population. For a discrete variable, X,

$$E(X) = \sum_{x} x f(x).$$

That is, the population mean consists of a weighted sum (see Section II.D in Appendix A for an explanation of weighted sums) of the X-values, where the weights are the densities, or probabilities, associated with each value. For a continuous variable, X,

$$E(X) = \int_x x f(x)\, dx.$$

In this case, the mean is the integral of x times $f(x)$ over the range of X-values. This is simply the continuous counterpart to the definition of $E(X)$ for discrete variables. In either case, $E(X)$ is just the population mean, or arithmetic average, of X.

Variance. The variance of X, denoted $V(X)$, is defined as $V(X) = E[X - E(X)]^2$. The variance is the mean-squared deviation of X from its expected value and indicates how spread out the X-values are around their mean. The standard deviation, denoted $SD(X)$, is the square root of the variance and is loosely interpreted as the average distance from the mean of the values in the distribution.

Covariance. The definition of covariance is given elsewhere (see Section III.A of Appendix A) but will be repeated here. The covariance of any two variables, X and Y, is $Cov(X,Y) = E(X - \mu_x)(Y - \mu_y)$, where $\mu_x = E(X)$ and $\mu_y = E(Y)$. The covariance is the average cross-product of X, deviated from its mean, times Y, deviated from its mean. The covariance measures the extent to which two variables vary together. Positive covariances suggest that higher values of X are associated with higher values of Y. Negative covariances, on the other hand, indicate that higher values of X are associated with lower values of Y.

Correlation. As a descriptive measure of joint variation, the covariance is inadequate, since its value depends on the units of measurement of X and Y. A better descriptive measure is the correlation coefficient, denoted ρ_{xy}. Its formula is

$$\rho_{xy} = \frac{Cov(X,Y)}{\sqrt{V(X)V(Y)}}.$$

The correlation is a "standardized" covariance, constructed so as to fall in the interval $[-1,1]$, with an absolute value of 1 indicating perfect correlation and a value of zero indicating no correlation. Another way to view the correlation is that it is the covariance between two standardized variables. Covariance and correlation are designed to detect linear association. Hence, either may be zero if two variables vary together in a systematic but nonlinear fashion.

Sampling and Sampling Distributions. Most of the science of statistics is concerned with making inferences about a population based on studying only a small subset of the population, the sample. Samples come in two major "flavors": probability and nonprobability. *Probability samples* are those in which each member of the population has some known a priori probability of being selected into the sample. *Nonprobability samples*, as the name suggests, are those in which the probability of given population members being selected into the sample cannot be specified ahead of time. Inferential statistics is based solely on probability sampling. It is the only type of sampling that lends itself to theoretical specification of the sampling distributions of statistics (discussed below), which form the basis of statistical inference. The simplest type of probability sample is the *simple random sample*, in which each member of the population has the same chance of being selected into the sample. If n cases are to be selected from a population of size N, each population member has a probability of selection of n/N.

Sample statistics such as the sample mean and variance of a variable ($\bar{y}$ and s^2, respectively) or a sample regression coefficient (b) indicating the effect of a predictor on an outcome are estimates of corresponding population parameters. Let θ denote any population parameter, and $\hat{\theta}$ the sample estimator of that parameter. In making inferences about θ based on the observed value of $\hat{\theta}$, we need to understand the nature of the relationship between the two. The *sampling distribution* of a statistic is critical to this enterprise: It is a probability distribution for a sample statistic. That is, it is an enumeration of all possible values of $\hat{\theta}$, together with their associated probabilities of occurrence, that would be obtained through an infinite repetition of collecting samples of size n from that population and recomputing $\hat{\theta}$. Although we collect only one sample and compute one value of $\hat{\theta}$ in practice, it is important to understand that the full distribution of $\hat{\theta}$ *could* be generated for any statistic via repeated sampling. The importance of this distribution is that it indicates the probability that $\hat{\theta}$ is within a specified "distance" from θ. It therefore places bounds on the degree to which we are in error in using $\hat{\theta}$ as an estimate of θ or in using $\hat{\theta}$ to test a hypothesis about θ.

Table 1A.1 presents a very simple illustration of the sampling distributions for the sample mean, $\bar{y}$, and the sample variance, s^2. As is evident in the table, the "population" consists of only five observations: A, B, C, D, and E. (The population is artificially small to keep the number of different samples manageable.) For each observation, a value is recorded for the variable Y. The mean of Y, or μ, for this population is 3 (as is easily verified), and the variance of Y, or σ^2, is 2. [This is also easily verified, keeping in mind that for the population, the variance of Y is $\sigma^2 = \sum(Y - \mu)^2/N$, where N is the population size.] If we draw samples from this population of $n = 3$, *without replacement*, there are 10 different possible samples that can be drawn. These are shown in the table along with the Y-values of the sample members and the sample mean and variance for each sample. The RF columns indicate the relative frequency of occurrence of each value of the sample mean and variance, respectively. These columns represent the sampling distributions of each statistic, since they indicate the probabilities associated with each different value of the sample statistics.

For the sample mean, it is clear that when drawing a sample of size 3 from this population, certain values of $\bar{y}$—such as 3.33, 3, and 2.67—are twice as likely as other values. Similarly, the most likely value for σ^2 is 2.33. We can also compute the average of the 10 sample means, denoted $E(\bar{y})$. We find that it is 3, the same as the population mean of Y. This is no accident, since it is always true that $E(\bar{y}) = \mu$. This means that the sample mean is an *unbiased estimator* of the population mean—its average value equals the population parameter. The average sample variance, or $E(s^2)$, is 2.5, which in this case is *not* equal to the population variance of 2. However, under ordinary sampling conditions with infinite (or approximately infinite) populations, it *is* the case that $E(s^2) = \sigma^2$. With finite populations we must apply a finite

Table 1A.1 Sampling Distribution for the Sample Mean and Sample Variance, Based on Repeated Sampling of Size $n = 3$ from a Population with Five Cases

Population Elements

Case	Y
A	2
B	3
C	1
D	5
E	4

Sampling Distribution (for n = 3)

Sample	Members	Y	$\bar{y}$	RF	s^2	RF
1	A,B,C	2,3,1	2	.1	1	.3
2	A,B,D	2,3,5	3.33	.2	2.33	.4
3	A,B,E	2,3,4	3	.2	1	.3
4	A,C,D	2,1,5	2.67	.2	4.33	.2
5	A,C,E	2,1,4	2.33	.1	2.33	.4
6	A,D,E	2,5,4	3.67	.1	2.33	.4
7	B,C,D	3,1,5	3	.2	4	.1
8	B,C,E	3,1,4	2.67	.2	2.33	.4
9	B,D,E	3,5,4	4	.1	1	.3
10	C,D,E	1,5,4	3.33	.2	4.33	.2

population correction to the sample variance to make it an unbiased estimator of σ^2. That correction is $(N - 1)/N$. Thus,

$$E\left(\frac{N-1}{N}s^2\right) = \frac{N-1}{N}\,E(s^2) = \sigma^2.$$

In this example, we see that

$$E\left(\frac{N-1}{N}s^2\right) = E\left(\frac{5-1}{5}s^2\right) = \frac{4}{5}\,E(s^2) = \frac{4}{5}(2.5) = 2 = \sigma^2.$$

Sampling from a Population vs. Sampling to a Population. Probability sampling from a target population allows inferences from the sample back to that target population. As we will see below, this is because probability sampling allows us to specify the sampling distributions of sample statistics and their relation to population parameters. What inferences can be made with nonprobability samples? Some would suggest that no inferences are possible. Nevertheless, as the reader is fully aware, researchers use nonprobability samples to make population inferences all the time. Data are frequently taken from convenience samples of students or members of voluntary organizations, analyses are performed, and results are discussed in terms of statistical significance or nonsignificance. Or, experimenters randomly assign volunteers to treatment and control groups and then outline which group differences are "significant." In either case, inferences are being made to some larger population of cases. The most conservative position would be that such "inferences" are meaningless since the data were not drawn in a probabilistic fashion from a known population. However, it is unlikely that any given sample, however haphazardly selected, is not representative of a larger group of cases. The question is: *What* larger group?

Some idea of the population represented by a nonprobability sample can be inferred from the concept of repeated sampling. For example, suppose that I wish to do an opinion survey about an important issue on my campus. To collect a sample, I go to the university library when it opens on Saturday morning and sample the first 30 students who come in. Clearly, this is not a probability sample from the population of students at the university, since only certain students patronize the library, especially on Saturday morning. Moreover, even if we identify the population of students who ever patronize the library on Saturday, this sample has not been selected randomly from that population. However, suppose that we repeat this sampling technique next Saturday, and the Saturday after that, and so on, indefinitely. Ultimately, this approach would generate a "population" of cases consisting of all unique students encompassed by the collection of all such samples. The first (and typically, only) sample we take could then be considered a random sample from *this* population. I refer to this scenario as sampling *to* a population rather than sampling *from* a population. This, of course, is a hypothetical population whose complexion could only be hinted at from the sociodemographic makeup of the sample. Nevertheless, it does represent a larger group to whom results from nonprobability samples might be generalized and for whom statistically "significant" results might apply.

Parameter Estimation and Statistical Inference

Sample statistics such as $\bar{y}$, s^2, or b are estimates of corresponding population parameters. In this section I discuss the desirable properties of estimators and common techniques for constructing estimators. I also consider inferential procedures for parameters, such as interval estimation and hypothesis tests. (Readers unfamiliar with differential calculus may want to review Section IV of Appendix A before continuing with this section.)

Least Squares. There are a number of techniques for constructing sample estimators of parameters, but one of the most commonly used is the technique of least squares. The *least squares estimator* of a parameter θ is that number, $\hat{\theta}$, that results in the smallest amount of prediction error when $\hat{\theta}$ is used in predicting the individual data values—the X's. Prediction error is measured by the squared distance of $\hat{\theta}$ [or $g(\hat{\theta})$, a function involving $\hat{\theta}$] from the data values (thus the appellation "least squares"). Hence, the least squares estimator of θ is the $\hat{\theta}$ that minimizes $\sum(x - \hat{\theta})^2$ (or, more generally, $\sum[x - g(\hat{\theta})]^2$). To illustrate, suppose that we draw a random sample from a normally distributed population of X-values whose population mean is μ. To estimate μ, we reason as follows. The best estimate should be the single number that is "closest" to the collection of sample X's. We use as our measure of closeness the squared distance from that number to each X-value. Thus, we take as our estimator of μ the value $\hat{\mu}$ that minimizes $\sum_{i=1}^{n} (x_i - \hat{\mu})^2$. The solution, from calculus, involves finding the value of $\hat{\mu}$ that makes the first derivative of this sum, with respect to $\hat{\mu}$, equal to zero:

$$\frac{d}{d\hat{\mu}}\left[\sum(x-\hat{\mu})^2\right] = -2\sum(x-\hat{\mu}).$$

$$-2\sum(x-\hat{\mu}) = 0$$

if

$$\sum(x-\hat{\mu}) = 0$$

or

$$\sum x - \sum\hat{\mu} = 0$$

or

$$\sum x = n\hat{\mu}$$

or

$$\hat{\mu} = \frac{\sum x}{n}.$$

Thus, the least squares estimator of the population mean is the sample mean, $\overline{X}$.

Maximum Likelihood. Another estimation technique is the one that we will employ for many of the models presented in this book. In estimation via *maximum likelihood*, we take $\hat{\theta}$ as the value of θ, among all possible values of θ, that would have rendered the sample data most likely to be observed. To illustrate this technique, I employ the following example. Suppose that X, a positive continuous variable, has an exponential distribution, with parameter λ, where $\lambda > 0$. The density function for the ith case is therefore $f(x_i) = \lambda e^{-\lambda x_i}$. If we sample n observations randomly from this population, the joint density function for the n observations on X is the product of the n individual density functions. This follows from the fact that the observations are independent, and the rule (see above) that the probability of independent events is the product of their individual probabilities (densities generally follow the same probability rules as probabilities). So if we let the vector $\mathbf{x}$ represent the collection of particular X-values observed in our sample, the joint density of $\mathbf{x}$ is

$$f(\mathbf{x}) = \prod_{i=1}^{n} \lambda e^{-\lambda x_i}.$$

Given the value of λ, the joint density of $\mathbf{x}$ allows us to calculate the density associated with any particular collection of X-values. On the other hand, given a particular collection of X-values, $f(\mathbf{x})$ is actually a function of λ. Viewed from this perspective, we can ask: Which value of λ makes $f(\mathbf{x})$ as large as possible and therefore makes the data most likely to have been observed? The function is then referred to as the *likelihood function* for the parameter, given the data, and denoted $L(\lambda \mid \mathbf{x})$. The value of lambda that maximizes it is called the *maximum likelihood estimate* of λ and denoted $\hat{\lambda}$.

Now it turns out that whatever maximizes the likelihood function also maximizes the log of the likelihood function. As the log-likelihood, denoted $\ln L(\lambda \mid \mathbf{x})$ or $\ell(\lambda \mid \mathbf{x})$, is more mathematically tractable than the likelihood, we seek to maximize this quantity. That is, we maximize

$$\ell(\lambda \mid \mathbf{x}) = \ln\left(\prod_{i=1}^{n} \lambda e^{-\lambda x_i} \right) = \ln\left(\lambda^n e^{-\lambda \sum x}\right) = n \ln\lambda - \lambda\sum x.$$

To find the λ that maximizes this quantity, we compute the first derivative of this function with respect to λ and then solve for the λ that makes it equal to zero. Thus,

$$\frac{d}{d\lambda}\left(\ell(\lambda \mid \mathbf{x})\right) = \frac{d}{d\lambda}\left(n \ln\lambda - \lambda\sum x \right) = \frac{n}{\lambda} - \sum x$$

and

$$\frac{n}{\lambda} - \sum x = 0 \quad \text{whenever} \quad \frac{n}{\lambda} = \sum x$$

or

$$\lambda = \frac{n}{\sum x}.$$

Thus, $n/\sum x$ is the maximum likelihood estimate (MLE) for λ. Now by the invariance property of MLEs (Bickel and Doksum, 1977), if $\hat{\theta}$ is the MLE for θ, then $g(\hat{\theta})$ is the MLE for $g(\theta)$, for any function $g(\cdot)$. In that the mean of an exponentially distributed variable is $1/\lambda$, the MLE for the mean of X in these data is

$$\hat{E}(X) = \frac{1}{\hat{\lambda}} = \frac{1}{n/\sum x} = \frac{\sum x}{n} = \bar{x}.$$

Here we see that the sample mean of the X's, $\bar{x}$, is the maximum likelihood estimator of the population mean for exponentially distributed data.

Asymptotic Properties of MLEs. Maximum likelihood estimators have several asymptotic, or large-sample, properties that make them especially desirable estimators (Bollen, 1989). First, they are *asymptotically unbiased*. What does this mean? The bias of an estimator $\hat{\theta}$ for a parameter θ, denoted $B(\hat{\theta})$, is $B(\hat{\theta}) = E(\hat{\theta}) - \theta$. An estimator $\hat{\theta}$ is *unbiased* for θ if $E(\hat{\theta}) = \theta$. That is, *an estimator is unbiased for a parameter if its average value, taken over its sampling distribution, equals the parameter.* MLEs have this property asymptotically; that is, as n tends to infinity, $E(\hat{\theta})$ converges to θ. Second, they are *consistent*. Consistency is a very important property of estimators. We say that $\hat{\theta}$ is consistent for θ if and only if $\lim_{n \to \infty} P(|\hat{\theta} - \theta| < \varepsilon) = 1$ for every $\varepsilon > 0$. This formulation says that a consistent estimator is one for which *the probability that it is arbitrarily close to the parameter converges to 1 as the sample size increases without bound*. Intuitively, this means that the sampling distribution of $\hat{\theta}$ becomes more and more concentrated over θ as n increases, so that ultimately, as n tends to infinity, the sampling distribution is entirely concentrated in a "spike" centered directly over the parameter. Consistent estimators are not necessarily unbiased in small samples, but the analyst is assured that as he or she uses a larger and larger sample, the estimator is getting closer and closer in value to the parameter of interest. Also, if $\hat{\theta}$ is consistent for θ, we say that the *probability limit*, or *plim*, of $\hat{\theta}$ is θ (Greene, 2003). An important theorem connected to the probability limit is the *Slutsky theorem*: *For a continuous function* $g(\hat{\theta})$ *that is not a function of n*, plim $g(\hat{\theta}) = g(\text{plim } \hat{\theta}) = g(\theta)$, *for* $\hat{\theta}$ *a consistent estimator of* θ (Greene, 2003). As an example, suppose that μ is the population mean of a continuous variable, Y, and we wish to estimate $\ln \mu$. In that $\bar{x}$ is consistent for μ, our consistent estimator of $\ln \mu$ is $\ln \bar{x}$, since plim $(\ln \bar{x}) = \ln (\text{plim } \bar{x}) = \ln \mu$. Third, MLEs are *asymptotically efficient*, which means that for large samples they achieve the smallest sampling variance among consistent estimators. Finally, they are *asymptotically normal*. This means that as n tends to infinity, their sampling distributions become more and more normal, enabling statistical inferences using the standard normal distribution.

Sampling Distributions of Estimators. We have seen a simple example above of sampling distributions for the sample mean and sample variance. These were generated by simple enumeration, since the population was artificially small. In practice, populations are infinite, or nearly so, and enumerating all possible samples is not feasible. However, various limit theorems in statistics can be drawn on to specify the sampling distributions of many sample statistics under certain conditions. For example, the central limit theorem (CLT) holds that *a weighted sum of random variables is normally distributed as n tends to infinity, regardless of the distributions of the original variables* (Hoel et al., 1971). That is, the CLT maintains that if X_i for $i = 1$, $2, \ldots, n$ are independent, identically distributed random variables each having mean μ and variance σ^2, and w_i, $i = 1, 2, \ldots, n$, constitute a set of constants, the sum $S_n = \sum_{i=1}^{n} w_i X_i$ converges in distribution to a normal distribution as n tends to infinity, regardless of the distribution of the X_i. Moreover, $E(S_n) = \mu \sum_{i=1}^{n} w_i$ and $V(S_n) = \sigma^2 \sum_{i=1}^{n} w_i^2$. As an example of the application of this theorem, consider the distribution of $\bar{X}$, the sample mean of the X_i, where $E(X_i) = \mu$ and $V(X_i) = \sigma^2$. Now $\bar{X} = \sum_{i=1}^{n} X_i/n$, which is in the form $\sum_{i=1}^{n} w_i X_i$, where $w_i = 1/n$ for all i. For large n (which in this case means an n of about 30 or more; see Agresti and Finlay, 1997), according to the CLT, $\bar{X}$ is approximately normally distributed, with mean equal to

$$\mu \sum_{i=1}^{n} w_i = \mu \sum_{i=1}^{n} \frac{1}{n} = \mu \frac{n}{n} = \mu.$$

Therefore, $\overline{X}$ is an unbiased estimator of μ. Moreover,

$$V(\overline{X}) = \sigma^2 \sum_{i=1}^{n} w_i^2 = \sigma^2 \sum_{i=1}^{n} \frac{1}{n^2} = \sigma^2 \frac{n}{n^2} = \frac{\sigma^2}{n}.$$

Thus, the standard deviation, or *standard error*, of the sample mean is $\sigma/\sqrt{n}$.

Knowing the shape and parameters (e.g., the mean and variance) of the sampling distribution of a statistic enables inferences to be made about the corresponding population parameter. Statisticians conduct two major types of inferences: interval estimation and hypothesis testing.

Interval Estimation. The sample mean, $\overline{X}$, is a *point estimate* of the population parameter, μ. However, we would not expect $\overline{X}$ to actually equal μ. In fact, given that $\overline{X}$ is a continuous variable in large samples (since it has a continuous density—the normal), we know that $P(\overline{X} = \mu) = 0$. On the other hand, we can use the sampling distribution of $\overline{X}$ to construct an interval of numbers within which we can be highly confident that μ falls. For example, a 95% *confidence interval* for μ is an interval of numbers of the form (a,b) that we are 95% confident contains the true value of μ. This means that if we were to engage in repeated sampling of size n from the population of X-values and construct this interval based on each sample, 95% of all such intervals would contain μ. The formula for the 95% large-sample confidence interval for μ is

$$\overline{X} \pm 1.96 \frac{s}{\sqrt{n}},$$

where s is the sample standard deviation of X. This formula has the following justification. Assuming a random sample of $n \geq 30$ from the population of X-values, we know by the CLT that $\overline{X}$ is normally distributed with mean μ and standard deviation $\sigma/\sqrt{n}$. This means that the variable $Z = (\overline{X} - \mu)/(\sigma/\sqrt{n})$ has the standard normal distribution. We also know that for any variable Z with a standard normal distribution, $P(-1.96 < Z < 1.96) = .95$. Putting these two facts together, we have that

$$P\left(-1.96 < \frac{\overline{X} - \mu}{\sigma/\sqrt{n}} < 1.96\right) = .95.$$

Notice that we have an inequality inside the probability statement. At this point, a brief word is appropriate regarding inequalities. Adding or subtracting the same number from all sides of an inequality does not change the inequality. Moreover, multiplying or dividing by the same positive number does not change the inequality. However, multiplying or dividing by a negative number reverses the inequality. With these principles in mind, we manipulate the inequality inside the probability statement above so as to isolate μ in the center of the inequality:

$$P\left(-1.96 < \frac{\overline{X} - \mu}{\sigma/\sqrt{n}} < 1.96\right) = .95$$

implies that

$$P\left(-1.96\frac{\sigma}{\sqrt{n}} < \overline{X} - \mu < 1.96\frac{\sigma}{\sqrt{n}}\right) = .95$$

or

$$P\left(1.96\frac{\sigma}{\sqrt{n}} > \mu - \overline{X} > -1.96\frac{\sigma}{\sqrt{n}}\right) = .95$$

or

$$P\left(\overline{X} + 1.96\frac{\sigma}{\sqrt{n}} > \mu > \overline{X} - 1.96\frac{\sigma}{\sqrt{n}}\right) = .95,$$

or, reorienting the expression,

$$P\left(\overline{X} - 1.96\frac{\sigma}{\sqrt{n}} < \mu < \overline{X} + 1.96\frac{\sigma}{\sqrt{n}}\right) = .95.$$

Thus, employing the CLT we have arrived at a formula for creating an interval of numbers that we can be 95% confident contains the true value of μ. In that s is a consistent estimator of σ, we substitute s for σ in this formula when computing the interval using a large sample.

Hypothesis Tests. Although confidence intervals are extremely informative, social and behavioral scientists tend to rely more on *hypothesis testing* to make inferences about population parameters. *A hypothesis is a tentative statement about the value of one or more population parameters.* In fact, we typically pose two competing hypotheses. The *research* or *alternative hypothesis* is the one the researcher usually believes to be true. We marshal evidence in favor of the research hypothesis by showing that the sample evidence is inconsistent with a contrary hypothesis, the *null hypothesis.* Together, the null and research hypotheses are mutually exclusive statements that cover the entire parameter space. For example, suppose we believe that the population mean of a continuous variable X is greater than 5. To test this idea, we pose the opposite: that the population mean is, at most, equal to 5. If we can show that the empirical (i.e., sample) evidence is inconsistent with this opposite position, we tend to accept the research hypothesis. Formally, the hypotheses are as follows. The null hypothesis is H_0: $\mu \leq 5$; the research hypothesis is H_1: $\mu > 5$. Notice that the hypotheses are mutually exclusive as well as exhaustive: They subsume all possible values that the parameter can take.

There are two ways to proceed with a test. The first is to decide in advance how willing we are to make an incorrect decision against H_0 when it is, in fact, true. Suppose that we are willing to take a 5% chance of being wrong when we reject H_0. This .05 probability of being wrong is called the *alpha level* for the test, denoted α. It is the probability of rejecting H_0 when it is true for a particular test. This would be an incorrect decision and is called a *type I error.* Suppose, further, that we have taken a random sample of 100 observations on X and have the following sample statistics: $\bar{x} = 5.8$, $s = 4$. Now since our sample is large, we know (from the CLT) that the sampling distribution of $\bar{X}$ is approximately normal and centered over μ and has a standard deviation of approximately $4/\sqrt{100}$, which equals .4. We therefore reason as follows. We want a rejection rule for H_0 such that P(reject $H_0 | H_0$ true) is only .05. If H_0 is true, then μ is, at most, 5. This value is the null-hypothesized value of μ, denoted μ_0. Let's suppose that H_0 is true and that μ is, in fact, 5. Then $\bar{X}$ has a normal distribution centered over 5, with a standard deviation of .4. We will want to reject H_0 based on the distance of our sample mean from this value. In particular, if our sample mean is too far above this value to be likely, given H_0, we reject H_0 as implausible. We choose a critical value of $\bar{X}$, denoted $\bar{X}_{cv}$, as our benchmark, so that we reject H_0 if $\bar{X} > \bar{X}_{cv}$. Since we want to take only a 5% chance of being wrong, we want to choose a critical value of the sample mean such that P($\bar{X} > \bar{X}_{cv} | H_0$ true) = .05. That is, we want

$$P\left(\frac{\bar{X} - \mu_0}{s/\sqrt{n}} > \frac{\bar{X}_{cv} - \mu_0}{s/\sqrt{n}}\right) = P\left(Z > \frac{\bar{X}_{cv} - \mu_0}{s/\sqrt{n}}\right) = .05.$$

Since P($Z > 1.645$) = .05, this implies that $(\bar{X}_{cv} - 5)/.4$ must equal 1.645, or that $\bar{X}_{cv}$ must equal $5 + 1.645(.4) = 5.658$. Hence, $\bar{X}_{cv} = 5.658$, and we reject H_0 if $\bar{X} > 5.658$, knowing that in doing so, there is only a 5% chance of rejecting an H_0 that is true.

The second way to proceed with the test is much simpler. We simply ask: *What is the probability of getting sample evidence at least as unfavorable to H_0 as we have obtained if H_0 were actually true?* This probability, called the *p-value* for the test, represents the *smallest α level at which H_0 could be rejected* for a given test. The *p*-value is therefore

P(sample evidence at least as unfavorable to H_0 as obtained | H_0 true)

$$= P(\bar{X} > 5.8 | H_0 \text{ true})$$

$$= P\left(\frac{\bar{X} - 5}{.4} > \frac{5.8 - 5}{.4}\right)$$

$$= P(Z > 2) = .0228.$$

We then compare the *p*-value to the α level. If $p \leq \alpha$, we reject H_0 and accept H_1. If not, we fail to reject H_0. In this case, if α is set at .05, then .0228 < .05, so we reject H_0 and conclude that the mean is greater than 5. By reporting the actual *p*-value instead of simply whether H_0 is rejected, however, we indicate the strength of the evidence against H_0. In the current case, the evidence against H_0 is sufficient to reject it but is not particularly strong. However, if p were, say, .001, the chances of getting the current sample mean, if H_0 is true, would only be 1 in 1000, constituting substantially stronger evidence against the null.

Power of Hypothesis Tests. When H_0 is rejected, it is because the sample evidence is considered very unlikely to occur were H_0 true. However, if we do not reject H_0, should we "accept" it as true? The answer is no. Generally, we do not accept H_0 despite failing to reject it. The primary reason for this is that failing to reject H_0 is more a statement about our ability to detect departures from the null-hypothesized value of the parameter (i.e., μ_0) than about the true state of nature. That is, H_0 may be false, but our test may not have enough *power* to detect it. *The power of a test is the probability that we will reject H_0 using the test when H_0 is false.* As an example, consider the fictitious sampling problem just discussed. Suppose that we had gotten a sample mean of 5.4 instead of 5.8. Then the *p*-value for the test would have been

$$p = P(\overline{X} > 5.4 \mid H_0 \text{ true})$$

$$= P\left(Z > \frac{5.4 - 5}{.4}\right)$$

$$= P(Z > 1) = .1587.$$

Employing conventional α levels, such as .05 or even .1, would still result in a failure to reject H_0. However, what if μ is really 5.2? Then, of course, the true sampling distribution of $\overline{X}$ is still normal with a standard deviation of .4, but it is centered over 5.2 instead of 5. However, we are operating under the assumption that μ is 5 (hypothesis tests are always conducted under the assumption that H_0 is true); hence our critical value of $\overline{X}$ is still 5.658. Let's consider, then, the probability of making a correct decision (i.e., reject H_0) in this case, using $\alpha = .05$:

$$P(\text{correct decision} \mid \mu = 5.2) = P(\text{decision is to reject } H_0 \mid \mu = 5.2)$$

$$= P(\overline{X} > 5.658 \mid \mu = 5.2)$$

$$= P\left(Z > \frac{5.658 - 5.2}{.4}\right) = P(Z > 1.145) = .1261.$$

Thus, the power of the test under this condition is only .1261. That is, there is only about a 13% chance of detecting that the null hypothesis is false in this case. If, on the other hand, μ were really 6.2, the probability of correctly rejecting H_0 would be

$$P(\overline{X} > 5.658 \mid \mu = 6.2) = P\left(Z > \frac{5.658 - 6.2}{.4}\right)$$

$$= P(Z > -1.355) = .9123.$$

In this case, there is a greater than 90% chance that we will detect that H_0 is false. As these extremes show, because we have no idea what the true value of the parameter is, it would be unwise to accept H_0 as true simply because we have failed to reject it.

More Complex Hypotheses. Hypotheses are always statements about the values of one or more population parameters. Although this definition sounds restrictive, it's not. Even complex hypotheses can be distilled down to statements about parameter values. For example, suppose I hypothesize that *open disagreements in marriage have a stronger positive effect on depressive symptomatology to the extent that individuals are less happily married.* How do we boil this down into a statement about a parameter? As we will see in Chapter 3, this hypothesis can be tested with the following regression model. If Y is a depressive symptomatology score, X_1 is open disagreement, and X_2 is marital happiness, all continuous variables measured for a sample of n cases, we can estimate the following regression model: $E(Y) = \beta_0 + \beta_1 X_1 + \beta_2 X_2 + \beta_3 X_1 X_2$, where $X_1 X_2$ is the product of X_1 with X_2. It turns out that our research hypothesis can be restated in terms of one model parameter, β_3. In particular, the research hypothesis is $\beta_3 < 0$.

NOTES

1. Finding the area, A, under a curve, $f(x)$, involves the use of integral calculus. The process of integration is, intuitively, as follows. To find the area under $f(x)$ between a and b, we divide the area into n equal-sized rectangles which are contained completely within the area between $f(x)$ and the X-axis. These are called *inscribed rectangles.* The area of a rectangle is the base of the rectangle times its height. The bases of each rectangle are the same: $(b - a)/n$. The height is the point at which the rectangle first touches the curve from beneath. If the X-value corresponding to that point is denoted x^*, the height of the rectangle at that point is $f(x^*)$. Hence, $f(x^*)$ times $(b - a)/n$ is the area of each inscribed rectangle. To approximate the area under the curve, we sum up all of the rectangles under $f(x)$ from a to b. Although this approximation is quite crude, we can get a better approximation by increasing n and repeating the procedure. This creates a more fine-grained subdivision of the area in question into narrower rectangles, which fill in more of the gaps under the curve. Once again, we compute the sum of the rectangles to refine our estimate of the area. If we repeat this procedure with ever-larger values of n, the resulting sum of the rectangles eventually converges to the area in question, as n tends to infinity.

2. *Asymptotically* refers to the behavior of a sample statistic as the sample size tends to infinity. Asymptotic results usually provide very good guidelines for what we can expect when n is large. How "large" n has to be for asymptotic results to be approximately correct must be established on a case-by-case basis, via simulation.

CHAPTER 2

Simple Linear Regression

CHAPTER OVERVIEW

This chapter explicates the simple linear regression (SLR) model and its application to social data analysis. I begin by discussing some real-data examples of linear relationships between a dependent variable, Y, and an explanatory variable, X. It turns out that many relationships between two continuous variables are approximately linear in nature, rendering linear regression an ideal approach to analysis, at least as a beginning. I then introduce the model formally and discuss its interpretation as well as the assumptions required for estimation via ordinary least squares (OLS). This is followed by an explanation of estimation using OLS, the most commonly used estimation technique. I then consider several theoretical properties of the slope estimate necessary for inferences about the relationship of X with Y. The chapter concludes with a discussion of model evaluation. In particular, I consider various means of checking regression assumptions as well as assessing the discriminatory power and empirical consistency of one's model. This chapter contains a fair amount of formal development of properties of the model. The reason for this is that such developments readily generalize to the case of multiple regression but are much simpler to do with only one predictor. This obviates the need to resort to matrix algebra, as would be the case with several predictors in the model. Nevertheless, readers with a limited math background will find it very helpful to review Sections I.P, II, and IV of Appendix A in tandem with reading this chapter. There you will find a tutorial on functions—in particular, linear functions—as well as tutorials on summation notation and derivatives. These skills will enhance your assimilation of the material in this chapter.

LINEAR RELATIONSHIPS

Recall the introductory statistics data described in Chapter 1. One of the questions frequently asked of an instructor in introductory statistics is: "How good does my

Regression with Social Data: Modeling Continuous and Limited Response Variables,
By Alfred DeMaris
ISBN 0-471-22337-9   Copyright © 2004 John Wiley & Sons, Inc.

math background have to be to do well in the class?" One answer to the question might be to examine how well math proficiency "predicts" or "explains" student performance on the first exam. Intuitively, given that statistics is a form of applied mathematics, one expects that students with stronger math skills should perform better on the first exam than those whose skills are weaker. Two approximately continuous variables have been recorded for 213 students who took the first exam. The first is their score on a math diagnostic quiz administered to all students on the first day of class. The quiz consists of 45 questions, with the score being the number of items answered correctly. Therefore, the score range is 0 to 45. This quiz roughly gauges a student's proficiency in math and is considered the explanatory variable, X, in this example. (The explanatory variable is also referred to as the *predictor*, the *regressor*, the *independent variable*, or the *covariate*.) The response variable (also called the *outcome*, the *criterion*, or the *dependent variable*) is the score on the first exam, which is administered during the sixth week of class. Its range is 0 to 102.

Figure 2.1 is a scatterplot of the (x,y) points for the 213 students, where X is *math diagnostic score* and Y is *score on exam 1*. There is a fairly clear linear upward trend in the scatter of points. Those with relatively low scores on the diagnostic (the minimum score was 28) tend to have relatively low scores on exam 1.

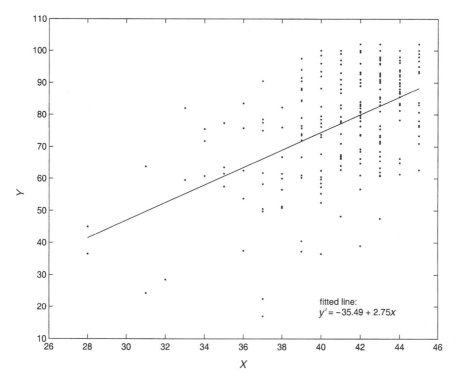

Figure 2.1 Scatterplot of Y (score on the first exam) against X (score on the math diagnostic) for 213 students in introductory statistics.

As the diagnostic takes on higher values, the scatter of points migrates upward. Those scoring in the middle 40s on the diagnostic, for example, have the highest exam performance, on average. This upward trend is nicely represented by the fitted line drawn through the center of the points. For now, the interpretation of this line is that it is the best-fitting straight line through the scatter of points were we to attempt to fit a straight line through them. (The meaning of *best fitting* is explained in more detail below.)

As another example we consider the *couples dataset*. Partners' attitudes about gender roles can sometimes be a source of friction for a couple, particularly when it comes to sharing housework or other necessary tasks. Family sociologists have therefore studied the factors that affect whether partners' attitudes are more traditional as opposed to more modern. One such factor, of course, is education. One would expect that couples with more years of schooling would tend to have more modern, or egalitarian, attitudes toward gender roles compared to others. A series of items asked in the NSFH allows us to tap gender-role attitudes. Four were included in the *couples dataset* and were asked of each partner of the couple. An example item asked for the extent of agreement with the statement: "It's much better for everyone if the man earns the main living and the woman takes care of the home and family." Response choices ranged from "strongly agree," coded 1, to "strongly disagree," coded 5. The other items were all similarly coded so that the high value represented the most modern, or egalitarian, response. To create a *couple modernism* score, I summed all eight items for both partners. The resulting scale ranged from 8 to 40, with the highest score representing couples with the most egalitarian attitude. Figure 2.2 shows a scatterplot of Y = *couple gender-role modernism* plotted against *male partner's years of schooling completed* (X) for the 416 couples in the dataset. The linear trend is again evident in the scatter of points, as highlighted by the fitted line.

As a third example, I draw on the *GSS98 dataset*. Among the questions asked of 1515 adult respondents in this survey was one about sexual activity. In particular, the question was: "About how often did you have sex during the last 12 months?" Response choices were coded 0 for "not at all," 1 for "once or twice," 2 for "about once a month," 3 for "2 or 3 times a month," 4 for "about once a week," 5 for "2 or 3 times a week," and 6 for "more than 3 times a week." Although this variable is not truly continuous, it is ordinal, and has enough levels—five or more is enough—to be treated as "approximately" continuous. This approximation is especially tenable if n is large, as it is for these data, and if the distribution of the variable is not too skewed. Regarding the latter condition, the percent of people giving each response—0, 1, 2, 3, 4, 5, 6— is 22.2, 7.5, 11.3,15.4, 18.3, 19.7, and 5.4, respectively. This represents an acceptable level of skew.

What predicts sexual activity? Several obvious determinants come to mind, such as age and health. But what about frequenting bars? There are several reasons why those who frequent bars might be more sexually active than others. Some reasons have to do with the selectivity of bar clientele and do not implicate bars as a cause of sexual activity per se. For example, sexually active couples often go to bars or night-clubs first before engaging in intimate activity. Also, those who go to bars more often are most likely younger and possibly looking for sexual partners to begin with.

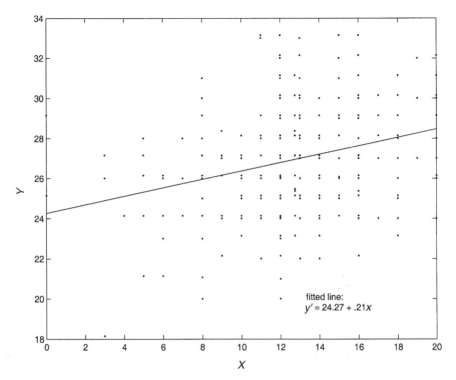

Figure 2.2 Scatterplot of Y (couple modernism score) against X (male's years of schooling) for 416 couples from the National Survey of Families and Households, 1987–1988.

Hence, it would not be surprising that bar clientele would tend to be more sexually active than others. On the other hand, frequenting bars might also play a causal role in enhancing sexual activity. First, regardless of whether one is looking for a sexual partner, bars tend to enhance one's opportunities for meeting potential sex partners. "Singles" bars are especially likely to draw a clientele that is in the "market" for sex, enhancing one's potential pool of willing partners. Then, too, the atmosphere of bars is especially conducive to romantic and/or sexual arousal. Alcohol is being freely consumed, the lights are usually dimmed, and erotic music is often playing in the background. For any of the foregoing reasons, I would expect a positive relationship between going to bars and having sex. Fortunately, it's easy to explore this hypothesis, since the GSS also asked respondents: "How often do you go to a bar or tavern?" with responses ranging from 1 for "never" to 7 for "almost every day." An examination of the distribution of this variable reveals that it is somewhat more skewed than sexual frequency (e.g., 49.2% of respondents never go to bars) but still at an acceptable level of skew for regression.

A scatterplot of $Y = frequency\ of\ having\ sex\ in\ the\ past\ year$ against $X = frequency\ of\ going\ to\ bars$ is shown in Figure 2.3. This scatterplot is somewhat deceptive. At first glance there is no visible linear trend. However the best-fitting line through these points does, indeed, have a slight upward slope, going from left to right. This indicates

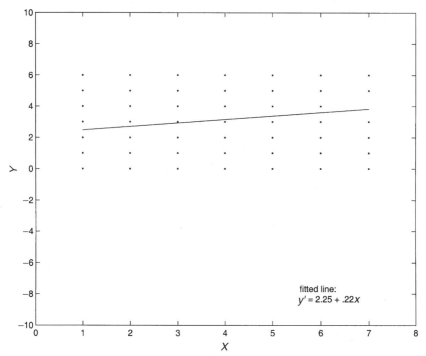

Figure 2.3   Scatterplot of Y (frequency of having sex in past year) against X (frequency of going to bars) for 1515 respondents from the General Social Survey, 1998.

that there is a positive relationship between X and Y. We will see (below) that this positive trend is also statistically significant.

SIMPLE LINEAR REGRESSION MODEL

The idea behind regression is quite simple. Again using the relationship in Figure 2.1 as a substantive referent, we imagine that a linear relationship between, say, *math diagnostic score* and *score on the first exam* characterizes the population from which the student sample was drawn. (In that this sample was not randomly drawn from any population, we have to imagine a hypothetical population of introductory statistics students that might be generated via repeated sampling.) That is, a straight line characterizes the relationship between *math diagnostic score* and *first exam performance* in the population, much as the fitted line in Figure 2.1 characterizes the relationship between these variables in the sample. Our task is, essentially, to find that line. The line is of the form $\beta_0 + \beta_1 X$, where β_0 is the Y-intercept of the line and β_1 is the slope of the line. That is, we want to estimate the parameters (β_0 and β_1) of that line, representing the "regression" of Y on X, and use them to further understand the relationship between the variables. The simple linear regression model is therefore nothing more than the equation for this line, plus a few additional assumptions.

Regression Assumptions

In particular, we assume (1) that the Y_i for $i = 1, 2, \ldots, n$ represent n independently sampled observations on a response variable, Y. We further assume (2) that Y for each observation has been "generated" (say, by nature) by a linear function applied to that case's value on an explanatory variable, X. The model for the ith observation on Y is therefore

$$Y_i = \beta_0 + \beta_1 X_i + \varepsilon_i. \tag{2.1}$$

It is assumed (3) that both Y and X are continuous variables. This is not to be taken too literally. As indicated above, regression is acceptable as long as Y and X are approximately continuous. A rule of thumb for *approximately continuous* to hold is that the variable should be at least ordinal and have at least five different levels, provided that n is large and that the distribution of the variable is not too skewed. In fact, X can also be dichotomous. This type of predictor, called a *dummy variable*, is introduced in Chapter 4 (as well as in Exercises 2.26 to 2.28). Moreover, Y can be dichotomous, although some problems emerge when using linear regression in such a case (more on this in Chapter 7). It is assumed (4) that X is fixed over repeated sampling. That is, the assumption is that observations on Y are sampled from subpopulations with values of X that are known ahead of time (later we will see that this assumption can be relaxed under certain conditions). It is also assumed (5) that X is measured with only negligible error (Myers, 1986).

The ε_i's, called the equation *errors* or *disturbances*, represent all influences on Y other than X. If Y were an *exact* linear function of X, that is, if Y_i were exactly equal to $\beta_0 + \beta_1 X_i$, called the *structural part of the model*, and all points lay right on the regression line, there would be no error term. However, this would represent a *deterministic relationship* between Y and X. That is, Y would be determined completely by X, such that increasing X by 1 unit would automatically increase Y by β_1 units. Such relationships are rarely—no, never—observed in the social sciences. Instead, social data tend to be characterized by *probabilistic relationships* in which increasing X increases the *likelihood* that Y will increase (or decrease, if $\beta_1 < 0$) (Lieberson, 1997), but doesn't guarantee it. The error term represents this *random* or *stochastic component* of the model. Manipulating equation (2.1), we see that $\varepsilon_i = Y_i - (\beta_0 + \beta_1 X_i)$. That is, the errors represent the departures of Y-values from the line $\beta_0 + \beta_1 X_i$. Typically, for any particular value x_i of X_i, there will be several different Y_i values among the observations with that x_i. Hence, there will be several different ε_i at any particular value of X.

The errors are also subject to some assumptions. It is assumed (6) that at each value of X, the mean of the errors is zero. The assumption is that the positive and negative errors tend to balance each other out, so that overall, the mean of the errors at any given X is zero. Additionally, it is assumed (7) that the errors all have the same variance at each X-value; this variance is denoted by σ^2. It is also assumed (8) that the errors are uncorrelated with each other. If the observations have been sampled independently and the data are cross-sectional, this is typically the case. Finally, it is assumed (9) that the errors are normally distributed in the population. The proper way to think about the regression assumptions is not: "Okay, let's assume that . . ." and then proceed with a regression analysis. Rather, the proper perspective is: "*If* the following assumptions hold, it makes sense to estimate a linear regression model." Below I discuss ways of assessing whether these assumptions are reasonable.

Interpreting the Regression Equation

Taking expectations of both sides of equation (2.1), we have

$$
\begin{aligned}
E(Y_i) &= E(\beta_0 + \beta_1 X_i + \varepsilon_i) \\
&= E(\beta_0 + \beta_1 X_i) + E(\varepsilon_i) \\
&= \beta_0 + \beta_1 X_i + 0 \\
&= \beta_0 + \beta_1 X_i.
\end{aligned}
\tag{2.2}
$$

(Note that the β's and X_i are constants at any particular X_i, and the mean of a constant is just that constant.) In other words, according to the model $E(Y_i)$, the mean of the Y_i (more properly, the *conditional mean* of the Y_i, since this mean is conditional on the value of X) is a linear function of the X_i. That is, the mean of the Y_i at any given X_i is simply a point on the regression line. Hence, using regression modeling assumes, at the outset, that the means of the dependent variable at each X-value lie on a straight line. To understand how to interpret β_0 and β_1, we manipulate equation (2.2) so as to isolate either parameter. For example, if $X_i = 0$, we have

$$
E(Y_i \mid X_i = 0) = \beta_0 + \beta_1(0) = \beta_0.
$$

Hence, β_0, the intercept, is the mean of Y when X equals zero. To isolate β_1, we consider the difference in the mean of Y for observations that are 1 unit apart on X. That is, we compute the difference $E(Y_i \mid x_i + 1) - E(Y_i \mid x_i)$. Notice that this is the difference in $E(Y)$ for a unit difference in X, regardless of the level of X at which that unit difference occurs. We have

$$
\begin{aligned}
E(Y_i \mid x_i + 1) - E(Y_i \mid x_i) &= \beta_0 + \beta_1(x_i + 1) - [\beta_0 + \beta_1 x_i] \\
&= \beta_0 + \beta_1 x_i + \beta_1 - \beta_0 - \beta_1 x_i = \beta_1.
\end{aligned}
$$

In other words, β_1 represents the difference in the mean of Y in the population for those who are a unit apart on X. If X is presumed to have a causal effect on Y, β_1 might be interpreted as the expected *change* in Y for a unit *increase* in X. In either case, I will refer to $E(Y_i \mid x_i + 1) - E(Y_i \mid x_i)$ as the *unit impact* of X in the model. The unit impact of X in any regression model—whether linear or not—can always be found by computing $E(Y_i \mid x_i + 1) - E(Y_i \mid x_i)$ according to the model.

The interpretation of β_1 can also be elucidated by taking the first derivative of $E(Y_i)$ with respect to X:

$$
\frac{d}{dX_i}[E(Y_i)] = \frac{d}{dX_i}[\beta_0 + \beta_1 X_i] = \beta_1.
$$

Thus, we see that β_1 is also the first derivative of $E(Y_i)$ with respect to X_i. Recall (from Section IV of Appendix A) that the first derivative, also called the *slope* of $E(Y_i)$ with respect to X_i, represents the instantaneous rate of change in $E(Y_i)$ with an increase in X_i at point x_i. It also represents the slope of the line tangent to the curve relating Y to X at point x. Because the curve in this case is really a straight line, the slope of the tangent line is just the slope of the regression line itself (there is no tangent line in this case, since it is impossible for a line to touch a straight line at only one point). Hence, the instantaneous rate of change in $E(Y)$ with increase in X is the same as the change in $E(Y)$ per unit increase in X. In other words, in regression models in which $E(Y)$ is a linear function of X, the first derivative and the unit impact are identical quantities. This is no longer the case for models in which $E(Y)$ is modeled as a nonlinear function of X (discussed in subsequent chapters).

ESTIMATION USING SAMPLE DATA

Estimation of the regression equation and associated parameters using sample data is most often accomplished by employing ordinary least squares estimation (OLS). The assumptions enumerated above are largely assumptions required for unbiased estimation of the regression parameters using OLS. Let's review the assumptions. They are:

1. Y is a linear function of X; that is, $Y_i = \beta_0 + \beta_1 X_i + \varepsilon_i$ for $i = 1, 2, \ldots, n$.
2. The observations are sampled independently.
3. X and Y are approximately continuous variables.
4. The X-values are fixed over repeated sampling and measured with only negligible error.
5. $E(\varepsilon_i) = 0$ at each x_i.
6. $V(\varepsilon_i) = \sigma^2$ at each x_i.
7. $Cov(\varepsilon_i, \varepsilon_j) = 0$ for $i \neq j$ (i.e., the errors are uncorrelated with each other). (Note that this is usually equivalent to assumption 2.)
8. The errors are normally distributed.

Assumption 5 also ensures the orthogonality condition that $Cov(X_i, \varepsilon_i) = 0$. The reason for this is straightforward. If there were a linear relationship between ε_i and X_i, reflected by a nonzero covariance, it would take the form $E(\varepsilon_i) = \gamma_0 + \gamma_1 X_i$. That is, the mean of the errors would be a linear function of X. If the mean of the errors is the same (in particular, it is zero) at each X, this implies that there is no linear relationship between the error and X. This, in turn, implies that $Cov(X_i, \varepsilon_i)$ is zero.

Rationale for OLS

To develop the rationale for OLS, we consider the following example. Table 2.1 presents fictitious data for five currently married persons; X is the number of years married and Y is a continuous measure of marital happiness. A scatterplot of these

Table 2.1 Visual Approximation versus Ordinary Least Squares Regression Lines Fit to Fictitious Data on Marital Happiness (Y) and Number of Years Married (X) for Five Persons

Person	Years Married	Marital Happiness	$\hat{y}_{VA}$	e_{VA}	$\hat{y}_{OLS}$	e_{OLS}
1	1	5	3.25	1.75	3.58	1.42
2	3	2	3.75	−1.75	3.96	−1.96
3	4	6	4.00	2.00	4.15	1.85
4	9	3	5.25	−2.25	5.10	−2.10
5	20	8	8.00	.00	7.19	.81
$\sum e$				−.25		.02
$\sum e^2$				15.1875		14.3466

Calculation of OLS Estimates

Case	X	Y	$(X - \bar{X})$	$(Y - \bar{Y})$	$[(X - \bar{X})(Y - \bar{Y})]$	$(X - \bar{X})^2$
1	1	5	−6.4	.2	−1.28	40.96
2	3	2	−4.4	−2.8	12.32	19.36
3	4	6	−3.4	1.2	−4.08	11.56
4	9	3	1.6	−1.8	−2.88	2.56
5	20	8	12.6	3.2	40.32	158.76
Sum					44.4	233.2

$$b_1 = \frac{44.4}{233.2} = .1904$$

$$b_0 = 4.8 - .1904(7.4) = 3.3911$$

(x,y) values is shown in Figure 2.4. There appears to be somewhat of a positive linear trend in the points, such that those married longer are, on average, more happily married. Assuming that there is a linear relationship between X and Y in the population, how do we go about estimating the population regression line representing this relationship? Well, suppose that we start with a visual approximation of the line that best fits the points. One possibility is the lightly drawn line in the figure, which touches the rightmost data point. The equation for this line is $\hat{y} = 3 + .25x$ ($\hat{y}$ is shown as "y'" in the figure). Here, $\hat{y}$, called the *fitted*, or *predicted*, *value* of Y, denotes the sample estimate of E(Y) (I have dropped the i subscript here for economy of expression). The sample regression equation is therefore $Y = 3 + .25x + e$, where e, called a *residual*, is our sample estimate of the equation error for a particular case. Notice that $e_i = y_i - \hat{y}_i$ is the difference between observed and fitted Y-values for the ith case. That is, e is the vertical distance from the point to the regression line for the ith data point. For example, the residual for the second observation is shown as e_2' in Figure 2.4. Table 2.1 shows the five fitted values of Y using the visual approximation (VA) approach, along with the residuals for each case.

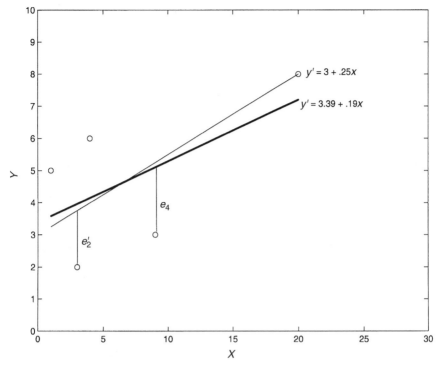

Figure 2.4 Scatterplot of Y (marital happiness) against X (number of years married) for five fictitious married persons.

Measuring Prediction Error. How well does this line fit the data? One way to tell is to examine the total prediction error made in using this equation to predict Y. Our first impulse might simply be to sum the e's for the five cases to assess total error. However, that doesn't work well, since large positive errors and large negative errors tend to cancel each other out. We might therefore end up with a small total error even with a poorly fitting line. Instead, the errors are first squared and then summed. The resulting quantity is called (fittingly) the *sum of squared errors* and denoted *SSE*:

$$SSE = \sum_{i=1}^{n} e_i^2 = \sum_{i=1}^{n} (y_i - \hat{y}_i)^2.$$

The idea behind ordinary least squares is to choose as our estimate of the population regression line that sample line that *minimizes SSE* (hence the resulting b_0 and b_1 are called the ordinary *least squares estimates*). The least squares line for these data is shown as the darker, thicker line in Figure 2.4 and is also the fitted line drawn through the point scatters in Figures 2.1 to 2.3. The least squares equation is $Y = 3.39 + .19x + e$. The residual for the fourth observation, based on the least squares line, is shown as e_4 in Figure 2.4. Table 2.1 also shows the fitted values and residuals based on the OLS regression equation. Notice that SSE_{OLS}, at 14.3466, is smaller than SSE_{VA}, which is 15.1875. In fact, the *SSE* based on the OLS estimates of b_0 and

b_1 is guaranteed to be smaller than *SSE* based on any other estimates. How is this condition satisfied?

Mathematics of OLS

Finding the b_0 and b_1 that minimize *SSE* is a minimization problem in two variables, which is readily solved using the techniques of differential calculus. First, we rewrite *SSE* to highlight the fact that it is a function of b_0 and b_1:

$$SSE = \sum_{i=1}^{n}(y_i - \hat{y}_i)^2 = \sum_{i=1}^{n}[y_i - (b_0 + b_1 x_i)]^2.$$

To find the b_0 and b_1 that minimize this function using calculus, we take the first derivative of *SSE* with respect to, first, b_0 and then b_1. We then set each of these expressions to zero and solve the resulting simultaneous equations for the b_0 and b_1 that satisfy the pair of equations. (Okay, this sounds complicated, but bear with me here. Of course, it will help immensely if you have read and absorbed Section IV of Appendix A before proceeding.) By a theorem in calculus, if *SSE* has a maximum or minimum at some value of (b_0, b_1), it occurs where these first derivatives equal zero (Anton, 1984).

Taking the first derivative of *SSE* with respect to b_0 gives us

$$\frac{\partial}{\partial b_0}\left[\sum_{i=1}^{n}(y_i - b_0 - b_1 x_i)^2\right] = -2\sum_{i=1}^{n}(y_i - b_0 - b_1 x_i),$$

which equals zero when

$$\sum_{i=1}^{n}(y_i - b_0 - b_1 x_i) = 0$$

or when

$$\sum_{i=1}^{n}y_i - nb_0 - b_1\sum_{i=1}^{n}x_i = 0$$

or when

$$nb_0 = \sum_{i=1}^{n}y_i - b_1\sum_{i=1}^{n}x_i.$$

Dividing both sides by n, we have

$$b_0 = \bar{y} - b_1\bar{x}. \tag{2.3}$$

So we have the solution for b_0, but of course it requires the solution for b_1 to compute. Therefore, we solve for b_1. Again we first take the derivative of *SSE* with respect to b_1:

$$\frac{\partial}{\partial b_1}\left[\sum_{i=1}^{n}(y_i - b_0 - b_1 x_i)^2\right] = -2\sum_{i=1}^{n}x_i(y_i - b_0 - b_1 x_i),$$

which equals zero when

$$\sum_{i=1}^{n} x_i(y_i - b_0 - b_1 x_i) = 0$$

or when

$$\sum_{i=1}^{n} x_i y_i - b_0 \sum_{i=1}^{n} x_i - b_1 \sum_{i=1}^{n} x_i^2 = 0.$$

At this point we can substitute the solution for b_0 from equation (2.3) for b_0, and after some mildly tedious algebra, we arrive at

$$b_1 = \frac{\sum_{i=1}^{n}(x_i - \bar{x})(y_i - \bar{y})}{\sum_{i=1}^{n}(x_i - \bar{x})^2}. \qquad (2.4)$$

Dividing both numerator and denominator of the right-hand side of equation (2.4) by $n - 1$, we see that the slope can also be expressed as

$$b_1 = \frac{\text{cov}(x,y)}{s_x^2}.$$

The equations $\sum_{i=1}^{n}(y_i - b_0 - b_1 x_i) = 0$ and $\sum_{i=1}^{n} x_i(y_i - b_0 - b_1 x_i) = 0$ are called the *normal equations*. Once we have found the values of b_0 and b_1 that satisfy these equations, we must then show that plugging these particular values into

$$SSE = \sum_{i=1}^{n}(y_i - \hat{y}_i)^2 = \sum_{i=1}^{n}(y_i - b_0 - b_1 x_i)^2$$

produces a minimum value for this function. Formally, by the second partials test (Anton, 1984), this can be shown using second partial derivatives of the function $f = \sum_{i=1}^{n}(y_i - b_0 - b_1 x_i)^2$. In particular, let f_{b_0,b_0} represent the second partial derivative of f with respect to b_0, f_{b_1,b_1} represent the second partial derivative of f with respect to b_1, and f_{b_0,b_1} represent the mixed second partial derivative of f with respect to b_0, followed by b_1. Then, by the second partials test, if the expression $D = (f_{b_0,b_0})(f_{b_1,b_1}) - f_{b_0,b_1}^2$ is greater than zero and f_{b_0,b_0} is greater than zero when these expressions are evaluated at a particular value of b_0 and b_1, then f has a minimum at those particular values of b_0 and b_1. In fact, f_{b_0,b_0} evaluated at the value in expression (2.3) is $2n$, f_{b_1,b_1} evaluated at the value in expression (2.4) is $2\sum x^2$, and f_{b_0,b_1} evaluated at expressions (2.3) and (2.4) is $2\sum x$. Hence,

$$D = 2n\left(2\sum x^2\right) - \left(2\sum x\right)^2 = 4n\left[\sum (x - \bar{x})^2\right],$$

which is greater than zero, as is f_{b_0,b_0}. Therefore, the OLS solutions to the normal equations are, in fact, the values that minimize *SSE*.

Intuitively, we can arrive at the same conclusion by simply regarding the function $f = \sum_{i=1}^{n}(y_i - b_0 - b_1 x_i)^2$. If there is an extreme value of this function, it *has* to be a minimum, since the maximum value is unbounded. That is, there is no finite maximum since we can make f as large as we wish simply by choosing b_0, b_1 values—and,

therefore a line—that is as far from the points as we want to make it. By the principles of calculus, the solutions to the normal equations show us the b_0 and b_1 that lead
to a critical value of the function. That critical value, by this reasoning, must therefore be a minimum.

The bottom panel of Table 2.1 illustrates the OLS computations for b_0 and b_1 for
our fictitious data. As shown, b_0, the estimate of the equation intercept, is 3.3911. This
implies that the average happiness level for newlyweds (i.e., those with zero years
married) is 3.3911. The intercept is not typically very meaningful whenever zero is
outside the range of observed X-values, as it is here. In fact, zero may not be a meaningful value for many explanatory variables, rendering the intercept uninterpretable
much of the time. The estimate of the equation slope, b_1, is .1904. This suggests that
those who are a year apart in marital duration are about .1904, or approximately two-
tenths of a unit, apart in marital happiness, on average. Or, in somewhat stronger
causal language, each additional year of marriage would be expected to increase marital happiness, on average, by about two-tenths of a point. Using the estimated equation, we can generate predicted marital happiness scores based on the number of years
that a person has been married. Hence, for someone married, say, 15 years, the predicted, or fitted, value of marital happiness is $\hat{y} = 3.3911 + .1904(15) = 6.2471$.
According to the model, this value is the estimated mean of marital happiness for all
those who have been married for 15 years. It is generally safe to generate predicted
scores on Y for X-values within the range observed in one's sample. However, we
would not want to try to predict Y for values outside that range—say, for someone
married 25 years in the current example. The reason for this is that the relationship
may or may not be linear beyond the range observed; hence the model may no longer
hold for more extreme levels of X.

OLS Estimates for the Examples. OLS estimates of the regression equations for the
regression of *exam performance* on *math diagnostic score*, the regression of *couple
modernism* on *male's education*, and the regression of *sexual frequency* on *going to
bars* are shown in Tables 2.2, 2.3, and 2.4, respectively. (Shown also are several other
statistics discussed below.)

**Table 2.2   Parameter Estimates from the Regression of $Y =$ Score on the First
Exam on $X =$ Math Diagnostic Score for 213 Students in Introductory Statistics**

Explanatory Variable	b	$\hat{\sigma}_b$	t	b^s	p
Constant	−35.494	12.840	−2.764		.006
Math diagnostic score	2.749	.313	8.788	.518	< .001

Model Summary Measure	Value
F	77.236
R^2	.268
R^2_{adj}	.265
$\hat{\sigma}^2$	213.756

Table 2.3 Parameter Estimates from the Regression of Y = Couple Modernism Score on X = Male's Years of Schooling for 416 Couples from the National Survey of Families and Households, 1987–1988

Explanatory Variable	b	$\hat{\sigma}_b$	t	b^s	p
Constant	24.274	.490	49.522		<.001
Male's years of schooling	.210	.038	5.586	.265	<.001

Model Summary Measure	Value
F	31.204
R^2	.070
R^2_{adj}	.068
$\hat{\sigma}^2$	5.751

Table 2.4 Parameter Estimates from the Regression of Y = Frequency of Having Sex in Past Year on X = Frequency of Going to Bars for 1515 Respondents from the General Social Survey, 1998

Explanatory Variable	b	$\hat{\sigma}_b$	t	b^s	p
Constant	2.252	.086	26.206		<.001
Frequency of going to bars	.224	.029	7.708	.194	<.001

Model Summary Measure	Value
F	59.417
R^2	.038
R^2_{adj}	.037
$\hat{\sigma}^2$	3.696

Exam Scores. The slope for the regression of *exam performance* on *math diagnostic score* is 2.749. This implies that those who are 1 unit apart on the diagnostic are, on average, about $2\frac{3}{4}$ points apart on exam performance. Moreover, someone with a perfect score of 45 on the diagnostic is expected to get a grade of $\hat{y} = -35.494 + 2.749(45) = 88.2$ on the first exam. Notice the uninterpretability of the intercept in this case, since it is impossible to have a negative score on the exam. Moreover, as the lowest observed score on the diagnostic was 28, trying to predict exam scores for those with diagnostic scores of zero—albeit a plausible value for the diagnostic—would be extrapolating beyond the range of observed X-values for these data. In all likelihood, the relationship between exam performance and *diagnostic score* becomes nonlinear at lower diagnostic scores, since zero on the exam is a "floor" below which no one can go.

Couple Modernism. From Table 2.3 we see that the estimated intercept for the regression of *couple modernism* on *male's years of schooling* is 24.274. In this case, the intercept is somewhat meaningful, since there is a case in which the male has zero years of schooling. In this instance, the predicted score for couple modernism is 24.274. Given that there is only one observation in which the male has zero years of schooling, however, the prediction of couple modernism at this level is probably not very robust. On the other hand, for males with four years of college—an X-value with several observations on Y—the predicted modernism score is $24.274 + .21(16) = 27.634$, a more robust estimate. The slope estimate of $.21$ suggests that each additional year of schooling for the male partner is expected to increase the couple's level of modernism by about two-tenths of a point.

Frequency of Sex. The estimates for the regression of *frequency of sex* on *going to bars* (Table 2.4) suggest that those who are 1 unit apart in frequenting bars are about .224 unit apart on sexual frequency during the previous year, on average. As neither Y nor X in this case is measured in a meaningful metric, the value of this slope is not very intuitive. Perhaps it suffices just to notice that there is, indeed, a positive relationship between going to bars and having sex. Notice, again, that the intercept is not particularly meaningful, since zero is not a possible value for the independent variable. Those who go to bars almost every day are predicted to have a sexual frequency during the past year, on average, of $2.252 + .224(7) = 3.82$. This corresponds roughly to having sex about once per week.

Estimating σ^2 and ρ^2. Two other quantities connected to the linear regression model are of interest to estimate. The first is σ^2, the variance of the error terms at each X-value. The second is denoted ρ^2 (pronounced "rho-squared") and is called the *coefficient of determination* of the regression equation. The latter is the primary index of discriminatory power for a regression model. To develop estimators of these quantities, let's consider partitioning the variability in Y, based on the assumption that it is a linear function of X in the population. Hence, if $Y = \beta_0 + \beta_1 X + \varepsilon$ (again, omitting the i subscript for simplicity), then

$$\begin{aligned} V(Y) &= V(\beta_0 + \beta_1 X + \varepsilon) = \text{Cov}[(\beta_0 + \beta_1 X) + \varepsilon, (\beta_0 + \beta_1 X) + \varepsilon] \\ &= \text{Cov}(\beta_0 + \beta_1 X, \beta_0 + \beta_1 X) + \text{Cov}(\beta_0 + \beta_1 X, \varepsilon) \\ &\quad + \text{Cov}(\varepsilon, \beta_0 + \beta_1 X) + \text{Cov}(\varepsilon, \varepsilon) \\ &= V(\beta_0 + \beta_1 X) + V(\varepsilon), \end{aligned}$$

since $\text{Cov}(X, \varepsilon) = 0$ by assumption [and, of course, $\text{Cov}(\beta_0, \varepsilon) = 0$, since the covariance of a constant with a variable is always zero]. What this shows is that the variability in Y can be neatly partitioned into two parts. The first, $V(\beta_0 + \beta_1 X)$, is the variability of the regression line itself, or the variance of the set of points that forms that line. This is the variance that is due to the linear function involving X, or the variance of the structural part of the model. The second, $V(\varepsilon)$, is the variance *around* that line, and reflects the variability in Y due to the stochastic part of the model—all of the presumably random influences on Y other than its linear relationship with X.

Now if we divide both sides of the equation $V(Y) = V(\beta_0 + \beta_1 X) + V(\varepsilon)$ by $V(Y)$, we have

$$\frac{V(Y)}{V(Y)} = 1 = \frac{V(\beta_0 + \beta_1 X)}{V(Y)} + \frac{V(\varepsilon)}{V(Y)} = \rho^2 + \frac{\sigma^2}{V(Y)}. \tag{2.5}$$

Equation (2.5) shows that the total variability in Y consists of two proportions: ρ^2 (which can also be written as $1 - \sigma^2/V(Y)$) is the proportion of variability in Y accounted for by the linear regression on X (i.e., by the structural part of the model) and $\sigma^2/V(Y)$ is the proportion accounted for by error. Thus, ρ^2 reflects our ability to account for variation in Y using a linear function of X, and as such, is our ideal measure of discriminatory power for linear regression. The value of ρ^2 ranges from 0, for the case in which X has absolutely no ability to account for Y, to 1.0, for the case in which Y is perfectly determined by X (i.e., all the points lie exactly on the regression line and there is no error). Typically, ρ^2 will range somewhere between these two extremes.

Decomposition of Variability in the Sample. The variability in Y among the sample observations can be decomposed in a similar manner. Here, however, the variability in Y is expressed in terms of the sum of squares in Y, or $\sum_{i=1}^{n}(y_i - \bar{y})^2$. This expression is the same as the numerator of the sample variance of Y and is referred to as the *total sum of squares in Y*, denoted *TSS*. It can be decomposed as follows:

$$\sum_{i=1}^{n}(y_i - \bar{y})^2 = \sum_{i=1}^{n}(y_i - \hat{y}_i + \hat{y}_i - \bar{y})^2.$$

Now, a word about this first step. You'll notice that I've added and subtracted $\hat{y}_i$ from the term inside the parentheses of *TSS*. This is a well-known mathematical "trick," if you will, that is equivalent to adding zero to an expression. Naturally, adding zero doesn't change anything. Now notice, however, what this facilitates. Continuing yields

$$= \sum_{i=1}^{n}[(\hat{y}_i - \bar{y}) + (y_i - \hat{y}_i)]^2$$

$$= \sum_{i=1}^{n}[(\hat{y}_i - \bar{y})^2 + (y_i - \hat{y}_i)^2 + 2(\hat{y}_i - \bar{y})(y_i - \hat{y}_i)]$$

$$= \sum_{i=1}^{n}(\hat{y}_i - \bar{y})^2 + \sum_{i=1}^{n}(y_i - \hat{y}_i)^2 + 2\sum_{i=1}^{n}(\hat{y}_i - \bar{y})(y_i - \hat{y}_i).$$

The last term in this expression is zero. Why? Well, first this expression can be written

$$2\sum_{i=1}^{n}(\hat{y}_i - \bar{y})(y_i - \hat{y}_i) = 2\sum_{i=1}^{n}[\hat{y}_i(y_i - \hat{y}_i) - \bar{y}(y_i - \hat{y}_i)]$$

$$= 2\sum_{i=1}^{n}\hat{y}_i(y_i - \hat{y}_i) - 2\bar{y}\sum_{i=1}^{n}(y_i - \hat{y}_i)$$

$$= 2\sum_{i=1}^{n}(b_0 + b_1x_i)(y_i - \hat{y}_i) - 2\bar{y}\sum_{i=1}^{n}(y_i - \hat{y}_i)$$

$$= 2b_0\sum_{i=1}^{n}(y_i - \hat{y}_i) + 2b_1\sum_{i=1}^{n}x_i(y_i - \hat{y}_i) - 2\bar{y}\sum_{i=1}^{n}(y_i - \hat{y}_i). \quad (2.6)$$

At this point, recall the normal equations $\sum_{i=1}^{n}(y_i - b_0 - b_1 x_i) = 0$ and $\sum_{i=1}^{n} x_i(y_i - b_0 - b_1 x_i) = 0$, or $\sum_{i=1}^{n}(y_i - \hat{y}_i) = 0$ and $\sum_{i=1}^{n} x_i(y_i - \hat{y}_i) = 0$. In that the OLS b_0 and b_1 solve these equations, they also make both equations true. Therefore, since $\sum_{i=1}^{n}(y_i - \hat{y}_i) = 0$ and $\sum_{i=1}^{n} x_i(y_i - \hat{y}_i) = 0$, expression (2.6) is also zero. Thus, we have the following decomposition of the total sum of squares in Y:

$$\sum_{i=1}^{n}(y_i - \bar{y})^2 = \sum_{i=1}^{n}(\hat{y}_i - \bar{y})^2 + \sum_{i=1}^{n}(y_i - \hat{y}_i)^2 \tag{2.7}$$

That is, TSS is the sum of two independent, or *orthogonal*, sums of squares. The first, $\sum_{i=1}^{n}(\hat{y}_i - \bar{y})^2$, is called the *regression sum of squares* and denoted RSS. This is the squared deviation of the regression line, represented by $\hat{y}_i$, around the overall sample mean of Y. This is the contribution of the linear relationship with X to the variability in Y. The second sum of squares is the *sum of squared errors*, or SSE. This is the squared deviation of the individual Y scores around the regression line. This is the component of variability in Y that is not accounted for by Y's linear relationship with X.

Dividing both sides of equation (2.7) by $\sum_{i=1}^{n}(y_i - \bar{y})^2$, we have

$$\frac{\sum_{i=1}^{n}(y_i - \bar{y})^2}{\sum_{i=1}^{n}(y_i - \bar{y})^2} = 1 = \frac{\sum_{i=1}^{n}(\hat{y}_i - \bar{y})^2}{\sum_{i=1}^{n}(y_i - \bar{y})^2} + \frac{\sum_{i=1}^{n}(y_i - \hat{y}_i)^2}{\sum_{i=1}^{n}(y_i - \bar{y})^2} \tag{2.8}$$

or

$$1 = \frac{RSS}{TSS} + \frac{SSE}{TSS}.$$

Hence, the decomposition of the variability in the sample Y-values mirrors the decomposition of the variability in Y in the population. Equation (2.8) shows that the sample variability in Y consists of two proportions, analogous to the proportions in equation (2.5). Our estimate of ρ^2 is therefore RSS/SSE, or $1 - (SSE/TSS)$, and is denoted r^2 (pronounced "r-squared"). Like its population counterpart ρ^2, r^2 ranges from 0 to 1. Also, our unbiased estimator of σ^2 is $\hat{\sigma}^2 = SSE/(n-2)$ and is typically denoted MSE for *mean-squared error*. This estimate is shown at the bottom of Tables 2.2 to 2.4.

Interpreting r^2. Let's consider the sample coefficient of determination, r^2, in more detail, as it is relied on heavily as a measure of how well the model performs. (r^2 is often referred to as a measure of model fit. However, assessing fit, or empirical consistency, is better accomplished via other approaches considered below.) The r^2 can be considered a *proportional reduction in error* measure, indicating the degree to which we improve our prediction of Y when using, as opposed to ignoring, the regression model. To see this, we reason as follows. First, we try to predict the individual Y scores while ignoring the model, assessing prediction error via squared deviation from the predicted value. Our best prediction, lacking any other information, is achieved by using the sample mean of Y as the predicted value for every case. Our total error is then TSS. Then, we try to predict Y using the regression model. Our total error is now SSE.

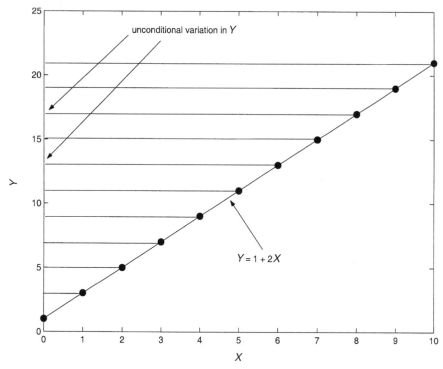

Figure 2.5 Regression of Y on X producing an r^2 of 1.0, showing how the regression completely "explains" the variation in Y.

The difference in the amount of error made in using, versus ignoring, the regression model is $TSS - SSE$. As a proportion of the original amount of error, it is

$$\frac{TSS - SSE}{TSS} = 1 - \frac{SSE}{TSS} = r^2.$$

Another way to view r^2 is to consider how it reflects the extent to which the linear function of X (i.e., the regression) explains the variation in Y. Figure 2.5 reveals this aspect of r^2. Depicted is a perfect linear relationship between Y and X, since Y is exactly determined by the line $1 + 2X$. The r^2 in this case is 1.0, since there is no error in predicting Y using the regression line. Suppose for the moment that we ignore X and simply examine the unconditional variation in Y. This is represented by the horizontal lines that map the Y values from the line $1 + 2X$ back to the Y axis. The Y values range from 1 to 21, exhibiting considerable variation. However, the reason for this variation is that Y is determined completely by its linear relationship with X. That is, the reason Y varies is that X varies from 0 to 10, and Y—being determined completely by X—varies accordingly. The conditional variation in Y given X is zero. That is, given any value x of X, Y is exactly $1 + 2x$, and there is no variability in Y at this x value. Here it is clear that the variation in Y is caused or *explained* totally by the regression on X, since Y exhibits no variance once X is held constant at any particular value. Even though r^2 is virtually never equal to 1.0, the closer it is to 1.0, the more the variation in Y is explained by the regression.

Adjusted r^2. Although r^2 provides a convenient comparative tool for assessing the relative discriminatory power of different models for Y, it has a tendency to overestimate the population ρ^2 (Stevens, 1986). If estimation of this parameter is one's goal, a better estimator of ρ^2 is the adjusted r^2 or r_{adj}^2. Its formula is

$$r_{adj}^2 = 1 - \frac{\hat{\sigma}^2}{s_Y^2},$$

or 1 minus the ratio of *MSE* to the sample variance of Y. Notice that this is the precise sample analog of $1 - \sigma^2/V(Y)$. In fact, r_{adj}^2 is a consistent estimator of r^2 because

$$\text{plim}(r_{adj}^2) = \text{plim}\left(1 - \frac{\hat{\sigma}^2}{s_Y^2}\right) = \text{plim}(1) - \frac{\text{plim}(\hat{\sigma}^2)}{\text{plim}(s_Y^2)} \qquad \text{(by the Slutsky theorem)}$$

$$= 1 - \frac{\sigma^2}{V(Y)} = \rho^2,$$

since the *MSE* and sample variance of Y are consistent for their respective population parameters. (Recall that the plim, consistency, and Slutsky's theorem were all discussed in the Chapter 1 appendix.)

Tables 2.2 to 2.4 show values of r^2 and r_{adj}^2 for the regressions of *exam 1* on *math diagnostic score*, *couple modernism* on *male's schooling*, and *frequency of sex* on *going to bars*. The values of r^2 and r_{adj}^2 for the same response are quite close. This will typically be the case in large samples, especially with only one regressor in the model. (I shall have more to say about the properties of r_{adj}^2 when we get to multiple regression models.) As is evident, the discriminatory power of the model for exam scores is fairly impressive, at least from a social science standpoint. Fully 26.8% of the variance in exam scores is accounted for by math diagnostic scores alone. Another interpretation of this value is that we would improve our ability to predict exam scores by 26.8% when using, as opposed to ignoring, its linear relationship to math diagnostic scores. The prediction of couple modernism using male's schooling is considerably less powerful. Only about 7% of the variance in couple modernism is accounted for by this predictor. And the predictive efficacy of frequenting bars as a determinant (or correlate) of having sex is less impressive still: A mere 3.8% of the variation in sexual frequency in the past year is explained by going to bars.

Limitations of r^2. The discriminatory power of one's model is certainly an important consideration to data analysts. However, we should be careful not to put too much weight on this dimension of regression. First, a high r^2 does not necessarily imply a correctly specified model. That is, high discriminatory power does not automatically imply authenticity. Second, high r^2 values are fairly easy to achieve if the regressor is conceptually close to the response. For example, a regression of someone's depressive symptomatology on his or her self-esteem is likely to produce an impressive r^2. But both attributes are typically tapped with paper-and-pencil scales that are affected by current mood. So, those who tend to be depressed are most likely also "down on themselves," suggesting that their self-esteem rating will also be diminished. Rather than this suggesting that self-esteem has high discriminatory power in predicting depression, or vice versa, a high r^2 value might only be indicating that people are consistent in answering mood-related items. Third, some criteria just lend themselves to better prediction than others. Exam scores of students in statistics, for example, can probably be nicely explained by college GPA, math aptitude, attitude toward the subject matter, and the number of other classes students are also taking. On the other hand, imagine trying to predict marital happiness. Certainly, there are a number of social factors that would be expected to affect this attribute: number of years married, number and ages of children, income level of the household, and so on. But ultimately, happiness is a fairly elusive feeling that may also have a large random component. One day we're happier with our spouse, one day we're not as happy with him or her, and some days we're just not sure. To expect a high r^2 value in predicting this type of variable is probably unrealistic unless one employs predictors such as satisfaction with communication in the marriage, which are themselves conceptually close to the response. In that case we might wonder whether we are really learning anything about the response variable, after all. In sum, the importance placed on r^2—and on discriminatory power, in general—depends on the researcher's goal. If the model is to be used for accurate prediction of responses for new observations, a high r^2 value is especially desirable. But if the central aim is to understand the nature of the "effect" of a regressor on the response in order to test theoretically driven hypotheses, discriminatory power may well be of secondary importance.

Correlation Coefficient and Standardized Slope. The r^2 is also the square of r, the Pearson correlation coefficient discussed in Section III.C of Appendix A. The correlation coefficient is calculated as

$$r = \frac{\text{cov}(x,y)}{\sqrt{s_X^2 s_Y^2}} = \frac{\sum_{i=1}^{n}(x_i - \bar{x})(y_i - \bar{y})}{\sqrt{\left[\sum_{i=1}^{n}(x_i - \bar{x})^2\right]\left[\sum_{i=1}^{n}(y_i - \bar{y})^2\right]}}.$$

The correlation coefficient ranges from -1.0 to $+1.0$ and is a measure of linear association between X and Y. When Y is a perfect linear function of X (i.e. all of the points lie exactly on the regression line), r is 1.0 in absolute value (this is proved in Section III.C of Appendix A). The correlation coefficient is also the *standardized slope* in simple linear regression. This terminology stems from the fact that if we standardize X and Y and then regress Y on X, the slope will be the correlation coefficient and the intercept will be zero (see also Exercises 2.17 and 2.18, where you're asked to verify this principle). The sample regression equation in standardized form is $Z_y = b_1^s Z_x + e^s$, where b_1^s is the standardized slope, or r. The standardized slope can be recovered from the unstandardized slope, b_1, via the formula

$$b_1^s = b_1 \frac{s_x}{s_y}.$$

The standardized slope has a rather cumbersome interpretation: it is the expected change in Y, in standard deviation units, for a 1-standard-deviation increase in X. Why? Well, if b_1 is the expected change in Y for each unit increase in X, $b_1 s_x$ is the expected change in Y for each 1-standard-deviation increase in X. Dividing again by s_y, $b_1 s_x/s_y = b_1(s_x/s_y)$ is the expected change in Y for each 1-standard-deviation increase in X, but now it is in terms of the number of standard deviations in Y. Standardized slopes for our three substantive regression examples are shown in Tables 2.2 to 2.4. They range in value from .518 for the regression of exam scores down to .194 for the regression of sexual frequency. No standardized value is shown for the intercept, since the intercept in the standardized equation is always zero.

INFERENCES IN SIMPLE LINEAR REGRESSION

In this section I discuss inferences connected with the simple linear regression model. In particular, I discuss three equivalent tests for the significance of the sample slope, as well as a test for the sample intercept. I also consider confidence intervals for the values of the population slope and intercept. Moreover, to justify inferences about the slope, I derive its expectation and variance, and its distribution, under general assumptions about the equation errors.

Tests about the Population Slope

Once we have estimated the parameters of the regression equation, the next question is: Is there a linear relationship between Y and X in the population? There *is*, provided that the value of the population slope is not zero. Hence, we wish to test the null hypothesis that β_1 equals zero against the alternative that β_1 is not zero. To do this, we need to find the sampling distribution of b_1, the sample estimator of the slope. It turns out that if n is sufficiently large, b_1 is normally distributed with mean equal to β_1 and variance equal to $\sigma^2/\sum_{i=1}^{n}(x_i - \bar{x})^2$, regardless of the distribution of ε. (In small samples it is necessary that ε be normally distributed with mean equal to zero and variance equal to σ^2 in order for b_1 to have the same distributional properties.) Given these properties, a t—or in large samples, z—test for H_0: $\beta_1 = 0$ against H_1: $\beta_1 \neq 0$ is $t = b_1/\hat{\sigma}_{b_1}$, where $\hat{\sigma}_{b_1} = \sqrt{\hat{\sigma}^2/\sum_{i=1}^{n}(x_i - \bar{x})^2}$ is the estimated standard deviation, or *standard error*, of the sample slope. Under the null hypothesis that the population slope is zero, this statistic has a t distribution with $n - 2$ degrees of freedom (or, equivalently, a z distribution) in large samples. As an example, the t test for the slope of the regression of *exam 1 score* on *math diagnostic score* is $t = 2.749/.313 = 8.788$. This t has a p-value less than .001 under the null hypothesis that the corresponding population slope is zero. In fact, t tests for all of the sample slopes in Tables 2.2 to 2.4 suggest significant linear relationships in the population. That is, we can reject the null hypothesis of zero population slope in each case and conclude that there are significant positive linear relationships between *math diagnostic score* and *exam 1 score*, between *male's schooling* and *couple modernism*, and between *going to bars* and *having sex*.

Justification. How does one derive the distribution of b_1? First, we rewrite the formula for b_1 to show that it is a weighted sum of the y_i:

$$b_1 = \frac{\sum_{i=1}^{n}(x_i-\bar{x})(y_i-\bar{y})}{\sum_{i=1}^{n}(x_i-\bar{x})^2}$$

$$= \frac{\sum_{i=1}^{n}[(x_i-\bar{x})y_i - (x_i-\bar{x})\bar{y}]}{\sum_{i=1}^{n}(x_i-\bar{x})^2}$$

$$= \frac{\sum_{i=1}^{n}(x_i-\bar{x})y_i}{\sum_{i=1}^{n}(x_i-\bar{x})^2} - \frac{\bar{y}\sum_{i=1}^{n}(x_i-\bar{x})}{\sum_{i=1}^{n}(x_i-\bar{x})^2}$$

$$= \frac{\sum_{i=1}^{n}(x_i-\bar{x})y_i}{\sum_{i=1}^{n}(x_i-\bar{x})^2}.$$

[The last step in this sequence is justified by exercise (1) in Section II.E of Appendix A.] Now, if we let $s_{xx} = \sum_{i=1}^{n}(x_i - \bar{x})^2$, we can write each person's weight as $w_i = (x_i - \bar{x})/s_{xx}$, and we can write b_1 as $b_1 = \sum_{i=1}^{n}w_i y_i$. This shows that b_1 is a weighted sum of the y_i. At this point we make use of the CLT, outlined in the Chapter 1 appendix, which states that a weighted sum of independent and identically distributed random variables is distributed approximately normal in large samples, with mean equal to $\mu\sum_{i=1}^{n}w_i$ and variance equal to $\sigma^2\sum_{i=1}^{n}w_i^2$, where μ and σ^2 are the mean and variance, respectively, of the random variables. In this case, the random variables, Y_i, are not exactly identically distributed, since although they all have the same variance of σ^2, their means are a function of X. In particular $E(Y_i) = \beta_0 + \beta_1 X_i$. No matter, however. This just means that the mean of the weighted sum is written as $\sum_{i=1}^{n}\mu_i w_i$ instead of $\mu\sum_{i=1}^{n}w_i$, where $\mu_i = \beta_0 + \beta_1 X_i$. At any rate, by the CLT, b_1 is normally distributed with mean

$$E(b_1) = \sum_{i=1}^{n}\mu_i w_i = \sum_{i=1}^{n}(\beta_0 + \beta_1 x_i)\frac{x_i - \bar{x}}{s_{xx}}$$

$$= \frac{1}{s_{xx}}\sum_{i=1}^{n}[\beta_0(x_i - \bar{x}) + \beta_1 x_i(x_i - \bar{x})]$$

$$= \frac{1}{s_{xx}}\beta_0\sum_{i=1}^{n}(x_i - \bar{x}) + \frac{1}{s_{xx}}\beta_1\sum_{i=1}^{n}(x_i^2 - x_i\bar{x})$$

$$= \frac{1}{s_{xx}}\beta_1\left(\sum_{i=1}^{n}x_i^2 - \bar{x}\sum_{i=1}^{n}x_i\right)$$

$$= \frac{1}{s_{xx}}\beta_1\left(\sum_{i=1}^{n}x_i^2 - n\bar{x}^2\right),$$

and recalling from Appendix A that

$$\sum_{i=1}^{n} x_i^2 - n\bar{x}^2 = \sum_{i=1}^{n} (x_i - \bar{x})^2 = s_{xx},$$

we have that

$$E(b_1) = \frac{1}{s_{xx}} \beta_1 s_{xx} = \beta_1.$$

It is therefore clear that b_1 is an unbiased estimator of β_1. Moreover, the variance of b_1 is

$$V(b_1) = \sigma^2 \sum_{i=1}^{n} w_i^2 = \sigma^2 \frac{\sum_{i=1}^{n}(x_i - \bar{x})^2}{s_{xx}^2} = \frac{\sigma^2}{\sum_{i=1}^{n}(x_i - \bar{x})^2},$$

which agrees with the result stated above. The point of this demonstration is that in large samples, the sample slope is normally distributed *regardless of the distribution of the errors*. This allows us to make inferences about the population slope using its sample counterpart, without the requirement of normally distributed errors. As mentioned previously, however, the normality of errors is a requirement in small samples. Provided that the errors are normally distributed, Y is also normally distributed. Hence the sample slope—shown above to be a weighted sum of the Y's—would then be normally distributed for any sample size.

Equivalent Tests for the Slope. Two other tests are available that are mathematically equivalent to the t test for the slope. The first is the t test for the correlation coefficient. The test statistic is

$$t = \frac{r}{\sqrt{(1 - r^2)/(n - 2)}}.$$

This test also has $n - 2$ degrees of freedom under the null hypothesis that the population slope is zero—or, equivalently, that the population correlation, ρ, between X and Y, is zero. In fact, this is exactly the same statistic as the test for the slope. For example, the coefficient for the correlation of *exam score* with *math diagnostic score* is .518. Testing this value, we get

$$t = \frac{.518}{\sqrt{(1-.518)/(213-2)}} = 8.797,$$

which, within rounding error, is the same value as that reported for the t test of the slope in Table 2.2 (both values round to 8.8).

The other test is an F test. This has the form $F = RSS/MSE$. Under the null hypothesis of zero population slope, this statistic follows the F distribution with 1 and $n - 2$ degrees of freedom. Again, for the regression of exam scores, the test is $F = 16509.689/213.756 = 77.236$, a highly significant result. In fact, this test statistic is also mathematically equivalent to the t test for the slope, as in simple linear regression, $F = t^2$. For example, the square root of 77.236 is 8.788, which agrees with the t test for the slope as reported in Table 2.2.

Testing the Intercept

As it is also a weighted combination of the Y_i (Neter et al., 1985), the sample estimate, b_0, of the equation intercept is also normally distributed in large samples (or
with any sample size provided that ε is normal). Its expected value is β_0, which means
that it is also unbiased for the population parameter. Its variance is (Neter et al., 1985)

$$V(b_0) = \frac{\sigma^2 \sum_{i=1}^{n} x_i^2}{ns_{xx}}.$$

Tests about the population intercept are almost never of interest, especially when zero
is not a meaningful value for X. Nonetheless, a t test of H_0: $\beta_0 = 0$ against H_1: $\beta_0 \neq 0$

takes the form $t = b_0/\hat{\sigma}_{b_0}$, where $\hat{\sigma}_{b_0} = \sqrt{\dfrac{MSE \sum_{i=1}^{n} x_i^2}{ns_{xx}}}$ is the estimated standard

error of the sample intercept. Under the null hypothesis that the population intercept
is zero, this statistic has a t_{n-2} (or equivalently, a z) distribution in large samples.
As is evident in Table 2.2, the t-test statistic for the intercept is $-35.494/$
$12.840 = -2.764$. Hence we would reject the null hypothesis that the population
intercept is zero at a p-value of .006. As indicated above, however, this test is not particularly meaningful, due to the potential nonlinearity of the relationship at diagnostic scores of zero.

Confidence Intervals for β_0 and β_1

The dominant vehicle for inferences in the social sciences appears to be the hypothesis
test. Yet we may, as well, be interested in trying to pin down the value of population
parameters as closely as possible. For this purpose, we can form confidence intervals
for values of β_0 or β_1. For example, a 95% confidence interval for β_1 is $b_1 \pm t_{(.025,n-2)}$
$(\hat{\sigma}_{b_1})$, where $t_{(.025,n-2)}$ is the 97.5th percentile of the t distribution corresponding to $n-2$
degrees of freedom. In the regression of exam scores, $n-2 = 211$ and $t_{(.025,211)}$ is 1.97.
For this regression, a confidence interval for the slope is $2.749 \pm 1.97(.313) = (2.132,$
$3.366)$. Similarly, a 95% confidence interval for β_0 is $b_0 \pm t_{(.025,n-2)}(\hat{\sigma}_{b_0})$. For the regression of *exam* 1, the interval is $-35.494 \pm 1.97(12.840) = (-60.789, -10.199)$. Only
the interval for the slope is particularly meaningful. It implies that we can be 95%
confident that those who are a unit higher on the math diagnostic score perform, on
average, between 2.13 and 3.37 points higher on the first exam in statistics compared to
others.

Additional Examples

Additional examples of simple linear regressions are shown in Table 2.5. Here, I
present the regression of *having sex* on *going to bars* separately by *gender* and
then separately by *marital status*. The reason for estimating the regression
separately within different subgroups of the population (male vs. female, or

Table 2.5   Parameter Estimates from the Regression of Y = Frequency of Having Sex in the Past Year on X = Frequency of Going to Bars for 1515 Respondents in the 1998 GSS, Separately by Gender and Marital Status

Explanatory Variable	b	$\hat{\sigma}_b$	t	p	R^2
A. *Males*					
Constant	2.682	.131	20.403	<.001	
Frequency of going to bars	.155	.041	3.824	<.001	.022
B. *Females*					
Constant	2.001	.113	17.645	<.001	
Frequency of going to bars	.254	.042	6.078	<.001	.041
C. *Married*					
Constant	3.053	.100	30.494	<.001	
Frequency of going to bars	.167	.040	4.215	<.001	.024
D. *Unmarried*					
Constant	1.185	.128	9.280	<.001	
Frequency of going to bars	.379	.039	9.807	<.001	.110

married vs. unmarried) is that it is reasonable to expect that frequenting bars would have a stronger or weaker association with having sex, depending on the subgroup.

For example, one might expect going to bars to have a stronger effect on sexual activity for women than for men. Why? Given that men are still expected to be the initiators of social contact in mixed-sex environments, the mere presence of women at bars is likely to garner more potential "suitors" for women than it is for men. That is, women can remain quite passive and still be approached by men at bars, whereas the passive male is less likely to end up with female companionship during the course of an evening. Also, given the rather sexualized atmosphere of many bars, such a venue may be especially selective of sexually willing females but not especially selective with respect to males. At any rate, I expect going to bars to bear a stronger association with sexual activity for women. This appears to be borne out by the results. Although the effect of going to bars on sexual frequency is significant for both genders, it is substantially larger for women, with a value of .254, compared to .155 for men.

In a similar vein, I expect the effect of going to bars to be stronger for the unmarried than for the married. Married people tend to have more regular sex than do the unmarried since they already have a presumably willing sex partner in their spouse. Going to bars should not, therefore, make as much difference in their sexual activity level as it might for single persons. Consistent with this notion, the effect of *going to bars* on *sexual frequency* is more than twice as great for the unmarried than for the married (slopes for the married and unmarried are .167 and .379, respectively).

The reader should notice that I'm comparing unstandardized rather than standardized slopes when comparing the impact of the same predictor on the same response across subpopulations. This is not just coincidental. For comparing across subpopulations in this manner, one should always use the unstandardized slope. The reason for this is that the standardized slope is a function not only of the unstandardized slope but also of the standard deviations of X and Y in each subpopulation. Therefore, the standardized slopes can be different for different subgroups solely because of differences in the distributions of X and Y in each group rather than because the effect of X is different in each group. Comparing the unstandardized slopes avoids this problem.

Once we have discovered that the effect of going to bars on having sex is different in samples from different subpopulations, a natural question is: Do these sample differences reflect real differences in the effect of X in the subpopulations themselves? We must be careful not to assume automatically that different slope estimates from different subsamples imply different effects in the corresponding subpopulations. What is needed is a statistical test to tell whether, say, the effect of going to bars is weaker for men than it is for women. That is, if β_{1m} is the effect for males and β_{1f} is the effect for females, we need a test for $H_0: \beta_{1m} = \beta_{1f}$ against $H_1: \beta_{1m} \neq \beta_{1f}$. This is a test for what is referred to as *statistical interaction*. If the effect of X depends on the level of another explanatory variable—in this case, *gender*—we say that *gender* and X *interact* in their effect on Y. We take up the issue of statistical interaction in regression in Chapter 3 and discuss such a test at that time.

ASSESSING EMPIRICAL CONSISTENCY OF THE MODEL

Above I discussed how we can assess the discriminatory power of a regression model using either r^2 or r_{adj}^2. What about empirical consistency? This issue is somewhat more complex. An empirically consistent model should conform to the model assumptions as well as accurately forecast the behavior of the response variable. In this section I discuss some informal and formal techniques for evaluating this dimension of regression.

Conforming to Assumptions

Recall that the equation disturbances, the ε_i's, are subject to some critical assumptions. First, in small samples, they should be normally distributed. In large samples this assumption is not as important. Second, they should have a mean of zero at each x. Third, they should have constant variance at each x. As our best estimates of the ε_i's, the sample residuals can be used to check the first and third of these assumptions. For example, the normality-of-errors assumption for the regression of exam scores in Table 2.2 was evaluated using SAS by saving the residuals from the regression and then requesting a test of normality via PROC UNIVARIATE. The test statistic, which was devised by Shapiro and Wilk (1965), had a p-value below .0001, resulting in rejection of the null hypothesis of normality. However, in that the sample size is relatively large, nonnormal errors are not of great concern here.

The constant variance assumption can be checked visually by plotting the residuals against X. Such a plot for the residuals from the regression of exam scores is shown in Figure 2.6. What we are looking for is a scatter of points of approximately

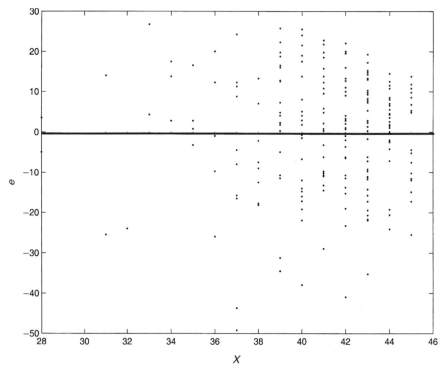

Figure 2.6   Plot of e (residuals) against X (math diagnostic score) from OLS regression of score on exam 1 on math diagnostic score for 213 students in introductory statistics.

constant width spread about the line $e = 0$—the dark line shown in the figure. The point scatter seems roughly to have this appearance. Nonconstant variance typically shows up as a point scatter in the shape of a wedge, in which the variance of e_i appears either to decrease or to increase with increasing X. As an example, Figure 2.7 shows a plot of residuals against X for the regression of *couples' coital frequency in the last month* on *male partner's age* from the *couples dataset*. *Coital frequency* is the average of the male's and female's report of how many times the couple "had sex" in the past month. As is evident here, the width of the band of points appears to taper down considerably with increasing age. In fact, a statistical test suggests that we should reject the null hypothesis of constant variance for these disturbances. The test is discussed in Chapter 6.

The assumption that the mean of the errors at each X is zero—the orthogonality condition—cannot be checked using the residuals. The reason for this is that it is an artifact of OLS estimation that the mean of the residuals is always zero, and moreover, that the covariance of the residuals with X is always zero. This is easy to show. Recall once again the normal equations $\sum_{i=1}^{n}(y_i - \hat{y}_i) = 0$ and $\sum_{i=1}^{n}x_i(y_i - \hat{y}_i) = 0$. Now,

$$\bar{e} = \frac{\sum_{i=1}^{n} e_i}{n} = \frac{\sum_{i=1}^{n}(y_i - \hat{y}_i)}{n} = \frac{0}{n} = 0.$$

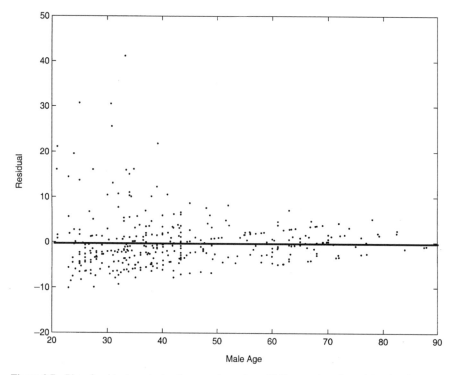

Figure 2.7 Plot of residuals against male partner's age from OLS regression of couple's coital frequency in past month on male partner's age for 416 couples from the NSFH.

Also, the covariance of the residuals with X is

$$\text{cov}(x_i, e_i) = \frac{\sum_{i=1}^{n}(x_i - \bar{x})(e_i - \bar{e})}{n-1}$$

$$= \frac{\sum_{i=1}^{n}(x_i - \bar{x})e_i}{n-1}$$

$$= \frac{\sum_{i=1}^{n}x_i e_i - \bar{x}\sum_{i=1}^{n}e_i}{n-1} = \frac{\sum_{i=1}^{n}x_i(y_i - \hat{y}_i) - \bar{x}\sum_{i=1}^{n}(y_i - \hat{y}_i)}{n-1}$$

$$= \frac{0-0}{n-1} = 0.$$

The orthogonality condition must be taken on faith. However, the best assurance of its veracity is to include in one's model any other determinants of Y that are also correlated with X. How to model Y as a function of multiple regressors is taken up in Chapter 3.

Nonlinearities. Plots of residuals against X are also useful for revealing potential nonlinearities in the relationship between Y and X. This would be suggested by a point scatter with a nonlinear shape around the line $e = 0$. No such pattern is evident in the point scatter in Figure 2.6.

Outliers. The diagnosis of unusual observations is taken up in detail in Chapter 6, but I also consider it briefly here. Although not officially an assumption of regression, it is ideal if the data are also free of *outliers*. These are observations that are noticeably "out of step" with the trend shown by the majority of the points. An outlier typically shows up as an unusually large (in absolute value) residual. Outliers are essentially observations that are not fit well by the model. At the least, we would like to identify them. They may show up in an ordinary residual plot, but a better strategy is to plot the standardized residuals against X.

Standardized Residuals. A residual is standardized by subtracting its mean and dividing by its standard deviation. For this purpose we recall that the mean of the residuals is zero and that the estimated standard deviation of the equation errors is $\hat{\sigma}$. The standardized residual is therefore $z_e = (e - 0)/\hat{\sigma} = e/\hat{\sigma}$. Figure 2.8 presents a plot of the standardized residuals against X for the regression of exam 1 performance on math diagnostic scores. If the equation errors were normally distributed, we might

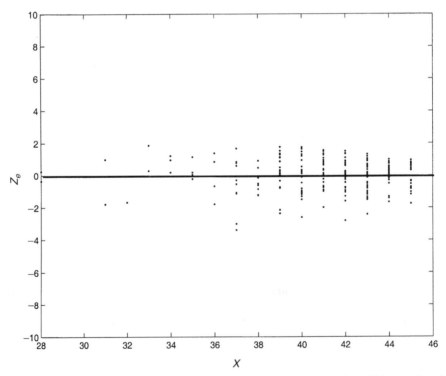

Figure 2.8   Plot of z_e (standardized residuals) against X (math diagnostic score) from OLS regression of score on exam 1 on math diagnostic score for 213 students in introductory statistics.

want to consider as outliers residuals that are ± 2 standard deviations or more on either side of the zero line. However, given that the errors may not be normally distributed, we should be more cautious. Neter et al. (1985) suggest four or more standard deviations as the benchmark for extreme residuals. Regarding Figure 2.8, it does not appear that any particular observations meet this criterion. Most of the z_e are well within this limit, with only two observations approaching it. Above all, however, none of the points appears dramatically out of step with the rest of the point scatter. (On the other hand, the observation with the largest magnitude z_e, in the lower middle of the plot, is somewhat of an outlier, as will be pointed out in Chapter 6.)

Formal Test of Empirical Consistency

A lack-of-fit test proposed by Neter et al. (1985) can be employed to formally test the empirical consistency of the model. In this section I discuss the test and then employ it to test the empirical consistency of the simple linear regression for exam scores in Table 2.2. This test assumes normality of the equation errors, in addition to zero mean and constant variance. The test, according to the authors, is for "ascertaining whether or not a linear regression function is a good fit for the data" (Neter et al., 1985, p. 123). The test generally requires repeated observations at each value of X, which is not a problem in the current example.

The test is based on a decomposition of SSE into *pure error* and *lack-of-fit components*. The first component, the *pure error sum of squares*, or $SSPE$, is the total variability of the Y-scores around their respective group means at each x. That is, suppose that there are c different X-values. In the regression of exam scores, for example, there are $c = 16$ different math diagnostic scores, ranging from 28 to 45. Let x_j denote the jth X-value and n_j denote the number of observations with $X = x_j$. Also, let y_{ij} denote the ith observation whose X-value is x_j, and let $\bar{y}_j$ stand for the mean of all of the y_{ij} whose X-value is x_j. Then the sum of squares of Y-values about their mean at a particular x_j is $\sum_{i=1}^{n_j}(y_{ij} - \bar{y}_j)^2$. Note that if there is only one observation at a particular x_j, then $y_{ij} = \bar{y}_j$, implying that $y_{ij} - \bar{y}_j = 0$. Thus, observations with unique values of X do not contribute any information to $SSPE$.

The total such variability across all of the X-values is

$$SSPE = \sum_{j=1}^{c} \sum_{i=1}^{n_j} (y_{ij} - \bar{y}_j)^2. \tag{2.9}$$

The *pure error mean square (MSPE)* is $SSPE$ divided by its degrees of freedom, which is $n - c$. Hence,

$$MSPE = \frac{SSPE}{n - c}.$$

It turns out that $MSPE$ is an unbiased estimator of the error variance, σ^2, regardless of the nature of the function relating X to Y. For this reason, it is dubbed an estimate of "pure error."

The second component of *SSE* is the *lack-of-fit sum of squares* (*SSLF*). Further denoting, by $\hat{y}_j$, the sample-regression fitted value at $X = x_j$, *SSLF* is

$$SSLF = \sum_{j=1}^{c} n_j (\bar{y}_j - \hat{y}_j)^2.$$

This is essentially a weighted (by the n_j) sum of squared deviations of the $\bar{y}_j$ from the regression line. The greater the deviation of the $\bar{y}_j$ from the regression line, the more evidence that a linear regression does not characterize the conditional mean of Y, given X (Neter et al., 1985). Conveniently, *SSLF* can also be calculated as *SSE* − *SSPE*. The degrees of freedom associated with *SSLF* is $c - 2$. Hence, the *lack-of-fit mean square* (*MSLF*), is

$$MSLF = \frac{SSLF}{c - 2}.$$

The test statistic for testing for a lack of fit is

$$F = \frac{MSLF}{MSPE}.$$

Under the null hypothesis that a linear regression model fits the data, or that the data are empirically consistent with a linear regression, F has the F distribution with $c - 2$ and $n - c$ degrees of freedom.

Implementing the Test. The lack-of-fit test is not currently implemented for the linear regression model in mainstream software—say in SPSS or SAS. However, it *can* be obtained in SAS, since this software implements the test for a response surface model via the program PROC RSREG. For a regression with one predictor, X, the response surface model is simply a model that adds X^2 as another regressor. At any rate, RSREG calculates and prints *SSPE* for this quadratic model. Nevertheless, *SSPE* is the same as in equation (2.9), since *SSPE* is always just the deviation of the Y_i around their group means at each x. For the regression of exam performance on math diagnostic scores, we have the following: *SSE* is 45102.50265 and *SSPE* is 2044.573675. Therefore, $SSLF = SSE - SSPE = 45102.50265 - 2044.573675 = 43057.928975$. As noted previously, $n = 213$ and $c = 16$. *MSPE* is thus $2044.573675/197 = 10.3785$, and $MSLF = 43057.928975/14 = 3075.5664$. The lack-of-fit test is therefore $F = 3075.5664/10.3785 = 296.34$. With 14 and 197 degrees of freedom, the result is highly significant ($p < .00001$). I should therefore reject the hypothesis that the data are empirically consistent with the model.

How troublesome is this finding? It should be noted, first, that in the quadratic model, it is only the linear effect that is significant. That is, the X^2 term is very nonsignificant, suggesting that there is no simple departure from linearity evident in the regression. Moreover, the scatterplot in Figure 2.1 and the residual plots in Figures 2.6 and 2.8 all seem to suggest that a linear regression adequately describes the relationship between *exam performance* and *math diagnostic score*. It is possible, particularly with larger sample sizes, that haphazard departures from linearity may result in the rejection of empirical consistency. Yet without a clear nonlinear trend that could be modeled, preserving the linear regression approach may be the best strategy, from the standpoint of parsimony. In sum, I accept the linear regression model as appropriate for the data in this example, despite having to reject the hypothesis of empirical consistency via a formal test.

Discriminatory Power vs. Empirical Consistency. As another illustration of the distinction between discriminatory power and empirical consistency, I conducted a regression simulation. First, I drew a sample of 200 observations from a true linear regression model in which Y was generated by the linear equation $3.2 + 1X + \varepsilon$, where ε was normally distributed with mean zero and variance equal to 22. Estimation of the simple linear regression equation with the 200 sample observations produced an r^2 of .21. The F statistic for testing empirical consistency was .2437. With 8 and 190 degrees of freedom, its p-value was .982. Obviously, the F test for lack of fit here suggests a good-fitting model. Note, however, that discriminatory power is modest, at best. Next I drew a sample of 200 observations from a model in which Y was generated as a nonlinear—in particular, a quadratic—function of X. Specifically, the equation generating Y was $1.2 + 1X + .5X^2 + \varepsilon$. Again, ε was a random observation from a normal distribution with mean zero. This time, however, the variance of ε was only 1.0. I then used the 200 sample observations to estimate a simple linear regression equation. That is, I estimated $Y = \beta_0 + \beta_1 X + \varepsilon$, a clearly misspecified model. The test for lack of fit in this case resulted in an F value of 339.389, a highly significant result (at $p < .00001$). Clearly, by a formal test, the linear model is rejected as empirically inconsistent. The r^2 for the linear regression, however, was a whopping .96! The point of this exercise is that contrary to popular conception, r^2 is not a measure of "fit" of the model to the data. It is a measure of discriminatory power. It's possible, as shown in these examples, for good-fitting models to have only modest r^2 values and for bad-fitting models to have very high discriminatory power. [See also Korn and Simon (1991) for another illustration of the distinction between these two components of model evaluation.]

Authenticity. The authenticity of a model is much more difficult to assess than is either discriminatory power or empirical consistency. Here we ask: Does the model truly reflect the real-world process that generated the data? This question usually does not have a statistical answer. We must rely on theoretical reasoning and/or evidence from experimental studies to buttress the veracity of our proposed causal link between X and Y. On the other hand, we *can* evaluate whether additional variables are responsible for the observed X–Y association, rendering the original causal model inauthentic. For example, I have attempted above to argue, theoretically, for the reasonableness of *math diagnostic score* as a cause of *exam performance*, for *years of schooling* as a cause of *couple modernism*, and for *frequenting bars* as a cause of *sexual frequency*. Objections to the authenticity of all of these models can be tendered. With respect to *exam performance*, it is certainly possible that academic ability per se is the driving force that affects performance on both the math diagnostic and the exam. In this case, the relationship between *diagnostic score* and *exam performance*, being due to a third, causally antecedent variable, would be spurious. *Couple modernism* may be determined strictly by the *female's years of schooling*. Its association with *male's years of schooling*, rather than reflecting a causal link, may simply be an artifact of the substantial correlation between each spouse's educational level. Finally, sexual activity and going to bars may be associated purely by virtue of the fact that younger people engage in both activities with greater frequency, rendering the correlation between sex and bar going, once again, spurious. Fortunately, each of these alternative explanations for the observed linear associations can be examined given that measures of the additional variables are available in the data sets. In Chapter 3 I introduce multiple linear regression, which allows us to examine the relationship between X and Y while controlling for additional factors. I can then assess each of these alternative hypotheses. For now, I leave these alternative models for the three outcomes as potential competitors for the models underlying the data.

STOCHASTIC REGRESSORS

The assumption of nonstochastic regressors, that is, regressor values that are fixed over repeated sampling, is quite restrictive. Generally, only in experiments does the researcher have the power to fix the different levels of X at particular values. With non-experimental data, such as are gathered via surveys of, say, the general population, what are actually sampled are random observations of (x,y) pairs. In this case, both Y and X are random, or stochastic, variables. In this situation, it is unreasonable to expect the X-values to remain constant over repeated sampling. The reason for this is that each sample may contain a somewhat different random sampling of X-values. Nevertheless, according to Neter et al. (1985), all results articulated earlier, pertaining to estimation, testing, and prediction, employing the model with fixed X, also apply under random X if the following two conditions hold: (1) "The conditional distributions of the Y_i, given X_i, are normal and independent, with conditional means $\beta_0 + \beta_1 x_i$ and conditional variance σ^2"; and (2) "The X_i are independent random variables, whose probability distribution $g(X_i)$ does not involve the parameters β_0, β_1, σ^2" (Neter et al., 1985, p. 84).

Regardless of whether these conditions hold, the fixed-X assumption is unnecessary if we are willing to make our results *conditional* on the sample values observed for the independent variable. This means that they are valid only for the particular set of X-values that we have observed in our sample. According to Wooldridge (2000), conditioning on the sample values of the independent variable is equivalent to treating the x_i as fixed over repeated sampling. Such conditioning then implies automatically that the disturbances are independent of X, which is the primary requirement for the validity of our model-based inferences.

ESTIMATION OF β_0 AND β_1 VIA MAXIMUM LIKELIHOOD

In this final section of the chapter, I show how to arrive at estimates of β_0 and β_1 using maximum likelihood rather than OLS estimation. I do this to provide some consistency in coverage of model estimation, since all models in this book other than the linear regression model typically rely on maximum likelihood as the estimation method.

In the appendix to Chapter 1, I discussed maximum likelihood estimation and provided a relatively simple illustration of the technique. Recall that the key mathematical expression is the log of the likelihood function for the parameters, given the data. To arrive at this, maximum likelihood estimation begins with an assumption about the density function of the observed response variable. In this case, we assume that the equation disturbances have a normal distribution with mean zero and variance σ^2 at each x_i. Given that the X's are fixed and that the conditional mean of y_i is $\beta_0 + \beta_1 x_i$, this means that the y_i are normal with mean equal to $\beta_0 + \beta_1 x_i$ and conditional variance equal to σ^2. Hence, the density function for each y_i is

$$f(y_i \mid x_i, \beta_0, \beta_1) = \frac{1}{\sqrt{2\pi\sigma^2}} \exp\left[-\frac{1}{2}\left(\frac{y_i - (\beta_0 + \beta_1 x_i)}{\sigma}\right)^2 \right],$$

which means that the joint density function for the vector **y** of sample responses, given the vector **x** of regressor values, is

$$f(\mathbf{y} \mid \mathbf{x}, \beta_0, \beta_1) = \prod_{i=1}^{n} \frac{1}{\sqrt{2\pi\sigma^2}} \exp\left[-\frac{1}{2}\left(\frac{y_i - \beta_0 - \beta_1 x_i}{\sigma}\right)^2 \right].$$

Given the data observed, this becomes a function only of the parameters; hence we change its symbol to $L(\beta_0, \beta_1 \mid \mathbf{y}, \mathbf{x})$, indicating that it is the likelihood function for the parameters, given the data, equal to

$$L(\beta_0, \beta_1 \mid \mathbf{y}, \mathbf{x}) = (2\pi\sigma^2)^{-n/2} \exp\left[\frac{-1}{2\sigma^2} \sum_{i=1}^{n} (y_i - \beta_0 - \beta_1 x_i)^2 \right].$$

Denoting the log of this function by $\ell(\beta_0, \beta_1 \mid \mathbf{y}, \mathbf{x})$, we have

$$\ell(\beta_0, \beta_1 \mid \mathbf{y}, \mathbf{x}) = -\frac{n}{2}\ln 2\pi\sigma^2 - \frac{1}{2\sigma^2} \sum_{i=1}^{n} (y_i - \beta_0 - \beta_1 x_i)^2. \qquad (2.10)$$

Recall from the earlier discussion of least squares estimation that finding the values of variables that minimize or maximize a function involves setting the first derivatives of the function, with respect to each of the variables, to zero and then solving the resulting simultaneous equations. In this case, with the variables b_0 and b_1 substituted for the parameters, we have

$$\frac{\partial}{\partial b_0}[\ell(\beta_0, \beta_1 \mid \mathbf{y}, \mathbf{x})] = \frac{1}{\sigma^2} \sum_{i=1}^{n} (y_i - b_0 - b_1 x_i)$$

and

$$\frac{\partial}{\partial b_1}[\ell(\beta_0, \beta_1 \mid \mathbf{y}, \mathbf{x})] = \frac{1}{\sigma^2} \sum_{i=1}^{n} x_i(y_i - b_0 - b_1 x_i).$$

Now, these expressions are zero whenever

$$\sum_{i=1}^{n}(y_i - b_0 - b_1 x_i) = 0 \quad \text{and} \quad \sum_{i=1}^{n} x_i(y_i - b_0 - b_1 x_i) = 0.$$

Notice that these are just the normal equations that were solved to find the least squares estimates. This shows that the OLS estimates, b_0 and b_1, are also the maximum likelihood estimates if the Y_i are normally distributed. Moreover, it is easy to convince ourselves that this solution, in fact, *maximizes* the log likelihood. Recall the argument supporting b_0 and b_1 as minimizing *SSE*. Now notice that $\ell(\beta_0, \beta_1 | \mathbf{y}, \mathbf{x})$ is generally a negative value, so to maximize it we have to find the values that make it *least negative*. Obviously, this occurs whenever *SSE* is *minimized*. In this case, the second negative expression on the right-hand side of equation (2.10), which is a function of *SSE*, reaches its least negative value, and therefore $\ell(\beta_0, \beta_1 | \mathbf{y}, \mathbf{x})$ attains its maximum.

EXERCISES

2.1 Using the computer, regress EXAM1 on COLGPA in the *students dataset*. Missing imputation: substitute the value 3.0827835 for missing data on COL-GPA. Then:

    (a) Interpret the values of b_0 and b_1.

    (b) Give the values of $\hat{\sigma}_{b_0}$ and $\hat{\sigma}_{b_1}$.

    (c) Give the F-value and its significance level.

    (d) Give the values of r^2 and r_{adj}^2.

    (e) Give the estimate of σ^2.

    (f) Give the predicted population mean of Y for those with a *college GPA* of 3.0.

    (g) Give a 95% confidence interval for β_1.

2.2 Using the computer, regress STATMOOD on SCORE in the *students dataset*. Missing imputation: Substitute the value 40.9358974 for missing data on SCORE. Then:

    (a) Interpret the values of b_0 and b_1.

    (b) Give the values of $\hat{\sigma}_{b_0}$ and $\hat{\sigma}_{b_1}$.

    (c) Give the F-value and its significance level.

    (d) Give the values of r^2 and r_{adj}^2.

    (e) Give the estimate of σ^2.

    (f) Give the predicted population mean of Y for those with a *math diagnostic score* of 43.

    (g) Give a 95% confidence interval for β_1.

2.3 Consider the regression of *coital frequency* on *male age* for our 416 couples. How reasonable is the orthogonality assumption? That is, are there any compelling theoretical reasons why you would expect the disturbances to be correlated with *male age* in the population?

2.4 Regard the following (x,y) values for 19 observations:

X	Y	X	Y	X	Y	X	Y
1	.5	2	1.7	4	1.3	10	1
1	1	3	.4	4	2	10	2.5
1	1.5	3	1.2	5	.6	10	4
2	.6	3	5	5	2	10	18
2	1	4	.7	5	2.2		

(a) Construct a scatterplot of Y against X.

(b) Estimate the regression of Y on X and draw the regression line on the plot.

(c) Omit the last observation (i.e., the point 10,18) and reestimate the regression on just the first 18 observations.

(d) How does the regression change, and what seems to account for the change?

2.5 Prove, for simple linear regression, that the point $\bar{x}, \bar{y}$ is on the regression line.

2.6 A centered variable, X^c, is defined as $X^c = X - \bar{X}$. Do the following:

(a) Prove that centering X leaves the slope unchanged in SLR.

(b) Prove that centering X makes the intercept equal to $\bar{y}$ in SLR.

(c) Interpret the intercept estimate in the centered-X model. Can you see an advantage to centering X in simple linear regression?

(d) Suppose that we center both X and Y; that is, let $Y^c = Y - \bar{Y}$ and $X^c = X - \bar{X}$. Now, for the regression of Y^c on X^c, show that the slope is again unchanged but that the intercept is now zero.

2.7 Using the computer, verify the properties associated with centering both Y and X, using the regression of COITFREQ on FEMAGE in the *couples dataset*.

2.8 A random sample of six students was drawn from a large university to determine if *depression* is related to GPA. *Depression* was measured with the Center for Epidemiologic Studies Depression Scale (CESD). This particular short version of the scale ranges from 0 to 84, with higher scores indicating more depressive symptomatology. The researchers think that students with lower GPAs suffer more from depression because they are more worried about their future job or graduate school prospects. Summary statistics are:

- Mean (GPA) = 3.
- Mean(CESD) = 50.
- Std Dev(GPA) = .894.
- Std Dev(CESD) = 28.284.
- cov(GPA,CESD) = − 20.

    (**a**) Give the regression equation for predicting CESD from GPA.

    (**b**) Give the correlation coefficient for the correlation of CESD with GPA.

    (**c**) If GPA for the first student is 2.0, give the CESD predicted for this student, based on the sample regression.

2.9 Regard the following data:

Student	SAT	GPA	Student	SAT	GPA	Student	SAT	GPA
1	4.8	2.4	5	3.8	2.7	9	7.2	3.4
2	6.6	3.5	6	5.2	2.4	10	6.0	3.2
3	5.9	3.0	7	6.6	3.0			
4	7.4	3.8	8	5.0	2.8			

To determine which incoming students should receive scholarships, a university admissions officer decided to study the relationship between a student's score on the *SAT verbal test* (taken in the final year of high school) and the student's *college GPA* at the end of the sophomore year. Ten student records for juniors at the university were examined. The data are shown above. The SAT scores have been divided by 100. The officer reasoned that if SAT scores are strongly related to GPA, an incoming student's SAT score gives a good indication of how well he or she will do in college.

(**a**) How strong is the relationship between GPA and SAT?

(**b**) Is this relationship significant?

(**c**) What is the prediction equation for predicting GPA from SAT?

(**d**) Using this equation, what is the predicted GPA for the first student?

(**e**) What is the residual for the first student?

2.10 The *cities dataset* consists of data compiled by the author on a random sample of 60 of the largest U.S. cities as of 1980. Several characteristics of these cities, based on 1980 census data, were measured, including:

- *Homicide rate*: number of homicide victims per 100,000 population.
- *Cost-of-living index*: the average market value of houses divided by average household income.
- *City growth rate*: percent increase in population from 1970 to 1980.
- *Population size*: the number of inhabitants of the city, in thousands of persons.
- *Reading quotient*: the total number of volumes, plus daily volume circulation, across all libraries in the city, divided by population size.

A simple linear regression of the *cost-of-living index* on the *city growth rate* produces the following sample statistics for 60 cities:

- $r = .18453$.
- $SSE = 34.8798$.
- $RSS = 1.2296$.

- $b_0 = 2.2299$.
- $b_1 = .0085$.
- $\hat{\sigma}_{b_1} = .0059$.

(a) What percent of variation in the *cost-of-living index* is accounted for by the *growth rate*?

(b) Give the sample estimate of the error (disturbance) variance.

(c) Compute the three different test statistics for testing whether the population slope is zero, and show that they are all the same value, within rounding error. What is the conclusion for this test?

(d) Interpret the values of b_0 and b_1.

(e) Give the value of r^2_{adj}.

2.11 For the *cities dataset*, let Y = *homicide rate* and X = *reading quotient*. Then for 54 cities we have:

- $\sum x = 384.7$.
- $\sum y = 409.3$.
- $\sum x^2 = 3069.29$.
- $\sum y^2 = 4564.81$.
- $\sum xy = 2663.45$.
- $\sum (y - \hat{y})^2 = 1268.58$ for the regression of Y on X.

(a) Find the OLS prediction equation for the linear regression of Y on X.

(b) Find the correlation between Y and X.

(c) Interpret the values of b_0 and b_1 in terms of the variables involved.

(d) Test whether or not there is a significant linear relationship between Y and X.

(e) What proportion of the variation in the *homicide rate* of a city is explained by its *reading quotient* here?

(f) What is the estimated average *homicide rate* for all cities in the population having a *reading quotient* of 7.1241?

2.12 Prove that for any given sample of X and Y values, the OLS estimates b_0 and b_1 produce the largest r^2 among all possible estimates for the linear regression of Y on X.

2.13 Using the computer, regress COLGPA on STUDYHRS in the *students dataset* and save the residuals. Then use software (e.g., SAS, SPSS) to test whether the errors are normally distributed in the population. Missing imputation: Follow the instructions for Exercise 2.1.

2.14 For the *cities dataset*, a regression of the *homicide rate* on the *1980 population size* for 60 cities produces the following statistics:

- Mean (*homicide rate*) = 7.471.
- Mean (*1980 population size*) = 759.618.

- Std Dev (*homicide rate*) = 5.151.
- Std Dev (*1980 population size*) = 1473.737.
- $r_{adj}^2 = .2190$.
- SSE = 1202.000.

(a) Give the OLS estimates of b_0 and b_1 for this regression. (*Hint*: Derive r from r_{adj}^2 by noting that $r_{adj}^2 = 1 - [(n-1)/(n-2)](SSE/TSS)$. Then use r to compute b_1.)

(b) Give the F statistic for testing whether H_0: $\beta_1 = 0$. Is it significant?

2.15 Regard the following data:

Obs.	X	Y	Obs.	X	Y
1	1	2	6	6	3
2	1	4	7	6	6
3	3	1.5	8	10	2
4	3	2	9	10	4
5	3	5	10	10	6.25

Estimate the simple linear regression of Y on X. In particular:

(a) Give the sample regression equation.
(b) Estimate the variance of the disturbances.
(c) Evaluate the discriminatory power of the model.
(d) Evaluate the empirical consistency of the model by performing the lack-of-fit test.
(e) Give r_{adj}^2.

2.16 For the 416 couples in the *couples dataset*, the regression of *coital frequency in the past month* on the *male partner's age* produces the following statistic: RSS = 1996.46395. Given that the sample variance of *coital frequency* is 37.1435 and the sample variance of *male age* is 215.50965, do the following:

(a) Find the absolute value of the correlation between *coital frequency* and *male age*.
(b) Find the absolute value of the slope of the regression of *coital frequency* on *male age*.
(c) Give the value of the standard error of the slope.
(d) Perform a test of H_0: $\beta_1 = 0$ against H_1: $\beta_1 \neq 0$ using the t test for the slope.

2.17 Let $Z_y = (y - \bar{y})/s_y$ and $Z_x = (x - \bar{x})/s_x$. Prove that an OLS regression of Z_y on Z_x results in an intercept estimate of zero and a slope estimate that is the correlation between X and Y.

2.18 Using the computer, verify that the regression of Z_y on Z_x gives $b_0 = 0$ and $b_1 = r$, using the example of the regression of COITFREQ on FEMAGE in the *couples dataset*.

2.19 Prove that for $n \geq 3$, r_{adj}^2 is always smaller than r^2. (*Hint*: Write out the formula for r_{adj}^2 from Exercise 2.14. Then subtract and add SSE/TSS from this formula and factor appropriately.)

2.20 Regard the following (x,y) values for 10 observations:

X	Y	X	Y	X	Y
1	1.5	5	7.5	9	8
1	1.75	5	7.75	9	8.25
3	4.5	7	8		
3	4.75	7	8.25		

Do the following:
(a) Estimate the simple linear regression of Y on X.
(b) Show that high discriminatory power can coexist with empirical inconsistency by comparing the r^2 for this regression with the results of the lack-of-fit test.

2.21 Changing metrics: If b_0 and b_1 are the intercept and slope, respectively, for the regression of Y on X, prove that the regression of $c_1 Y$ on $c_2 X$, where c_1 and c_2 are any two arbitrary constants, results in a sample intercept, b_0', equal to $c_1 b_0$, and a sample slope, b_1', equal to $(c_1/c_2)b_1$.

2.22 Using the computer, change the metrics of COLGPA and STUDYHRS in the *students dataset* as follows. Let MCOLGPA $= 100($COLGPA$)$ and MSTUDHRS $=$ STUDYHRS$/24$. Then verify the principles of Exercise 2.21 by regressing MCOLGPA on MSTUDHRS and comparing the coefficients to those from the regression of COLGPA on STUDYHRS. Missing imputation: Follow the instructions for Exercise 2.1.

2.23 Using the computer, regress EXAM1 on STATMOOD in the *students dataset* and save the residuals. Then correlate the residuals with STATMOOD, EXAM1, COLGPA, and SCORE. Also, correlate $\hat{y}$ with STATMOOD, the residuals, and EXAM1. Explain the value of each of these correlations. Missing imputation: Substitute 3.0827835 for missing values on COLGPA and 40.9358974 for missing values on SCORE.

2.24 Rescaling: If b_0 and b_1 are the intercept and slope, respectively, for the regression of Y on X, prove that the regression of $Y + c_1$ on $X + c_2$ results in a sample intercept, b_0^*, equal to $b_0 + c_1 - b_1 c_2$, and a sample slope, b_1^*, equal to b_1. Then let $c_1 = -\bar{y}$ and $c_2 = -\bar{x}$, and show that this principle subsumes Exercise 2.6(d) as a special case.

2.25 Recall the decomposition of the population variance of Y, assuming that

$$Y = \beta_0 + \beta_1 X + \varepsilon: V(Y) = V(\beta_0 + \beta_1 X + \varepsilon) = V(\beta_0 + \beta_1 X) + V(\varepsilon).$$

Note that this equals $\beta_1^2 V(X) + V(\varepsilon)$. Show an analogous decomposition for the sample variance of Y based on the fact that $Y = b_0 + b_1 X + e$.

2.26 Dichotomous independent variables can be used in regression if they are dummy-coded. This involves assigning a 1 to everyone in the category of interest (the *interest category*) and a 0 to everyone who is in the other category on X (the *contrast category*, or *omitted group*, or *reference group*). Show that in the population, the regression of a continuous variable Y on a dummy variable X results in β_0 being the conditional mean of Y for those in the group coded 0 on X, and $\beta_0 + \beta_1$ being the conditional mean of Y for those in the group coded 1 on X (i.e., the interest category).

2.27 Prove that the sample regression of a continuous Y on a dummy variable, X, has b_0 equal to the mean of Y for the contrast category, and b_1 equal to the mean of Y for the interest category minus the mean of Y for the contrast category. (*Hint*: Let:

- $\bar{y}_0$ = mean of Y for those in the contrast category
- $\bar{y}_1$ = mean of Y for those in the interest category
- $\hat{\pi}$ = proportion of the sample in the interest category
- n_1 = number of sample cases in the interest category
- n_0 = number of sample cases in the contrast category

and note that:

- $\hat{\pi} = \sum x / n$
- $\sum x = n\hat{\pi} = n_1$
- $\bar{y} = (\hat{\pi})\bar{y}_1 + (1 - \hat{\pi})\bar{y}_0$
- $\sum xy = \sum(y | x = 1)$, that is, the sum of Y for those in the interest category)

2.28 Using the computer, verify the properties associated with the dummy coding of X. That is, show that $b_0 = (\bar{y} | x = 0)$ and $b_1 = (\bar{y} | x = 1) - (\bar{y} | x = 0)$. As the example, use the regression of CONFLICT on PRESCHDN in the *couples dataset*.

CHAPTER 3

Introduction to Multiple Regression

CHAPTER OVERVIEW

In this chapter I introduce the multiple linear regression (MULR) model, building on the theoretical foundations established in Chapter 2. I begin with an artificial example that illustrates the primary advantages of multiple regression over simple linear regression (SLR). A fundamental concept, statistical control for another regressor, is then explicated in some detail. I then formally develop the model and its assumptions, and discuss estimation via ordinary least squares as well as inferential tests in MULR. I then explain the problem of *omitted-variable bias* and define the three major types of bias that can occur when key regressors are omitted from the model: *confounding, suppression*, and *reversal*. I also consider the phenomenon of *mediation* by omitted variables, which, although not a form of bias, is central to causal modeling. Next I discuss *statistical interaction* and the related issue of comparing models across discrete groups of cases. Finally, I return to the issue of assessing empirical consistency and evaluate a MULR model for scores on the first exam for 214 students in introductory statistics.

EMPLOYING MULTIPLE PREDICTORS

Advantages and Rationale for MULR

Suppose that we have several potential predictors, $X_1, X_2, \ldots, X_K$, for a given response variable, Y. Certainly, we could assess the impact of each X_k on Y via a series of SLRs. In fact, if the X's are all orthogonal—that is, mutually uncorrelated—the impact, b_k, of a given X_k will be no different in an SLR of Y on X_k than in a MULR of Y on X_k and all other $K - 1$ predictors. Moreover, R^2 from the multiple regression

Regression with Social Data: Modeling Continuous and Limited Response Variables,
By Alfred DeMaris
ISBN 0-471-22337-9   Copyright © 2004 John Wiley & Sons, Inc.

would simply be the sum of all the individual r^2's from the K different SLRs of Y on each X_k. However, unless the data were gathered under controlled conditions (e.g., via an experiment), the regressors in social data analysis will rarely be orthogonal. Most of the time they are correlated—sometimes highly so.

With this in mind, there are four primary advantages of MULR over SLR. First, by including several regressors in the same model, we can counteract omitted-variable bias in the coefficient for any given X_k. Such bias would be present were we to omit a regressor that is correlated with both X_k *and* a determinant of Y. Second, we can examine the discriminatory power achieved when employing the collection of regressors simultaneously to model Y. When the X's are correlated, R^2 is no longer the simple sum of r^2's from the SLRs of Y on each X_k. R^2 can be either smaller or larger than that sum. The first two advantages are germane only when regressors are correlated. However, the next two advantages apply even if the regressors are all mutually orthogonal. The third advantage of MULR is its ability to model relationships between Y and X_k that are nonlinear, or to model statistical interaction among two or more regressors. Although interaction is discussed below, a consideration of nonlinear relationships between the X's and Y is postponed until Chapter 5. The final advantage is that in employing MULR, we are able to obtain a much more precise estimate of the disturbance variance than is the case with SLR. By "precise" I mean an estimate that is as free as possible from systematic influences and that comes as close as possible to representing purely random error. The importance of this, as we shall see, is that it makes tests of individual slope coefficients much more sensitive to real regressor effects than would otherwise be the case.

Example

Figure 3.1 presents a scatterplot of Y against X for 26 cases, along with the OLS fitted line for the linear regression of Y on X. It appears that there is a strong linear impact of X on Y, with a slope of 1.438 that is highly significant ($p < .0001$). For this regression, the r^2 is .835, and the estimate of the error variance is 3.003. However, the plot is somewhat deceptive in suggesting that X has such a strong impact on Y. In truth, this relationship is driven largely by a third, omitted variable, Z. Z is strongly related to both Y ($r_{zy} = .943$) and X ($r_{zx} = .891$). If Z is a cause of both Y and X, or even if Z is a cause of Y but only a correlate of X, the SLR of Y on X leads us to overestimate the true impact of X on Y, perhaps by a considerable amount. What is needed here is to control for Z and then reassess the impact of X on Y.

Controlling for a Third Variable

What is meant in this instance by *controlling for Z* or *holding Z constant*? I begin with a mechanical analogy. Figure 3.2 illustrates a three-variable system involving X, Y, and Z. Actually, there are four variables—counting ε, the disturbance—but only three that are observed. Suppose that this system is the true model for Y. Further, imagine that the circles enveloping X, Y, Z, and ε are gears, and the arrows and

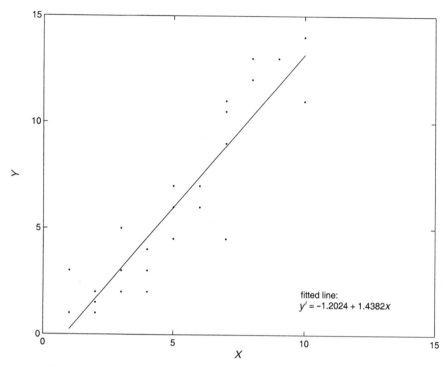

Figure 3.1 Scatterplot of (x,y) values, along with OLS regression line, depicting the linear relationship between Y and X for 26 cases.

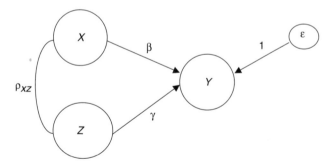

Figure 3.2 Regression model for Y showing confounding of X–Y relationship due to a third variable, Z.

curved lines connecting these variables are drive shafts. Running the SLR of Y on X alone is tantamount to hiding the shaft, γ, that runs from Z to Y. When you turn the X gear, the Y gear also turns—maybe quite a bit—because X is also connected to Z, and Z also turns Y. However, since the shaft from Z to Y is hidden, we are misled into thinking that X's power to move Y is all realized through the shaft β. What we really want to know, of course, is how much X turns Y, if at all, *just on its own*. In such a system, one simple solution, of course, is simply to disconnect γ. Then when we turn X, if Y also turns, we know that this is strictly due to β, not to the other connection through Z. What this solution accomplishes is to *stop Z from turning along with X*, so that if turning X turns Y, this can only be due to β. The analogy to variables and their relationships is made by equating "turning" with variation. We want to know whether varying X also induces Y to vary, and we don't want X's effect to be confounded with Z's effect on Y. Hence, we must stop Z from *varying along with X* in order to isolate X's effect on Y.

The principle of holding Z constant and observing how much Y varies with X alone is illustrated, in the current example, in Figures 3.3 to 3.5. The variable Z, measured for all 26 cases, takes on three values: 1, 2, and 3. Figure 3.3 shows a plot of Y against X for all cases whose Z-value is 1. Notice that we are literally holding Z constant in this plot, since all cases have the same value of Z (i.e., Z is no longer varying here). Also notice that the slope of the regression of Y on X for just these cases is now much shallower than before, with a value of only .444, about a third of its previous value achieved when Z was ignored. Figure 3.4 similarly shows a plot of Y against X for cases whose Z-value is 2. The slope of the regression here is .595. Finally, Figure 3.5 shows Y plotted against X for those with Z-values of 3. The slope of the regression in this case is .544.

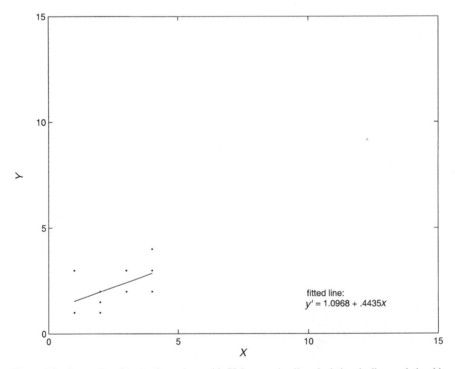

Figure 3.3 Scatterplot of (x,y) values, along with OLS regression line, depicting the linear relationship between Y and X for the 10 cases with Z-values equal to 1.

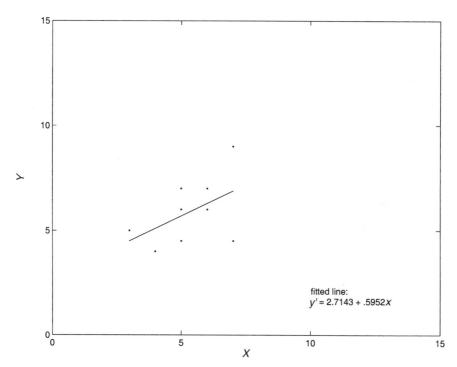

Figure 3.4 Scatterplot of (x,y) values, along with OLS regression line, depicting the linear relationship between Y and X for the nine cases with Z-values equal to 2.

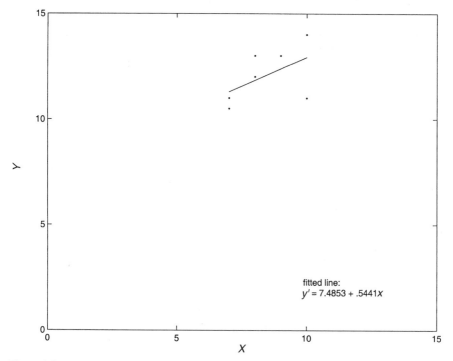

Figure 3.5 Scatterplot of (x,y) values, along with OLS regression line, depicting the linear relationship between Y and X for the seven cases with Z-values equal to 3.

A simple summary of X's impact on Y when controlling for Z might be obtained by averaging the three individual X–Y slopes from each level of Z, since their values are quite similar. This is not exactly the formula used to calculate the effect of X in the multiple regression of Y on X and Z (see Exercise 3.17 for the correct formula), but it is a good heuristic approximation. The average of the three slopes is .528. Hence, we could say that holding Z constant, each unit increase in X is expected to increase Y by about .528 unit, on average. This is substantially smaller than the slope of X's effect on Y when Z is not controlled, which was 1.438. Clearly, failing to control for Z results in an overestimate of the impact of X. Statistically, controlling for Z is accomplished by including it, along with X, in the equation for Y and estimating the MULR model. In the MULR, the slope for X is actually .564 instead of .528; nonetheless, the reduction in value compared to 1.438 is clear. Additionally, the estimate of the disturbance variance in the MULR is 1.614, or about half that in the SLR. The error variance in the SLR was obviously inflated by not removing the systematic variance accounted for in Y by Z. Finally, R^2 in the MULR is .915, compared to .835 in the SLR. In that X and Z are highly correlated, the R^2 in the MULR is much less than the sum of the individual r^2's explained by X and Z in isolation (.835 and .889, respectively).

A final and more formal way of understanding the meaning of statistical control is as follows. Suppose that we regress X on Z and arrive at the estimated equation: $X = c + dZ + u$. Now, notice that $u = X - (c + dZ)$. That is, u is the part of X that is not a linear function of Z. Recall from Chapter 2 that u is also uncorrelated with Z. Then suppose that we perform the SLR of Y on u and get $Y = a + bu + e$. Then b from this SLR is exactly the same as the slope of X in the MULR of Y on X and Z. Here we see that control, once again, is achieved by preventing Z from varying along with X. In this case, it is evident that such control is achieved by using a form of X from which the linear association with Z has been removed. Since that form of X is now orthogonal to Z, b represents the impact of X on Y while preventing Z from varying along with X, or holding Z constant. (Exercise 3.8 asks you to verify this principle using the computer to analyze a four-variable model from the *students dataset*.)

MULR Model

Now that we understand the need and rationale for MULR, it is time to present the model and assumptions for estimation via OLS. The MULR model for the ith response, Y_i, is

$$Y_i = \beta_0 + \beta_1 X_{i1} + \beta_2 X_{i2} + \cdots + \beta_K X_{iK} + \varepsilon_i. \tag{3.1}$$

Or, the model for the conditional mean of Y, given the X's, is

$$E(Y_i | X_{i1}, X_{i2}, \ldots, X_{iK}) = \beta_0 + \beta_1 X_{i1} + \beta_2 X_{i2} + \cdots + \beta_K X_{iK},$$

or, in more compact notation,

$$E(Y_i) = \sum_k \beta_k X_{ik}, \tag{3.2}$$

where the index k ranges from 0 to K, and X_0 equals 1 for all cases. [From here on, for economy of presentation, I denote the conditional mean in the regression model as $E(Y_i)$ instead of $E(Y_i|X_{i1}, X_{i2}, \ldots, X_{iK})$.] The right-hand side of equation (3.2) called the *linear predictor*, represents the structural part of the model. As in SLR, the conditional mean is assumed to be a linear function of the equation parameters— the β's. Now, however, the points $(x_1, x_2, \ldots, x_K, y)$ no longer lie on a straight line. Rather, they lie on a *hyperplane* in $(K + 1)$-dimensional space. A particular combination of values, $x_1, x_2, \ldots, x_K$, for the X's is called a *covariate pattern* (Hosmer and Lemeshow, 2000). $E(Y_i)$ is therefore the mean of the Y_i's for all cases with the same covariate pattern. As before, the equation disturbances, the ε_i's, represent the departure of the individual Y_i's at any given covariate pattern from their mean, $E(Y_i)$.

The other two quantities that are of importance in MULR are σ^2 and P^2 (rho-squared). As before, σ^2 is $V(\varepsilon)$, the variance of the equation errors, which is assumed to be constant at each covariate pattern. Because the variance in Y at each covariate pattern is assumed to be due to random error alone, σ^2 is also the conditional variance of Y. P^2 is the coefficient of determination for the MULR model, and as in SLR, is the primary index of the model's discriminatory power. P^2 can be understood, once again, by expressing the variance of Y in terms of the model:

$$V(Y_i) = V\left(\sum_k \beta_k X_{ik} + \varepsilon_i\right) = V\left(\sum_k \beta_k X_{ik}\right) + V(\varepsilon_i).$$

Dividing through by $V(Y_i)$, we have

$$1 = \frac{V\left(\sum_k \beta_k X_{ik}\right)}{V(Y_i)} + \frac{V(\varepsilon_i)}{V(Y_i)},$$

or

$$1 = P^2 + \frac{\sigma^2}{V(Y_i)}.$$

P^2 is therefore the proportion of the variance in Y that is due to variation in the linear predictor—that is, to the structural part of the model.

Interpretation of the Betas. The interpretation of the betas is, as in SLR, facilitated by manipulating equation (3.2) so as to isolate β_0 and then each β_k. By setting all of the X's to zero, we see that β_0 is the expected value of Y when all of the X's equal zero. The interpretation of any β_k, say β_1, can be seen by considering increasing X_1 by 1 unit while holding all other X's constant at specific values. If we let $\mathbf{x}_{-1}$ represent all of the X's other than X_1, the change in the mean of Y is

$$E(Y\,|\,x_1 + 1, \mathbf{x}_{-1}) - E(Y\,|\,x_1, \mathbf{x}_{-1})$$
$$= \beta_0 + \beta_1(x_1 + 1) + \beta_2 X_2 + \cdots + \beta_K X_K - (\beta_0 + \beta_1 x_1 + \beta_2 X_2 + \cdots + \beta_K X_K)$$
$$= \beta_1(x_1 + 1 - x_1) = \beta_1.$$

Here it is clear that β_1 represents the change in the mean of Y for a unit increase in X_1, controlling for the other regressors in the model. Or, in language with fewer causal connotations, β_1 is the expected difference in Y for those who are 1 unit apart on X_1, controlling for the other regressors in the model. As in SLR, β_1 (and β_k generally) is both the unit impact of X_1 (or X_k) on $E(Y)$ as well as the first partial derivative of $E(Y)$ with respect to X_1 (or X_k).

Assumptions of the Model. The assumptions of MULR relevant to estimation via OLS mirror those for SLR with perhaps a couple of exceptions:

1. The relationship between Y and the X's is *linear in the parameters*; that is,

   $$Y_i = \beta_0 + \beta_1 X_{i1} + \beta_2 X_{i2} + \cdots + \beta_K X_{iK} + \varepsilon_i \qquad \text{for } i = 1, 2, \ldots, n.$$

   What does this mean? It means that the parameters enter the equation in a *linear* fashion or that the equation is a weighted sum of the *parameters*, where the weights are now the X's. An example of an equation that is nonlinear in the parameters is $Y_i = \beta_0 X_{i1}^{\beta_1} X_{i2}^{\beta_2} \varepsilon_i$. Here, since β_1 and β_2 enter as exponents of the X's, the right-hand side of this equation clearly cannot be described as a weighted sum of the β's. On the other hand, this equation is easily made into one that is linear in the parameters by taking the log of both sides: $\ln Y_i = \ln \beta_0 + \beta_1 \ln X_{i1} + \beta_2 \ln X_{i2} + \ln \varepsilon_i$. In contrast, there is no simple way to linearize the equation $Y_i = \beta_0 + X_{i1}^{\beta_1} + X_{i2}^{\beta_2} + \varepsilon_i$.
2. The observations are sampled independently.
3. Y is approximately interval level, or binary (although the ideal procedures when Y is binary are probit or logistic regression, described in Chapter 7). The X's are approximately interval-level, or dummy variables. *Dummy variables* are binary-coded X's that are used to represent qualitative predictors or predictors that are to be treated as qualitative (more about this in Chapter 4).
4. The X's are fixed over repeated sampling. As in the case of SLR, this requirement can be waived if we are willing to make our results conditional on the observed sample values of the X's.
5. $E(\varepsilon_i) = 0$ at each covariate pattern. This is the orthogonality condition.
6. $V(\varepsilon_i) = \sigma^2$ at each covariate pattern.
7. $\text{Cov}(\varepsilon_i, \varepsilon_j) = 0$ for $i \neq j$, or the errors are uncorrelated with each other. Again, this is equivalent to assumption 2 if the data are cross-sectional.

8. The errors are normally distributed.

9. None of the X_k is a perfect linear combination, or weighted sum, of the other X's in the model. That is, if we regress each X_k on all of the other X's in the model, no such MULR would produce an R^2 of 1.0. Should we find an R^2 of 1.0 for the regression of one or more X's on the remaining X's, we say that there is an *exact collinearity* among the X's; that is, at least one of the X's is completely determined by the others. Under this condition, the equation parameters are no longer identified, and there is no unique OLS solution to the normal equations. This is almost never a problem, although once in a while the unsuspecting analyst will try to use X's that are exactly collinear in a regression. One situation that produces this condition is when one tries to model an *M*-category qualitative variable using all *M* dummy variables that can be formed from the categories. I postpone discussion of dummies until Chapter 4. Another scenario resulting in perfect collinearity occurs when someone tries to enter, say, *age at marriage*, *marital duration*, and *current age*—all in years of age—into a model. In that *current age = age at marriage + marital duration*, these variables are exactly collinear. If this happens, it is immediately obvious from software output. There are no regression results, and an error message appears warning the analyst that the "XTX matrix" is "singular" or "has no inverse." (In Chapter 6, where I discuss the matrix representation of the regression model, these concepts will be clearer.) Although perfect collinearity is rare, a somewhat more common problem occurs under conditions of near-collinearity among the X's. This occurs when the regression of a given X_k on the other X's produces an R^2 close to 1.0, say .98. This situation is termed *multicollinearity*. Unlike the case of exact collinearity, it does not violate an assumption of regression. The normal equations can still be solved and parameter estimates produced. However, the estimates and their standard errors may be quite "flawed." The symptoms, consequences, diagnosis, and remedies for multicollinearity are taken up briefly below, and with substantially greater rigor in Chapter 6.

Estimation via OLS. Estimation of the MULR model proceeds in a fashion similar to estimation of the SLR model. The idea, once again, is to find the $b_0, b_1, b_2, \ldots, b_K$ that minimize *SSE*, where

$$SSE = \sum_{i=1}^{n}(y_i - \hat{y}_i)^2 = \sum_{i=1}^{n}\left[y_i - \left(\sum_k b_k x_{ik}\right)\right]^2$$

For any given sample of data values, this expression is clearly only a function of the b_k's. To find the b_k's that minimize it, we take the first partial derivative of *SSE* with respect to each of the b_k's and set each resulting expression to zero. This produces a set of simultaneous equations representing the multivariate version of the normal equations. These are then solved to find the OLS estimates of the b_k's (the solution, in matrix form, is shown in Chapter 6).

As in SLR, we are also interested in estimating σ^2 and P^2. In MULR, as in SLR, the estimate of σ^2 is *SSE* divided by its degrees of freedom, which is $n - K - 1$.

Thus,

$$\hat{\sigma}^2 = MSE = \frac{SSE}{n - K - 1}.$$

R^2, the estimate of P^2, has the same basic formula in MULR as in SLR:

$$R^2 = 1 - \frac{SSE}{TSS}.$$

In MULR, R^2 is the proportion of variation in Y that is accounted for by the *collection* of X's in the model, as represented by the linear predictor. R^2 is also the squared correlation of Y with its model-fitted value. That is,

$$R^2 = [\text{corr}(y, \hat{y})]^2.$$

Example. Table 3.1 shows the results of three different regression analyses of students' scores on the first exam in introductory statistics, for 214 students. The first model is just an SLR of *exam1 score* on the *math diagnostic score* and is essentially a replication of the analysis in Table 2.2. The second model adds *college GPA* as a predictor of exam scores, whereas the third model adds *attitude toward statistics* (a continuous variable ranging from -10 to 20, with higher values indicating a more positive attitude), *class hours in the current semester* (number of class hours the

Table 3.1   Regression Models for Score on the First Exam for 214 Students in Introductory Statistics

	Model 1		Model 2		Model 3	
Regressor	b	b^s	b	b^s	b	b^s
Intercept	$-35.521**$	.000	$-57.923***$	.000	$-65.199***$	.000
Math diagnostic score	$2.749***$	.517	$2.275***$	.428	$2.052***$	.386
College GPA			$13.526***$	.398	$13.659***$	.402
Attitude toward statistics					.376*	.122
Class hours in the current semester					.819*	.115
Number of previous math courses					1.580	.094
RSS	16509.326		25803.823		27762.947	
MSE	212.919		169.878		162.909	
F	77.538***		75.948***		34.084***	
R^2	.268		.419		.450	
R^2_{adj}	.264		.413		.437	

$*p < .05. \ **p < .01. \ ***p < .001.$

student is currently taking), and *number of previous math courses* (the number of previous college-level math courses taken by the student).

Recall that when discussing the authenticity of the first model in Chapter 2, I suggested academic ability as the real reason for the association of math diagnostic performance with exam scores. Should this be the case, we would expect that with academic ability held constant, diagnostic scores would no longer have any impact on exam performance. That is, if academic ability is Z in Figure 3.2 and diagnostic score is X, this hypothesis suggests that there is no connection from X to Y, but rather, it is the connection from X to Z and from Z to Y that causes Y to vary when X varies. (Instead of a curved line connecting X with Z, we now imagine a directed arrow from Z to X, since Z is considered to cause X as well as Y. The mathematics will be the same, as shown below in the section on omitted-variable bias.) The measure of academic ability I choose in this case is *college GPA*, since it reflects the student's performance across all classes taken prior to the current semester and is therefore a proxy for academic ability.

The first model shows that exam performance is, on average, 2.749 points higher for each point higher that a student scores on the diagnostic. Model 2, with *college GPA* added, shows that this effect is reduced somewhat but is still significant. (Whether this reduction itself is significant is assessed below.) Net of academic ability, exam performance is still, on average, 2.275 points higher for each point higher a student scores on the diagnostic. It appears that academic ability does not explain all of the association of diagnostic scores with exam scores, contrary to the hypothesis. *College GPA* also has a substantial effect on exam performance. Holding the *diagnostic score* constant, students who are a unit higher in GPA are estimated to be, on average, about 13.5 points higher on the exam. The model with two predictors explains about 41% of the variance in exam scores. Adding *college GPA* apparently enhances the proportion of explained variation by $.419 - .268 = .151$. This increment to R^2 resulting from the addition of *college GPA* is referred to as the *squared semipartial correlation coefficient* between exam performance and *college GPA*, controlling for *diagnostic score*. The *semipartial correlation coefficient* between *college GPA* and exam performance, controlling for *diagnostic score*, is the square root of this quantity, or .389. Although the increment to R^2 is a meaningful quantity, the semipartial correlation coefficient is not particularly useful. A more useful correlation coefficient that takes into account other model predictors is the *partial correlation coefficient*, explained below. Our estimate of σ^2 for model 2 is *MSE*, which is 169.878.

The third model adds the last three predictors. This model explains 45% of the variance in exam scores. Of the three added predictors, two are significant: *attitude toward statistics* and *class hours in the current semester*. Each unit increase in *attitude* is worth about a third of a point increase in exam performance, on average. Each additional hour of *classes taken during the semester* is worth about eight-tenths of an additional point on the exam, on average. This last finding is somewhat counterintuitive, in that those with a greater class burden have less time to devote to any one class. Perhaps these students are especially motivated to succeed, or perhaps these students have few other obligations, such as jobs or families, which allows them to devote more time to studies. The intercept in all three models is clearly uninterpretable, since the only predictor that can take on a value of zero is *number of previous math courses*. The model might be useful for forecasting exam scores for prospective students, since its discriminatory power is moderately strong. Suppose that a prospective student scores 43 on the diagnostic, has a 3.2 GPA, has a 12 on attitude toward statistics, is taking 15 hours in the current semester, and has had one previous math course. Then his or her predicted score on the first exam would be $\hat{y} = -65.199 + 2.052(43) + 13.659(3.2) + .376(12) + .819(15) + 1.58(1) = 85.1228$.

Discriminatory Power, Revisited. Although R^2 is used to tap discriminatory power, it is an upwardly biased estimator of P^2. This is evident in the fact that (Stevens, 1986)

$$E(R^2 \mid P^2 = 0) = \frac{K}{n-1}.$$

That is, even if P^2 equals zero in the population, a regression model with, say, 49 predictors and 50 cases will show an R^2 of 1.0, regardless of the real utility of the explanatory variables in explaining Y. A better estimator of P^2 is the adjusted R^2, whose formula, as given in Chapter 2, is

$$R_{adj}^2 = 1 - \frac{\hat{\sigma}^2}{s_y^2}.$$

R_{adj}^2 is typically smaller in magnitude than R^2, adjusting for the latter's tendency to "overshoot" P^2. For model 3 in Table 3.1, for example, R^2 is .45 whereas R_{adj}^2 is .437.

Additionally, R^2 has the rather unpleasant property that it can never decrease when variables are added to a model, no matter how useless they are in explaining Y. This could entice one to add ever more variables to a model in order to increase its discriminatory power. The drawback to this strategy is that we end up with a model that maximizes R^2 in a given sample but has little replicability across samples. Another advantage to R_{adj}^2 is that it reflects whether or not "junk" is being added to a regression model. Rewriting R_{adj}^2 as

$$R_{adj}^2 = 1 - \frac{SSE/(n-K-1)}{TSS/(n-1)} = 1 - \frac{n-1}{n-K-1}\frac{SSE}{TSS},$$

we see that as K increases—that is, as we add more and more predictors to the model—the factor $(n-1)/(n-K-1)$ also increases. Now TSS is constant for any given set of Y values. So for R_{adj}^2 to increase as we add predictors, SSE must be reduced correspondingly. That is, each predictor must be removing systematic variation from the error term in order for us to experience an increase in R_{adj}^2. Hence, R_{adj}^2 is a more effective barometer for gauging whether an additional variable should be entered into a model than is R^2.

Standardized Coefficients and Elasticities. Frequently, we wish to gauge the relative importance of explanatory variables in a given model. The unstandardized coefficients, that is, the b_k's, are not suited to this purpose, since their magnitude depends on the metrics of both X_k and Y. A better choice is the *standardized partial regression coefficient*, denoted b_k^s in this volume. The formula for b_k^s is

$$b_k^s = b_k \frac{s_{x_k}}{s_y}.$$

That is, the standardized coefficient is equal to the unstandardized coefficient times the ratio of the standard deviation of X_k to the standard deviation of Y. This standardization removes the dependence of b_k on the units of measurement of X_k and Y. Its interpretation, however, is cumbersome: It represents *the estimated standard-deviation difference in Y expected for a standard-deviation increase in X_k, net of other model predictors*. Why? Well, if b_k is the change in the mean of Y for a unit increase in X_k, then $b_k s_{xk}$ is the change for a 1-standard-deviation increase in X_k. Dividing this by s_y gives us the change in the mean of Y expressed in standard deviations of Y, which is, of course, b_k^s. Typically, we are not interested in interpreting b_k^s but rather, in comparing the magnitudes of the b_k^s's. For example, in model 2 the b_k's suggest that *college GPA* has a substantially stronger effect on exam performance than that of *math diagnostic score*. However, the b_k^s's paint a different picture, suggesting that *math diagnostic score* has a stronger impact than *college GPA*. In the last model, on the other hand, the picture changes again. In the presence of the other model predictors, *college GPA* has the strongest effect on exam performance, followed by *math diagnostic score, attitude toward statistics, class hours in the current semester*, and *number of previous math courses*.

Although the b_k^s's are useful for gauging the relative importance of predictors in a given model, they have drawbacks with respect to other model comparisons. Because they are a function of the sample standard deviations of the predictors and of the response, they should not be used to compare coefficients across samples. For example, they should not be used to compare the effects of predictors in, say, two different populations, based on samples from each population. The reason for this is that the effect of a given predictor, as assessed by the unstandardized coefficient, might be the same in each population. But the standard deviations of the predictor and/or the response might be different in each population, resulting in different standardized coefficients. The same reasoning suggests that we should not use the b_k^s's to compare coefficients in different samples from the same population—say, taken in different time periods. In these cases, the b_k's are preferable for making comparisons.

A unitless measure that can be used for comparing the relative importance of regressor effects both within and *across* samples is the variable's *elasticity*, denoted E_k for the kth variable (Pindyck and Rubinfeld, 1981). This statistic is less susceptible to sampling variability than is the standardized coefficient. For the kth predictor, the population elasticity is calculated as

$$E_k = \beta_k \frac{\overline{x}_k}{\overline{y}}.$$

The elasticity is the *percentage change in Y that could be expected from a 1% increase in X_k* (Hanushek and Jackson, 1977). To see how this interpretation arises, consider the change in the mean of Y for a 1% increase in X_k, holding the other X's constant:

$$E(Y\,|\,x_k + .01x_k, \mathbf{x}_{-k}) - E(Y\,|\,x_k, \mathbf{x}_{-k})$$
$$= \beta_0 + \beta_1 X_1 + \cdots + \beta_k(x_k + .01x_k) + \cdots + \beta_K X_K$$
$$\quad - (\beta_0 + \beta_1 X_1 + \cdots + \beta_k x_k + \cdots + \beta_K X_K)$$
$$= .01\beta_k x_k.$$

Dividing by Y gives us $.01\beta_k x_k/Y$, which is the proportionate change in $E(Y)$—as a proportion of Y—for a .01 increase in X. Finally, multiplying by 100 results in $\beta_k(x_k/Y)$, which is then the percent change in $E(Y)$ resulting from a percent increase in X_k. This change depends on the levels of both X_k and Y, but it is customary to evaluate it at the means of both variables. For model 3 in Table 3.1, the elasticities are 1.091 for the *diagnostic score*, .549 for *college GPA*, .024 for *attitude toward statistics*, .158 for *class hours*, and .026 for the *number of previous math courses*. According to the elasticities, *math diagnostic score* clearly has the strongest impact. Each 1% increase in the diagnostic score is expected to increase exam performance by 1.09%.

Inferences in MULR

Several inferential tests are available to test different types of hypotheses about variable effects in MULR. The first test we might want to consider is a test for the model as a whole. Here we ask the question: Is the model of any utility in accounting for variation in Y? We can think of the null hypothesis as H_0: $P^2 = 0$, and the research hypothesis as H_1: $P^2 > 0$. That is, if the model is of any utility in explaining Y, there is some nonzero proportion of variance in Y in the population that is explained by the regression. Another way of stating these hypotheses is as follows: H_0: $\beta_1 = \beta_2 = \cdots = \beta_K = 0$ versus H_1: at least one $\beta_k \neq 0$. Actually, this is a narrow representation of the hypotheses. A more global expression of the hypotheses is: H_0: all possible linear combinations of the β_k's $= 0$ versus H_1: at least one linear combination of the β_k's $\neq 0$ (Graybill, 1976). The connection between these two (latter) statements of H_0 and H_1 is that $\beta_1, \beta_2, \ldots, \beta_K$ each represents linear combinations, or weighted sums, of the β_k's. To see that, say, β_1 is a weighted sum of the β_k's, we simply write β_1 as $1(\beta_1) + 0(\beta_2) + \cdots + 0(\beta_K)$. Here it is evident that the first weight is 1 and all the rest of the weights equal zero. The F statistic, which is used to test the null hypothesis in all three cases here, is actually a test of the third null hypothesis. The reason for highlighting this is that occasionally, the null hypothesis will be rejected and none of the individual coefficients turns out to be significant. One reason for this is that it is some other linear combination of the coefficients that is nonzero, but perhaps not a combination that makes any intuitive sense. (Another reason for this phenomenon is that multicollinearity is present among the X's; multicollinearity is considered below.)

At any rate, the test statistic for all of these incarnations of H_0 is the F statistic, where

$$F = \frac{RSS/K}{SSE/n - K - 1} = \frac{MSR}{MSE}$$

and MSR denotes mean-squared regression, or the regression sum of squares divided by its degrees of freedom, K. Under the null hypothesis—that is, if the null hypothesis is true—this statistic has the F distribution with K and $n - K - 1$ degrees of freedom. Table 3.1 displays RSS, MSE, and F for all three models. All F tests are highly significant, suggesting that each model is of some utility in predicting exam scores.

If the F test is significant, it is then important to determine which β_k's are not equal to zero. The individual test for any given coefficient is a t test, where

$$t = \frac{b_k}{\hat{\sigma}_{b_k}}$$

has the t distribution with $n - K - 1$ degrees of freedom under the null hypothesis that $\beta_k = 0$. All coefficients in all models with the exception of the effect of *number of previous math courses*, in model 3, are significant. As in simple linear regression, we may prefer to form confidence intervals for the regression coefficients. A 95% confidence interval for β_k is $b_k \pm t_{(.025, n-K-1)}\hat{\sigma}_{b_k}$.

Nested F. Often, it is of interest to test whether a nested model is significantly different from its parent model. *A model, B, is nested inside a parent model, A, if the parameters of B can be generated by placing constraints on the parameters in A.* The most common constraint is to set one or more parameters in A to zero. For example, model 2 is nested inside model 3 because the parameters in model 2 can be generated from those in model 3 by setting the β's for *attitude toward statistics, class hours in the current semester*, and *number of previous math courses*, in model 3, all to zero. We can therefore test whether a significant loss in fit is experienced when setting these parameters to zero, or, alternatively, whether a significant improvement in fit is experienced when adding these three parameters. In reality, we are testing H_0: $\beta_3 = \beta_4 = \beta_5 = 0$ versus H_1: at least one of $\beta_3, \beta_4, \beta_5 \neq 0$. That is, the test is a test of the validity of the constraints in H_0. In general form, the nested F-test statistic is

$$F = \frac{(RSS_A - RSS_B)/\Delta df}{MSE_A},$$

where Δdf is the number of constraints imposed on model A to derive model B, or the difference in the number of parameters estimated in each model. An alternative but equivalent form of the test statistic is

$$F = \frac{(R_A^2 - R_B^2)/\Delta df}{(1 - R_A^2)/(n - K - 1)},$$

where K is the total number of regressors in the parent model. In the current example, $K = 5$ and $\Delta df = 3$, since we have constrained three parameters in model 3 to zero to arrive at model 2. If H_0 is true—that is, the constraints are valid—this statistic has the F distribution with Δdf and $n - K - 1$ degrees of freedom. In the current example, the test statistic is

$$F = \frac{(27762.947 - 25803.823)/3}{162.909} = 4.009.$$

With 3 and 208 degrees of freedom, the attained significance level is .0083. At conventional significance levels (e.g., .05 or .01), we would reject H_0 and conclude that at least one of the constrained parameters is nonzero. Individual t tests suggest that two of the parameters are nonzero: the effect of *attitude toward statistics* and the effect of *class hours in the current semester*. If only one parameter is hypothesized to be zero, say β_k, the nested F is just the square of the t test for the significance of b_k in the model containing that parameter estimate.

Nesting: Another Example. There are other means of constraining parameters that do not involve setting them to zero. As an example, Table 3.2 presents two different regression models for *couple modernism*, based on the 416 intimate couples in the *couples dataset*. Model 1 shows the results of regressing *couple modernism* on male

Table 3.2   Regression Models for Couple Modernism for 416 Intimate Couples, Showing Nesting to Test Equality of Coefficients

Regressor	Model 1		Model 2	
	b	b^s	b	b^s
Intercept	23.368***	.000	23.481***	.000
Male's schooling	.110*	.139	.160***	.335
Female's schooling	.211***	.237	.160***	.335
Male's church attendance	−.136*	−.147	−.067**	−.131
Female's church attendance	.020	.021	−.067**	−.131
Male's income	−.015	−.097	.004	.030
Female's income	.048***	.189	.004	.030
RSS	473.121		348.285	
MSE	5.104		5.370	
F	15.450***		21.621***	
R^2	.185		.136	
R^2_{adj}	.173		.130	

Partial variance–covariance matrix for model 1

	Male's Schooling	Female's Schooling
Male's schooling	.001990	−.001004
Female's schooling	−.001004	.002340

*$p < .05$. **$p < .01$. ***$p < .001$.

and female partner's *schooling* (years of schooling completed), *church attendance* (interval-level predictor ranging from 1 = "never" to 9 = "more than once a week") and *income* (in thousands of dollars). Recall the discussion of the authenticity of a SLR of modernism in Chapter 2. There I suggested that the effect of *male's schooling* might be only an artifact of its association with *female's schooling*. In that case, *male's schooling* might not have any independent effect on *modernism*, and the relationship between these two variables would be spurious. In model 1 in Table 3.2 we see that controlling for *female's schooling*, *male's schooling* still has a significant effect on *modernism*, albeit substantially reduced compared to its effect in the SLR (see Table 2.3). However, we might now ask whether *male's schooling* and *female's schooling* have *equal* effects on *couple modernism*. In fact, we might wonder whether there is any difference in the impact of males' versus females' *schooling, church attendance*, or *income* on *couple modernism*.

The effects in model 1 seem to suggest that there is. The effect for *female's schooling* is close to twice as large as the effect for *male's schooling*. *Church attendance* has opposite effects for males and females, although only *male's church attendance* is significant. Similarly, male and female *incomes* have opposite effects, with only *female's income* significant. Nevertheless, sample coefficients can be different from each other due entirely to sampling error, even were there no difference between males' and females' effects in the population. We can use a nested F test to test whether the impact of males' and females' characteristics is the same. The parent model is model 1, which can be represented in the population as

$$\text{E}(Y) = \beta_0 + \delta_1 X_1 + \delta_2 X_2 + \gamma_1 X_3 + \gamma_2 X_4 + \lambda_1 X_5 + \lambda_2 X_6, \tag{3.3}$$

where Y is *couple modernism*, X_1 is *male's schooling*, X_2 is *female's schooling*, X_3 is *male's church attendance*, X_4 is *female's church attendance*, X_5 is *male's income*, and X_6 is *female's income*. The null hypothesis that we want to test is that the parameters for males' and females' characteristics are equal. That is, we test H_0: $\delta_1 = \delta_2 = \delta$, $\gamma_1 = \gamma_2 = \gamma$, $\lambda_1 = \lambda_2 = \lambda$ against H_1: at least one pair of parameters is not equal. Under the null hypotheses, the model becomes

$$\text{E}(Y) = \beta_0 + \delta X_1 + \delta X_2 + \gamma X_3 + \gamma X_4 + \lambda X_5 + \lambda X_6$$
$$= \beta_0 + \delta(X_1 + X_2) + \gamma(X_3 + X_4) + \lambda(X_5 + X_6). \tag{3.4}$$

Notice that equation (3.4) is now nested inside equation (3.3) because of the constraints in H_0. There are three constraints being imposed here. The nature of each constraint is that a given parameter is being set equal in value to another parameter. Therefore, H_0 is tested by performing a nested F test to compare these two models. Notice also that model (3.4) can be estimated by summing male and female scores on each of the variables representing *schooling, church attendance*, and *income*, and entering these three sums as the regressors in the model. These results are shown in model 2 in Table 3.2. The coefficient for the sum of male and female schooling, .160, is shown as the common coefficient for male and female schooling in the model. The other coefficients are similarly portrayed. The test statistic for H_0 is then

$$F = \frac{(473.121 - 348.285)/3}{5.104} = 8.153.$$

Under the null hypothesis, this statistic has the F distribution with 3 and 409 degrees of freedom and a significance level of $p < .001$. Apparently, we should reject H_0 and conclude that at least one pair of coefficients is not equal.

Testing Coefficient Differences. Differences between particular pairs of coefficients can be tested with individual t tests. The estimated difference between the jth and kth coefficients is $b_j - b_k$. To find the standard error of this difference, we first find its variance. Using covariance algebra, we have

$$\begin{aligned}
V(b_j - b_k) &= \text{Cov}(b_j - b_k, b_j - b_k) \\
&= \text{Cov}(b_j, b_j) - \text{Cov}(b_j, b_k) - \text{Cov}(b_k, b_j) + \text{Cov}(b_k, b_k) \\
&= V(b_j) + V(b_k) - 2\text{Cov}(b_j, b_k).
\end{aligned} \tag{3.5}$$

This variance can be estimated by requesting the *variance–covariance matrix of parameter estimates* as part of one's regression output. The entries in this matrix make sense if one considers the possibility of repeatedly taking random samples of the same size (i.e., 416) from the population of intimate couples, running model 1 each time, and recording the estimates from each run. The diagonal entries of the matrix represent the estimated variances of the coefficients across all of the samples, and the off-diagonal entries represent the estimated covariances between each pair of coefficients. The relevant terms from this matrix can then be substituted into (3.5).

The bottom panel of Table 3.2 shows the relevant part of that matrix for testing whether male and female *schooling* have equal effects on *couple modernism*, based on estimating model 1. The variance of the difference in coefficients for male versus female *schooling* is

$$v(d_1 - d_2) = .00199 + .00234 - 2(-.00104) = .006338.$$

The standard error is the square root of this quantity, .0796. The test statistic for the equality of these coefficients is then

$$t = \frac{.110 - .211}{.0796} = -1.269.$$

Under the null hypothesis that the coefficients for *schooling* are equal, this statistic has the t distribution with 409 degrees of freedom. With $p > .1$, we would fail to reject the null hypothesis. Apparently, there is not enough evidence to conclude that *female schooling* has a different effect on *modernism* than *male schooling* in the population of intimate couples. The other pairs of coefficients in model 1 could be tested in a similar fashion were we so inclined.

Testing Coefficient Changes. Researchers are frequently interested in whether the effect of a given variable changes after other variables are held constant. For example, let us return to the analyses in Table 3.1. We might hypothesize that in the population, part of the impact of *math diagnostic score* on *exam score* is accounted for by the association of *college GPA* with both of these variables. We notice that the effect of math diagnostic score is indeed reduced when *college GPA* is added to the model for *exam score*, which appears to support the hypothesis. But the observed coefficient change in the math diagnostic effect across models might be due only to chance. The question is: Is there a *significant* reduction in the coefficient for *math diagnostic score* when *college GPA* is added to the model? What is needed is a test for the *difference* in the effect of *math diagnostic score* between models 1 and 2. The testing approach is complicated by the fact that the coefficients in the initial model (model 1) are not independent of the same coefficients after variables have been added (Clogg et al., 1995).

The test proposed by Clogg et al. (1995), which accounts for the dependence of the coefficients, is as follows. We suppose that the baseline model with p parameters is

$$Y = \alpha + \sum \beta_p^* X_p + \varepsilon, \tag{3.6}$$

whereas the full model with q added parameters is

$$Y = \alpha + \sum \beta_p X_p + \sum \gamma_q Z_q + \upsilon. \tag{3.7}$$

Now, let $\delta_k = \beta_k^* - \beta_k$ be the difference in the kth parameter after the additional variables—the Z's—have been added. The sample estimate of δ_k is $d_k = b_k^* - b_k$, the difference between the corresponding sample coefficients. We test $H_0: \delta_k = 0$ against $H_1: \delta_k \neq 0$ using the test statistic

$$t = \frac{d_k}{\hat{\sigma}_{d_k}},$$

which is distributed as a t random variable with $n - (p + q) - 1$ degrees of freedom under H_0. The estimated variance of d_k is

$$\hat{\sigma}_{d_k}^2 = s_{b_k}^2 - s_{b_k*}^2 \frac{\hat{\sigma}_\upsilon^2}{\hat{\sigma}_\varepsilon^2},$$

where $s_{b_k}^2$ is the sample variance of b_k in model (3.7), $s_{b_k*}^2$ is the sample variance of b_k^* in model (3.6), and the *MSE*'s from the models are used to estimate their respective error variances. The sample variance of the coefficient for *diagnostic score* in model 1 is .097455, while the sample variance for the same coefficient in model 2 is .081867. In the current case, then, we have $\hat{\sigma}_{d_k}^2$ for math diagnostic score as

$$\hat{\sigma}_{d_k}^2 = .081867 - .097455 \left(\frac{169.878}{212.919} \right) = .004.$$

The standard error of the coefficient difference is therefore .0632, and the test statistic is

$$t = \frac{2.749 - 2.275}{.0632} = 7.5.$$

With 211 degrees of freedom, this result is highly significant at $p < .00001$. We conclude that *college GPA* accounts for some of the effect of *math diagnostic score* on *exam score*. One final comment about tests involving nested models is in order: *Models are nested only if the sample size remains the same in each model.* If the n's change as variables are added or constraints are imposed, nesting no longer holds.

Omitted-Variable Bias

Earlier I alluded to the problem of omitted-variable bias. In this section I elaborate on the problem. The estimate of a given coefficient is biased whenever a regressor is omitted from the model that is a determinant of Y and is also correlated with the predictor(s) of interest. The nature of the bias depends on both the correlation between the included and omitted regressors and the effect of the omitted regressor on the response. Figure 3.6 presents models exhibiting three different types of omitted-variable bias, which I refer to as *confounding, spuriousness*, and *mediation*. Without loss of generality, I assume that all variables have means of zero, obviating the need for an equation intercept in any model. In each case the focus variable is X, the omitted variable is Z, and the response is Y. Also, in each case the parameter that we are trying to estimate without bias is β. I employ simple models involving only three variables to avoid excessively tedious mathematics. In Chapter 6 I revisit this issue with a more complex model. As scalar algebra becomes unwieldy with more than three variables in the model, I employ matrix algebra for that explication.

Panel (a) of Figure 3.6 illustrates the confounding scenario, in which the effect of X on Y is *confounded* with Z. That is, part of X's effect on Y is realized via X's connection (correlation) with Z. In this case, we are unwilling to specify the nature of the causal relationship between Z and X. Or, perhaps Z and X are not causally related but simply correlated for reasons that remain unanalyzed. Hence, Z is depicted as falling into the same causal sequence as X by my showing Z even with X in the figure. The correlation between Z and X is symbolized by ρ, where

$$\rho = \frac{\mathrm{Cov}(X,Z)}{\sigma_x \sigma_z},$$

which implies that $\mathrm{Cov}(X,Z) = \rho \sigma_x \sigma_z$. The model depicted is the true model for X, Y, and Z. Mathematically, it is $Y = \beta X + \gamma Z + \varepsilon$, with $\mathrm{Cov}(X,\varepsilon) = \mathrm{Cov}(Z,\varepsilon) = 0$. Instead, suppose that we estimate the model $Y = \beta X + \varepsilon'$, where $\varepsilon' = \gamma Z + \varepsilon$ is now no longer uncorrelated with X, since X and Z are correlated. Thus, the orthogonality condition no longer holds. What are the consequences? Recall from Chapter 2 that the OLS estimate of b in this SLR can be written

$$b = \frac{\mathrm{cov}(X,Y)}{s_x^2}.$$

(a) Confounding

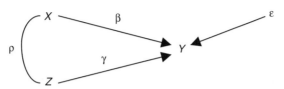

(b) Spuriousness

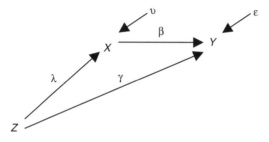

(c) Mediation

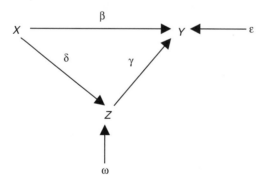

Figure 3.6 Three-variable models exhibiting varieties of omitted-variable bias.

Now we ask: What parameter is b consistent for? To answer this, we find plim b:

$$\text{plim } b = \text{plim}\left(\frac{\text{cov}(X,Y)}{s_x^2}\right) = \frac{\text{plim}(\text{cov}(X,Y))}{\text{plim } s_x^2} \qquad \text{(by the Slutsky theorem)}$$

$$= \frac{\text{Cov}(X,Y)}{\sigma_x^2}$$

(because the sample variance and covariance are consistent estimators of their respective population parameters)

$$= \frac{\text{Cov}(X, \beta X + \gamma Z + \varepsilon)}{\sigma_x^2} = \frac{\beta \sigma_x^2 + \gamma \text{Cov}(X,Z)}{\sigma_x^2}$$

$$= \beta + \frac{\gamma \rho \sigma_x \sigma_z}{\sigma_x^2} = \beta + \gamma \rho \frac{\sigma_z}{\sigma_x}. \tag{3.8}$$

This result shows that b is consistent for $\beta + \gamma \rho(\sigma_z/\sigma_x)$ rather than β. The bias in the estimation of β is the term $\gamma \rho(\sigma_z/\sigma_x)$. The nature of the bias depends on β, the true effect of X on Y, γ, the nature of the impact of Z on Y, and ρ, the nature of the correlation between X and Z. (Since σ_z/σ_x is always positive, it has no effect on the direction of the bias.) If the sign of the product $\gamma \rho$ is the same as the sign of β, we will *overestimate* the magnitude of X's impact on Y when Z is excluded from the model. But if $\gamma \rho$ is opposite in sign to β, the effect of X is *suppressed*, or underestimated, when Z is excluded. If $\gamma \rho$ is opposite in sign to β and its magnitude is greater than β's, the effect of X is *reversed* when Z is excluded. This last phenomenon is also known as *Simpson's paradox* (Agresti, 2002).

Panel (b) illustrates *spuriousness*, in which the effect of X on Y is partially (or completely) spurious, due to Z. That is, part or all of X's "effect" on Y is due to the fact that Z causes both X and Y. In that Z is causally prior to both X and Y, Z is pictured as the leftmost variable in the diagram. Once again, the true model for Y is $Y = \beta X + \gamma Z + \varepsilon$, with $\text{Cov}(X,\varepsilon) = \text{Cov}(Z,\varepsilon) = \text{Cov}(Z,\upsilon) = \text{Cov}(\upsilon,\varepsilon) = 0$. The model for X is $X = \lambda Z + \upsilon$. As before, if we estimate the model $Y = \beta X + \varepsilon'$, where $\varepsilon' = \gamma Z + \varepsilon$, we find that ε' is no longer orthogonal to X, since X is correlated with Z by virtue of Z's effect on X. Moreover, for the OLS estimate of β we have

$$b = \frac{\text{cov}(X,Y)}{s_x^2}$$

and by virtue of the same mathematics, plim b is, again,

$$\beta + \gamma \rho \frac{\sigma_z}{\sigma_x},$$

and the remarks about the nature of the bias are the same as before. In this case, plim b can be rewritten by taking advantage of a reexpression of ρ. Recall that in SLR, the correlation coefficient is the standardized slope, which equals the unstandardized slope times the ratio of the standard deviation of X to the standard deviation of Y. In panel (b), Z plays the role of X, while X plays the role of Y in this calculation. With λ as the unstandardized slope, we have

$$\rho = \lambda \frac{\sigma_z}{\sigma_x},$$

and therefore,

$$\beta + \gamma\rho\frac{\sigma_z}{\sigma_x} = \beta + \gamma\left(\lambda\frac{\sigma_z}{\sigma_x}\right)\frac{\sigma_z}{\sigma_x}$$

$$= \beta + \gamma\lambda\frac{\sigma_z^2}{\sigma_x^2}.$$

Here it is evident that the bias in b as an estimate of β is a function of the effect of Z on X and the effect of Z on Y.

Panel (c) illustrates *mediation*. We say that the impact of X on Y is *mediated* by Z if Z is partly or completely the *mechanism* by which X's effect on Y is realized. In this case, X causes Z, and Z, in turn, causes Y. Hence, Z is pictured as lying between X and Y in the figure. As before, the true model for Y is $Y = \beta X + \gamma Z + \varepsilon$, with $\text{Cov}(X,\varepsilon) = \text{Cov}(Z,\varepsilon) = \text{Cov}(X,\omega) = \text{Cov}(\omega,\varepsilon) = 0$. The model for Z is now $Z = \delta X + \omega$. Suppose, once again, that we estimate the model $Y = \beta X + \varepsilon'$, where $\varepsilon' = \gamma Z + \varepsilon$. We find that ε' is no longer orthogonal to X, since X is correlated with Z by virtue of X's effect on Z. By virtue of the, by now, very familiar mathematics of consistency, plim b is, again,

$$\beta + \gamma\rho\frac{\sigma_z}{\sigma_x}.$$

Once again, plim b can be rewritten by taking advantage of a reexpression of ρ. In that

$$\rho = \delta\frac{\sigma_x}{\sigma_z},$$

we have that

$$\beta + \gamma\rho\frac{\sigma_z}{\sigma_x} = \beta + \gamma\left(\delta\frac{\sigma_x}{\sigma_z}\right)\frac{\sigma_z}{\sigma_x}$$

$$= \beta + \gamma\delta.$$

This last expression suggests that the bias in b as an estimate of β is due to the product of X's effect on Z with Z's effect on Y. In path analysis, this is called the *indirect effect* of X on Y via Z, while β itself is called the *direct effect* of X on Y. The *total effect* of X on Y is the sum of direct and indirect effects, or $\beta + \gamma\delta$ [see, e.g., Bollen (1989) for a thorough discussion of path analysis]. From this perspective, mediation is arguably not a form of bias, simply a conflation of the indirect and direct effects of one variable on another into one omnibus effect.

Example: Deciphering Omitted-Variable Bias. Recall the positive association between *frequenting bars* and *having sex* that was documented in Chapter 2. In the discussion of model authenticity in that chapter I also suggested that this association might be spurious, due to the fact that younger people go to bars more frequently and are also more sexually active. Let me now expand that argument. I propose that part of the reason for the positive association between *frequenting bars* and *having sex* is due to three background factors that affect both phenomena: *age, religiosity,* and

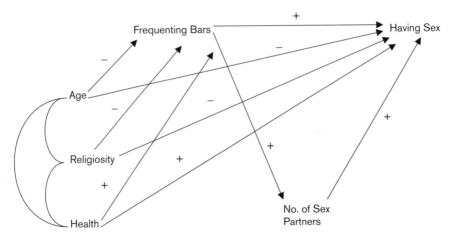

Figure 3.7 Model for frequency of having sex showing potential spurious, mediated, and direct effects for frequenting bars.

health. That is, I suggest that younger people, those who are less religious, and those in better health are more likely both to frequent bars and to have sex. To the extent that this is the case, the relationship between *frequenting bars* and *having sex* is spurious. However, if *frequenting bars* also exerts a causal influence on sexual activity, as was the argument in Chapter 2, it should still have a significant effect on sexual activity even net of these three background factors. Moreover, I made the argument in Chapter 2 that one means by which frequenting bars might enhance sexual activity is by widening one's pool of potential sex partners. If this is the case, the *number of sex partners* should mediate the remaining impact of *frequenting bars* on sexual activity. Figure 3.7 presents the conceptual model representing these hypothesized relationships. Notice that all of the compound "paths" of the form $\gamma\lambda$ (for spuriousness) or $\gamma\delta$ (for mediation) linking *frequenting bars* with *having sex* via *age, religiosity, health,* or *number of sex partners* are positive. This suggests that the effect of *frequenting bars* should diminish once these other variables are controlled.

Table 3.3 presents three regression models for *having sex,* which allow us to evaluate the model in Figure 3.7. The first model is the SLR for *having sex* as a function of *frequenting bars,* and is similar to the results in Table 2.4 except that all analyses in Table 3.3 are based on 2,320 respondents. We see that the effect of frequenting bars is significant and positive, with a value of .224. Model 2 adds *age* (in years), *religiosity* (continuous predictor that is the sum of five standardized measures; higher scores indicate those who are more religious), and *self-assessed health* (ranging from 1 = "poor" to 4 = "excellent"). The effect of *frequenting bars,* although still significant, has been reduced by a proportion of $(.224 - .058)/.224 = .74$, or about 74%, once the background factors are held constant. This suggests that about 74% of the "effect" of *frequenting bars* on *having sex* is noncausal and due to the antecedents of both variables. Nevertheless, *frequenting bars* still has a nonzero effect on *having sex* (the sample coefficient is .058). Model 3 adds the respondent's *number of sex partners in the past*

Table 3.3 Regression Models for Frequency of Having Sex for 2320 Respondents from the General Social Survey

Regressor	Model 1		Model 2		Model 3	
	b	b^s	b	b^s	b	b^s
Intercept	2.270***	.000	3.984***	.000	3.225***	.000
Frequenting bars	.224***	.157	.058*	.042	.001	.001
Age			−.047***	−.401	−.038***	−.326
Religiosity			−.009	−.011	.008	.010
Self-assessed health			.249***	.103	.265***	.110
No. partners in past 5 years					.276***	.218
RSS	218.446		1815.340		2152.511	
MSE	3.747		3.062		2.918	
F	58.301***		148.220***		147.558***	
R^2	.025		.204		.242	
R^2_{adj}	.024		.203		.240	

* $p < .05$. ** $p < .01$. *** $p < .001$.

five years (censored count variable ranging from 0 = "no partners" to 8 = "more than 100 partners") as a potential mediator of the impact of *frequenting bars* on *having sex*. We see now that the effect of *frequenting bars* is diminished to approximately zero in this last model (the coefficient is .001). This suggests that the remaining effect of *frequenting bars* on *having sex* may be causal, since it appears to be mediated or "explained" by one's *number of sex partners*, which was hypothesized to be enhanced by *frequenting bars*.

At this point a word of caution is in order in regard to the model's authenticity. Although the causal priority of *age, religiosity*, and *health* over *frequenting bars* is a reasonable assumption, it is not so clear that *frequenting bars* is causally prior to *number of sex partners*. The difficulty is the retrospective time frame associated with each variable. *Frequenting bars* was measured with the question "How often do you go to a bar or tavern," while *number of sex partners* refers to the last five years. Temporally speaking, therefore, it could be argued that *number of sex partners* is prior to *frequenting bars*. On the other hand, it is theoretically reasonable that *frequenting bars* enhances one's *number of sex partners*. But it is not particularly reasonable that one's *number of sex partners* should determine how often one goes to bars. In fact, if anything, those with more sex partners might go to bars less often, since they already have what many bar patrons are looking for. This would imply a negative relationship between *frequenting bars* and *number of sex partners*, which is contrary to the results. Since the positive effect of *frequenting bars* is *reduced* when *number of sex partners is added*, and since *number of sex partners* is *positively* related to *having sex*, the path from *frequenting bars* to *number of sex partners must* be positive. At any rate, I argue, based on these theoretical considerations, that *frequenting bars* is causally prior to *number of sex partners*. Finally, to ascertain whether *number of sex partners* indeed mediates the remaining effect of *frequenting bars* on *having sex*, we must test whether the reduction in the impact of *frequenting bars* from model 2 to model 3 is significant. Here, again, we draw on the test proposed by Clogg et al. (1995), which was introduced earlier. As the result is highly significant ($t = 10.76$; $p < .00001$), I conclude that part of the impact of *frequenting bars* on *having sex* is due to the influence of *frequenting bars* on one's pool of potential sex partners.

Modeling Interaction Effects

Recall, in the analyses in Figures 3.3 to 3.5, that the slope of the regression of Y on X was approximately the same at each level of Z. Therefore, it made sense to report the "average" of these three slopes as the effect of X on Y when holding Z constant. But what if the effect of X *depends* on the level of Z? In this case, it no longer makes sense to speak of *the* impact of X controlling for Z, since there may be many different impacts of X depending on which level of Z is being considered. When the slope of the regression of Y on X changes significantly across levels of Z, we say that X and Z *interact* in their effects on Y. Or, we say that the impact of X on Y is *moderated by* or *conditioned on* the level of Z, and Z is referred to as a *moderator variable*. In this section of the chapter I discuss the modeling of interaction effects. I also discuss a related notion, the idea that the model as a whole might differ across groups of cases.

Interaction Model. When X and Z interact in their effects on Y, this is captured by including a cross-product term—representing the product of X with Z—as an additional regressor in the model. The model is

$$E(Y) = \beta_0 + \beta_1 X + \beta_2 Z + \gamma XZ.$$

Since the partial slope of the regression with respect to X is the parameter that multiplies X in the model, let's factor this equation to isolate the common multipliers of X:

$$E(Y) = \beta_0 + \beta_2 Z + (\beta_1 + \gamma Z)X. \tag{3.9}$$

Equation (3.9) shows how the cross-product term captures the interaction effect: The impact of X—its partial slope—is now a function of the level of Z. In fact, the partial slope of X, denoted ps_x, is a simple linear regression model of the form $ps_x = \beta_1 + \gamma Z$. That is, ps_x increases (if γ is positive) or decreases (if γ is negative) linearly with increases in Z. Interaction effects are symmetric. We could also factor the equation so as to isolate the common multipliers of Z. We would then find that the partial slope for Z is $\beta_2 + \gamma X$. In other words, the impact of Z is a simple linear function of X, too. This type of interaction effect, in which the effect of X (Z) varies in a simple linear fashion with the level of Z (X), is the one most commonly modeled. However, non-linear interaction effects can also be modeled, as demonstrated in Chapter 5.

The interaction effect modeled in equation (3.9) is called a *first-order interaction effect*. This means that the effect of X varies only according to the values of one other variable. If the effect of X varies according to the *combination* of values of two other variables, we have a *second-order interaction effect*, which is much more complicated.

A second-order interaction effect among, say, X, Z, and W in their effects on Y would be modeled by including the following predictors in the model: X, Z, W, XZ, XW, ZW, and XZW. The equation would then be

$$E(Y) = \beta_0 + \beta_1 X + \beta_2 Z + \beta_3 W + \beta_4 ZW + \gamma XZ + \lambda XW + \delta XZW$$

Factoring the common multipliers of X, we have

$$E(Y) = \beta_0 + \beta_2 Z + \beta_3 W + \beta_4 ZW + (\beta_1 + \gamma Z + \lambda W + \delta ZW)X$$

Now the partial slope for X is $\beta_1 + \gamma Z + \lambda W + \delta ZW$. This is an interaction model in the variables Z and W. Thus, the model suggests that the impact of X on Y is a function of the interaction of Z with W; or, the extent to which Z moderates the impact of X on Y itself depends on the level of W. It should be apparent that higher-order interactions become increasingly unwieldy to understand, let alone model. As one rarely encounters the modeling of interactions any higher than first-order [see MacDonald and DeMaris (2002) for an exception, however], they will not be considered further here [but for a lucid discussion of higher-order interaction effects, see Aiken and West (1991)]. The partial slope of X in the first-order interaction model has a unit-impact interpretation, just like the effect of X in the *no-interaction*, or *main effects*, *model*. It reflects the expected change in Y for a unit increase in X (in Exercise 3.19 I ask you to prove this). It is also the first partial derivative of equation (3.9) with respect to X.

Ordinal versus Disordinal Interaction. Let's consider some examples of interaction models. Suppose that the equation is $E(Y) = 5 + .2X + 1.5Z + .05XZ$, where X ranges from 0 to 10 and Z ranges from 0 to 5. Then the partial slope for X is $.2 + .05Z$. Thus, when Z is 0, the partial slope for X is .2. When Z is 2.5, the partial slope for X is $.2 + .05(2.5) = .325$. When Z is 5, the partial slope for X is $.2 + .05(5) = .45$. In this case, X always has a positive effect on Y, but the magnitude of the effect is stronger at higher values of Z. This type of interaction is called an *ordinal interaction* (Kerlinger, 1986): The effect of X changes in magnitude, but not direction, with changing values of Z. But suppose that the equation is $E(Y) = 5 + .2X + 1.5Z - .1XZ$, with X and Z being the same as before. In this equation, the partial slope for X is $.2 - .1Z$. Now, when Z is 0, the partial slope for X is .2. However, when Z is 2, the partial slope for X is $.2 - .1(2) = 0$; and when Z is 5, the partial slope for X is $.2 - .1(5) = -.3$. In this case, the effect of X changes *direction* over values of Z, resulting in a *disordinal interaction* (Kerlinger, 1986). Note that the only real way to ascertain whether the interaction is ordinal or disordinal within the observed range of the moderator variable is to calculate some sample values of ps_x over different values of Z, and see whether ps_x changes sign across these values. (Technically, the terms *ordinal* and *disordinal* refer to whether the slopes for the regression of Y on X cross when plotted according to different values of Z. However, even if the lines do not cross within the observed range of Z, I refer to the interaction as disordinal if the slope changes from positive to negative, or vice versa, over values of Z.)

Centering, Revisited. In Exercise 2.6 I introduced the term *centered variable*, refer-ring to a variable that is deviated from its mean. Centered variables are particularly useful in interaction models. Consider equation (3.9) again. Suppose that zero is not a legitimate value for Z. Then the t test for b_1—the main effect of X—is not particu-larly meaningful since it refers to the effect of X when Z is zero. Similarly, the test for b_2 refers to the effect of Z when X is zero, which, again, may not be a meaningful value for X. However, if Z and X have first been centered, and the cross-product XZ is constructed using the centered variables, the main effects of X and Z are always mean-ingful. The reason for this is that a centered variable, say $Z^c = Z - \bar{Z}$, has a mean of zero. Therefore, the main effect of X, β_1, in the centered-variable interaction model is the effect of X when Z^c is zero or when Z is at its mean. The same interpretation applies to the main effect of Z if X is also centered: It is the effect of Z when X is at its mean. Hence, centering in interaction models renders the main effects of the inter-acting variables interpretable. Another advantage of centering variables involved in interactions has to do with the problem of multicollinearity (discussed below). Recall that multicollinearity arises because one variable is highly correlated with another variable or with a linear combination of the other variables. Cross-product terms of the form XZ are highly correlated with their component variables—X and Z—and therefore introduce collinearity problems into the model. It turns out that centering variables before creating cross-product terms brings about a substantial reduction in this collinearity [see Aiken and West (1991) for the mathematics behind this].

Example. Table 3.4 presents a MULR analysis of faculty salary for 725 faculty members at Bowling Green State University (BGSU) for the academic year 1993–1994. The dependent variable is the *nine-month salary in dollars*. The independent variables are the *number of years of prior experience* (job experience prior to start-ing at BGSU), the *number of years at the university*, the *number of years in rank*, and a continuous variable tapping the *marketability of one's discipline*. This marketabil-ity factor is the ratio of average academic-year salary of full-time faculty in a partic-ular discipline to average academic-year salary of all full-time faculty. The variables *years at the university* and *years in rank* are both centered. Model 1 is the main effects, or additive model, the model without any interaction effects. All variables except *years in rank* have significant effects on salary. The directions of effects suggest that *years of prior experience*, *years at the university*, and *marketability of the discipline* are all positively associated with salary. Although the marketability variable appears to have the largest unstandardized effect, the standardized coefficients suggest that *years at the university* has the strongest impact on salary.

Model 2 investigates the interaction of *years at the university* with *years in rank* in their effects on salary. Therefore, model 2 adds to the main effects model the cross-product of centered *years at the university* with centered *years in rank*. The coefficient for the interaction effect is significant at $p < .05$, suggesting that the impact of *years at the university* is a function of *years in rank*. To ascertain the nature of the interac-tion, we examine the partial slope for *years at the university*. Its value is $(1008.267 - 15.481$ *years in rank*$)$. In that the main effect of *years at the university* is significant, we see that *years at the university* has a significant positive effect on salary for those

Table 3.4 Regression Models for Faculty Salary for 725 Faculty Members, Showing First-Order Interaction of Years at the University with Years in Rank

Regressor	Model 1		Model 2	
	b	b^s	b	b^s
Intercept	12219.000***	.000	13241.000***	.000
Years of prior experience	924.893***	.317	906.418***	.311
Years at the university[a]	1072.730***	.759	1008.267***	.713
Years in rank[a]	−121.466	−.061	37.996	.019
Marketability of discipline	35001.000***	.372	34948***	.372
Years at the university × years in rank[b]			−15.481*	−.074
RSS	84691222609.000		85141325670.000	
MSE	79234093.911		78718281.718	
F	267.218***		216.319***	
R^2	.598		.601	
R^2_{adj}	.595		.598	

[a] Centered variable.
[b] Cross-product of centered variables.
* $p < .05$. ** $p < .01$. *** $p < .001$.

who are average in the number of years in rank. In particular, for these faculty, each year longer that they have been at the university is estimated to be worth $1008.27 additional salary, on average. However, this effect grows slightly weaker the longer someone has been in rank. For example, for someone who is 1 standard deviation, 7.043 years, above mean years in rank, the effect of *years at the university* is 1008.267 − 15.481(7.043) = 899.234. Thus, for someone this long in rank, each additional year at the university is only worth an additional $899.23 of salary. The most likely explanation of this effect is that *years in rank* is an inverse proxy for productivity. That is, the most productive faculty tend to be promoted sooner, all else equal. Therefore, a greater number of years in rank tends to be associated with lower productivity. Finding that *years at the university* has a weaker effect the longer one has been in rank suggests that seniority has a weaker effect on salary for the relatively less productive.

As mentioned, the fact that the main effect of *years at the university* is significant suggests that this factor is significant for those who are average in *years in rank*. Suppose that we wish to know whether *years at the university* is significant *among* all those who are 1 standard deviation above mean *years in rank*. At this point, some clarification is in order. In the sample regression equation, $\hat{y} = b_0 + b_1X + b_2Z + gXZ$, a significant coefficient, g, for the cross-product XZ, does not imply that the impact of X is significant *at* a particular level of Z. In fact, the effect of X may not be significant at *any* level of Z, even though g is significant. The significance of g means only that we can reject the null hypothesis that $\gamma = 0$. In that γ captures the variation in ps_x over levels of Z, rejection of this null hypothesis only suggests that ps_x *varies* with Z, not that ps_x is different from zero at any particular level of Z. To answer the latter question, we must test the significance of $b_1 + gZ$ at a given level, z, of Z. The test statistic is

$$t = \frac{b_1 + gz}{\hat{\sigma}_{b_1 + gz}},$$

where $\hat{\sigma}_{b_1 + gz}$ is the estimated standard error of $b_1 + gz$. Under H_0: $\beta_1 + \gamma z = 0$, this statistic has the t distribution with $n - K - 1$ degrees of freedom. The standard error is just the square root of the variance of $b_1 + gz$. The expression for the variance of $b_1 + gz$ can be found using covariance algebra. Assuming that Z is fixed over repeated sampling, we have

$$
\begin{aligned}
V(b_1 + gz) &= \text{Cov}(b_1 + gz, b_1 + gz) \\
&= \text{Cov}(b_1, b_1) + z\text{Cov}(b_1, g) + z\text{Cov}(b_1, g) + z^2\text{Cov}(g, g) \\
&= V(b_1) + 2z\text{Cov}(b_1, g) + z^2 V(g).
\end{aligned}
\tag{3.10}
$$

In model 2, let X be *years at the university*, Z be *years in rank*, b_1 be the main effect of *years at the university*, and g be the coefficient for the interaction of *years at the university* with *years in rank*. From the variance–covariance matrix of parameter estimates for model 2 (not shown), the relevant estimates are 5145.182 for $V(b_1)$, 41.913 for $V(g)$, and 174.527 for $\text{Cov}(b_1, g)$. At one standard deviation (7.043 years) above mean *years in rank*, the estimated variance of ps_x is therefore $5145.182 + 2(7.043)(174.527) + 7.043^2(41.93) = 9682.615$. The estimated standard error of ps_x is the square root of this, 98.4. The test statistic for the significance of *years at the university* at this level of *years in rank* is therefore

$$t = \frac{899.234}{98.4} = 9.139.$$

With 719 degrees of freedom, this is a highly significant result ($p < .00001$).

Problems with Cross-Product Terms. In addition to the collinearity created by cross-product terms, two other difficulties can arise when investigating interaction effects. First, researchers examining interaction effects using nonexperimental data may often find these terms to be nonsignificant. Or, if significant, they may turn out to account for very little variance in the criterion. McClelland and Judd (1993) discuss the reasons for this. They point out that it is the residual variation in XZ—the unique variance not shared with the other predictors in the model—that determines the statistical power of the test for addition of the XZ term. Moreover, the residual variance of XZ is determined entirely by the joint distribution of X and Z. The more correlated X is with Z, the greater the power for detecting interaction effects. What is needed, in particular, is an "optimal" distribution of X and Z in which "extreme values of X co-occur with similarly extreme values of Z" (McClelland and Judd, 1993, p. 384). Controlled experimentation allows the researcher to arrange X and Z so that this type of distribution occurs. However, with nonexperimental data, there is no guarantee that the joint distribution of X and Z will be optimal. McClelland and Judd (1993) therefore advise nonexperimental data analysts to temper their expectations regarding the sizes of interaction effects in their analyses, and to regard more modest effects as equally important. This should be the case especially when interactions are guided by strong theory.

The other difficulty is that nonlinearitiy in the relationship between a regressor and the criterion can frequently be confused with an interaction effect. In particular, when the relationship between X and Y is quadratic in nature—the true model has Y as a function of X and X^2 (Chapter 5 considers such models at length)—such an effect may masquerade as an interaction. The reason for this is that the reliability of X^2 is lower than the reliability of XZ when X and Z are highly correlated (McCallum and Mar, 1995). Since unreliability attenuates true effects, it is likely that spurious moderator effects will override real quadratic effects in the data when the true model is quadratic. McCallum and Mar (1995) demonstrate this phenomenon using extensive simulations. They advise that the best strategy is to have a compelling theoretical or substantive rationale for preferring one type of model over the other. In a similar vein, Ganzach (1997) argues that researchers investigating interaction effects of the form XZ should always include X^2 and Z^2 in the model, and vice versa. He demonstrates using empirical examples that if this strategy is not followed, the nature of interaction or quadratic effects found in the sample will often be biased. Again, the best strategy is to have a sound theoretical rationale before investigating any such more complex effects in the data.

Multicollinearity. I indicated above that the creation of cross-product terms—or quadratic terms, for that matter—induces collinearity problems. Collinearity that results from the creation of special terms to capture interaction or nonlinear effects is known as *nonessential ill-conditioning*. Collinearity due to high correlations among naturally occurring variables, on the other hand, is referred to as *essential ill-conditioning* (Aiken and West, 1991). In either case, multicollinearity can pose problems in data analysis that the researcher needs to be aware of. The two major consequences that interfere with good parameter estimation are (1) the variances of parameter estimates become greatly inflated, causing wide fluctuations in the values of estimates from sample to sample; and (2) the magnitudes of parameter estimates are substantially inflated, making it appear that variables have much stronger effects than they really have. There are some well-known symptoms associated with collinearity, which the researcher should learn to recognize. One symptom is a very significant F test or high R^2 value in combination with the finding that no individual coefficients are significant. However, as noted above, this may also be due to the fact that some complex linear combination of the parameters is what is driving the F or R^2 results. Other symptoms are parameter estimates that are unreasonably large in magnitude or have counterintuitive signs, standard error estimates that are especially large, or standardized coefficients that are outside the range of $[-1, +1]$.

The best single diagnostic for detecting multicollinearity is the *variance inflation factor* for the kth coefficient, denoted VIF_k. This measure is defined as

$$VIF_k = \frac{1}{1 - R^2_{X_k}|\mathbf{x}_{-k}}, \qquad (3.11)$$

where $R^2_{X_k}|\mathbf{x}_{-k}$ is the R^2 for the regression of X_k on all other X's in the model. The denominator of equation (3.11), called the kth variable's *tolerance*, represents the proportion of variation in X_k not shared with all other X's. Tolerances smaller than .1, or equivalently, VIF's greater than about 10 suggest that collinearity is beginning to be a problem in one's analysis (Myers, 1986). This is also somewhat dependent on sample size, with larger samples better able than smaller ones to tolerate collinearity. As an example, the VIF's for model 2 in Table 3.4 range from 1.007 for marketability of discipline to 5.822 for years in rank. What can be done about collinearity? One remedy for nonessential collinearity is to center variables before creating cross-products or polynomial terms, as indicated above. This reduces the estimated standard errors for all terms except the highest-order cross-product or polynomial term (Aiken and West, 1991). For collinearity among naturally occurring variables, there are some simple options. First, if two variables are very highly correlated, consider dropping one of them from the model. If most of the variance of one variable is shared with the other, not much more information is gained by including the second one in the model. Or if two or more items are highly correlated because they are measuring the same underlying construct, consider incorporating them into a single scale. Another remedy when X and Z are highly correlated is to substitute $\ln X$ and $\ln Z$ for X and Z in the model since the nonlinear transformation of the natural log reduces the degree of correlation (which only taps linear association). More elaborate remedies are discussed in Chapter 6.

Comparing Models across Groups. A variant on the interaction theme is the situation in which the model as a whole might be different for different groups. For example, does the same salary model characterize both male and female faculty? One could argue that they should not, for either of two reasons. First, in the United States, women have historically been paid less for doing the same work. One might therefore expect that such factors as the *number of years at the university* or *marketability of the discipline* may not have as strong an effect on salary for women as they do for men. This would be the *gender bias hypothesis.* On the other hand, given the concern with gender equity in pay that has characterized the American workplace over the past 25 years or so, one might expect the opposite. That is, women may be more highly rewarded for seniority or for marketability of discipline than men are, to make up for past injustices. This would be the *reverse discrimination hypothesis.* Table 3.5 explores this issue by presenting salary model 2 from Table 3.4 separately for male and female faculty.

There are some notable differences. To begin, the intercepts are different, with the male intercept being about $2500 higher than the female intercept. The intercepts are not entirely interpretable because zero is not a plausible value for *marketability of the discipline.* Nevertheless, this difference suggests that for those with no prior experience, who are average in years at the university and years in rank, and who are in a discipline with zero marketability, males make $2500 more than females. Additionally, *years of*

Table 3.5 Regression Models for Faculty Salary for 511 Male Faculty Members and 214 Female Faculty Members

Regressor	Male Model		Female Model	
	b	b^s	b	b^s
Intercept	17402.000***	.000	14954.000***	.000
Years of prior experience	1020.899***	.370	370.365**	.156
Years at the university[a]	929.251***	.674	1100.884***	.896
Years in rank[a]	93.937	.049	−245.221	−.142
Marketability of discipline	33303.000***	.367	26883.000***	.346
Years at the university × years in rank[b]	−14.936	−.068	−22.298*	−.164
SSE	40991619682.000		10930510277.000	
F	135.154***		45.185***	
R^2	.572		.521	

[a] Centered variable using separate means by gender.

[b] Cross-product of centered variables.

$*p < .05.$ $**p < .01.$ $***p < .001.$

experience and *marketability of discipline* have stronger effects on salary among males, as suggested by the gender bias hypothesis, whereas *years at the university* has a stronger effect among females, as suggested by the reverse discrimination hypothesis. Also, the interaction of *years at the university* with *years in rank* is only significant for females. [No substantive implications should be drawn from this analysis, since it excludes several key factors in the determination of salary. For a complete assessment of potential gender discrimination in salary at BGSU, see Balzer et al. (1996) or Boudreau et al. (1997).]

To assess whether the models are truly different for each group, we can perform a Chow test (Chow, 1960). If we let p equal $K + 1$, the number of parameters in the model, c denote the model estimated for the combined sample (i.e., model 2 in Table 3.4), and m and f denote models for males and females, respectively, the test statistic has the form

$$F = \frac{[SSE_c - (SSE_m + SSE_f)]/p}{(SSE_m + SSE_f)/(n - 2p)}.$$

Under the null hypothesis that the same model applies to each group, this statistic has the F distribution with p and $n - 2p$ degrees of freedom. Now, SSE for the combined analysis is 56598444555. Hence, for the current problem we have

$$F = \frac{[56598444555 - (40991619682 + 10930510277)]/6}{(40991619682 + 10930510277)/(725 - 12)} = 10.703.$$

With 6 and 713 degrees of freedom, this result is quite significant ($p < .00001$). Apparently, the dynamics of salary determination work differently for males than they do for females.

EVALUATING EMPIRICAL CONSISTENCY

In this final section of the chapter, I discuss ways of evaluating the empirical consistency of the MULR model. As an example, I evaluate the full model (model 3) of exam scores for 214 students, which was presented in Table 3.1. Recall that in chapter 2 I presented a formal test for empirical consistency for SLR: Neter et al.'s (1985) lack-of-fit test. This test was based on the ratio of $MSLF$, the mean of the sum of squares for lack of fit, to $MSPE$, the mean of the sum of squares for pure error. The same test applies in MULR, except that $SSPE$ is based on the sum of squared deviations of the individual Y-values around the group mean of the Y's at each covariate pattern in the data. As before, $SSLF = SSE - SSPE$. The test again employs an F statistic of the form

$$F = \frac{MSLF}{MSPE}.$$

The degrees of freedom for this statistic are now $n - c$ and $c - p$, where $p = K + 1$ is the number of parameters in the model and c is the number of different covariate patterns in the data. $SSPE$ can still be recovered in SAS using PROC RSREG. However, a problem in MULR with continuous independent variables is that there will usually be as many different covariate patterns as there are cases. In this event, $SSPE$ is necessarily zero, and the test is of no utility. This is, in fact, the case for model 3 in Table 3.1. There are 214 cases and 214 different covariate patterns in the data. (The number of different covariate patterns can easily be counted by counting the number of different predicted values there are in the data, since each different covariate pattern results in a unique predicted value.)

Examination of Residuals

Lacking a formal test of empirical consistency, we can always resort to more informal methods. One such technique employed in Chapter 2 was an examination of the raw and standardized residuals for potential outliers, nonconstant variance, or nonlinearities. Figure 3.8 shows a plot of the raw residuals against $\hat{y}$ for model 3. The plot appears to have the desired shape, a band of points spread evenly around the line $e = 0$. There does not appear to be any noticeable nonlinearity in the relationship between e and $\hat{y}$, nor does the trend in the points suggest any dramatic variation in the variance of the residuals across $\hat{y}$-values. For the detection of outliers, it is better to plot the standardized residuals against $\hat{y}$. Figure 3.9 shows such a plot. The plot looks pretty much the same except that the values of z_e are much smaller than those of e. Outliers would be indicated by standardized residuals greater than about 4 in absolute value. There is perhaps one such data point in the

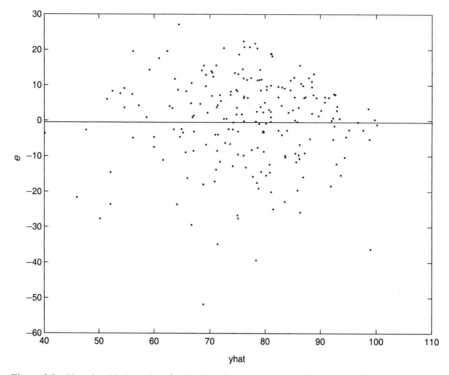

Figure 3.8 Plot of residuals against fitted values for the regression of score on the first exam on math diagnostic score, college GPA, attitude toward statistics, class hours in the current semester, and number of previous math courses.

middle lower part of the plot, although its value is right around -4. Were we concerned about this data point, it could be investigated further. However, perhaps more important, no single observation stands out as being dramatically different than the others.

Partial Regression Leverage Plots

In Chapter 2 we examined scatterplots of Y against a single X to ensure that the X–Y relationship was linear. In MULR the assumption is that the relationship between Y and *each* X_k is linear at fixed levels of the other predictors. This assumption is obvious from the regression equation itself. Assume, for the moment, that all X's other than X_k are fixed at the values $x_1, x_2, \ldots, x_{k-1}, x_{k+1}, x_{k+2}, \ldots, x_K$. Then the MULR equation is

$$
\begin{aligned}
E(Y) &= \beta_0 + \beta_1 x_1 + \beta_2 x_2 + \cdots + \beta_{k-1} x_{x-1} + \beta_k X_k + \cdots + \beta_K x_K \\
&= \beta_0 + \beta_1 x_1 + \beta_2 x_2 + \cdots + \beta_{k-1} x_{k-1} + \beta_{k+1} x_{k+1} + \cdots + \beta_K x_K + \beta_k X_k \\
&= \alpha' + \beta_k X_k,
\end{aligned}
$$

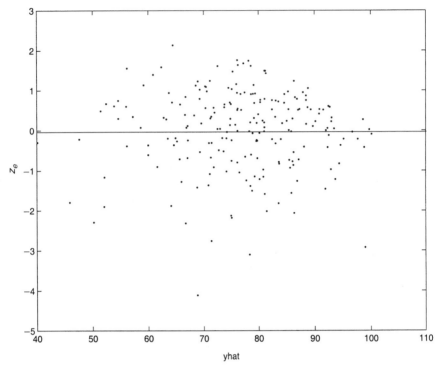

Figure 3.9 Plot of standardized residuals against fitted values for the regression of score on the first exam on math diagnostic score, college GPA, attitude toward statistics, class hours in the current semester, and number of previous math courses.

where

$$\alpha' = \beta_0 + \beta_1 x_1 + \beta_2 x_2 + \cdots + \beta_{k-1} x_{k-1} + \beta_{k+1} x_{k+1} + \cdots + \beta_K x_K.$$

This shows that at fixed levels of all other regressors, Y is a simple linear function of X_k. If this is not the case, the appropriate model should contain either transformations of X_k that make the relationship linear, or additional terms to model the nonlinearity (Chapter 5 takes up these issues in greater detail).

One way to examine whether the relationship between Y and each X_k is linear, controlling for all other regressors, is to look at the *partial regression leverage plots*, or *partial plots*, of Y with each X in the model. These are scatterplots of Y against each X_k, where the influence of all other regressors has been partialed out of each variable. *Partialing out* all other regressors involves removing the linear association between the given variable and all other regressors. How is this accomplished? Suppose that we wish to look at the partial plot for X_1, controlling for $X_2, X_3, \ldots, X_K$. First, we regress Y on $X_2, X_3, \ldots, X_K$ and save the residuals. Call these $e_{y|x-1}$, representing the residuals for the regression of Y on all X's except (minus) X_1. Then regress X_1 on $X_2, X_3, \ldots, X_K$. Save these residuals and call them, correspondingly,

$e_{x_1|\mathbf{x}-1}$. Then a scatterplot of $e_{y|\mathbf{x}-1}$ against $e_{x_1|\mathbf{x}-1}$ is the partial plot for X_1. In that OLS residuals are always uncorrelated with all regressors in the same equation, $e_{y|\mathbf{x}-1}$ and $e_{x_1|\mathbf{x}-1}$ represent versions of Y and X_1, respectively, that are uncorrelated with all other model regressors. The same procedure is used to generate plots for the other $K - 1$ regressors.

Figures 3.10 to 3.14 present the partial plots for the regression of exam score in introductory statistics. The plots are for model 3 in Table 3.1. Shown, in order, are the plots for *math diagnostic score, college GPA, attitude toward statistics, class hours in the current semester*, and *number of previous math courses*. The fitted line on each plot is the OLS line for the simple linear regression of each $e_{y|\mathbf{x}-k}$ on each $e_{x_k|\mathbf{x}-k}$. None of the relationships depicted in the plots suggest noticeable non-linear associations between $e_{y|\mathbf{x}-k}$ and $e_{x_k|\mathbf{x}-k}$. This confirms the acceptability of modeling each of the predictors as having a linear relationship to exam scores. Also, note that the simple correlation between $e_{y|\mathbf{x}-k}$ and $e_{x_k|\mathbf{x}-k}$ is called the *partial correlation coefficient* between Y and X_k, controlling for all other regressors, and is denoted pr_k in this book. This represents the correlation between Y and X_k after removing from each variable the linear association with all other regressors in the model. The *squared partial correlation*, or pr_k^2, can be used as a measure of the

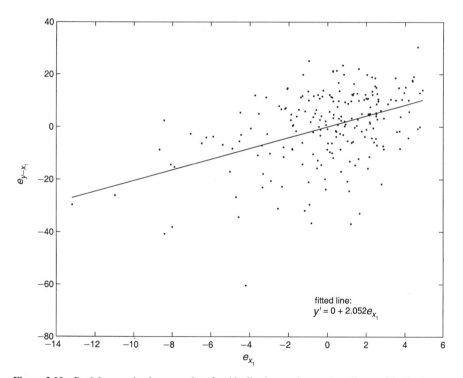

Figure 3.10 Partial regression leverage plot of residualized exam 1 score (ey-x1) on residualized score on the math diagnostic (ex1).

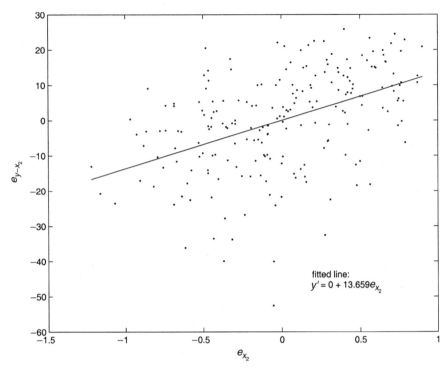

Figure 3.11 Partial regression leverage plot of residualized exam 1 score (ey-x2) on residualized college GPA (ex2).

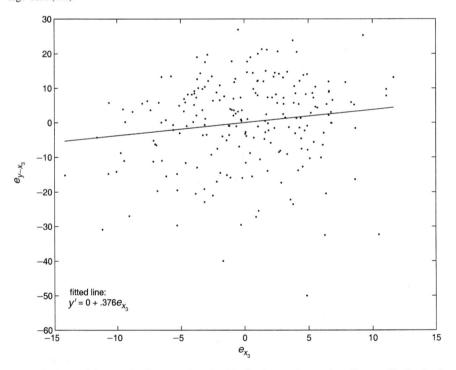

Figure 3.12 Partial regression leverage plot of residualized exam 1 score (ey-x3) on residualized attitude toward statistics score (ex3).

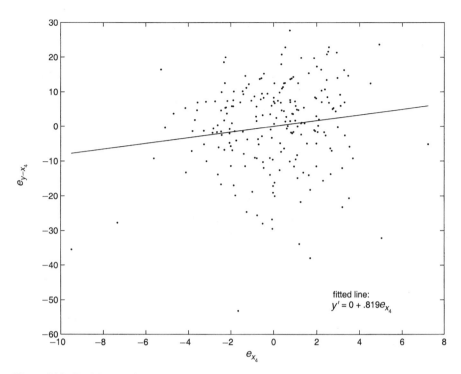

Figure 3.13 Partial regression leverage plot of residualized exam 1 score (ey-x4) on residualized number of current semester hours (ex4).

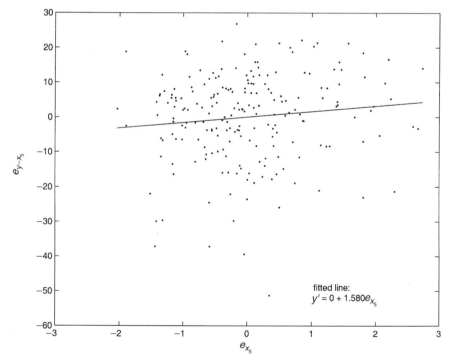

Figure 3.14 Partial regression leverage plot of residualized exam 1 score (ey-x5) on residualized number of previous math courses (ex5).

unique variance in Y accounted for by X_k after the influence of all other variables has been removed (McClelland and Judd, 1993).

Using this measure of *effect size*—the size of the effect of a given regressor on Y—the most unique variance in exam scores is accounted for by *college GPA*. Its squared partial correlation with *exam score* is .21. The second largest effect size is for *math diagnostic score*, with a squared partial correlation of .196. The least influential factor, net of the others, is the *number of previous math courses taken*. This variable accounts for only 1.4% of unique variance in exam scores. Unlike the standardized regression coefficient, which can fall outside the range $[-1, 1]$ in the presence of strong collinearity among the X's, the partial correlation coefficient *is* confined within this range under all conditions. A simple (although not trivial to prove) formula for pr_k^2 is (from Greene, 2003)

$$pr_k^2 = \frac{t_k^2}{t_k^2 + dfE},$$

where t_k is the t test for X_k in the corresponding MULR model, and dfE is the error degrees of freedom from this regression, which is $n - K - 1$, assuming a total of K regressors in the model. Using this formula, we assign to pr_k the sign of the coefficient for X_k in the MULR. As an example of this calculation, the t-value for *math diagnostic score* (variable 1) in model 3 in Table 3.1 is 7.127. The squared partial correlation of *exam score* with *math diagnostic score*, controlling for *college GPA, attitude toward statistics, class hours in the current semester*, and *number of previous math courses* is then equal to

$$pr_1^2 = \frac{7.127^2}{7.127^2 + 208} = .196.$$

EXERCISES

3.1 In the sample regression equation $\hat{y} = b_0 + b_1 X_1 + b_2 X_2$, show how holding X_2 constant at the respective values 3, 6, and 9 changes the intercept for the regression of Y on X_1, but not the slope.

3.2 In the sample regression equation $\hat{y} = b_0 + b_1 X_1 + b_2 X_2 + g X_1 X_2$, show how holding X_2 constant at the respective values 3, 6, and 9 changes the intercept *and* slope for the regression of Y on X_1.

3.3 The squared partial correlation between Y and X_2, controlling for X_1, is $pr_2^2 = (R_{Y|X_1 X_2}^2 - R_{Y|X_1}^2)/(1 - R_{Y|X_1}^2)$. That is, pr_2^2 is the additional variation in Y accounted for by X_2, over and above the variance accounted for by X_1, as a proportion of the variance unaccounted for by X_1.

(a) Using this formula for pr_2^2, find the partial correlation between *college GPA* and *exam 1 score*, while controlling for *math diagnostic score*, for the 214 students represented in Table 3.1.

    (b) Also, find the same partial correlation using Greene's (2003) formula if the t statistic for the effect of *college GPA* in model 2 is 7.397.

3.4 The regression of CMODERN (*couple modernism*) on MEDUC (*male's education*) and FEDUC (*female's education*) for the 416 couples in the *couples dataset* produces the following results:

Predictor	b	t
Intercept	22.6853	
FEDUC	.2497	5.139
MEDUC	.0898	2.069
RSS	322.58233	
SSE	2237.94588	

Standard deviations:

CMODERN	2.4839
FEDUC	2.8
MEDUC	3.1

    (a) Give R^2 and R^2_{adj}.
    (b) Give the F-value and significance for the test of the model's utility.
    (c) Give the standardized coefficients for FEDUC and MEDUC.
    (d) Give the partial correlation between CMODERN and each predictor, controlling for the other predictor.
    (e) Interpret the intercept and the unstandardized coefficients.

3.5 To the model in Exercise 3.4 are added FCHATT, DURYRS, MCHATT, MALEAGE, and FEMAGE, producing the following result: $R^2 = .16672$. Test whether the addition of these five variables produces an improvement in the model (i.e., test the null hypothesis that the five additional coefficients are zero in the population).

3.6 Regard model 3 in Table 3.1. The regression of each predictor on the other four predictors in the model produces the following R^2's: .1007 for *math diagnostic score*, .0953 for *college GPA*, .2102 for *attitude toward statistics*, .0942 for *class hours in the current semester*, and .1291 for *number of previous math courses*. Give the VIF's for each coefficient in the model. Does multicollinearity appear to be a problem?

3.7 The regression of STATMOOD (*attitude toward statistics*) on COLGPA (*college GPA*), SCORE (*score on the math diagnostic*), HOURS (*class hours in* the current semester), and PREVMATH (number of previous math courses), for 214 students in introductory statistics, produces the following sample equation (standard errors of coefficients are in parentheses underneath the coefficients) and results:

$$\hat{y} = -6.748 + 1.597\ colgpa + .296\ score - .490\ hours + 1.516\ prevmath.$$

$$(4.981)\ (.604)\qquad\quad (.110)\qquad (.145)\qquad (.344)$$

$$TSS = 6505.32710,\ R^2_{adj} = .195.$$

(a) Test the null hypothesis that all regression coefficients (except the intercept) in the population equal zero. (Hint: Since

$$R^2_{adj} = 1 - \frac{SSE/(n-K-1)}{TSS/(n-1)},$$

use this to solve for MSE, then compute the F statistic.)
(b) Give the value of R^2.
(c) Give the estimate of σ^2.
(d) Regardless of your answer to part (a), test the significance of each regression coefficient in the equation.
(e) Give the predicted STATMOOD score for someone with a GPA of 3.15, a diagnostic score of 40, who's taking 12 hours this semester, and who has had two previous math courses.
(f) Is the intercept meaningful? Why or why not?

3.8 Using the students dataset, verify that statistical control for the variables SCORE and COLGPA when looking at the effect of STATMOOD on EXAM1 is accomplished by partialing out of STATMOOD its association with a linear combination of SCORE and COLGPA. (Hint: You need to regress STATMOOD on SCORE and COLGPA among only those with valid EXAM1 scores and save the residuals. Then regress EXAM1 on these residuals and verify that the regression coefficient is the same as the coefficient for STATMOOD in the MULR of EXAM1 on STATMOOD, SCORE, and COLGPA.) Missing imputation: Substitute 40.9358974 for missing data on SCORE and 3.0827835 for missing data on COLGPA.

3.9 Using the students dataset, use residual and partial regression leverage plots to judge the empirical consistency of the MULR of STATMOOD on COLGPA, SCORE, PREVMATH, HOURS, STUDYHRS, and TVHRS for the 235 students. Missing imputation: Substitute the parenthetical values for missing data on each variable indicated: COLGPA (3.0827835), SCORE (40.9358974), PREVMATH (1), STUDYHRS (16.7844828), TVHRS (9.5128755). Also, recode values greater than 4 on PREVMATH to the value 4. What do you conclude?

3.10   The regression of STATMOOD on COLGPA, SCORE, PREVMATH, HOURS, STUDYHRS, and TVHRS for 235 students in the *students dataset* produces the following results:

Variable	Coefficient	Variable Mean	Variable Standard Deviation
STATMOOD		5.0383	5.4915
INTERCEPT	−6.7106		
COLGPA	1.1721	3.0828	.5012
SCORE	.3222	40.9359	3.1133
PREVMATH	1.4474	1.2383	1.0308
HOURS	−.5075	14.8426	2.3362
STUDYHRS	.0662	16.7845	10.1542
TVHRS	−.0444	9.5129	7.5350

Calculate both the standardized coefficients and the elasticities for all regressors. Is there any difference in relative importance of the predictors when the E_k's are used to judge this rather than the b_k^*'s? Interpret the elasticities for SCORE and for HOURS.

3.11   The regression for STATMOOD in Exercise 3.10 has an R^2 of .2256 and a *TSS* of 7056.65532.
   (a)   Give the value of R^2_{adj}. [*Hint*: See the hint for Exercise 3.7(a).]
   (b)   Test whether the model is of any utility in predicting STATMOOD.
   (c)   Give the correlation between STATMOOD and its predicted value from the MULR.
   (d)   Give the estimate of σ^2.

3.12   Suppose that the true model for Y is $.5X − 1.2Z + \varepsilon$, but you estimate $Y = \beta X + \varepsilon'$. Assuming that the correlation between X and Z is $−.75$ and $\sigma_x = 1.75$ while $\sigma_z = 2.25$, what parameter value is your sample regression coefficient consistent for? Repeat this computation assuming that the true coefficients for X and Z are $−2, −4.148$, and again if they are $−2, 2.75$. In each case, describe the nature of the bias in estimating the true effect of X on Y.

3.13   For the 416 couples in the *couples dataset*, the regression of CONFLICT (*couple verbal conflict*) on MCHATT, FCHATT (*male and female church attendance*), MALEAGE, FEMAGE (*male and female age*), MEDUC, FEDUC (*male and female education*), IHTOT2 (*total household income*), and DURYRS (*duration of the relationship in years*) produces the following results:

$\hat{y} = 2.842 − .001667\ mchatt − .007571\ fchatt − .006181\ maleage − .001743$ $femage − .002062\ meduc − .007608\ feduc − .003551\ ihtot2 − .000858$ $duryrs.$

Partial variance–covariance matrix of parameter estimates:

	MCHATT	FCHATT
MCHATT	.000165	−.000103
FCHATT	−.000103	.000162

	MALEAGE	FEMAGE
MALEAGE	.000015	−.000010
FEMAGE	−.000010	.000017

	MEDUC	FEDUC
MEDUC	.000113	−.000056
FEDUC	−.000056	.000131

Use *t* tests to test the differences between the effects of male and female characteristics on couple conflict.

3.14 Use the *GSS98 dataset* to run the following: regress ABORTION (*attitude toward abortion*, the sum of the seven abortion-attitude items coded 1 = "yes," 0 = "no," where a high score indicates a liberal attitude toward obtaining abortions) on EDUCAT, INCOME, PAED, MAED, and RESPAGE for the 1868 respondents who are nonmissing on ABORTION. Then, add the variables RELOSITY, PARTNRS5, and CONSERV to the model. Missing imputation: Substitute the parenthetical values for missing data on each variable indicated: EDUCAT (13.2824919), INCOME (13.2024490), CONSERV (4.0702152).

(**a**) Test whether the addition of the last three variables results in a significant improvement in the model's utility.

(**b**) Test whether the effects of PAED and MAED in the full model are significantly different, using the *t* test for $b_{PAED} - b_{MAED}$.

(**c**) Interpret all model coefficients in the full model.

3.15 Using the *kids dataset*, estimate the model for the regression of ADVENTRE on PERMISIV, MSEXATT, FSEXATT, MVALUES, FVALUES, MSTYLE1, FSTYLE1, MSTYLE2, and FSTYLE2. Then conduct an omnibus test for the equality of effects of male versus female parents' sex attitudes, values, and parenting styles on the focal child's *sexual adventurism* (ADVENTRE).

3.16 Using the *students dataset*, estimate the following models of STATMOOD:

(**a**) The regression of STATMOOD on COLGPA, PREVMATH, HOURS, STUDYHRS, and TVHOURS. Evaluate the hypothesis that the *number of previous math courses* elevates *attitude toward statistics* because it improves math proficiency as measured by the *math diagnostic score*. That is, add SCORE to the model and test whether this results in a significant *reduction* in the effect of PREVMATH.

(**b**) Finally, add to the model interactions between SCORE and COLGPA and between HOURS and COLGPA. Test whether this block of two interaction terms is significant. Regardless of significance, interpret the effect of COLGPA in this model.

Missing imputation: Follow the instructions for Exercise 3.9. Also, center COLGPA, SCORE, and HOURS in all models, and be sure to create the cross-product terms from the centered variables.

3.17 In the model $E(Y) = \beta_0 + \beta_1 X_1 + \beta_2 X_2$, the OLS estimate of β_1 is (Hanushek and Jackson, 1977)

$$b_1 = \frac{r_{yx_1} - r_{x_1 x_2} r_{yx_2}}{1 - r_{x_1 x_2}^2} \frac{s_y}{s_{x_1}}.$$

Use this formula to demonstrate why:

(a) b_1 is no different from the value for b in the SLR of Y on X_1 alone if $r_{x_1 x_2} = 0$.

(b) Multicollinearity tends to inflate the magnitudes of MULR coefficients.

3.18 Demonstrate the mathematical equivalence of the two forms of the nested F test shown in the text.

3.19 Show that in the interaction model

$$E(Y) = \beta_0 + \beta_1 X_1 + \beta_2 X_2 + \cdots + \beta_k X_k + \cdots + \beta_K X_K + \gamma X_k X_j,$$

a unit increase in X_k results in a change in $E(Y)$ of $\beta_k + \gamma X_j$.

3.20 Verify that $\partial E(Y)/\partial X_k$ in the interaction model in Exercise 3.3 is $\beta_k + \gamma X_j$.

3.21 Based on the results for model 2 in Table 3.4, compute the estimated partial slope for *years at the university* for those with 2 years below, 2 years above, and 9 years above mean years in rank. Then give the estimated mean salary for these faculty members, assuming that they have been at the university 3 years more than average, that they have 3 years of prior experience, and that their marketability factor is 1.1.

3.22 Although R^2 is confined to the range $[0,1]$, is it also true that R_{adj}^2 is so confied? That is, can R_{adj}^2 ever be < 0 or > 1? If so, under what conditions? (*Hint:* Set $R_{adj}^2 < 0$ and $R_{adj}^2 > 1$ and see what these inequalities imply.)

3.23 In the *kids dataset*, a MULR of offspring's *sexual adventurism* (ADVENTRE) on CPERMISS (a centered version of offspring's *sexual permissiveness*), FCAGE2 (offspring's *age at time 2*), MONITOR (*parental monitoring at time 1*), and the interaction of CPERMISS with MONITOR for 357 cases produces the following sample equation:

$$\hat{y} = .0197 + .1625 \, cpermiss + .0001 \, fcage2 + .0155 \, monitor$$
$$+ .024 \, cpermiss * monitor.$$

Significant coefficients are in boldface type. MONITOR is a scale consisting of the sum of six standardized items and therefore has a mean of zero. Also, letting a represent the coefficient for CPERMISS and d the coefficient for CPERMISS * MONITOR, we have $v(a) = .00157$, $v(d) = .00014$, and $cov(a,d) = .000016$.

(a) Interpret the main effect of CPERMISS in the model.

(b) Show the partial slope for CPERMISS in the model, and give its value at 1 standard deviation, 3.4 units, above mean *parental monitoring*.

(c) Test whether the effect of CPERMISS on ADVENTRE is significant at 1 standard deviation above mean *parental monitoring*, as well as at 2 standard deviations below mean *parental monitoring*.

3.24 Using the *students dataset*, test whether the following model for EXAM1 score is any different for males versus females in the population, using the Chow test (use only the 214 cases with valid scores on EXAM1):

$$exam1 = \beta_0 + \beta_1 \, colgpa + \beta_2 \, score + \beta_3 \, prevmath + \beta_4 \, hours$$
$$+ \beta_5 \, studyhrs + \beta_6 \, statmood + \varepsilon.$$

Missing imputation: Follow the instructions for Exercise 3.9. What do you conclude?

3.25 For the *students dataset*, a regression of STATMOOD on COLGPA, SCORE, PREVMATH, HOURS, STUDYHRS, and TVHRS, for 235 students, gives $SSE = 5464.69347$, while separate regressions of STATMOOD on the same predictors give $SSE = 3415.49428$ for the 163 female students and $SSE = 1451.03996$ for the 72 male students. Conduct a Chow test to discern whether there is any difference in the population between the models for males versus females.

3.26 Refer to Exercise 3.25. A variable called MALE is coded 1 for males and 0 for females. This type of binary variable is called a *dummy, design*, or *indicator* variable. If it is added to the model for STATMOOD for the combined sample of 235 males and females in Exercise 3.25, it allows the intercept for males to be different from the intercept for females, even though the coefficients of regressors for males versus females are constrained to be the same.

(a) Why does adding MALE allow the intercepts to differ for males and females?

(b) If SSE for this model in the combined sample is now 5018.55625, test whether just the regressor effects (but not the intercepts) are different for males versus females, using the Chow test. (*Hint*: The degrees of freedom in the numerator of the test is no longer simply $p =$ the number of parameters in the model. Rather, the df is the difference in the number of parameters estimated in the combined-sample model versus the total number of parameters estimated in the models for males and females.) What do you conclude?

3.27 Suppose that we want to test whether $\beta_1 = \beta_2$ in model A: $E(Y) = \beta_0 + \beta_1 X_1 + \beta_2 X_2 + \beta_3 X_3$. We have seen that this can be done via a nested F test of this model against model B: $E(Y) = \beta_0 + \gamma_1(X_1 + X_2) + \gamma_2 X_3$, as well as via a t test of the form $t = (b_1 - b_2)/\hat{\sigma}_{b_1 - b_2}$, using estimates from model A. Woodridge (2000) suggests yet a third way of performing this test. Let $\theta = \beta_1 - \beta_2$, implying that $\beta_1 = \theta + \beta_2$. Then model A can be expressed as $E(Y) = \beta_0 + (\theta + \beta_2)X_1 + \beta_2 X_2 + \beta_3 X_3$, or $E(Y) = \beta_0 + \theta X_1 + \beta_2 X_1 + \beta_2 X_2 + \beta_3 X_3$, or $E(Y) = \beta_0 + \theta X_1 + \beta_2(X_1 + X_2) + \beta_3 X_3$. A test of $\theta = 0$ for this model is then a test of H_0: $\beta_1 = \beta_2$ in model A. Notice that θ is just the coefficient for X_1 in a model that uses X_1, X_3, and $X_1 + X_2$ as the three regressors. Use the *couples dataset* to verify that these three ways of testing $\beta_1 = \beta_2$ are the same, with $Y = \text{WIFHAP}$, $X_1 = \text{MFIGHTS}$, $X_2 = \text{FFIGHTS}$, and $X_3 = \text{DURYRS}$.

CHAPTER 4

Multiple Regression with Categorical Predictors: ANOVA and ANCOVA Models

CHAPTER OVERVIEW

Often, the explanatory variables in a regression model are *categorical*, that is, variables with only a few discrete values. These values may not even be ordered, but may instead represent categories of a purely qualitative variable such as *religious affiliation* or *ethnic identification*. In this chapter I discuss how to incorporate such variables into a regression model. When all of the predictors are categorical, MULR is equivalent to the *analysis of variance* (ANOVA). When both categorical and continuous variables are present in the model, the procedure is equivalent to the *analysis of covariance* (ANCOVA). Although MULR is equivalent to these procedures, researchers typically reserve the terms ANOVA and ANCOVA for analyses in which the emphasis is on group comparisons of means on a dependent variable. More generally, in the regression context, a categorical predictor may simply be one of a set of important explanatory variables in which the researcher is interested. I begin by outlining two systems of coding for categorical variables: dummy coding and effect coding. I then discuss one-way and two-way ANOVA via regression and illustrate interaction between categorical predictors. Regression with both categorical and continuous predictors is then taken up, along with the issue of multiple comparisons of group means. Finally, I discuss models in which continuous and categorical predictors are allowed to interact with one another, and end the chapter by showing the equivalence between the Chow test and a model in which a categorical predictor is allowed to interact with all other covariates in the model. The example used throughout is drawn from the *faculty salary dataset*.

Regression with Social Data: Modeling Continuous and Limited Response Variables,
By Alfred DeMaris
ISBN 0-471-22337-9   Copyright © 2004 John Wiley & Sons, Inc.

MODELS WITH EXCLUSIVELY CATEGORICAL PREDICTORS

Dummy Coding

Categorical predictors cannot simply be entered as is into a regression equation. One obvious reason is that the values may not convey any real quantitative information, as in the case of a nominal variable. Even with a quantitative variable, however, its relationship with Y may not be linear. What is needed is a system of coding that is invariant to both the qualitative nature of a covariate's values and to the functional form of its relationship with Y. One such system is called *dummy coding*. The name comes from the fact that the codes—ones and zeros—only represent whether or not a case is in a given category of the variable, and otherwise convey no quantitative meaning. As an example, regard Table 4.1, which presents average academic-year salaries for 725 faculty members at Bowling Green State University (BGSU) according to college and to whether they are on graduate faculty. Suppose that we wish to regress *academic year salary* on whether or not someone is *on graduate faculty* (a status that depends on research productivity and when conferred, allows one to teach graduate classes). We create a variable, GRAD, coded 1 if the person is on graduate faculty and 0 otherwise. This is called a *dummy variable*. Letting $Y = academic\ year\ salary$, the model is $E(Y) = \beta_0 + \delta\ \text{GRAD}$ (I like to use deltas to denote the coefficients of dummy variables). How is this interpreted? Well, for those who are not on graduate faculty, the mean salary is $E(Y) = \beta_0 + \delta(0) = \beta_0$. Thus, the intercept is the mean of Y for those in the group coded 0, which is called the *contrast*, *reference*, or *omitted* group. The mean salary for those on graduate faculty is $E(Y) = \beta_0 + \delta(1) = \beta_0 + \delta$. I refer to this group as the *interest* category. The difference in means between these two groups is $E(Y|\text{on graduate faculty}) - E(Y|\text{not on graduate faculty}) = \beta_0 + \delta - \beta_0 = \delta$. A test of whether or not this mean difference is significant is a test of $H_0: \delta = 0$. This is just the usual test for the significance of a regression coefficient, consisting of the parameter estimate, d, divided by its estimated standard error. Least squares estimates of the parameters are obtained in the usual fashion—by minimizing *SSE* with respect to the parameters. The least squares estimate of β_0 is the sample mean for the omitted group, while the least squares estimate of δ is the difference in sample means for the interest and omitted groups. The estimated regression equation in this case is $\hat{y} = 39582 + 11393\ \text{GRAD}$. From Table 4.1 it is evident that the intercept here is just the mean salary for those not on graduate faculty, and the slope is the difference in mean salaries for the two groups: $50975.061 - 39581.895 = 11393.166$. The test statistic for the slope (not shown) is a t value of 10.552, which is highly significant ($p < .0001$). Recall that the regression model assumes equal error variance, implying equal Y variance, at each covariate pattern. There are only two covariate patterns here, 1 and 0. The assumption, therefore, is equal Y variance in each group—those on graduate faculty and those not on graduate faculty—in the population. In other words, in this case, regression accomplishes a test for the difference between group means under the assumption of equal Y variance and is therefore equivalent to the two-sample t test.

Table 4.1   Mean Academic Year Salaries for 725 Faculty Members by College and Whether on Graduate Faculty

| College | On Graduate Faculty? | | Overall |
	Yes (n)	No (n)	(n)
Arts and Sciences	51471.122	40592.507	49188.688
	(290)	(77)	(367)
Firelands	55411.000	40106.206	40543.486
	(1)	(34)	(35)
Business	60250.923	36498.769	54196.452
	(76)	(26)	(102)
Education	47028.545	38294.214	44311.198
	(62)	(28)	(90)
Other	44500.859	40137.915	43268.577
	(94)	(37)	(131)
Overall	50975.061	39581.895	46302
(n)	(523)	(202)	(725)

Multicategory Variables. Suppose now that the categorical variable has more than two categories. For example, the variable *college* in Table 4.1 has five categories: "arts and sciences," "business," "education," "other," and "firelands" (actually, a branch campus of BGSU being treated as a college here). In general, for an M-category variable, we need to create $M - 1$ dummy variables to represent it in a regression model. Hence, we need to create four dummy variables to represent *college*. I will let "arts and sciences" be the contrast group and will call the dummies FIREL, BUSINESS, EDUCATN, and OTHER. Table 4.2 shows the coding of these dummy variables for faculty members from the five colleges. As is evident, each dummy variable takes on the value of 1 if a faculty member is in a particular college, and 0 otherwise. If someone is in "arts and sciences," the contrast category, all dummies equal 0. Notice the naming convention for the dummies that I follow here: Each dummy is named after the interest category for that dummy. Hence FIREL takes on the value 1 if someone is in the "firelands" college, and 0 otherwise, and so on. This makes it very easy to identify what the interest category is for a particular dummy. Why don't we need another dummy for the category "arts and sciences"? Recall the assumption for MULR that no predictor is an exact linear combination of the other predictors. Suppose that we add one more dummy called ARTSCI, coded 1 if someone is in "arts and sciences," and 0 otherwise. Then it is easy to verify that the following linear equation perfectly identifies ARTSCI for each case:

$$ARTSCI = 1 - FIREL - BUSINESS - EDUCATN - OTHER.$$

For example, a faculty member in "arts and sciences" has ARTSCI as $1 - 0 - 0 - 0 - 0 = 1$. Someone in the "firelands" college has ARTSCI as $1 - 1 - 0 - 0 - 0 = 0$.

Table 4.2 Dummy Variable Coding to Represent the Variable College for Faculty Members in Each of the Colleges at BG

Faculty Member Is In:	Dummy Variable			
	FIREL	BUSINESS	EDUCATN	OTHER
Arts and Sciences	0	0	0	0
Firelands	1	0	0	0
Business	0	1	0	0
Education	0	0	1	0
Other	0	0	0	1

(The reader can verify that ARTSCI is similarly determined by the values of the other four dummies for the remaining classifications of *college*.) In this case, since ARTSCI is a perfect linear combination of the other dummies, the no-exact-collinearity assumption is violated, and the regression parameters are no longer identified. Intuitively, it is also evident that the pattern of ones and zeros for the four dummies conveys all of the information required regarding group membership in each of the five categories. The pattern in which all dummies equal zero identifies membership in the omitted group.

The model now becomes

$$E(Y) = \beta_0 + \delta_1 \text{ FIREL} + \delta_2 \text{ BUSINESS} + \delta_3 \text{ EDUCATN} + \delta_4 \text{ OTHER}. \quad (4.1)$$

This is equivalent to a one-way analysis of variance (one-way ANOVA). It is called "one-way" since there is only one factor, or one independent variable, in the model. Mean salary for "arts and sciences" faculty is

$$E(Y) = \beta_0 + \delta_1(0) + \delta_2(0) + \delta_3(0) + \delta_4(0) = \beta_0.$$

Thus, once again, the intercept is the mean of Y for the omitted group. Each regression coefficient (i.e., each δ) is the difference in means for the dummied category (the interest category) and the reference category. For example, δ_1 is the difference in mean salary between "firelands" faculty and "arts and sciences" (A&S) faculty, as can be seen by

$$E(Y|\text{firelands}) - E(Y|\text{A\&S}) = \beta_0 + \delta_1(1) + \delta_2(0) + \delta_3(0) + \delta_4(0)$$
$$- [\beta_0 + \delta_1(0) + \delta_2(0) + \delta_3(0) + \delta_4(0)] = \delta_1.$$

(Again, the reader can verify using the model that the other deltas represent mean contrasts for each college with "arts and sciences.") With a multicategory predictor, however, the deltas, individually, do not capture all of the potential mean contrasts between pairs of categories. For an M-category predictor there are a total of $M(M-1)/2$ nonredundant contrasts that can be evaluated. The deltas capture, in the current model, the contrasts between each of the dummied colleges and "arts and sciences." What about the contrast between, say, "firelands" and "business"? It turns out that the differences between the deltas capture the other contrasts. In the case of "firelands" vs. "business," we have that

$$E(Y|\text{firelands}) - E(Y|\text{business}) = \beta_0 + \delta_1 - (\beta_0 + \delta_2) = \delta_1 - \delta_2.$$

Similarly,

$$E(Y|\text{firelands}) - E(Y|\text{education}) = \beta_0 + \delta_1 - (\beta_0 + \delta_3) = \delta_1 - \delta_3.$$

The reader can again verify that $\delta_1 - \delta_4$, $\delta_2 - \delta_3$, $\delta_2 - \delta_4$, and $\delta_3 - \delta_4$ capture the rest of the contrasts, for a total of $5(4)/2 = 10$ possible contrasts. Least squares estimates for the model in equation (4.1) are shown as model 1 in Table 4.3.

Once again, the least squares estimate of β_0 is the sample mean of Y for the group omitted. Thus, the intercept in model 1 is the sample mean salary for faculty in "arts and sciences," or 49189 (it's 49188.688 in Table 4.1). The OLS estimates of the deltas are, again, differences in mean salaries, compared to "arts and sciences," for each college. For example, the coefficient for *firelands*, -8645.202, is the mean salary for "firelands" faculty (40543.486 in Table 4.1) minus the mean for "arts and sciences" faculty (49188.688). The difference is -8645.202. The global F test for the model is significant, which means that at least one of the deltas is nonzero. We see, in fact, that t statistics for the coefficients suggest that all of the deltas are nonzero. Recall that the F test actually tests the null hypothesis that every linear combination of the parameters is zero. In dummy variable regression, it is especially important to keep this more general character of the F test in mind, since linear combinations of the parameters

Table 4.3 Models for Academic Year Salary Regressed on College and Whether on Graduate Faculty

Predictor	Model 1[a]	Model 2[b]	Model 3[a]	Model 4[a]
Intercept	49189.000***	46302.000***	40348.000***	40593.000***
Firelands	−8645.202***	−5758.194***	−123.946	−486.301
Business	5007.765***	7894.772***	5512.277***	−4093.737
Education	−4877.490**	−1990.482	−3744.090*	−2298.292
Other	−5920.111***	−3033.103**	−5107.463***	−454.592
Grad faculty			11188.000***	10879.000***
Grad faculty × Firelands				4426.179
Grad faculty × Business				12874.000***
Grad faculty × Education				−2144.285
Grad faculty × Other				−6515.672*
F	14.415***	14.415***	32.930***	22.096***
ΔF			99.142***	7.151***
R^2	.074	.074	.186	.218

[a] Uses dummy coding.
[b] Uses effect coding.
* $p < .05$. ** $p < .01$. *** $p < .001$.

now make sense. In fact, every difference of the form $\delta_i - \delta_j$, comparing mean differences between interest categories, is a linear combination of the parameters that we are interested in. Hence, it may happen that the global F test is significant but none of the individual dummy coefficients is. This is entirely reasonable, since it may be one of the other contrasts—*between* interest categories—that is nonzero. The test statistic for these other contrasts is of the form

$$t = \frac{d_i - d_j}{\hat{\sigma}_{d_i - d_j}},$$

where the numerator of the test is the difference between sample estimates of the dummy coefficients, and the denominator of the test is the estimated standard error of that difference. We could compute these contrasts by hand, employing the variance–covariance matrix of parameter estimates to obtain the standard errors of the differences. But the equivalent and much simpler procedure is simply to change the contrast category and rerun the regression. In the present case, I reran the regression with, alternately, "firelands," "business," and "education" as the contrast categories to obtain the other six contrasts. The other significant contrasts, using an α of .05, were "business" versus "firelands," "education" versus "business," and "other departments" versus "business."

Effect Coding

Another type of coding that can be quite useful for categorical variables is *effect coding*. Rather than contrasting a given group's mean with that of a single other group, we might want to contrast it with a kind of overall average, across groups, on the dependent variable. Effect coding allows comparisons of each interest category's mean of Y to a "grand mean" of Y across groups. In effect coding, we once again require $M - 1$ variables for an M-category predictor. The interest categories of effect-coded indicators are, once again, coded 1 and 0 for being, vs. not being, in the category of interest. However, this time, instead of taking the value of zero on each indicator, the contrast category is coded "-1" on each. Table 4.4 shows effect coding for the colleges at BGSU, with "arts and sciences" once again serving as the omitted group.

Table 4.4 Effect Coding to Represent the Variable College for Faculty Members in Each of the Colleges at BG

Faculty Member Is In:	Effect Variable			
	FIREL	BUSINESS	EDUCATN	OTHER
Arts and Sciences	-1	-1	-1	-1
Firelands	1	0	0	0
Business	0	1	0	0
Education	0	0	1	0
Other	0	0	0	1

With this type of coding, the intercept is now the unweighted average of all five group means. Why? Let's let the letters A, F, B, E, and O represent the colleges "arts and sciences," "firelands," "business," "education," and "other departments," respectively. Then the model using effect coding of colleges is

$$E(Y) = \mu = \beta_0 + \delta_1 F + \delta_2 B + \delta_3 E + \delta_4 O.$$

The population mean salary for each college is then

$$\mu_F = \beta_0 + \delta_1(1) + \delta_2(0) + \delta_3(0) + \delta_4(0) = \beta_0 + \delta_1,$$

$$\mu_B = \beta_0 + \delta_1(0) + \delta_2(1) + \delta_3(0) + \delta_4(0) = \beta_0 + \delta_2,$$

$$\mu_E = \beta_0 + \delta_1(0) + \delta_2(0) + \delta_3(1) + \delta_4(0) = \beta_0 + \delta_3,$$

$$\mu_O = \beta_0 + \delta_1(0) + \delta_2(0) + \delta_3(0) + \delta_4(1) = \beta_0 + \delta_4,$$

$$\mu_A = \beta_0 + \delta_1(-1) + \delta_2(-1) + \delta_3(-1) + \delta_4(-1) = \beta_0 - (\delta_1 + \delta_2 + \delta_3 + \delta_4).$$

Now consider the unweighted average of the group means:

$$\bar{\mu}_j = \frac{\sum_j \mu_j}{5} = \frac{\beta_0 + \delta_1 + \beta_0 + \delta_2 + \beta_0 + \delta_3 + \beta_0 + \delta_4 + \beta_0 - (\delta_1 + \delta_2 + \delta_3 + \delta_4)}{5}$$

$$= \frac{5\beta_0}{5} = \beta_0.$$

Hence β_0 is clearly the unweighted mean of the group means, or the *grand mean*, and each delta is the difference between the mean of a given college's salary and the grand mean of all colleges' salaries. For example, the difference between Firelands' average salary and the grand mean is $\beta_0 + \delta_1 - \beta_0 = \delta_1$, and so on.

Model 2 in Table 4.3 shows *academic year salary* regressed on *college*, now coded using effect coding. The grand mean of all colleges' salaries is 46302. According to the estimate of δ_1, "firelands" average salaries are 5758.194 below the overall average salary, a difference that is quite significant. On the other hand, the Business College's average salary is 7894.772 above the grand mean, and the "other departments" category of departments has an average salary that is 3033.103 lower than the grand mean. These are also significant differences. Although the mean salary for the College of Education is 1990.482 below the grand mean in the sample, this is not a significant difference. Hence, there is not enough evidence to suggest that average salaries for the College of Education are any different than average salaries across all colleges at BGSU. Another way to phrase this is that there is not enough evidence to conclude that Education departments' salaries are any different than the university's average salary for professors. If the difference in average salary between the College of Arts and Sciences and the grand mean is of interest, the coding must be changed to make a different college the reference category. As with dummy coding, the mean difference in Y between different interest categories is captured by differences between the deltas. For example, the difference in mean salaries between "business" and "education" is $E(Y|\text{business}) - E(Y|\text{education}) = \beta_0 + \delta_2 - (\beta_0 + \delta_3) = \delta_2 - \delta_3$. According to model 2, the estimated difference is $7894.772 - (-1990.482) = 9885.254$. From Table 4.1 we can verify this figure by calculating the difference between the two sample mean salaries: $54196.452 - 44311.198 = 9885.254$. The reader is invited to consult Hardy (1993) or McClendon (1994) for other ways of coding categorical variables. For most of the book, I employ dummy coding exclusively. Dummy coding is by far the most common form of coding for categorical variables in regression models.

Two-Way ANOVA in Regression

Model 3 in Table 4.3 adds the variable GRAD, representing membership on the graduate faculty, to the model for *academic year salary*. The theoretical model is now

$$E(Y) = \beta_0 + \delta_1 \text{ FIREL} + \delta_2 \text{ BUSINESS} + \delta_3 \text{ EDUCATN}$$
$$+ \delta_4 \text{ OTHER} + \delta_5 \text{ GRAD}. \tag{4.2}$$

This is equivalent to a *two-way* ANOVA model, since there are now two categorical factors in the model: *college* and *graduate faculty status*. The model posits that salary is a purely additive function of *college* and *graduate faculty status*. Therefore, controlling for *college*, being on the graduate faculty is estimated to result in an increase of 11188 in average academic year salary, a very significant increment. Also, controlling for being on graduate faculty, being in, say, the College of Education is worth a reduction in mean salary of 3744.09 compared to being in the College of Arts and Sciences—a significant decrement. The model also allows us to predict mean salary based on *college* and *graduate faculty status*. Thus, for those in, say, the College of Education, who are on graduate faculty, the estimated mean salary is

$$\hat{y} = b_0 + d_1(0) + d_2(0) + d_3(1) + d_4(0) + d_5(1) = b_0 + d_3 + d_5$$
$$= 40348 - 3744.09 + 11188 = 47791.91.$$

However, this time the predictions do not equal the sample means shown in Table 4.1. The average salary for those in the College of Education who are members of the graduate faculty is actually 47028.545, according to Table 4.1. Why the discrepancy? The additive model assumes that there is no interaction between the categorical predictors in their effects on Y. This means, for example, that being on graduate faculty is worth the same salary increment, regardless of college. Or, it means that the difference between any two colleges in average salary is the same, regardless of whether someone is on graduate faculty or not. In the current example, as we shall see, this is not particularly realistic.

Interaction between Categorical Predictors

The model that allows interaction between *college* and *graduate faculty status* is

$$E(Y) = \beta_0 + \delta_1 \text{ FIREL} + \delta_2 \text{ BUSINESS} + \delta_3 \text{ EDUCATN}$$
$$+ \delta_4 \text{ OTHER} + \delta_5 \text{ GRAD} + \gamma_1 \text{ GRAD} * \text{FIREL} + \gamma_2 \text{ GRAD} * \text{BUSINESS}$$
$$+ \gamma_3 \text{ GRAD} * \text{EDUCATN} + \gamma_4 \text{ GRAD} * \text{OTHER}. \quad (4.3)$$

There are two ways to interpret this model, depending on which is the *focus* variable (the variable whose effect varies over levels of the other variable) and which is the *moderator* variable (the variable whose levels condition the effect of the focus variable). If GRAD (*graduate faculty status*) is the focus and *college* is the moderator, we factor equation (4.3) so that the common multipliers of GRAD are all collected in one partial effect. The result is

$$E(Y) = \beta_0 + \delta_1 \text{ FIREL} + \delta_2 \text{ BUSINESS} + \delta_3 \text{ EDUCATN} + \delta_4 \text{ OTHER}$$
$$+ (\delta_5 + \gamma_1 \text{ FIREL} + \gamma_2 \text{ BUSINESS} + \gamma_3 \text{ EDUCATN} + \gamma_4 \text{ OTHER}) \text{ GRAD}.$$

Here it is clear that the effect of GRAD, controlling for *college*, is

$$\delta_5 + \gamma_1 \text{ FIREL} + \gamma_2 \text{ BUSINESS} + \gamma_3 \text{ EDUCATN} + \gamma_4 \text{ OTHER}$$

This implies that the effect of being on graduate faculty depends on membership in a particular college. For example, the effect of being on graduate faculty for those in "arts and sciences" is $\delta_5 + \gamma_1(0) + \gamma_2(0) + \gamma_3(0) + \gamma_4(0) = \delta_5$. This gives meaning to the main effect of GRAD in equation (4.3)—it's the expected difference in salary for those on, versus not on, graduate faculty among all those in the College of Arts and Sciences. For faculty in, say, the Business College, the effect of GRAD is $\delta_5 + \gamma_1(0) + \gamma_2(1) + \gamma_3(0) + \gamma_4(0) = \delta_5 + \gamma_2$. Hence, $\delta_5 + \gamma_2$ is the expected difference in salary for those on, versus not on, graduate faculty among all those in the Business College. Should γ_2 prove to be equal to zero, the *effect* of GRAD would not be different in "arts and sciences" than in "business." The other gammas are interpreted in a similar fashion: Each is the difference in the *impact* of being on graduate faculty (as opposed to not being on graduate faculty) for the given college compared to "arts and sciences."

On the other hand, let's say that *college* is the focus variable and GRAD is the moderator. Then factoring the common multipliers of each college dummy in equation (4.3), we have

$$E(Y) = \beta_0 + \delta_5 \text{ GRAD} + (\delta_1 + \gamma_1 \text{ GRAD}) \text{ FIREL} + (\delta_2 + \gamma_2 \text{ GRAD}) \text{ BUSINESS}$$
$$+ (\delta_3 + \gamma_3 \text{ GRAD}) \text{ EDUCATN} + (\delta_4 + \gamma_4 \text{ GRAD}) \text{ OTHER}.$$

In this factoring of the equation, it becomes clear that the impact of being in a particular college, compared to being in "arts and sciences," is dependent on whether someone is on graduate faculty. For example, the expected difference in salary for someone in "firelands" versus someone in "arts and sciences" is $\delta_1 + \gamma_1$ GRAD. Therefore, it is δ_1 if someone is not on graduate faculty and $\delta_1 + \gamma_1$ if they are. Once again, if γ_1 equals zero, the difference in mean salaries between "firelands" and "arts and sciences" is the same whether or not one is on the graduate faculty.

Test for Interaction. If all of the gammas in equation (4.3) equal zero, there is no interaction between *college* and *graduate faculty status* in their effects on mean salary. This can be tested using the nested F test, since setting all of the gammas to zero in equation (4.3) results in the simplified model of equation (4.2). Hence, the model in (4.2) is nested inside the model in (4.3). Model 4 in Table 4.3 presents sample estimates for the model in equation (4.3), along with the results of the nested F test. As is evident in the table, the nested F statistic (shown as "ΔF" in the table) is 7.151. With 4 and 715 degrees of freedom, this is a highly significant result ($p < .0001$). It appears from tests of the individual gammas that there are two significant interactions: between GRAD and BUSINESS and between GRAD and OTHER. Let's interpret the overall interaction effect using GRAD and *college* alternately, as the focus variables. With GRAD as the focus, its partial effect is

$$10879 + 4426.179 \text{ FIREL} + 12874 \text{ BUSINESS}$$
$$- 2144.285 \text{ EDUCATN} - 6515.672 \text{ OTHER.}$$

The overall interpretation of this effect is that being on the graduate faculty seems to be worth an increase in average salary for all faculty, but the extra amount is greatest for the Business College and smallest for those in "other departments." For example, the increment due to being on graduate faculty for the Business College is $10879 + 12874 = 23753$, whereas for those in "other" departments, it is only $10879 - 6515.672 = 4363.328$.

If college is the focus, the partial effects are:

Firelands: $-486.301 + 4426.179$ GRAD.

Business: $-4093.737 + 12874$ GRAD.

Education: $-2298.292 - 2144.284$ GRAD.

Other: $-454.592 - 6515.672$ GRAD.

Once again, the overall interpretation is that Firelands and Business faculty make, on average, lower salaries than Arts and Sciences faculty if they do not have graduate faculty status, but more if they do. On the other hand, those in "education" and those in "other departments" make, on average, less than Arts and Sciences faculty if they are not on the graduate faculty, and *considerably* less if they are. We must keep in mind, however, that only two of the gammas are significant, so some of these apparent differences do not necessarily hold in the population.

In contrast to the estimates in model 3 of Table 4.3, those for model 4 will perfectly (within rounding error) reproduce the cell means in Table 4.1. Why? Notice that there are 10 cells in Table 4.1, each containing a mean salary that is independent of the other means. Hence there are 10 independent pieces of information, or degrees of freedom, in this table. Model 4 uses 10 parameters, an intercept and nine regression coefficients, to explain these 10 observations. In terms of the means (but not the individual faculty members), model 4 is saturated. That is, there are as many parameters as observations. Whenever this occurs, the model will perfectly reproduce the sample observations, which in this case refer to the cell means. As an example, consider using models 3 and 4 in Table 4.3, the main effect and interaction models, to predict average salary for those in the Business College with graduate faculty status. According to Table 4.1, the sample mean salary for these faculty is 60250.92. Model 3's prediction is $40348 + 11188 + 5512.28 = 57048.28$, which is off by about \$3000. Model 4's prediction is $40593 + 10879 - 4093.737 + 12874 = 60252.26$, which, within rounding error, is the correct value.

MODELS WITH BOTH CATEGORICAL AND CONTINUOUS PREDICTORS

Typically, in regression models we have a mix of both categorical and continuous predictors. If there is no interaction between the categorical and continuous regressors, these models are equivalent to ANCOVA models. The idea in ANCOVA is to examine group differences in the mean of Y while adjusting for differences among groups on one or more continuous variables, called *covariates*, that also affect the dependent variable. We want to see how much group membership "matters," in the sense of affecting the response variable, after taking account of group differences on the covariates. "Taking account," of course, means holding the covariates constant, or treating the groups as though they all had the same means on the covariates.

For simplicity's sake, let's assume that we have a categorical predictor, Z, with categories A, B, and C, and one continuous covariate, X. Then letting C be the reference group for Z, and letting A and B be dummy variables for being in categories A and B, respectively, the regression model for Y is

$$E(Y) = \beta_0 + \delta_1 A + \delta_2 B + \beta_1 X. \tag{4.4}$$

Figure 4.1 shows how this model can be interpreted. To begin, the model can be expressed for each group separately, by substituting the values for each dummy into equation (4.4). Hence, for group C the equation is

$$E(Y) = \beta_0 + \delta_1(0) + \delta_2(0) + \beta_1 X = \beta_0 + \beta_1 X, \tag{4.5}$$

for group A we have

$$E(Y) = \beta_0 + \delta_1(1) + \delta_2(0) + \beta_1 X = \beta_0 + \delta_1 + \beta_1 X, \tag{4.6}$$

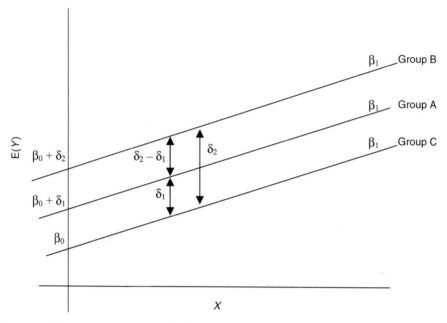

Figure 4.1   Regression model $E(Y) = \beta_0 + \delta_1 A + \delta_2 B + \beta_1 X$, depicting the absence of interaction between the categorical variable, Z, and the continuous variable, X.

and for group B we have

$$E(Y) = \beta_0 + \delta_1(0) + \delta_2(1) + \beta_1 X = \beta_0 + \delta_2 + \beta_1 X. \qquad (4.7)$$

These three equations are essentially simple linear regressions of Y on X in which the equation for each group has a different intercept but a common slope of β_1. Due to the common slope, the regression lines are parallel. This condition is depicted in the figure. For group C, the regression of Y on X has intercept β_0 and slope β_1. For group A, the intercept is $\beta_0 + \delta_1$, and the slope is β_1. For group B, the intercept is $\beta_0 + \delta_2$ and the slope is β_1. Hence, controlling for Z, the impact of X is constant; that is, the effect of X is the same in each group. On the other hand, δ_1, δ_2, and $\delta_2 - \delta_1$ represent constant differences in the mean of Y across groups regardless of the level of X. This can easily be verified by taking differences between equations (4.5) to (4.7). For example, the difference in the mean of Y for groups A and C is

$$E(Y|A) - E(Y|C) = \beta_0 + \delta_1 + \beta_1 X - (\beta_0 + \beta_1 X) = \delta_1,$$

for groups B and C we have

$$E(Y|B) - E(Y|C) = \beta_0 + \delta_2 + \beta_1 X - (\beta_0 + \beta_1 X) = \delta_2,$$

and for groups A and B we have

$$E(Y|B) - E(Y|A) = \beta_0 + \delta_2 + \beta_1 X - (\beta_0 + \delta_1 + \beta_1 X) = \delta_2 - \delta_1.$$

In other words, the deltas and their difference represent group differences in the mean of Y after adjusting for the covariate. In the figure, these differences are depicted as two-headed arrows. Notice that these group differences in $E(Y)$ are the same all along the range of X. This is an artifact of the model, since it does not allow for interaction between X and Z.

Adjusted Means

When interest centers on differences in the mean of Y across groups, we often want to examine mean differences both before and after adjusting for one or more covariates. Actually, the *adjusted mean differences* themselves—the differences between the adjusted means—are represented by the deltas in equation (4.4). But we may want to present both unadjusted (disregarding any covariates) and *adjusted* (controlling for covariates) means, in order to highlight variability in group means both before and after controlling for one or more continuous predictors. The idea behind adjusted means is that we statistically "force" the groups to all have the same mean on the covariate(s). How? We simply choose a value for each covariate and evaluate the model at that value or values. It is customary to choose the sample mean of the covariate as the control value. Thus, for the model in equation (4.4), the equation for the adjusted mean of Y is

$$E(Y) = \beta_0 + \delta_1 A + \delta_2 B + \beta_1 \bar{x}.$$

The equations for the adjusted means of each group are

$$\text{C:} \quad E(Y) = \beta_0 + \beta_1 \bar{x}.$$
$$\text{A:} \quad E(Y) = \beta_0 + \delta_1 + \beta_1 \bar{x}.$$
$$\text{B:} \quad E(Y) = \beta_0 + \delta_2 + \beta_1 \bar{x}.$$

By taking differences among these three equations, the reader can verify that the adjusted mean differences are, in fact, δ_1, δ_2, and $\delta_2 - \delta_1$. The method is easily extended to the case of multiple covariates: We simply substitute each covariate's mean into the equation and then evaluate the equation for the different groups to produce the adjusted means for each group.

Faculty Salary Example. Table 4.5 presents ANCOVA models for faculty salary regressed on the dummies representing *college*, plus four centered covariates: *years in rank, years at BG* (referred to in Chapter 3 as "years at the university"), *prior experience*, and *marketability*. In Table 3.4 we saw that these covariates are important predictors of salary. Model 1 in that table is reproduced as model 1 in Table 4.5

Table 4.5   Models for Academic Year Salary Regressed on College Plus Covariates for Number of Years in Rank, Number of Years at BG, Prior Experience, and Marketability for 725 Faculty Members

Predictor	Model 1[a]	Model 2[a]	Model 3[b]	Adjusted Mean Salary
Intercept	47801.000***	47950.000***	46993.000***	47950.000
Firelands		−6098.382***	−5142.106***	41851.618
Business		4444.926***	5401.202***	52394.926
Education		−1507.845	−551.569	46442.155
Other		−1620.079	−663.803	46329.921
Years in rank[c]	−121.466	−137.728	−137.728	
Years at BG[c]	1072.730***	1093.968***	1093.968***	
Prior experience[c]	924.893***	952.136***	952.136***	
Marketability[c]	35001.000***	27020.000***	27020.000***	
SSE	57048547616	54259263744	54259263744	
F	267.218***	144.298***	144.298***	
R^2	.598	.617	.617	

[a] Uses dummy coding.
[b] Uses effect coding.
[c] Centered variable.
* $p < .05$. ** $p < .01$. *** $p < .001$.

(although the intercepts are different, since not all of the covariates were centered in Table 3.4). It shows the results of salary regressed on the four centered covariates. Although *years in rank* is not significant, the other three covariates all have significant, positive effects on salary. If the distributions of these covariates are substantially different across colleges, it is important to adjust for them in any analysis of the effect of location in a particular college of the university on salary. Otherwise, we will have a misleading picture of the extent to which salary differences across colleges are due to college per se rather than to key covariates which happen to differ for each college. Table 4.6 shows the means on the four covariates both overall and by each college.

The variability in the means reveals that the distributions on these four covariates do, in fact, change considerably across colleges. Compared to the overall mean for *years in rank*, "arts and sciences," "education," and "other departments" are above average on years in rank, while "firelands" and "business" are below average. In that *years in rank* is not a significant predictor of salary, however, these differences may not be of consequence. Differences on the other three covariates, on the other hand, could be important. "Arts and sciences" and "other departments" are above average in *years at BG*, whereas "firelands," "business," and "education" are below average. In terms of *prior experience*, "firelands" and "education" stand out as being substantially above average, while "business" has the lowest mean on this factor. Finally, *marketability* shows an interesting pattern in which the mean for the Business College is not only above average but is substantially higher than the *marketability* scores for

Table 4.6 Means for the Covariates Years in Rank, Years at BG, Prior Experience, and Marketability: Overall and by College of the University for 725 Faculty Members at BGSU

| College | Covariate | | | |
	Years in Rank	Years at BG	Prior Experience	Marketability
Arts and Sciences	7.880	13.777	2.973	.935
Firelands	5.486	10.514	3.143	.955
Business	5.843	9.304	2.069	1.159
Education	7.456	11.778	3.656	.865
Other	7.725	12.603	2.756	.830
Overall	7.397	12.530	2.899	.940

any of the other colleges. In that *marketability* has a strong impact on salary, one would expect that controlling for this covariate should reduce the gap in salary between "business" and the other colleges.

Model 2 in Table 4.5 shows the ANCOVA model for salary regressed on the college dummies plus the four covariates. Is college's effect on salary significant after controlling for the covariates? That is, does the addition of the four dummies representing *college* make a significant contribution to the model? We answer this question using the nested F test, since model 1 is nested inside model 2. The test statistic, using the R^2 values from the two models, is

$$F = \frac{(.6172 - .5975)/4}{(1 - .6172)/716} = 9.212.$$

With 4 and 716 degrees of freedom, this is a very signficant result ($p < .00001$). We can conclude that at least one of the dummy coefficients is nonzero. Or more generally, we can conclude that at least one of the mean contrasts in salary between colleges is nonzero. We see that two coefficients are significant: the mean difference between "firelands" and "arts and sciences" and the mean difference between "business" and "arts and sciences." The other two coefficients, contrasting "education" and "other departments" with "arts and sciences," are nonsignificant. Recall from model 1 in Table 4.3 that all of the coefficients reflecting contrasts with "arts and sciences" were significant when the covariates were ignored. Thus, some of the effects of college on salary have been accounted for by adjusting for differences across colleges in key predictors of salary. This is highlighted further by the fact that the dummy coefficients representing mean differences across colleges have all been reduced in magnitude compared to model 1 in Table 4.3. For example, the unadjusted mean difference in salary between "firelands" and "arts and sciences" is −8645.202. The adjusted mean difference is now −6098.382. So a little over $2500 in the average salary difference between "firelands" and "arts and sciences" is due to differences between these colleges in average *years in rank, years at BG, prior experience,* and *marketability*. The same comment can be made regarding the other coefficient reductions.

Model 3 in Table 4.5 is the same as model 2 except that effect coding is used to represent colleges. In comparison to model 2 in Table 4.3, we see that the departures of each college's mean salary from the grand mean are smaller than was the case without controlling for covariates. Moreover, whereas three of these departures were significant before, only two are significant now. To further foreground the closing of the salary gaps across colleges, the last column of Table 4.5 shows the adjusted mean salaries for each college after accounting for the four key covariates. Because the covariates are centered and their means are therefore all zero, calculating adjusted means is quite straightforward. One just ignores the coefficients for the covariates in model 2 in Table 4.5 and uses the intercept and dummy coefficients to calculate the means. For example, the adjusted mean for "arts and sciences" is just the intercept in model 2—47950. The adjusted mean for "firelands" is $47950 - 6098.382 = 41851.618$, and so on. It is evident that compared to the unadjusted means for each college in the last column of Table 4.1, the adjusted means exhibit less variability.

Mean Contrasts with an Adjusted Alpha Level. Recall that with five categories of the variable college, there are 10 possible mean salary contrasts between pairs of colleges that can be tested. Up until now, I have been conducting these tests without controlling for the increased risk of type I error—or capitalization on chance—that accrues to making multiple tests. There are several procedures that accomplish this control; here I discuss one, the *Bonferroni comparison procedure*. The Bonferroni technique is advantageous because of its great generality. It is not only limited to tests of mean contrasts. It can be used to adjust for capitalization on chance whenever multiple tests of hypothesis are conducted, regardless of whether or not they are the same type of test. The rationale for the procedure is quite simple. Suppose that I were making 10 tests and I wanted my overall chance of making at least one type I error to be .05 for the collection of tests. That is, I want the probability of rejecting at least one null hypothesis that is, in fact, true, to be no more than .05 over all tests. If I make each test at an α level of .05, the probability of making a type I error on each test is .05. This means that the probability of *not* making a type I error on any given test is .95, and the probability of not making any type I errors across all 10 tests is therefore $(.95)^{10} = .599$. This implies that the probability of making at least one type I error in all these tests is $1 - .599 = .401$. In other words, we have about a 40% chance of declaring one H_0 to be false when it is not. The Bonferroni solution in this case is to conduct each test at an α level of $.05/10 = .005$. This way, the probability of not making a type I error on any given test is .995, and the probability of making at least one type I error across all 10 tests is $1 - (.995)^{10} = .049$. In general, if one is making K tests and one wants the probability of making at least one type I error to be held at α across all tests, the Bonferroni procedure calls for each test to be made at an α level of $\alpha' = \alpha/K$.

Although the Bonferroni procedure has the advantages of simplicity and flexibility, it tends to be somewhat low in power. Holland and Copenhaver (1988) discuss several modifications of the Bonferroni procedure that result in enhanced power to detect false null hypotheses. The one I focus on here is due to Holm (1979). The Bonferroni–Holm approach is to order the attained p-values for the K tests from smallest to largest. One then compares the smallest p-value to α/K. If $p < \alpha/K$, one rejects the corresponding null hypothesis and moves to the test with the next smallest p-value. This is compared to $\alpha/(K-1)$. If this p is less than $\alpha/(K-1)$, one rejects the corresponding null hypothesis and moves to the test with the next smallest p-value, which is compared to $\alpha/(K-2)$. We continue in this fashion, each time comparing the next smallest p-value to $\alpha/(K-3)$, then to $\alpha/(K-4)$, and so on, until the largest p-value is compared to $\alpha/1 = \alpha$. As long as the p-value for the given test is smaller than the relevant adjusted α level, the corresponding null hypothesis is rejected and we continue to the next test. If, at any point, we fail to reject the null, testing stops at that point and we fail to reject all of the remaining null hypotheses.

Table 4.7 presents tests for mean salary differences between colleges after adjusting for *years in rank, years at BG, prior experience*, and *marketability*, using Bonferroni–Holm adjusted α levels (shown as α' in the table). The tests are based on the dummy coefficients in model 2 in Table 4.5. The adjusted α levels are .05/10 = .005, .05/9 = .0056, .05/8 = .0063, and so on, so that the last adjusted α level is .05. The first seven contrasts are all significant since the p-value for each of the tests is less than the adjusted α level. However, we fail to reject equality of mean salaries for "other departments" versus "arts and sciences," hence we fail to reject equality of the last two contrasts as well. The outcomes of these tests are the same as would have been realized if we had simply used the .05 α level for each test. However, this procedure, unlike that simpler one, holds the overall α level down to .05. Notice that the unmodified Bonferroni procedure, using $\alpha = .005$ for each test, would have failed to reject equality for the "other departments" versus "firelands" and the "education" versus "firelands" contrasts. That they are rejected under Bonferroni–Holm reveals the greater power of the latter procedure.

Table 4.7 Tests for Mean Salary Differences between Colleges after Adjusting for Years in Rank, Years at BG, Prior Experience, and Marketability, Using Bonferroni–Holm Adjusted α Levels (Based on Model 2 in Table 4.3)

Contrast	Mean Salary Difference	p	α'
Business–Firelands	10543.000	<.0001	.0050
Other–Business	−6065.005	<.0001	.0056
Firelands–A&S	−6098.382	<.0001	.0063
Education–Business	−5952.772	<.0001	.0071
Business–A&S	4444.926	.0002	.0083
Other–Firelands	4478.303	.0086	.0100
Education–Firelands	4590.537	.0092	.0125
Other–A&S	−1620.079	.0850	.0167
Education–A&S	−1507.845	.1508	.0250
Other–Education	−112.233	.9254	.0500

Interaction between Categorical and Continuous Predictors

The ANCOVA model, together with adjusted means, makes sense only if the differences in group means are the same at different values of the covariate (or at different values of the covariate *patterns* exhibited by the continuous regressors). In this case it makes sense to speak of group mean differences that exist after adjusting for the covariates—treating group differences as constant regardless of the levels of the covariates. If this condition is not met, we have statistical interaction between the continuous and categorical predictors. In this section of the chapter, I discuss this type of interaction and show how it can be interpreted with, alternately, the continuous and the categorical variable as the focus.

Interaction Model. For simplicity, once again, I explicate interaction with a model that has one categorical predictor, Z, with three categories—A, B, and C (the reference category)—and one continuous predictor, X. This time, I allow Z to interact with X in its impact on Y. The model is

$$E(Y) = \beta_0 + \delta_1 A + \delta_2 B + \beta_1 X + \gamma_1 AX + \gamma_2 BX. \qquad (4.8)$$

To interpret the interaction effect, let's write the simple linear regression of Y on X separately for each group. For group C the equation is

$$E(Y) = \beta_0 + \beta_1 X, \qquad (4.9)$$

for group A it is

$$E(Y) = \beta_0 + \delta_1 + (\beta_1 + \gamma_1)X, \qquad (4.10)$$

and for group B we have

$$E(Y) = \beta_0 + \delta_2 + (\beta_1 + \gamma_2)X. \qquad (4.11)$$

We can see immediately that the regression of Y on X has a different slope in each group. Thus, if X is the focus, its impact is β_1 in group C, $\beta_1 + \gamma_1$ in group A, and $\beta_1 + \gamma_2$ in group B. That is, the effect of X changes across groups. If the gammas both turn out to be zero, there is no difference in the effect of X in each group. The Y-intercepts for each group are, once again, β_0 for group C, $\beta_0 + \delta_1$ for group A, and $\beta_0 + \delta_2$ for group B. Figure 4.2 illustrates the model for the three groups.

If group is the focus, we ask how group differences might depend on the level of X. To ascertain this, we take differences between equations (4.9) to (4.11). The difference in the mean of Y for group A vs. group C is (4.10) − (4.9), or

$$E(Y|A) - E(Y|C) = \beta_0 + \delta_1 + \beta_1 X + \gamma_1 X - \beta_0 - \beta_1 X = \delta_1 + \gamma_1 X.$$

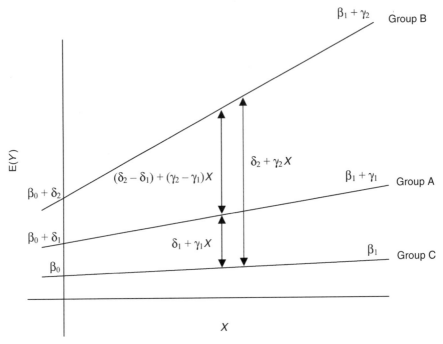

Figure 4.2 Regression model $E(Y) = \beta_0 + \delta_1 A + \delta_2 B + \beta_1 X + \gamma_1 AX + \gamma_2 BX$, depicting ordinal interaction between the categorical variable, Z, and the continuous variable, X.

The difference in means for groups B and C is (4.11) − (4.9), or

$$E(Y|B) - E(Y|C) = \beta_0 + \delta_2 + \beta_1 X + \gamma_2 X - \beta_0 - \beta_1 X = \delta_2 + \gamma_2 X.$$

Finally, the difference in means for groups B and A is (4.11) − (4.10), or

$$E(Y|B) - E(Y|A) = \beta_0 + \delta_2 + \beta_1 X + \gamma_2 X - \beta_0 - \delta_1 - \beta_1 X - \gamma_1 X$$
$$= (\delta_2 - \delta_1) + (\gamma_2 - \gamma_1)X.$$

It is evident here that group differences in the mean of Y are no longer constant as they were in the no-interaction model. Instead, they are a function of the level of X. Figure 4.2 depicts this situation by showing that the gap between any two groups' means increases as we go from lower to higher values of X. This particular model is designed to exhibit interaction that is ordinal in both X and Z. With respect to the effect of X, this means that the direction of its impact (which in this example is positive) is the same in each group but that the magnitude of its effect changes across groups. Similarly, that the interaction is ordinal in Z is illustrated by the fact that although the gap between groups' means gets larger with increasing X, the ordering of group means is always the same: The mean for group B is always higher than the mean for group A, which is always higher than the mean for group C.

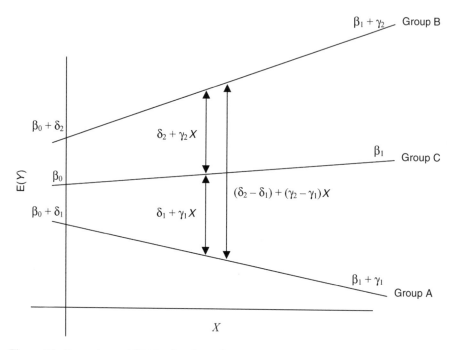

Figure 4.3 Regression model $E(Y) = \beta_0 + \delta_1 A + \delta_2 B + \beta_1 X + \gamma_1 AX + \gamma_2 BX$, depicting interaction that is disordinal in the effect of the continuous variable, X, but ordinal in the effect of the categorical variable, Z.

Figures 4.3 and 4.4 depict various types of disordinal interaction. Figure 4.3 shows interaction that is disordinal in X but ordinal in Z. That is, the effect of X changes direction across groups: positive in groups B and C but negative in group A. Nevertheless, with respect to Z, the ordering of group means is always the same, with the mean for group B being the highest and the mean for group A being the lowest, regardless of the level of X. Hence the effect of Z is of the same nature, regardless of X, but varies in magnitude over the values of X. Figure 4.4 illustrates interaction that is disordinal in both X and Z. The effect of X changes direction across groups since, again, it's positive in groups B and C but negative in group A. The "direction" of Z's effect also changes over X. This is shown by the fact that although the means for groups A and B are always higher than the mean for group C, group A's mean is higher than group B's mean at lower values of X, but lower than B's mean at higher values of X. In other words, the nature of Z's effect changes over levels of X. The only way to tell whether the interaction is ordinal or disordinal in either variable involved in the interaction is to substitute into the equation some sample estimates of the main-effect and interaction coefficients. After evaluating the equation for the different groups, or at different sample values of X, it should be relatively easy to discern the nature of the interaction effect.

Interaction Models for Faculty Salary. Table 4.8 shows the results of estimating a model for faculty salary that includes the interaction of *college* with *marketability* in

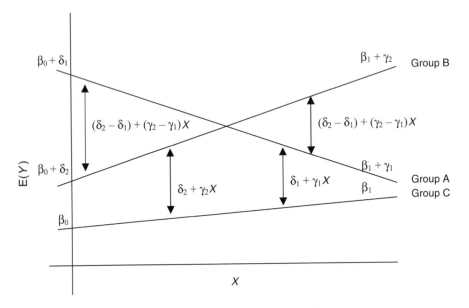

Figure 4.4 Regression model $E(Y) = \beta_0 + \delta_1 A + \delta_2 B + \beta_1 X + \gamma_1 AX + \gamma_2 BX$, depicting interaction that is disordinal both in the effect of the continuous variable, X, and in the effect of the categorical variable, Z.

Table 4.8 Main Effect and Interaction Models for the Regression of Academic Year Salary on College Plus Covariates for Number of Years in Rank, Number of Years at BG, Prior Experience, and Marketability for 725 Faculty Members

Predictor	Model 1	Model 2
Intercept	47950.000***	47958.000***
Firelands	−6098.382***	−5639.779***
Business	4444.926***	2201.368
Education	−1507.845	−767.187
Other	−1620.079	−2000.977
Years in rank[a]	−137.728	−132.748
Years at BG[a]	1093.968***	1088.270***
Prior experience[a]	952.136***	956.563***
Marketability[a]	27020.000***	27820.000***
Marketability[a] × Firelands		−32394.000***
Marketability[a] × Business		9392.054
Marketability[a] × Education		9321.962
Marketability[a] × Other		−4197.697
F	144.298***	98.524***
ΔF		3.267*
R^2	.617	.624

[a] Centered variable.

* $p < .05$. ** $p < .01$. *** $p < .001$.

their effects on salary. For comparison purposes, the main effects (ANCOVA) model without interaction, shown as model 2 in Table 4.5, is first reproduced here as model 1. Model 2 adds the cross-products of *marketability* with the four *college* dummies. First we must investigate whether the interaction is significant. As it involves four additional terms, we perform the nested F test to compare models 1 and 2. The nested F is 3.267, which is just significant ($p < .05$). Evidently, the only significant individual coefficient is for the cross-product of *marketability* with the "firelands" group. Nevertheless, for didactic purposes, I consider all of the coefficients when interpreting the interaction effect. First, if *marketability* is the focus, its partial slope is

$$27820 - 32394 \text{ FIREL} + 9392.054 \text{ BUSINESS}$$
$$+ 9321.962 \text{ EDUCATION} - 4197.697 \text{ OTHER}.$$

Here it is evident that the interaction is disordinal in *marketability*. Although *marketability* has a positive effect on salary for "arts and sciences," "business," "education," and "other departments," its effect is estimated to be negative for "firelands." With a value of $27820 - 32394 = -4574$, the effect suggests that a unit increase in marketability in the Firelands college actually lowers one's salary by $4574, on average. This seemingly counterintuitive result may make more sense when viewed from the perspective of *marketability* as the moderator variable, as we do next.

With *college* as the focus, the dummy coefficients turn out to be:

Firelands: $-5639.779 - 32394$ *marketability.*

Business: $2201.368 + 9392.054$ *marketability.*

Education: $-767.187 + 9321.962$ *marketability.*

Other: $-2000.977 - 4197.697$ *marketability.*

Each coefficient reflects the difference in mean salary between the indicated college and "arts and sciences." As is evident, this difference depends on the marketability value for that faculty member's discipline. The range of marketability values is .58 to 1.33, with a mean of .94 and a standard deviation of .149. Here it is important to remember that *marketability* is centered. So the main effect of any given dummy represents the difference in mean salary between the indicated college and "arts and sciences," at average marketability of faculty disciplines. Consider the mean salary difference between "firelands" and "arts and sciences," for example. At average marketability, it is -5639.779. That is, for faculty members with average marketability, Firelands faculty make about $5640 less than Arts and Sciences faculty. However, at 2 standard deviations below average marketability, or a value of $-.298$, the effect of "firelands" is $-5639.779 - 32394(-.298) = 4013.633$. That is, for those in less marketable disciplines, it is more advantageous, salarywise, to be in the Firelands college, compared to being in the College of Arts and Sciences. At 2 standard deviations above average marketability, or a value of .298, the effect is $-5639.779 - 32394(.298) = -15293.191$. The effect is clearly disordinal in college here, since

Firelands faculty make a higher average salary than Arts and Sciences faculty at lower marketability but a lower average salary at higher marketability. Similar changes in direction of effects can be seen for the Education and Business colleges. For "education," the contrast with "arts and sciences" at -2 standard deviations of marketability is -3545.132. At $+2$ standard deviations of marketability it is 2010.758. For "business," the numbers are -597.464 and 5000.200, respectively. For "other departments," however, average salary is always lower than for "arts and sciences," regardless of marketability levels observed in the sample. At any rate, sample estimates suggest that overall, the interaction is disordinal in both *college* and *marketability*.

Comparing Models across Groups, Revisited

The interaction of *all* model regressors with a dummy variable (or variables) representing group membership suggests that the dynamics of a given model are different for each group. In Chapter 3 we saw that we could use a Chow test to test whether the impact of a set of regressors on a response was different in different groups. We can also accomplish the same task using cross-product terms. In this last section of the chapter I show the equivalence of these two ways of addressing the same issue.

Another Look at the Chow Test. Recall that the Chow test is an omnibus test for whether a model differs across groups: meaning whether the effect of at least one explanatory variable is different across groups. It turns out that an assumption for this test, as well as the equivalent test using cross-product terms, is that *the error variance in each group is the same*. That is, if A and B represent the two groups in question, and σ^2, as usual, represents $V(\varepsilon)$, the assumption is that $\sigma_A^2 = \sigma_B^2$. In Chapter 3 I simply assumed equal error variances in the equations for *academic year salary* for male and female faculty and proceeded to test whether the factors affecting salary had the same effects for each gender. However, to ensure that the test is valid, we must first test whether, in fact, the assumption of equal error variances is reasonable. The following test assumes that the error variance is normally distributed in each group. The test statistic is

$$F = \frac{MSE_A}{MSE_B},$$

assuming that group A has the larger of the two error variances. If the error variance in group B is larger, the test statistic is

$$F = \frac{MSE_B}{MSE_A}.$$

Under the null hypothesis that the error variances are the same, this statistic follows the F distribution with numerator degrees of freedom equal to the error degrees of freedom (df_E) for the model whose MSE is in the numerator of the test, and denominator degrees of freedom equal to df_E for the model whose MSE is in the denominator (Hardy, 1993). As an example, let's test whether the error variances for the salary models in Table 3.5 are the same for male versus female faculty. For the male faculty, the MSE is

81171524.123, and the df_E is 505. For the female faculty, the MSE is 52550530.18, and the df_E is 208. The test statistic, therefore, is

$$F = \frac{81171524.123}{52550530.18} = 1.545.$$

With 505 and 208 degrees of freedom, this result is quite significant ($p < .001$). This suggests that the Chow test for this problem, which found the salary model to differ for males and females, was not valid. In Chapter 6, in which weighted least squares is introduced, we will see how to conduct a test for equality of models across two groups under the condition of unequal error variance.

In that *graduate faculty status* appears to play a key role in determining faculty salary, one question worth investigating is whether model 1 in Table 4.8 differs by graduate faculty status. That is, do the effects of *college, years in rank, years at BG, prior experience*, and *marketability* have different effects on salary for those who are on the graduate faculty, as opposed to those who are not? Table 4.9 presents several models pertinent to this question. Model 1 shows the estimated effects of these factors on salary for the 202 faculty members not on graduate faculty, while model 2 shows the results for the 523 faculty members on graduate faculty. It seems that all of the regression coefficients are quite different in each model, with some differences more pronounced than others. Particularly noticeable are the effects of *being in the business college, prior experience*, and *marketability*, all of which have substantially greater positive effects on salary for those on graduate faculty. Moreover, the effects of *being in the business college* and *prior experience* are significant only for those on graduate faculty. We should keep in mind, however, that any time we run separate analyses of the same model in different subgroups, the coefficient estimates will differ to some extent purely because of sampling error. And even though a given effect is significant in one group but not the other, this is not enough evidence to conclude that the effects are significantly *different* in each group.

Before we can test for model differences, we must again test for the equality of error variances in each group. The estimated error variance, or MSE, for those on graduate faculty status is 53934137.987, with 514 df. The MSE for those not on graduate faculty is 50674043.62, with 193 df. Therefore, the test statistic is

$$F = \frac{53934137.987}{50674043.62} = 1.064.$$

With 514 and 193 df, this is not a significant result ($p > .3$). The assumption of equal error variance in this instance appears reasonable.

In Chapter 3 the Chow test was performed by constraining all model coefficients, including the intercept, to be the same in each group under the null hypothesis. This may not always be desirable. In this particular example, the intercept represents the salary of faculty members in "arts and sciences" who are average in *years in rank, years at BG, prior experience*, and *marketability*. It may well be that average salary for these faculty members differs according to *graduate faculty status*. That is, being on graduate faculty may add some increment to salary. But the *impact* on salary of *college, years in rank, years at BG, prior experience,* or *marketability* may be the same, regardless of whether or not one is on the graduate faculty. That is, in this instance, we may want the null hypothesis to allow the intercept to differ by *graduate faculty status* while constraining the effects of the other regressors. We perform the test both ways, so that the reader can see the difference in the outcomes.

The test that constrains the intercept as well as the other regressors is performed by comparing model 1 in Table 4.8, the combined-sample model, with models 1 and 2 in Table 4.9. The test statistic is

$$F = \frac{[54259263744 - (9780090418.7 + 27722146925)]/9}{(9780090418.7 + 27722146925)/707} = 35.101.$$

Table 4.9 Regression Models Pertaining to the Interaction of Predictors of Academic Year Salary with Whether on Graduate Faculty

Predictor	Model 1	Model 2	Model 3	Model 4
Intercept	40698.000***	50023.000***	39891.000***	40698.000***
Graduate faculty			10318.000***	9324.678***
Firelands	1409.370	803.314	1960.409	1409.370
Business	714.042	6957.555***	5547.657***	714.042
Education	−188.553	−1486.229	−1000.544	−188.553
Other	−1757.682	−1543.441	−1575.448	−1757.682
Years in rank[a]	−279.113	246.232**	195.092*	−279.113
Years at BG[a]	1183.076***	863.967***	877.906***	1183.076***
Prior experience[a]	227.027	1010.841***	845.124***	227.027
Marketability[a]	9660.048*	27146.000***	22442.000***	9660.048*
Graduate faculty × Firelands				−606.056
Graduate faculty × Business				6243.512**
Graduate faculty × Education				−1297.677
Graduate faculty × Other				214.241
Graduate faculty × years in rank[a]				525.345*
Graduate faculty × years at BG[a]				−319.109*
Graduate faculty × prior experience[a]				783.814***
Graduate faculty × marketability[a]				17486.000***
SSE	9780090418.7	27722146925	41978431466	37502237344
F	55.884***	145.241***	188.799***	115.595***
R^2	.699	.693	.704	.735

[a] Centered variable.
* $p < .05$. ** $p < .01$. *** $p < .001$.

With 9 and 707 *df*, this result is very significant ($p < .00001$), resulting in rejection of the null hypothesis. This suggests that the salary models do differ according to *graduate faculty status*. However, to what extent is this result an artifact of constraining the intercepts to be equal? Model 3 in Table 4.9 is the same salary model as models 1 and 2, except that it is performed on the combined sample and adds a dummy representing *graduate faculty status*. Notice that the effect of *graduate faculty status* is quite pronounced: All else equal, being on the graduate faculty adds, on average, \$10,318 to academic year salary. Model 3 is an ANCOVA model that allows for different intercepts according to *graduate faculty status*. For example, for those not on graduate faculty, the mean salary is

$$E(Y) = 39891 + 1960.409 \text{ } firelands + 5547.657 \text{ } business - 1000.544 \text{ } education$$
$$- 1575.448 \text{ } other \text{ } departments + 195.092 \text{ } years \text{ } in \text{ } rank$$
$$+ 877.906 \text{ } years \text{ } at \text{ } BG + 845.124 \text{ } prior \text{ } experience + 22442 \text{ } marketability.$$

For those on graduate faculty, the mean salary is

$$E(Y) = (39891 + 10318) + 1960.409 \text{ } firelands + 5547.657 \text{ } business$$
$$- 1000.544 \text{ } education - 1575.448 \text{ } other \text{ } departments + 195.092 \text{ } years \text{ } in \text{ } rank$$
$$+ 877.906 \text{ } years \text{ } at \text{ } BG + 845.124 \text{ } prior \text{ } experience + 22442 \text{ } marketability.$$

Here, it is evident that the intercept for those not on graduate faculty is 39891, while for those on graduate faculty it is $39891 + 10318 = 50209$. The Chow test allowing the intercepts to differ is performed by comparing model 3 to models 1 and 2. Notice, however, that there is a change in the numerator degrees of freedom, representing the difference in the number of parameters estimated. With the intercept constrained, this difference is just the number of parameters in the model, since the model is simply being duplicated in each group. With the intercept unconstrained, there is one more parameter being estimated in the combined model than before. Now the difference is 18 parameters in models 1 and 2 minus 10 parameters in model 3 = 8 numerator degrees of freedom. The test statistic is

$$F = \frac{[41978431466 - (9780090418.7 + 27722146925)]/8}{(9780090418.7 + 27222146925)/707} = 10.55.$$

With 8 and 707 degrees of freedom, the result is very significant ($p < .00001$), but the test statistic is substantially smaller compared to the result of the constrained-intercept test. Nevertheless, we would conclude that the impact of at least one model regressor is different for those who are on, versus not on, the graduate faculty.

Model Comparison Using Cross-Product Terms. One limitation of the Chow test is that although it allows us to conclude that models are different across groups, it doesn't tell us which regressors have different effects. Performing model comparison tests using cross-product terms rectifies that limitation. Model 4 in Table 4.9 is model 3 with the addition of the cross-products of *graduate faculty status* with all other model regressors. To compare models across levels of *graduate* *faculty status* with the intercept constrained, we can perform a nested F test comparing model 1 in Table 4.8 ($RSS = 87480506481$), the ANCOVA model for faculty salary without the dummy for *graduate faculty status*, with model 4 in Table 4.9 ($RSS = 104237532881$, $MSE = 53044182.948$), the interaction model that includes this dummy along with the cross-product terms. The test statistic is

$$F = \frac{(104237532881 - 87480506481)/9}{53044182.948} = 35.101,$$

which agrees with the result of the first Chow test above. To compare models without constraining the intercept, the more usual practice, we perform a nested F test comparing models 3 and 4 in Table 4.9. This is essentially a test for the significance of the block of cross-product terms capturing the interaction of *graduate faculty status* with all other regressors in model 3. The R^2 for model 3 is .7038 and the R^2 for model 4 is .7354. Hence, the test statistic is

$$F = \frac{(.7354 - .7038)/8}{(1 - .7354)/707} = 10.55,$$

which also agrees with the second Chow test above. In either case, we would prefer the interaction model over the model that does not allow interaction between graduate faculty status and the other regressors.

Model 4 also indicates which regressor effects differ by graduate faculty status via the t tests for the cross-product terms. Apparently, the factors whose effects differ according to graduate faculty status are being in the Business College as opposed to being in Arts and Sciences, *years in rank, years at BG, prior experience*, and *marketability*. Interpretation of the model is facilitated by examining the partial effect of each of these variables as a function of graduate faculty status. The impact of being in the Business College compared to being in Arts and Sciences is $714.042 + 6243.512$ *graduate faculty*. This suggests that being on the graduate faculty adds about \$6243 to the average salary gap between the Business College and the Arts and Sciences faculty. The effect of *years in rank* is $-279.113 + 525.345$ *graduate faculty*. Thus, a greater number of years in rank reduces salary for those not on the graduate faculty but enhances salary for graduate faculty members. The effect of *years at BG* is $1183.076 - 319.109$ *graduate faculty*. Working at BG a year longer raises average salary more for those who are not on the graduate faculty than for those who are. The effect of *prior experience* is $227.027 + 783.814$ *graduate faculty*, while the effect of *marketability* is $9660.048 + 17386$ *graduate faculty*. Both of these regressors have stronger positive effects on salary for those who are on the graduate faculty than for those who are not.

Generalizing the Chow Test. Although the Chow test examples I have used involve only two groups, the test can be generalized to any number of groups. For example, suppose that we wish to compare a model across three groups, denoted A, B, and C. First, we run the model on the combined sample, either constraining the intercepts for the groups to be equal by excluding group membership from the model (giving us SSE_c) or by allowing the intercepts to be unconstrained by including two dummies in the model representing group membership (giving us SSE_u). Next, we run the model in each group separately (which gives us SSE_A, SSE_B, and SSE_C). If there are J parameters in the model (including the intercept), the test statistic with intercept constrained is

$$F = \frac{[SSE_C - (SSE_A + SSE_B + SSE_C)]/2J}{(SSE_A + SSE_B + SSE_C)/(n-3J)},$$

which under the null hypothesis of no model difference across groups has the F distribution with $2J$ and $n-3J$ degrees of freedom. The test statistic with the intercept unconstrained is

$$F = \frac{[SSE_u - (SSE_A + SSE_B + SSE_C)]/(2J-2)}{(SSE_A + SSE_B + SSE_C)/(n-3J)}.$$

The combined-sample model in this case has $J + 2$ parameters. Under the null hypothesis of no difference across groups in the effects of the $J-1$ regressors in the model, this statistic has the F distribution with $2J-2$ and $n-3J$ degrees of freedom. Extension to more than three groups follows in a similar fashion. Examples are given in the exercises.

Generalizing the Variance Homogeneity Test. The Chow test for model equivalence across multiple groups assumes, as before, that the error variance is the same in each group. Once again, this assumption can be tested. This time, however, an appropriate test is Barlett's test (Neter et al., 1985). This test is based on the assumption that the error variance is normally distributed in each group. Let $\hat{\sigma}_1^2, \hat{\sigma}_2^2, \ldots, \hat{\sigma}_G^2$ be the estimated error variances (i.e., MSE's) for G different groups, with error degrees of freedom equal to df_g, for $g = 1, 2, \ldots, G$. Then the weighted average of the error variances is

$$MSE^* = \frac{1}{df_T} \sum_{g=1}^{G} df_g \hat{\sigma}_g^2,$$

where df_T is the sum of the G error degrees of freedom. That is,

$$df_T = \sum_{g=1}^{G} df_g.$$

Barlett's test statistic is then

$$B = \frac{1}{C}\left(df_T \log MSE^* - \sum_{g=1}^{G} df_g \log \hat{\sigma}_g^2 \right),$$

where "log" refers to the natural logarithm, and

$$C = 1 + \frac{1}{3(G-1)}\left[\left(\sum_{g=1}^{G} \frac{1}{df_g} \right) - \frac{1}{df_T} \right]$$

For large sample sizes in each group, and under the null hypothesis that the G error variances are equal, this test statistic has approximately a chi-squared distribution with $G - 1$ degrees of freedom. Two points should be noted. First, Barlett's test is very sensitive to departures from normality, so if the errors are not normally distributed, the test may not be accurate. Second, C is always greater than 1, so if the term inside the parentheses in the expression for B is already under the critical chi-squared value for testing at level α, it is not necessary to calculate C—the null hypothesis will not be rejected either way. If the term inside the parentheses is greater than the critical chi-squared value, it is necessary to calculate C and do the complete computation for B (Neter et al., 1985). Examples are given in the exercises.

EXERCISES

4.1 In the *couples dataset*, 189 couples have no children, and 227, or 54.57%, of couples have children. For all 416 couples, the mean *couple-conflict* score is 1.79, with a standard deviation of .601384. Couples without children have a mean *conflict* score of 1.5658 with a standard deviation of .480. Couples with children have a mean *conflict* score of 1.9759, with a standard deviation of .629. If a dummy variable, PRESCHDN, is created with those having children the interest category and those without children the contrast group, then:

(a) Give the sample equation for the SLR of *couple conflict* on PRESCHDN.
(b) Give r and r^2 for the SLR of *couple conflict* on PRESCHDN.
(c) Test whether there is a significant relationship between *couple conflict* and having children using the two-sample t test, the test for the significance of r, and the test for the dummy coefficient (i.e., $t = d/\hat{\sigma}_d$). In so doing, you should find that all three t tests are equivalent. (*Hint*: See Exercise 4.25 for helpful ideas on this problem.)

4.2 For the 416 couples in the *couples dataset*, a regression of *couple conflict* on a dummy variable OWNKID, representing all children in the household being the natural children of both partners, and a dummy variable STEPKID, representing at least one child being a stepchild ("no children" is the omitted group) produced the following equation:

$$\hat{y} = 1.5723 + .4023 \text{ OWNKID} + .3866 \text{ STEPKID}.$$

The variance–covariance matrix of parameter estimates is:

	OWNKID	STEPKID
OWNKID	.00365	
STEPKID	.00171	.00711

RSS is 16.3622 and *SSE* is 133.7280.

(a) Give the mean *couple conflict* for the three groups of couples.
(b) Test whether there is a significant relationship between *child type* and *couple conflict*.

(c) Which *child type* categories are significantly different from each other in average couple conflict?

4.3 For the *students dataset*, CLASSF is recoded into the categories "sophomore," "junior," "senior," and "postgraduate/graduate" (there are no freshmen). The means (*n*'s) on the first exam for these groups are "sophomore" 81.595 (21); "junior" 74.681 (70); "senior" 74.884 (88); "postgraduate/graduate" 84.064 (35).

(a) Show how the categories of CLASSF should be effect coded, naming the effect variables SOPH, JUN, and POST (with "senior" as the reference group).

(b) Using only hand calculations, give the sample regression equation for regressing EXAM1 on SOPH, JUN, and POST.

4.4 Suppose that a continuous variable Y is regressed on two categorical variables: Z_1 with categories A and B, and Z_2 with categories C, D, and E. If Z_1 is dummied with B as the interest category and Z_2 is dummied with C and D as the interest categories, the sample regression equation is

$$\hat{y} = 4 + 6B - 2C - D - 4BC - 2BD.$$

Fill in the following table with each of the cell means, based on this equation:

	A	B
C		
D		
E		

4.5 Using the *GSS98 dataset*, perform a two-way ANOVA for the effect of *gender* and *marital status* on *attitude toward cohabitation* (COHABTN) using dummy-variable regression. In particular:

(a) Test whether *marital status* affects *cohabitation attitude*, controlling for *gender*.

(b) Test whether there is a significant interaction effect between *gender* and *marital status*.

(c) Use the interaction regression model to display the 10 sample cell means for the cross-classification of *gender* with *marital status*.

4.6 In the two-way ANOVA model without interaction in Exercise 4.5, use Bonferroni–Holm to test all 10 mean contrasts in *cohabitation attitude* by marital-status categories.

4.7 In DeMaris (2002b) I present several regression analyses of the frequency of sexual activity in the past month for 2997 couples in the NSFH. A regression of *sexual frequency* on a dummy variable reflecting *cohabiting status* (1 = "cohabiting unmarried," 0 = "married") produces the following equation: $\hat{y} = 8.171 + 3.314$ *cohabiting*. A second analysis adds the continuous variables *female age* in years (mean = 33.43) and *relationship duration* in months (mean = 92.63) to the model. The second equation is

$$\hat{y} = 13.969 + 2.322 \text{ cohabiting} - .144 \text{ female age}$$
$$- .01 \text{ relationship duration.}$$

(a) Give the unadjusted mean sexual frequency for cohabiting vs. married couples.
(b) Give the adjusted mean sexual frequency for cohabiting vs. married couples, after adjusting for *female age* and *relationship duration*. Comment on the differences between the unadjusted and adjusted means.

4.8 In the analyses in Exercise 4.7, *MSE* for the first equation is 43.183, *MSE* for the second equation is 40.631, and the standard error of the cohabiting effect is .201 in each equation. Test whether *female age* and *relationship duration* account for a significant part of the cohabitation effect on sexual frequency, using the Clogg et al. (1995) test discussed in Chapter 3.

4.9 For the *couples dataset*, the variable MPERCENT is the *percent of total weekly housework hours contributed by the male partner*, calculated as

$$100\left(\frac{MMHOURS}{MMHOURS + FFHOURS}\right).$$

This was regressed on MPARTIME and MUNEMP, dummies for male partners who work "part time" or are "unemployed," respectively (working "full time" is the reference category); FPARTIME and FUNEMP, dummies for female partners who work "part time" or are "unemployed," respectively (working "full time" is the reference category); and MINCOME and FINCOME (male and female *income*, respectively). The sample equation is

$$\hat{y} = 23.766 + 2.104 \text{ MPARTIME} - .177 \text{ MUNEMP} - 6.043 \text{ FPARTIME}$$
$$- 4.738 \text{ FUNEMP} - .020 \text{ MINCOME} + .286 \text{ FINCOME.}$$

(a) Interpret the intercept and each of the dummy coefficients.
(b) Give the adjusted mean percentages of housework done by male partners in each of the nine cross-classifications of male employment with female employment. (*Note*: Mean MINCOME = 23.5, mean FINCOME = 9.04.)

4.10 For the *couples dataset*, the regression of *couple happiness* (the mean of HUSHAP and WIFHAP) on PRESCHDN (a dummy representing the *presence* *of children* in the household), CONFLICT (*couple conflict*), and their interaction (PRESCONF) produces the following equation:

$$\hat{y} = 6.997 + .185 \text{ PRESCHDN} - .479 \text{ CONFLICT} - .157 \text{ PRESCONF.}$$

Interpret the interaction effect with, alternately, *presence of children* and *conflict* as the focus variable. Note that the *couple conflict* scale ranges from 1 ("no conflict") to 6 ("maximum conflict").

4.11 For the 230 main-respondent females in the *couples dataset*, the following three regression equations were produced, where Y = the Center for Epidemiological Studies Depression Scale [the 12-item version (CESD)], and the regressors are some combination of OWNKID, STEPKID, and CONFLICT (significant coefficients are shown in boldface type):

(1) $\hat{y} = 13.2178 + .6332 \text{ OWNKID} + 3.0679 \text{ STEPKID}; R^2 = .0043.$

(2) $\hat{y} = -\mathbf{6.9947} + \mathbf{11.8270} \text{ CONFLICT}; R^2 = .2062.$

(3) $\hat{y} = -\mathbf{7.0448} - \mathbf{5.0846} \text{ OWNKID} - .4777 \text{ STEPKID}$
$\qquad + \mathbf{13.0701} \text{ CONFLICT}; R^2 = .2279.$

(a) Test whether *child type* is a significant predictor of *women's depressive symptomatology* after controlling for *couple conflict*.
(b) Interpret the coefficient for OWNKID in the last equation.
(c) Explain why the effects of OWNKID and STEPKID change signs from equation 1 to equation 3.

4.12 Refer to Exercise 4.3. A regression of EXAM1 on SCORE, COLGPA, and PREVMATH for the 214 students with valid grades produces an R^2 of .4312. When SOPH, JUN, and POST (coded as effect variables) are added to the model, the R^2 goes up slightly, to .4383.
(a) Test whether *student classification* is a significant predictor of exam scores once SCORE, COLGPA, and PREVMATH have been accounted for.
(b) If the sample equation for the more complete model is

$$\hat{y} = -55.122 + 1.728 \text{ SOPH} - 1.429 \text{ JUN} + 1.603 \text{ POST}$$
$$+ 2.2 \text{ SCORE} + 13.139 \text{ COLGPA} + 1.805 \text{ PREVMATH}$$

and the means for SCORE, COLGPA, and PREVMATH are, respectively, 40.925, 3.091, and 1.257, give the estimated average EXAM1 scores for sophomores, juniors, seniors, and postgraduate/graduate students with average values of SCORE, COLGPA, and PREVMATH.

4.13 Refer to Exercise 4.12. A regression of EXAM1 on SCORE, COLGPA, and PREVMATH and *dummy* variables representing *student classification* and named SOPHM ("sophomore"), JUNR ("junior"), and POSTG ("postgrad/grad"), with seniors as the contrast group, produced the following sample equation:

$$\hat{y} = -57.025 + 3.63 \text{ SOPHM} + .474 \text{ JUNR} + 3.506 \text{ POSTG}$$
$$+ 2.2 \text{ SCORE} + 13.139 \text{ COLGPA} + 1.805 \text{ PREVMATH}.$$

Give the adjusted mean EXAM1 scores for sophomores, juniors, seniors, and postgrad/grads. Compare your answer to part (b) of Exercise 4.12.

4.14 One analysis of faculty salary for the 725 faculty members at BGSU looks at the interaction of *gender* with *college* ("firelands" is the group omitted here) in their effects on *faculty salary*. The estimated equation, including several controls (contained in the vector **CONTROLS**), is

$$\hat{y} = 36959 + 702.006 \text{ FEMALE} + 1201.846 \text{ ARTSCI} + 7738.588 \text{ BUSINESS}$$
$$+ 499.644 \text{ EDUCATN} + 2900.191 \text{ OTHER}$$
$$- 1301.887 \text{ FEMALE} * \text{ARTSCI} - 4817.316 \text{ FEMALE} * \text{BUSINESS}$$
$$- 1388.864 \text{ FEMALE} * \text{EDUCATN} - 1058.814 \text{ FEMALE} * \text{OTHER}$$
$$+ \mathbf{g'} \text{ CONTROLS}.$$

There are a total of 20 regressors in this model, and the R^2 is .8382. The model without the interaction terms has an R^2 of .8363.

(a) Test whether the interaction is significant.

(b) Interpret the intercept and all of the main effects of *gender* and *college* in the model.

(c) Interpret the interaction effect, regardless of its significance level, with *gender* and *college* alternating as the focus variables.

4.15 Using the *faculty salary dataset*, estimate the regression of AYSALARY on *rank* (captured by the dummies R1 for "full professor," R2 for "associate professor," and R4 for "instructor/lecturer," with "assistant professors" as the reference group), centered versions of PRIOREX, YRBG, YRRANK, and SALFAC, plus the dummies TERMDEG, GRAD, ADMIN, and FIRELAND.

(a) Interpret the *rank* effects.

(b) Give the adjusted mean AYSALARY for those in each rank in the baseline group (i.e., no terminal degree, not on graduate faculty, not in an administrative position, on the main campus).

4.16 Using the *GSS98 dataset*, conduct a one-way ANOVA via dummy-variable regression, where $Y =$ ABORTION and $X = $ *religious affiliation* ("Protestant," "Catholic," "Jewish," "None," "Other"). Use Bonferroni–Holm to test all 10 mean contrasts. Then add the covariates EDUCAT and CONSERV to the model and once again use Bonferroni–Holm to test the 10 religious-group contrasts on abortion attitude. Missing imputation: Follow the instructions for Exercise 3.14.

4.17 For the 416 couples in the *couples dataset*, an analysis was conducted to replicate DeMaris (1997). That article examined the impact of *coital frequency* on *women's depressive symptomatology* and how this might be moderated by the male's *physical aggression* against the female partner. The analysis regressed the CESD on *coital frequency* (COITFREQ), dummies for male and female *violence* (MALEHIT and FEMAHIT, respectively), the interaction between COITFREQ and each *violence* dummy (MHITCOIT and FHITCOIT, respectively), and a vector of control variables (**CONTROLS**). The model has 15 parameters. The results for all 416 couples produced an $SSE = 81596.74241$. The model was then estimated separately for the 186 couples in which the main respondent was male, and the 230 couples in which the main respondent was female. For the male sample, $SSE = 39246.83241$, $MSE = 229.51364$, $df_E = 171$; for the female sample, $SSE = 39607.02799$, $MSE = 184.21873$, $df_E = 215$.

The females' equation was

$$\hat{y} = .3344 + \mathbf{g'}\ \mathbf{CONTROLS} - .0183\ \text{COITFREQ} - 28.4758\ \text{MALEHIT}$$
$$- 3.0606\ \text{FEMAHIT} + 1.2076\ \text{MHITCOIT} - .2841\ \text{FHITCOIT}.$$

(**a**) Test whether the error variances in the equations for males and females are equal.

(**b**) Regardless of your answer to part (a), test whether the same model (including the intercept) holds in the population of partnered males versus partnered females.

(**c**) In the female equation, interpret the effect of COITFREQ as moderated by *partner violence*.

4.18 Using the *GSS98 dataset*, estimate the regression of ABORTION on MALE, RELOSITY, CONSERV, and EDUCAT for the 1868 respondents with valid ABORTION scores. Then perform the Chow test with both constrained and unconstrained intercepts to test model invariance across categories of RACE ("White," "Black," "Other"). First, use Bartlett's test to test the homogeneity of error variance across the racial groups; but do the Chow test regardless of this outcome. Missing imputation: Follow the instructions for Exercise 3.14.

4.19 Retest the model's equivalence across racial groups in Exercise 4.18 using the cross-product method. Again, use both constrained- and unconstrained-intercepts approaches. You should demonstrate that this produces the same results as in Exercise 4.18. Missing imputation: Follow the instructions for Exercise 3.14.

The following information is to be used for Exercises 4.20 and 4.21.

- For the 725 faculty members in the *faculty salary dataset*, the ANCOVA model in Exercise 4.15, when estimated using the entire sample, has: $RSS = 112828737391$, $SSE = 28911032834$, $MSE = 40548433.147$, and $df_E = 713$.
- If a dummy for *gender* (FEMALE) is added, the results are $RSS = 112913052400$, $SSE = 28826717824$, $MSE = 40486963.236$, and $df_E = 712$.
- If the cross-product of FEMALE with all 11 other covariates is then added to the model, the results are $RSS = 113515775771$, $SSE = 28223994454$, $MSE = 40262474.256$, and $df_E = 701$.
- If the model in Exercise 4.15 is estimated for the 511 male faculty members, the results are $RSS = 72214388564$, $SSE = 23630713768$, $MSE = 47356139.816$, and $df_E = 499$.
- If the model in Exercise 4.15 is estimated for the 214 female faculty members, the results are $RSS = 18209728848$, $SSE = 4593280685.4$, $MSE = 22739013.294$, and $df_E = 202$.

4.20 Test the model equivalence of the model in Exercise 4.15 for males versus females using the Chow test. Perform the test with both constrained- and unconstrained-intercept versions. First, test homogeneity of error variance in each group; then do the Chow test regardless of this outcome.

4.21 Redo Exercise 4.20 using the cross-product approach.

4.22 For the 214 students with valid EXAM1 scores in the *students dataset*, EXAM1 was regressed on COLGPA, STATMOOD, and SCORE. The results were $RSS = 26652.37864$, $SSE = 34995.71120$, $MSE = 166.64624$, and $df_E = 210$.

- For the 69 "sociology" majors, the same model produced $RSS = 12969.83816$, $SSE = 7162.26656$, $MSE = 110.18872$, and $df_E = 65$.
- For the 43 "other social science" majors, the same model produced $RSS = 3070.59860$, $SSE = 5257.61617$, $MSE = 134.81067$, and $df_E = 39$.
- For the 102 "other fields" majors, the same model produced $RSS = 11694.08078$, $SSE = 21442.49456$, $MSE = 218.80096$, and $df_E = 98$.

Use Barlett's test to test for homogeneity of error variance across the three groups of majors. Regardless of the outcome, conduct a Chow test for model equivalence across the three groups of majors, constraining the intercept to be the same across groups.

4.23 Suppose that a model for Y is estimated as a function of a categorical variable, Z, with categories A, B, C, and D (the reference category), and a continuous variable X (range: -10 to $+10$). Suppose that the sample equation is

$$\hat{y} = 2 + .5A + .8B - .2C + 1.5X + .05AX - .15BX - .25CX.$$

Determine whether the interaction effect in X and in Z is ordinal or disordinal, with, alternately, Z and then X as the focus variables.

4.24 Suppose that a model for Y is estimated as a function of a categorical variable, Z, with categories A, B, C, and D (the reference category), and a continuous variable X (range: -10 to $+10$). Suppose that the sample equation is

$$\hat{y} = 3.5 - A - 4B - 2C - .5X + .5AX + .25BX + .75CX.$$

Determine whether the interaction effect in X and in Z is ordinal or disordinal, with, alternately, Z and then X as the focus variables.

4.25 Prove that the t test for d (the dummy coefficient) in the SLR of Y on a dummy variable X is equivalent to the two-sample t test. Note that for two independently sampled groups, denoted group 0 and group 1, with sample sizes n_0 and n_1, respectively, and means $\bar{y}_0$ and $\bar{y}_1$, respectively, the two-sample t test is

$$t_{(n_0+n_1-2)} = \frac{\bar{y}_1 - \bar{y}_0}{\hat{\sigma}\sqrt{(1/n_0)+(1/n_1)}},$$

where $\hat{\sigma}^2$ is the pooled estimate of the common population variance of Y, with the formula

$$\hat{\sigma}^2 = \frac{(n_0-1)s_0^2 + (n_1-1)s_1^2}{n_0 + n_1 - 2}.$$

(a) First, prove that $\hat{\sigma}^2$ in the two-sample t test equals MSE in the dummy-variable SLR. [*Hint*: Start with $MSE = \sum(y - \hat{y})^2/(n - 2)$; then use the fact that $\hat{y} = \bar{y}_0$ for $X = 0$, and $\hat{y} = \bar{y}_1$ for $X = 1$, plus the fact that $n_0 + n_1 = n$.]

(b) Then prove that the t tests are equivalent. [*Hint*: Start with $\hat{\sigma}^2_{\bar{y}_1 - \bar{y}_0}$, which equals $\hat{\sigma}^2((1/n_0) + (1/n_1))$. Then use the fact that $(1/n_0) + (1/n_1) = (n_1 + n_0)/n_0 n_1$; $n_0 = n(1 - \hat{\pi})$, $n_1 = n\hat{\pi}$; and $\sum(x - \bar{x})^2 = n\hat{\pi}(1 - \hat{\pi})$.]

CHAPTER 5

Modeling Nonlinearity

CHAPTER OVERVIEW

Until now the models under discussion have been strictly linear. By this is meant that (1) Y is held to be a linear, additive function of the explanatory variables, or, if interaction is present, it is of a linear nature; and (2) the model is linear in the parameters (i.e., it is a weighted sum of regressors times parameters). (More formal definitions of linearity in terms of the explanatory variables or the model as a whole are given below.) In this chapter I introduce models in which Y is held to be a nonlinear function of either the explanatory variables or the model parameters. I begin by defining these concepts and giving examples of models illustrating various nonlinear relationships between Y and the structural component of the model. I then illustrate some of the more common nonlinear functions of explanatory variables that may be helpful in modeling the response variable. This discussion segues into a lengthy illustration of the use and interpretation of the quadratic model, perhaps the most commonly employed nonlinear function in social data analysis. Here I also define and discuss nonlinear interaction effects. Finally, I introduce the reader to nonlinear regression, the technique employed when one's model is nonlinear in the parameters and cannot be converted to a convenient linear form. Readers unfamiliar with derivatives from calculus are advised to peruse Section IV of Appendix A before tackling this chapter. Because derivatives and partial derivatives are central to an understanding of nonlinearity, they figure prominently in the subject matter that follows.

NONLINEARITY DEFINED

I begin by making a distinction between nonlinearity in the *functional form of the relationship* between Y and X, versus nonlinearity of the *model* for Y. Nonlinearity

Regression with Social Data: Modeling Continuous and Limited Response Variables,
By Alfred DeMaris
ISBN 0-471-22337-9   Copyright © 2004 John Wiley & Sons, Inc.

in the functional form of the relationship between Y and X is determined by the first partial derivative of the model with respect to X. In any model in which Y is a function of X, *if the first partial derivative of Y with respect to X is a function of X, the model is nonlinear in X; otherwise, the model is linear in X.* For example, suppose that model A is

$$Y = \alpha + \beta X + \gamma Z + \varepsilon.$$

Since the first partial derivative of Y with respect to X, or $\partial Y/\partial X$, is β, which is not a function of X, the model is linear in X. However, in model B,

$$Y = \alpha + \beta X + \delta X^2 + \gamma Z + \varepsilon,$$

the first partial derivative is $\partial Y/\partial X = \beta + 2\delta X$. Since this is a function of X, the model is nonlinear in X. In particular, this model, called a *quadratic* model, or *curvilinear* model, in X, describes a parabolic curve (or part of a parabolic curve) relating Y to X at any given value of Z. We can also define nonlinearity in the functional form relating Y to X using the second partial derivative. *If the second partial derivative of Y with respect to X is not zero, the model for Y is nonlinear in X; if it is zero, the model is linear in X.* In model A, $\partial^2 Y/\partial X^2 = \partial(\beta)/\partial X = 0$ showing again that the model is linear in X. On the other hand, in model B, $\partial^2 Y/\partial X^2 = 2\delta$, which is nonzero provided that δ is not zero. This once again reveals that model B is nonlinear in X. Intuitively, the first derivative measures the change in Y with change in X at the point x. As long as this is not a function of X, that change is constant over levels of X. This condition means that the relationship between Y and X can be represented by a straight line, which is characterized by a constant slope (see Section I.P of Appendix A). If, on the other hand, the first derivative is a function of X, this means that the *rate* at which Y changes with change in X *is itself changing* with levels of X, describing some type of curve instead of a straight line. Moreover, if the first partial derivative of Y with respect to X is not a function of X, the second partial derivative of Y with respect to X is necessarily zero. Note that conventional interaction models such as model C,

$$Y = \alpha + \beta X + \delta Z + \gamma XZ + \varepsilon,$$

are not nonlinear in X, since $\partial Y/\partial X = \beta + \gamma Z$ is not a function of X.

The linearity or nonlinearity of the *model* as a whole is determined by the first partial derivative of Y with respect to the model's *parameters*. Denote each of the P parameters of any model by θ_p, for $p = 1, 2, \ldots, P$ (e.g., in linear regression we typically have $P = K + 1$ parameters). *If the first partial derivative of Y with respect to at least one of the θ_p is a function of any of the model parameters, the model for Y is nonlinear* (Ratkowsky, 1990). Model B, which is nonlinear in X, is nevertheless not a nonlinear model, since the first partial derivatives of Y with respect to, alternately, α, β, δ, and γ are 1, X, X^2, and Z. Notice that none of these terms involves any of the model parameters. On the other hand, consider model D: $Y = \alpha + X^\beta + \varepsilon$. Now, finding the first partial derivative of Y with respect to β involves finding the first derivative of X^β with respect to β. To do this we write X^β as $\exp(\log X^\beta)$. Then

$$\frac{d}{d\beta}[\exp(\log X^\beta)] = \frac{d}{d\beta}[\exp(\beta \log X)] = \exp(\beta \log X)\log X = X^\beta \log X.$$

Since this expression involves β, model D is nonlinear for Y. Notice that it is also nonlinear in X, since $\partial Y/\partial X = \beta X^{\beta-1}$ is also a function of X. (Most models that are nonlinear for Y are also nonlinear in X.) Some other examples of nonlinear models are the exponential model with additive error term (Neter et al., 1985),

$$Y = \gamma_0 e^{\gamma_1 X} + \varepsilon,$$

the exponential model with multiplicative error term (Fox, 1997),

$$Y = \gamma_0 e^{\gamma_1 X} e^\varepsilon,$$

the logistic population-growth model (Fox, 1997),

$$Y = \frac{\beta_1}{1 + \exp(\beta_2 + \beta_3 X)} + \varepsilon,$$

and the gravity model of migration (Fox, 1997),

$$Y_{ij} = \alpha \frac{P_i^\beta P_j^\gamma}{D_{ij}^\delta} \varepsilon_{ij},$$

where Y_{ij} is the number of migrants moving from city i to city j, D_{ij} is the distance between cities i and j, and P_i and P_j are the respective population sizes of cities i and j.

Employing the terminology of Fox (1997) and Neter et al. (1985), I make a further distinction among three types of models. *Linear models* are those that are linear in the parameters, such as models A through C above. *Intrinsically linear models* (Neter et al., 1985) are those that are linear in the parameters after applying some kind of transformation to the response and/or explanatory variables. An example is the exponential model with multiplicative errors. If we transform Y by taking its natural logarithm, the model becomes $\log Y = \log \gamma_0 + \gamma_1 X + \varepsilon$. Defining α as $\log \gamma_0$, the model is $\log Y = \alpha + \gamma_1 X + \varepsilon$, which is now linear in the parameters, and therefore a linear model. Similarly, logging Y_{ij} in the gravity model of migration produces

$$\log Y_{ij} = \alpha' + \beta \log P_i + \gamma \log P_j + \delta(-\log D_{ij}) + \log \varepsilon_{ij},$$

where $\alpha' = \log \alpha$. This model is, again, linear in the parameters. Provided that the errors in each model (ε and $\log \varepsilon_{ij}$, respectively) are independent and normally distributed with zero mean and constant variance, these models can be estimated using ordinary least squares. *Essentially nonlinear models* (Fox, 1997) are nonlinear models that cannot be made linear by any transformations of the response or explanatory variables. The logistic population-growth model, the exponential model with additive errors, and model D are all essentially nonlinear models. Estimating these models requires the use of nonlinear least squares, which also depends on the assumption that model errors are independent and normal, with zero mean and constant variance (Fox, 1997; Greene, 2003; Neter et al., 1985) This technique is discussed in the final section of this chapter.

COMMON NONLINEAR FUNCTIONS OF X

Figures 5.1 to 5.4 present several functions of X that are useful in modeling a nonlinear relationship with Y. These are all curves with one "bend" for X in the range $[0,\infty)$, and represent the most frequently encountered patterns of nonlinearity. I identify the curves by linking them to the shapes of the corresponding segments of a circle that has been quartered. Figure 5.5 illustrates this idea. That is, if we divide a circle in half by running a line through its middle from left to right, and then divide it in quarters by running a line through its middle from top to bottom, the circle is separated into four segments. Segment I is the upper right-hand quarter of the circle, segment II is the lower right-hand quarter, segment III is the lower left-hand quarter,

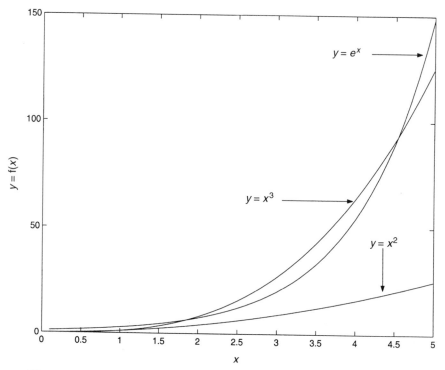

Figure 5.1 Nonlinear relationships exemplified by the functions $y = x^2$, $y = x^3$, and $y = e^x$.

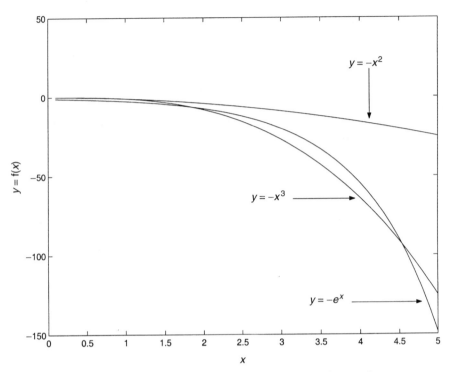

Figure 5.2 Nonlinear relationships exemplified by the functions $y = -x^2$, $y = -x^3$, and $y = -e^x$.

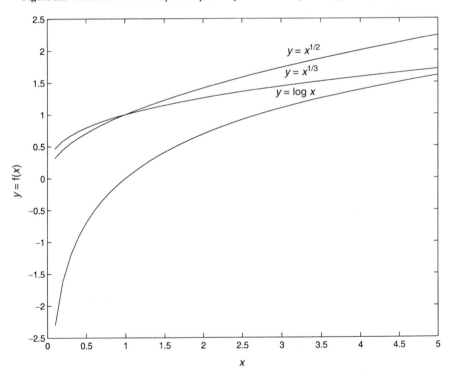

Figure 5.3 Nonlinear relationships exemplified by the functions $y = x^{1/2}$, $y = x^{1/3}$, and $y = \log x$.

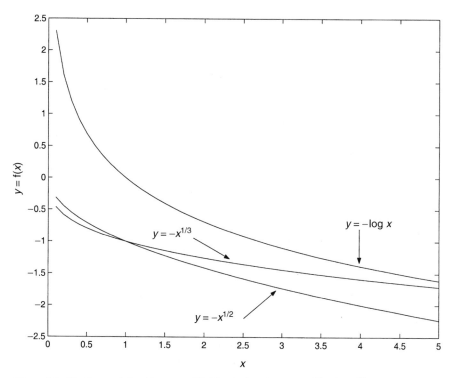

Figure 5.4 Nonlinear relationships exemplified by the functions $y = -x^{1/2}$, $y = -x^{1/3}$, and $y = -\log x$.

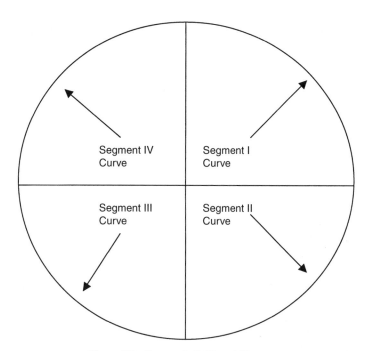

Segment IV
Curve

Segment I
Curve

Segment III
Curve

Segment II
Curve

Figure 5.5 Segment I, II, III, and IV curves.

and segment IV is the upper left-hand quarter. The curves in Figures 5.1 to 5.4 resemble these circle segments.

Figure 5.1 presents three functions of X representing segment II curves. These are all curves with positive slopes, or first derivatives, that become more positive with increasing X. The curve with the most rapidly increasing slope is $y = e^x$, whose derivative is e^x. A slightly less dramatic increase in slope is seen in the curve $y = x^3$. Its derivative is $3x^2$. Finally, a curve with a relatively gentle increase in slope with increasing X is $y = x^2$, whose derivative is $2x$. Figure 5.2 shows some segment I curves. These can be obtained by simply "reflecting," or changing the sign of, the functions in the second segment. Notice that all of these curves have negative slopes that become more negative with increasing X. Figure 5.3 illustrates three segment IV curves. All of these have the characteristic of having positive slopes that become less positive with increasing X. This is also easily seen by examining the slopes themselves. The most dramatic change, at least in the range of X shown here (.1 to 5) is exhibited by $y = \log x$. Its slope is $1/x$, which ranges from $1/.1 = 10$ to $1/5 = .2$ in the figure. The curve represented by $y = x^{1/3}$ appears to "flatten out" most quickly. Its derivative is $1/3x^{2/3}$. Its slope ranges from 1.55 when X is .1 to .11 when X is 5. The curve $y = x^{1/2}$ shows a similar pattern but with a more gentle reduction of the slope with increasing X. Finally, Figure 5.4 presents segment III curves. These curves all have negative slopes that become less negative with increasing X. Once again, they can be produced by reflecting the three segment IV curves.

Quadratic Functions of X

Although Figures 5.1 to 5.4 present a variety of functions of X that can be used to fit curvilinear relationships with Y, an especially useful function is the quadratic function, $y = x^2$. Actually, the correct functional form of a quadratic model for Y is

$$Y = \beta_0 + \beta_1 X + \beta_2 X^2 + \varepsilon. \tag{5.1}$$

This model includes the linear component of X, $\beta_1 X$, along with the curvilinear component, $\beta_2 X^2$. As Aiken and West (1991) explain, models with higher-order functions of X should always contain all lower-order components of the higher-order terms. As these components are often highly correlated with the higher-order terms, omitting them introduces a type of bias into the equation. That is, the regression of Y on X^2 alone might produce a significant coefficient for X^2 due to the correlation between X and X^2 along with a significant effect of X. In this case, the supposed quadratic effect would be "driven" by a significant linear trend that is correlated with X^2. A true test of whether there is a curvilinear component to the X–Y relationship is achieved only when we control for X simultaneously. In Aiken and West's words: ". . . higher order terms actually represent the effects they are intended to represent *if and only if* all lower order terms are partialled from them . . . " (1991, p. 110; emphasis in original).

For equation (5.1), $d(Y)/dx = \beta_1 + 2\beta_2 X$. In this expression, β_1 encapsulates the linear component of the curve while β_2 captures the departure from linearity. In fact, if $\beta_2 = 0$, there *is* no departure from linearity, and we are left, once again, with a linear relationship between Y and X. The signs of β_1 and β_2 typically reveal the nature of the curvilinear relationship between Y and X. For example, let's assume that X is always ≥ 0. If the X–Y relationship is characterized by the shape of a segment II curve, both β_1 and β_2 should be positive. This indicates that the slope starts out positive (when $X = 0$) and becomes increasingly positive with increasing X. A segment I curve would be indicated by both β_1 and β_2 being negative. That is, the slope starts out negative and becomes increasingly so with increase in X. For a segment IV curve, we would expect β_1 to be positive and β_2 to be negative. That is, the slope is initially positive but becomes less so as X increases. Hence the $2\beta_2X$ component of the slope adds an increasingly negative number to β_1 to bring the overall size of the slope down ever further with increasing X. Finally, the segment III curve would be indicated by the opposite pattern: β_1 should be negative and β_2 should be positive. The slope starts out negative, but we add an increasingly positive number ($2\beta_2X$) to β_1, which has the effect of making the slope less and less negative with increasing X.

Figures 5.1 to 5.4 do not represent the only possible curvilinear patterns that the X–Y relationship might exhibit. In particular, they represent relationships between Y and X that are *monotonic* in nature. That is, Y is always either increasing (in segments II and IV) or decreasing (in segments I and III) with X in each case. Fitting a model that is linear in X in these situations does not lead one too far astray, since each of these curves could be—at least roughly—approximated by a straight line. The correlation between Y and X in all cases should be significantly nonzero. Not so with the curves in Figure 5.6. These are U-shaped (bottom curve) and inverted U-shaped (top curve) curvilinear relationships that are not monotonic. In each case, Y is increasing with X over part of X's range, and decreasing with X over the rest of X's range. Fitting a model that is linear in X in these cases is likely to be very misleading, producing a correlation close to zero. However, a quadratic model nicely captures this type of curve. As shown in the figure, each curve was, in fact generated by a quadratic model. The U-shaped curve has a negative β_1 and a positive β_2, while the inverted U-shaped curve shows the opposite pattern. The only way the analyst can tell whether the data evince the pattern in Figure 5.6, as opposed to the patterns in Figures 5.3 and 5.4 (segment IV and III curves), is either to graph the fitted values from the model against X or to plug some sample values of X into the expression $\beta_1 + 2\beta_2X$.

A quadratic model is just a special case of a polynomial model in X. The Jth-order polynomial model in X is

$$E(Y) = \beta_0 + \beta_1X + \beta_2X^2 + \beta_3X^3 + \cdots + \beta_JX^J.$$

A Jth-order polynomial will fit any curve with $J - 1$ bends. As we have seen, the second-order polynomial, the quadratic equation, will fit any curve with one bend. For a curve with two bends, we could try the third-order polynomial, or cubic equation:

$$E(Y) = \beta_0 + \beta_1X + \beta_2X^2 + \beta_3X^3. \tag{5.2}$$

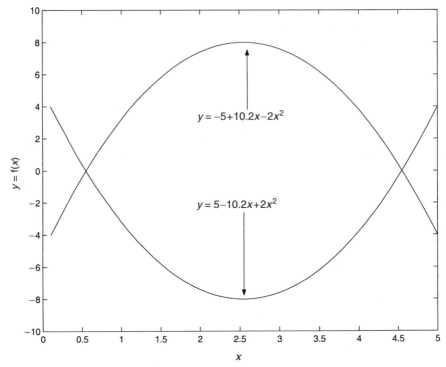

Figure 5.6 U-shaped and inverted U-shaped functions of the form $y = a + bx + cx^2$.

Polynomials higher than second order are rarely used, however. Adding increasingly higher powers of X tends to produce severe collinearity problems, making estimation quite unreliable (more about multicollinearity in Chapter 6). Moreover, higher-order models become increasingly cumbersome to interpret. For example, for equation (5.2), $d(Y)/dx = \beta_1 + 2\beta_2 X + 3\beta_3 X^2$. This suggests that the slope of the X–Y relationship is itself a quadratic function of X. Not only is this somewhat difficult to grasp, but it is hard to imagine many theories precise enough to allow the forecasting of such a trend in advance. Most of the time, the quadratic function of X will be sufficient to capture the types of nonlinear relationships found in the social sciences. (An exception is discussed below, however.)

Applications of the Quadratic Model

To illustrate the fitting and interpretation of models that are nonlinear in X, I consider the quadratic model in some detail. The *GSS98 dataset* contains information about a respondent's *sexual frequency in the last year* (coded 0 = "not at all" to 6 = "more than 3 times a week"), his or her *age* (in years), and his or her *health* (coded 1 = "poor" to 4 = "excellent"). [Note that although *age* is approximately continuous, *sexual frequency* and *health* are not. Nevertheless, they will be treated as such, since Bentler and Chou (1988) suggest that as long as a quantitative variable has at least four categories, it is safe in analyses to treat it as continuous.] In earlier work with a different sample (DeMaris, 2002b) I have found that there is a nonlinear relationship between *age* and *sexual frequency* which can be captured by a quadratic model. Intuitively, one would expect that sexual activity would decline with advancing age but that this decline would not be linear. Age should have a relatively modest negative impact on *sexual activity* until after age 50 or so, at which time a decline in testosterone levels, particularly in men, should accelerate the decline in sexual activity with advancing age. That is, I expect the relationship between *age* and *sexual frequency* to exhibit a segment I curvilinear pattern such as those in Figure 5.2.

One means of examining relationships for nonlinearity is via the scatterplot. (In a multivariate context, one might prefer to examine partial plots, in which other variables have been controlled.) Let's begin by examining a scatterplot of *sexual frequency* with respondent's *age*, which is depicted in Figure 5.7. A problem that arises, particularly with discrete response variables having relatively few categories, is that the scatterplot is not very informative. As is evident, it is difficult to discern any pattern in the relationship between the two variables. An alternative strategy is to partition a continuous *X* into intervals and then to plot the mean of *Y* for cases in each interval against the interval number. In the current case, I partitioned age according to deciles of its distribution, with each decile containing approximately 10% of the 2320 respondents. Within each decile, I computed the mean of respondents' sexual

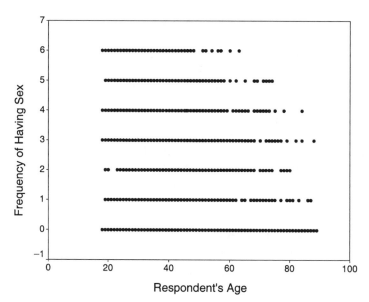

Figure 5.7 Scatterplot of frequency of having sex with respondent age for 2320 respondents in the 1998 GSS.

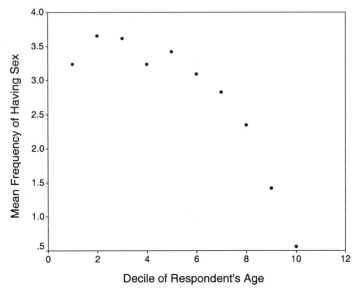

Figure 5.8   Scatterplot of mean frequency of having sex with decile of respondent's age.

frequencies. A plot of mean sexual frequency with decile of respondent's age is shown in Figure 5.8. As expected, the plot resembles the types of curves found in Figure 5.2: There is a nonlinear decline in mean sexual frequency with advancing age, with the decline getting ever steeper as age increases. The decline is particularly swift after the sixth decile, which represents ages 42–47. The curvilinear pattern also poses some irregularities compared to the curves in Figure 5.2. For one thing, there appears to actually be an increase in sexual frequency with age at the lowest ages, followed by a decrease from the third decile (ages 30–34) on. However, this pattern is easily handled with a quadratic model. Additionally, the fourth decile (ages 34–38) appears to represent a departure from the general trend of increase followed by accelerating decline. Whether this is worth taking into account in the model depends on whether there is strong enough theoretical underpinning to support it. In the current case, I treat it as a sample anomaly and assume that a quadratic function is most appropriate.

Testing Departures from Linearity

As a first step in modeling sexual frequency as a function of age, I employ a model that will fit any pattern in the relationship between these two variables. I refer to it as the *unconstrained model*, since it does not constrain mean sexual frequency to follow any particular pattern with respect to age. The variable *age decile*, representing deciles of respondent age, is coded 1 for the first decile to 10 for the tenth decile. If we dummy up this variable, omitting the first decile (i.e., the youngest age group, aged 18–25), the unconstrained model simply regresses *sexual frequency* on the nine resulting dummies. The model is

$$E(Y) = \delta_1 + \delta_2 D_2 + \delta_3 D_3 + \cdots + \delta_{10} D_{10}, \tag{5.3}$$

where $D_2, D_3, \ldots, D_{10}$ denote the nine dummies that represent the second through the tenth deciles of age. The results are shown as model 1 in Table 5.1. The pattern of the dummy coefficients confirms the trend shown in Figure 5.7. There is an initial increase in sexual frequency from the first to the second age decile, since those in the second decile have mean sexual frequency that is .416 higher than those in the first decile. After this, there is a steady decline in sexual frequency, which becomes more and more pronounced at later ages. Again it is clear that the fourth decile is not quite in sync with this general trend.

At any rate, is the departure from linearity in the *age–sexual frequency* relationship, as revealed in model 1, significant? That is, might a linear model nevertheless fit the trend in the data adequately? We can test this with the *test for linearity*, a nested F test based on comparison of model 1 with a model utilizing a linear effect for *age decile*. This test takes advantage of the fact that a linear version of model (5.3) of the form

$$E(Y) = \alpha + \beta \; age \; decile \tag{5.4}$$

is nested inside model (5.3). Why? Recall from the definition of nesting in Chapter 3 that model (5.4) is nested inside model (5.3) if its parameters can be generated by

Table 5.1 Dummy Variable, Linear, and Quadratic Models for the Regression of Frequency of Sex on Age for 2320 Respondents in the 1998 GSS

Regressor	Model 1	Model 2	Model 3
Intercept	3.238***	2.808***	3.335***
Age decile		−.277***	−.260***
(Age decile)^2			−.065***
Second age decile	.416**		
Third age decile	.378*		
Fourth age decile	−.002		
Fifth age decile	.182		
Sixth age decile	−.148		
Seventh age decile	−.413**		
Eighth age decile	−.888***		
Ninth age decile	−1.817***		
Tenth age decile	−2.676***		
RSS	2002.350	1448.369	1954.210
MSE	2.988	3.216	2.999
R^2	.225	.163	.220

Note: Age decile is centered, and the quadratic term for age is based on its centered version.
*$p < .05$. ** $p < .01$. *** $p < .001$.

placing constraints on the parameters in model (5.3). In fact, model (5.4) is produced from model (5.3) by placing the following constraints on the deltas in model (5.3): $\delta_1 = \alpha + \beta$; $\delta_j = (j-1)\beta$ for $j = 2, \ldots, 10$. To see that this works, let's take a couple of examples. The intercept, δ_1, in model (5.3) is the mean sexual frequency for those in the first age decile. According to model (5.4), that mean is $E(Y) = \alpha + \beta(1) = \alpha + \beta$. The parameter δ_8 in model (5.3) is the difference in mean sexual frequency between those in the eighth age decile and those in the first. According to model (5.4), that difference is $\alpha + \beta(8) - [\alpha + \beta(1)] = \alpha + 8\beta - \alpha - \beta = 7\beta$, or $(j-1)\beta$ for $j = 8$. The linear model in equation (5.4) is much more parsimonious than equation (5.3) since it uses only two parameters instead of 10 to model the data. The difference in the model degrees of freedom for these two models is, therefore, 8. The results of estimating equation (5.4) are shown as model 2 in Table 5.1. The test statistic for whether the linear model is adequate is

$$F_{(8,2310)} = \frac{(2002.350 - 1448.369)/8}{2.988} = 23.175,$$

which is quite significant ($p < .00001$). Apparently, a linear model is not adequate to fit the data here.

Next, I try the quadratic model:

$$E(Y) = \alpha + \beta \, age \, decile + \gamma \, (age \, decile)^2. \tag{5.5}$$

We can test whether the pattern of nonlinearity in the data is adequately captured by the quadratic model with a nested test for model (5.5) versus model (5.3), since model (5.5) is also nested inside model (5.3). The constraints placed on model (5.3) to produce the model in (5.5) are $\delta_1 = \alpha + \beta + \gamma$; $\delta_j = (j-1)\beta + (j^2-1)\gamma$ for $j = 2, \ldots, 10$. Again a couple of examples reveal why this works. The mean sexual frequency for those in the first decile, according to equation (5.3), is δ_1. According to equation (5.5), it is $\alpha + \beta(1) + \gamma(1^2) = \alpha + \beta + \gamma$; and δ_5 in model (5.3) is the difference in mean sexual frequency between the fifth and first deciles. According to model (5.5), that difference is $\alpha + \beta(5) + \gamma(5^2) - (\alpha + \beta + \gamma) = 4\beta + 24\gamma$, or $(j-1)\beta + (j^2-1)\gamma$, where $j = 5$. As the quadratic model replaces the 10 parameters in equation (5.3) with only three parameters, the difference in model degrees of freedom between these two models is 7. The test statistic for adequacy of the quadratic model is

$$F_{(7,2310)} = \frac{(2002.350 - 1954.210)/7}{2.988} = 2.302.$$

This is just significant at $p < .03$, which suggests that the quadratic model is not quite adequate to capture all of the nonlinearity in the *age–sexual frequency* relationship. However, in terms of discriminatory power, the quadratic model is a substantial improvement over the linear model (the change in R^2 is .057) and only slightly less efficacious than the unconstrained model (the change in R^2 is .0054). Adding perhaps a cubic term might produce a nonsignificant nested F test, but I do not believe the increase in complexity would be worthwhile. Hence, I choose the quadratic model as exhibiting maximum parsimony while capturing most of the nonlinearity in the data. Additionally, the quadratic model is most consistent with theoretical expectation.

Centering. For models 2 and 3 in Table 5.1, I employ the centered version of *age decile* and then form the quadratic term by squaring this centered variable. This has two advantages. First, it renders the main effect of *age decile* interpretable in the quadratic model. The partial slope for *age decile* in model (5.5), which captures the "effect" of *age decile* on *sexual frequency*, is $\beta + 2\gamma$ *age decile*. The main effect of *age decile*, or β, is the effect when *age decile* is zero, since in this case the "2γ *age decile*" term disappears. If *age decile* is uncentered, this effect has no meaning, since *age decile*, which begins at the value 1, cannot possibly take on the value zero. However, if *age decile* is centered, it is zero whenever *age decile* is at its mean. Thus, β is the effect of *age decile* at its mean. Moreover, the test of significance of b, the sample estimate of β, is a test for whether the impact of increasing *age* (in deciles) is significant at its mean value.

The second advantage of centering has to do with collinearity. Recall from Chapter 3 that it was important to center the continuous variables involved in cross-product terms in order to reduce potential collinearity problems. For the same reason, we want to center X before creating higher-order powers of X (e.g., X^2, X^3) to include in the model. As an example, without centering, *age* and *age-squared* are correlated .9737, producing *VIF*'s (not shown) of 19.26 for each coefficient in the quadratic model. After centering, the correlation is reduced to .1, and the *VIF*'s for each coefficient are only 1.01. At the least, collinearity inflates the sampling variance of one's estimators, which tends to reduce the power of tests for the coefficients. In this case, it has relatively little effect on the estimates, however, and both are quite significant in the uncentered version of the model as well. In this model (not shown), the coefficient for *age decile*, or b, is .422, while the coefficient for $(age\ decile)^2$, or g, is $-.065$. Although $\hat{\sigma}_g$ is no different than for the model using the centered variables, $\hat{\sigma}_b$ is about four times larger in the uncentered, versus the centered, model. The quadratic effect, $-.065$, is the same in both models. The main effects in the centered and uncentered models are not directly comparable, since the value of .422 in the uncentered model is the effect when *age decile* is zero. To make them comparable, consider the effect of *age decile* at its mean for the uncentered model. As the mean of *age decile* is 5.279, the effect is $.422 + 2(-.065)(5.279) = -.264$, compared to $-.26$ in the centered model. Both models apparently give rise to comparable estimates of *age decile*'s effect on *sexual frequency*.

Interpreting Quadratic Models

The use of *age decile* in place of the continuous variable, *age*, was necessary for testing various alternatives to the unconstrained model, since it allowed for the creation of nested models. However, once the quadratic model was chosen, it was reestimated using continuous *age* in place of *age decile*. Again, *age* was centered prior to taking its square. The results are shown as model 1 in Table 5.2. The effects of both *age* and

Table 5.2 Curvilinear and Interaction Models for the Regression of Frequency of Sex on Age and Health for 2320 Respondents in the 1998 GSS

Regressor	Model 1	Model 2	Model 3	Model 4
Intercept	3.113***	3.109***	2.834***	3.128***
Age	−.040***	−.037***	−.048***	−.039***
Age^2	−.001***	−.001***		−.0011***
Health		.241***	.225***	.410***
Age × health			.008**	.012***
Age^2 × health				−.0007***
RSS	1981.836	2066.429	1841.966	2142.310
MSE	2.987	2.952	3.049	2.922
R^2	.223	.232	.207	.241

Note: Age and health are centered, and all higher-order terms involving these variables use their centered versions.

*p < .05. ** p < .01. *** p < .001.

age^2 are significant, and both are negative, again suggesting a segment I curve, the pattern exhibited in Figure 5.8. The partial slope for age is $-.04 - 2(.001)\, age$. (The effects are smaller than for the $age\ decile$ model, since the units of age are now single years instead of deciles.) Thus, at mean age ($age = 44.455$) the partial slope is $-.04$; at 1 standard deviation above mean age ($age = 61.229$) it is $-.04 - 2(.001)$ $(16.774) = -.074$; and at 2 standard deviations above mean age ($age = 78.003$) it is $-.04 - 2(.001)(33.548) = -.107$. These calculations suggest an accelerating decline in sexual frequency with advancing age, as was expected.

Unit Impact versus Partial Derivative. Recall from Chapter 2 the distinction between the unit impact of X [the change in E(Y) for a unit increase in X, at x] and the partial derivative with respect to X [the instantaneous change in E(Y) with change in X, at x]. As mentioned in Chapter 2, these are identical in linear models. However, in nonlinear models they are different quantities. The partial derivative for the quadratic model [e.g., model (5.1)] at a particular x is, as noted, $\beta_1 + 2\beta_2 x$. The unit impact, however, is

$$E(Y|x + 1) - E(Y|x) = \beta_0 + \beta_1(x + 1) + \beta_2(x + 1)^2 - (\beta_0 + \beta_1 x + \beta_2 x^2)$$
$$= \beta_0 + \beta_1 x + \beta_1 + \beta_2 x^2 + 2\beta_2 x + \beta_2 - \beta_0 - \beta_1 x - \beta_2 x^2$$
$$= \beta_1 + \beta_2 + 2\beta_2 x.$$

How much difference does this really make? Actually, it doesn't make much difference as long as a unit change in X is a relatively small change. In model 1 in Table 5.2, for example, the difference is $-.04 - 2(.001)x = -.04 - .002x$ for the partial derivative, versus $-.04 - .001 - 2(.001)x = -.041 - .002x$ for the unit impact. So, for example, at 1 standard deviation above mean age the partial derivative with respect to age is $-.074$ and the unit impact is $-.075$. In this case to interpret the partial derivative as the approximate change in average sexual frequency for a year's (i.e., a unit of *age*) increase in *age* is pretty accurate.

But if a unit increase represents a relatively large change in X, approximating the unit impact with the partial derivative can be misleading. As an example, Lennon and Rosenfield (1994) investigated the impact of *perceived fairness in the household division of labor* on *depressive symptomatology* among employed wives, using a quadratic effect of *fairness*. *Perceived fairness* was coded -1 for "very unfair to me" (i.e., the wife), $-.5$ for "unfair to me," 0 for "fair to both of us," .5 for "unfair to my spouse," and 1 for "very unfair to my spouse." Actually, I prefer to refer to this as a scale of *overbenefit*, since the highest score represents maximum overbenefit, whereas maximum fairness occurs in the middle of the scale (see also Longmore and DeMaris, 1997). According to equity theory (Walster et al., 1978) people experience distress when they are either underbenefited or overbenefited. With this in mind, the authors expected that *depressive symptomatology* would exhibit a U-shaped relationship with *overbenefit*. Increases in the scale away from underbenefit and in the direction of fairness should reduce depressive symptoms, whereas increases away from fairness and in the direction of overbenefit should again increase depressive symptoms. The coefficients for *overbenefit* and its square were, respectively, .086 and .188. Therefore, the partial derivative is .086 + 2(.188) *overbenefit*, while the unit impact is .086 + .188 + 2(.188) *overbenefit*. Using the partial derivative to approximate a change in average *depressive symptomatology* for a unit increase in *overbenefit* at the value of "very unfair to me" on the scale, we get a value of .086 + 2(.188)(−1) = −.29. In actuality, a unit increase in the scale is associated with a change of .086 + .188 + 2(.188)(−1) = −.102. At the value of "fair to both," the partial derivative estimates the change as .086, whereas the unit impact is actually .086 + .188 = .274. In this case, the partial derivative is not a particularly good approximation to the unit impact. Whether this is a problem depends on how important it is to be accurate about the estimated effect of a "unit increase" in the explanatory variable.

NONLINEAR INTERACTION

In Chapters 3 and 4 I discussed various interaction models. However, all of these involved linear interactions, in which the impact of X might change over levels of Z, but the relationship between Y and X at each z was always linear. In this section I *consider nonlinear interaction models*, in which the nonlinear relationship between Y and X varies as a function of Z. To motivate interest in the topic, let's reexamine the relationship between *age* and *sexual frequency*. Prior research on *sexual frequency* (e.g., Rao and DeMaris, 1995) suggests that those in better health have sex more often. I would go one step beyond this and suggest that *health* and *age* should interact in their effects on *sexual frequency*. I have already shown that there is a nonlinear decline in sexual frequency with age. I further hypothesize that *the nonlinear decline in sexual frequency with advancing age will be less pronounced for those in better, as opposed to poorer, health*. This is a nonlinear interaction effect, and we must first consider how to model it.

Once again, I rely on the first partial derivative to define nonlinear interaction in
X. *In a model for Y containing the variables X and Z, if the first partial derivative of
Y with respect to X is a function of both X and Z, the model is said to be characterized by a nonlinear interaction in X. If the first partial derivative of Y with respect to
X is only a function of Z but not X, the interaction is linear in X.* Let's examine a
series of quadratic models containing X and Z to see how this definition applies. In
model (5.6), Z is added purely as another covariate:

$$E(Y) = \beta_0 + \beta_1 X + \beta_2 X^2 + \beta_3 Z. \tag{5.6}$$

The effect of X (i.e., $\partial[E(Y)]/\partial X$) is $\beta_1 + 2\beta_2 X$. As this is not a function of Z, there is
no interaction effect here. Thus the curves relating Y to X at each level of Z are par-
allel. Moreover, the curves are identical in shape since, in particular, the departure
from linearity, β_2, is not a function of Z.

Model (5.7) adds the cross-product of X with Z:

$$E(Y) = \beta_0 + \beta_1 X + \beta_2 X^2 + \beta_3 Z + \beta_4 XZ. \tag{5.7}$$

The effect of X is $\beta_1 + 2\beta_2 X + \beta_4 Z$. This is a nonlinear interaction effect, since the
term is a function of both X and Z. This means that the curves relating Y to X over
levels of Z will not be parallel. However, they will be similar in shape since the
departure from linearity is still β_2, which is not a function of Z. Notice that with
respect to Z, the interaction is linear because $\partial[E(Y)]/\partial Z$ is $\beta_3 + \beta_4 X$. This is just a
conventional linear interaction, as this effect is not also a function of Z.

Model (5.8) adds the cross-product of X^2 with Z:

$$E(Y) = \beta_0 + \beta_1 X + \beta_2 X^2 + \beta_3 Z + \beta_4 XZ + \beta_5 X^2 Z. \tag{5.8}$$

First, you should notice the hierarchical nature of these successive equations. For
every higher-order term in a given equation, all the lower-order components are also
in the equation. This is especially important to remember in equation (5.8), as it is
not immediately obvious that the XZ term must be in the equation. But if $X^2 Z$ is in
the equation, we must also have all of its component regressors: X, X^2, Z, and XZ. To
continue, the effect of X is $\beta_1 + 2\beta_2 X + \beta_4 Z + 2\beta_5 XZ$. This can be further expressed
as $(\beta_1 + \beta_4 Z) + 2(\beta_2 + \beta_5 Z)X$. This makes clear that the effect is of the form
$\beta + 2\gamma X$, typical of the quadratic model, except that now β, the linear component of
the curve, is $(\beta_1 + \beta_4 Z)$, and γ, the departure from linearity, is $(\beta_2 + \beta_5 Z)$. Obviously
this is a nonlinear interaction effect, since this expression is a function of both X and
Z. But in this case, not only are the X–Y curves not parallel, they are also not of the
same shape, since the departure from linearity is now also a function of Z. The effect
of Z, on the other hand, is $\beta_3 + \beta_4 X + \beta_5 X^2$. This, again, is a conventional linear
interaction in Z, since the expression is not a function of Z. In other words, the Z–Y
relationship at each level of X is still linear.

Finally, model (5.9) adds the cross-product of X^2 with Z^2, plus the additional lower-order components necessitated by this term—Z^2 and XZ^2:

$$E(Y) = \beta_0 + \beta_1 X + \beta_2 Z + \beta_3 X^2 + \beta_4 Z^2 + \beta_5 XZ + \beta_6 XZ^2 + \beta_7 X^2 Z + \beta_8 X^2 Z^2. \quad (5.9)$$

In this equation, the effect of X can be written $(\beta_1 + \beta_5 Z + 2\beta_6 Z^2) + 2(\beta_3 + \beta_7 Z + \beta_8 Z^2)X$, while the effect of Z can be written $(\beta_2 + \beta_5 X + \beta_7 X^2) + 2(\beta_4 + \beta_6 + \beta_8 X^2)Z$. In this case, then, we have a nonlinear interaction effect in both X and Z. For either interaction, the curves relating Y to X (Z) are neither parallel nor of the same shape across levels of Z (X).

Because model (5.8) is central to testing the hypothesis regarding the *age–health* interaction posed above, let's consider some additional issues concerning this model. In particular, we can perform global tests of both curvilinearity and interaction by comparing this model with other selected models (Aiken and West, 1991). If we factor equation (5.8) as

$$E(Y) = (\beta_0 + \beta_3 Z) + (\beta_1 + \beta_4 Z)X + (\beta_2 + \beta_5 Z)X^2.$$

Then a test of curvilinearity tests the null hypothesis that the quadratic effect, which in this case is $\beta_2 + \beta_5 Z$, is zero. Assuming that Z is not uniformly zero, this will be true if β_2 and β_5 are both zero. Hence a global test of curvilinearity is a nested F test comparing model (5.8) with the usual linear interaction model,

$$E(Y) = \beta_0 + \beta_1 X + \beta_2 Z + \beta_3 XZ. \quad (5.10)$$

To understand the global test of interaction, we factor (5.8) as

$$E(Y) = \beta_0 + \beta_1 X + \beta_2 X^2 + \beta_3 Z + (\beta_4 + \beta_5 X)XZ.$$

The test for interaction involves the null hypothesis that $\beta_4 + \beta_5 X$ equals zero (i.e., that $\beta_4 = \beta_5 = 0$). The appropriate test is the nested test comparing model (5.8) with model (5.6).

Returning to the *GSS98 dataset*, where $Y = sexual\ frequency$, if we let $X = age$ and $Z = health$, model 2 in Table 5.2 estimates equation (5.6), model 3 estimates equation (5.10), and model 4 estimates equation (5.8). As the table note indicates, *age*, *health*, and all of their higher-order relatives are based on the centered versions of these variables. It is clear that *health* has a significantly positive linear effect on *sexual frequency* in each model, as one would expect. It also looks like all interactions involving *age* and *health* as well as all quadratic terms in *age* are significant. Nevertheless, I begin by performing the global tests just outlined. The global test of curvilinearity compares model 4 with model 3:

$$F_{(2,2314)} = \frac{(2142.310 - 1841.966)/2}{2.922} = 51.394.$$

The global test of interaction compares model 4 with model 2:

$$F_{(2,2314)} = \frac{(2142.310 - 2066.429)/2}{2.922} = 12.984.$$

Both test statistics are highly significant, with $p < .00001$. Hence, the model for *sexual frequency* appears to be characterized by both curvilinearity in the relationship with *age* and an interaction between *age* and *health* in their effects on the response.

To evaluate my hypothesis completely, it is necessary to examine the *age–health* interaction effect more closely. From the discussion above, the effect of *age* is of the form $(\beta_1 + \beta_4 Z) + 2(\beta_2 + \beta_5 Z)X$, which as pointed out earlier, is a nonlinear interaction. Here β_1 is the coefficient of *age*, β_4 is the coefficient of the *age* * *health* cross-product term, β_2 is the coefficient of *age²*, and β_5 is the coefficient of the *age²* * *health* cross-product term. In terms of the coefficient estimates, the effect of *age* is

$$(-.039 + .012 \ health) + 2(-.0011 - .0007 \ health) \ age.$$

It is instructive to examine the impact of *age* at particular values of *health*. At 1 standard deviation (.815) below mean *health*, the effect of *age* is

$$[-.039 + .012(-.815)] + 2[-.0011 - .0007(-.815)] \ age = -.049 - .0011 \ age.$$

At mean *health*, the effect of *age* is

$$[-.039 + .012(0)] + 2[-.0011 - .0007(0)] \ age = -.039 - .0022 \ age.$$

At 1 standard deviation above mean *health*, the effect of *age* is

$$[-.039 - .012 \ (.815)] + 2[-.0011 - .0007(.815)] \ age = -.029 - .0033 \ age.$$

A more complete picture of the impact of *age* is obtained by evaluating each of these three partial slopes at selected values of *age*. In this case I evaluated them at 1 standard deviation below mean *age*, at mean *age*, and at 1 standard deviation above mean *age*. The results are shown in Table 5.3. Each number in the table is the impact of *age* on *sexual frequency* at selected settings of *age* and *health*. The pattern that emerges is somewhat unusual. Among younger respondents, *age* apparently has an adverse effect on *sexual activity* that is more pronounced for those in poorer *health* (reading down the first two columns). However, among the oldest respondents (1 standard deviation above mean *age*), this effect is reversed, and *age* has the most adverse effect among the healthiest respondents (reading down the third column). In other words, the decline in *sexual activity* with *age* is least dramatic for those in better *health*, but only up to a certain age; after that, the decline is more rapid for those in better *health*. This is much more easily discerned in Figure 5.9, which graphs the *age–sexual frequency* relationship at the three levels of *health*. As is evident in the figure, those in poorest health experience a nearly linear decline in sexual frequency

Table 5.3   Partial Derivatives with Respect to Age for the Regression of Sexual Frequency on Age and Health at Selected Values of Age and Health

	Value of Age		
Value of Health	One Standard Deviation Below Mean	Mean	One Standard Deviation Above Mean
One standard deviation below mean	−.031	−.049	−.067
Mean	−.002	−.039	−.076
One standard deviation above mean	.026	−.029	−.084

Note: Calculations are based on model 4 in Table 5.2.

with age; for those in average health, the decline is much more gradual; and for those in good health, there is an initial increase in sexual activity with age, followed by a gradual decline that becomes ever steeper after about age 55 (i.e., 10 years beyond the mean of 44.455). As the decline in sexual activity with age tends to be most pronounced for the unhealthiest respondents, I consider my hypothesis to be largely supported.

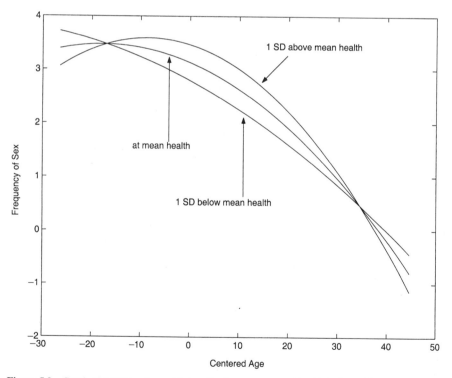

Figure 5.9   Graph of nonlinear (in age) interaction between age and health in their effects on frequency of sex.

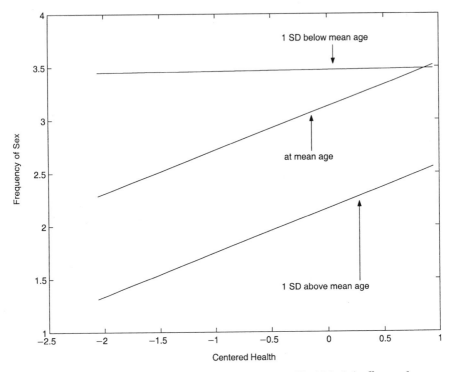

Figure 5.10 Graph of linear (in health) interaction between age and health in their effects on frequency of sex.

Using the same approach, we can recover the partial slopes for the impact of *health* at the three settings of *age* employed above. (This is left as an exercise for the reader.) Recall that in this type of model, the interaction is linear in *health*. This is easily revealed by Figure 5.10, in which the relationship between *sexual frequency* and *health* is plotted separately for those who are 1 standard deviation below mean *age*, at mean *age*, and 1 standard deviation above mean *age*. It appears that *health* has little impact on *sexual frequency* for the youngest respondents; their sexual activity is uniformly high regardless of health status. However, for those of average age or older, *health* appears to have a substantial positive impact, with more sexual activity being reported by those in better health.

Nonlinear Interaction: Another Example. Another example of a nonlinear interaction, one that does not involve the quadratic model, can be found in Mirowsky and Hu (1996). They investigated the relationship between *income* and *physical impairment* in two national probability samples. Based on an initial investigation of the bivariate relationship between *income* and *physical impairment*, they decided that a cube-root transformation of income in thousands would best represent the relationship. They also allowed this function of income to interact with *education*. Their estimated equation takes the form

$$\hat{y} = a + \mathbf{g}' \textbf{ controls} + bX^{1/3} + cZ + dZX^{1/3}, \tag{5.11}$$

where $\mathbf{g}'$**controls** is a weighted sum of coefficients times control variables, X is *income* in thousands of dollars, and Z is *education* in years of completed schooling. The first partial derivative of this equation with respect to X is

$$\frac{\partial \hat{y}}{\partial X} = \frac{1}{3}bX^{-2/3} + \frac{1}{3}dZX^{-2/3} = \frac{b + dZ}{3X^{2/3}}.$$

As this expression is a function of both X and Z, it is clearly a nonlinear interaction (also made very clear by their graphs of the *income–physical impairment* relationship at different levels of *education*; see Mirowsky and Hu, 1996, Fig. 4).

A sense of the nature of the interaction can be obtained by evaluating the impact of income at, say, *income* = $10,000, at three levels of *education*. The coefficients are $b = -.038$ and $d = .021$. In this analysis, Z (i.e., *education*) is centered, and although its standard deviation is not given, I estimate it as 3, based on the range of *education*. At 1 standard deviation below mean *education*, the impact of *income* is

$$\frac{-.038 + .021(-3)}{3(10^{2/3})} = -.00725.$$

At mean *education*, the impact of *income* is

$$\frac{-.038 + .021(0)}{3(10^{2/3})} = -.00273.$$

At 1 standard deviation above mean *education*, the impact of *income* is

$$\frac{-.038 + .021(3)}{3(10^{2/3})} = .0018.$$

It is instructive to repeat these computations for *income* = $25,000. At 1 standard deviation below mean *education*, the impact of *income* is

$$\frac{-.038 + .021(-3)}{3(25^{2/3})} = -.0039.$$

At mean *education*, the impact of *income* is

$$\frac{-.038 + .021(0)}{3(25^{2/3})} = -.0015.$$

At 1 standard deviation above mean *education*, the impact of *income* is

$$\frac{-.038+.021(3)}{3(25^{2/3})} = .00097.$$

What these computations show is that *income* appears, in general, to reduce physical impairment. But income has a stronger impact on reducing impairment for those with less education than for those with more education. Additionally, this trend (i.e., the differential impact of *income* according to education level) becomes less and less evident at higher levels of *income*. As a final note, the partial derivative of (5.11) with respect to Z (i.e., *education*) is $c + dX^{1/3}$. Since this is not a function of Z, the interaction is linear in *education*.

NONLINEAR REGRESSION

In this final section of the chapter, I consider the estimation of essentially nonlinear models. These are referred to as *nonlinear regression models* and are typically estimated using *nonlinear least squares*. To avoid excessive mathematical complexity, I restrict attention to a relatively simple model yet one that illustrates all of the key facets of this technique. Table 5.4 presents measurements on the *diagnostic quiz* and the score on the *final exam* for 11 students in my beginning graduate-level statistics class. The quiz is given during the seventh week of the semester to gauge how well students are assimilating the material. The final exam is given during the sixteenth week. Figure 5.11 shows a scatterplot of these variables, with *final exam* scores plotted against *diagnostic quiz* scores. There is clearly a strong positive relationship between the variables but one that does not appear to be linear. In fact, it very much resembles the pattern exhibited by the function $y = e^x$ in Figure 5.1. Is this reasonable? Assuming that the diagnostic quiz taps statistical aptitude, the trend in the figure suggests that an increase in aptitude has an accelerating positive effect on performance in the class. That is, an increase in aptitude has a stronger effect on performance for those who already have considerable aptitude than for those whose skills are more rudimentary. If this is true, one appropriate model for these data, with

Table 5.4   Score on Diagnostic Quiz and Score on Final Exam for 11 Students in Graduate Statistics

Diagnostic Quiz	Final Exam	Diagnostic Quiz	Final Exam
32	36.50	84	75.25
60	49.75	84	103.00
60	36.75	92	108.00
76	61.75	92	89.25
84	92.25	100	91.50
84	61.00		

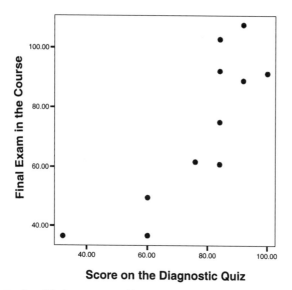

Figure 5.11 Scatterplot of final exam score with score on the diagnostic quiz for 11 students in graduate statistics.

$Y = final\ exam$ score and $X = diagnostic\ quiz$ score, is the exponential model with additive errors, a nonlinear model. As noted earlier, the model is

$$Y = \gamma_0 e^{\gamma_1 X} + \varepsilon. \tag{5.12}$$

Before discussing the estimation of this model, I digress briefly and consider instead the exponential model with multiplicative errors, also shown earlier:

$$Y = \gamma_0 e^{\gamma_1 X} e^{\varepsilon}. \tag{5.13}$$

Note that the only difference between these two models is the error structure: that is, the manner in which error enters the model. Because of this difference, each model suggests that the processes generating Y in either case are slightly different. For example, let's assume that $\gamma_0 = 1.5$, $\gamma_1 = .25$, $X = .75$, and $\varepsilon = -2$. Then according to model (5.12), the value of Y is

$$Y = 1.5 e^{.25(.75)} - 2 = -.1907,$$

while model (5.13) has the value of Y as

$$Y = 1.5 e^{.25(.75)} e^{-2} = .2449.$$

More generally, however, if ε is characterized by constant variance, model (5.12) implies that the variance of Y is constant with increasing Y. The multiplicative nature of the error term in model (5.13), on the other hand, implies that the variance of Y increases with increasing Y (Ratkowsky, 1990; the reader is also asked to verify this in the exercises). Unless there is a compelling theoretical reason for choosing one error structure over another ahead of time, it will not be a simple matter to decide which is more reasonable. Based on the scatterplot in Figure 5.11, it does not appear that a multiplicative error structure is warranted. There does not appear to be a pronounced trend for the variance in Y to increase with Y, with the possible exception of the large scatter of points at $X = 84$. An examination of model residuals plotted against fitted values might also reveal which of model (5.12) or model (5.13) is the correct choice. In the current case, both residual plots (not shown) appeared reasonable. However, 11 data points are simply too few to allow for a reliable empirical check on model plausibility. For this reason, I estimate both models, as well as the simple linear regression model, and compare the results. First I consider the estimation of model (5.13).

Estimating the Multiplicative Model

As noted earlier, if we transform Y by taking its log, model (5.13) becomes a linear model:

$$\log Y = \alpha + \gamma_1 X + \varepsilon, \tag{5.14}$$

where $\alpha = \log \gamma_0$. If we can assume that ε is normally distributed with zero mean and constant variance, σ^2, this model can be estimated via OLS. Panel (a) of Table 5.5 shows the results of estimating this model, called the *log y model* in the table, as well

Table 5.5   Regression Models for the Regression of Final Exam Score on Diagnostic Quiz

(a) *Regression Results*

Parameter Estimate	Linear Model	Log y Model	Exponential Model
Intercept	−11.190	2.868***	17.129*
Beta	1.094**	.018***	.018***
R^2	.690	.718	.717

* $p < .05$. ** $p < .01$. *** $p < .001$.

(b) *Iteration History for Exponential Model*

Iteration	Intercept	Beta	SSE
0	17.5983	.0176	1893.8412
1	17.0852	.0182	1854.9534
2	17.1307	.0182	1854.9443
3	17.1294	.0182	1854.9443

Note: All R^2s are calculated as $r^2_{y,\hat{y}}$.

as estimating the SLR model for Y (the *linear model* in the table). Both models show a significant relationship between *diagnostic quiz* and *final exam* scores, with the $\log y$ model having slightly greater discriminatory power. The estimates are, of course, not directly comparable, but the fitted values will be similar. For example, the linear model predicts final exam scores of 54.45 for $X = 60$ and 89.458 for $X = 92$. Calculating the fitted values, $\hat{y}$, or estimated $E(Y)$, for the $\log y$ model is not quite as straightforward as it might seem. The estimated log of y would be given by $\log \hat{y} = 2.868 + .018X$. But according to Wooldridge (2000), if we simply take $\hat{y} = \exp(\log \hat{y})$, we will systematically underestimate $E(Y)$. The reason for this is that for this model, $E(Y)$ is given by

$$E(Y|X) = \exp\left(\frac{\sigma^2}{2}\right) \exp(\alpha + \gamma_1 X), \tag{5.15}$$

where σ^2 is the variance of ε. Therefore, the correct fitted values are given by

$$\hat{y} = \exp\left(\frac{\hat{\sigma}^2}{2}\right) \exp(\log \hat{y}),$$

where the *MSE* in the regression for $\log Y$ would be used as the measure of $\hat{\sigma}^2$. Wooldridge (2000) explains that these fitted values are not unbiased, but they are consistent. The development in equation (5.15), however, relies on the normality of the error term. With an *MSE* of .03981 in the current example, the fitted values for X at scores of 60 and 92 for the $\log y$ model are 52.874 and 94.057, respectively. Notice that these are slightly closer to the means of the actual Y values at these X values (43.25, 98.63) than are the fitted values for the linear model.

Interpreting γ_1. Interpreting the effect of X in the $\log y$ model is facilitated by considering the ratio of $E(Y|x+1)$ to $E(Y|x)$. That is, we examine the ratio of mean responses for those who are 1 unit higher on X than others. Let $c = \exp(\sigma^2/2)$. Then, according to equation (5.15), this ratio is

$$\frac{E(Y|x+1)}{E(Y|x)} = \frac{ce^{\alpha}e^{\gamma_1(x+1)}}{ce^{\alpha}e^{\gamma_1 x}} = \frac{e^{\gamma_1 x}e^{\gamma_1}}{e^{\gamma_1 x}} = e^{\gamma_1},$$

which implies that

$$E(Y|x+1) = e^{\gamma_1}E(Y|x).$$

In other words, $\exp(\gamma_1)$ is the multiplicative impact on the mean of Y for each unit increase in X. Or, $100(e^{\gamma_1} - 1)$ is the percent change in the mean for each unit increase in X. In the current example, each point higher the student scores on the diagnostic quiz is estimated to increase the average final exam score by $100(e^{.018} - 1) = 1.8\%$. Moreover, an increase of 10 points on the diagnostic quiz should raise the average final exam score by $100(e^{10(.018)} - 1) = 19.7\%$.

Discriminatory Power. As is the case with the fitted values, calculation of R^2 is also affected by equation (5.15). Wooldridge (2000) recommends calculating R^2 using an approach that depends only on the orthogonality of the errors with X (i.e., the orthogonality condition), but not on normality. This approach is as follows. Regress the log of Y on X, using model (5.14), and obtain $\log \hat{y}$ (i.e., the fitted values from this regression). Create $M = \exp(\log \hat{y})$ for each observation. Then regress Y on M without an intercept; this is known as a *regression through the origin*. Obtain the fitted values from this regression and denote them as $\hat{y}'$. Then R^2 is calculated as the squared correlation between y and $\hat{y}'$. Using this procedure resulted in the value of .718 shown in Table 5.5.

Estimating the Nonlinear Model

Estimation of the additive exponential model is complicated by the fact that the normal equations for this model have no *closed-form solution* or algebraic formula that can be applied in one step. This can be seen by first considering the SSE for this model. Writing the sample equation as

$$Y = g_0 e^{g_1 X} + e,$$

we see that the residual, e, is

$$e = Y - g_0 e^{g_1 X}.$$

The criterion to be minimized by the least squares estimates, in this case, is

$$SSE = \sum (Y - g_0 e^{g_1 X})^2.$$

The normal equations are found by taking the first partial derivatives of SSE with respect to, alternately, g_0 and g_1, setting them to zero, and solving for g_0 and g_1. The normal equations in this case are (Neter et al., 1985)

$$\sum_{i=1}^{n} y_i e^{g_1 X_i} - g_0 \sum_{i=1}^{n} e^{2g_1 X_i} = 0,$$

$$\sum_{i=1}^{n} y_i X_i e^{g_1 X_i} - g_0 \sum_{i=1}^{n} X_i e^{2g_1 X_i} = 0.$$

Since these equations are nonlinear in the parameter estimates, they have no closed-form solution and must be solved through iterative methods.

Actually, a more efficient procedure is to find the estimates via a direct search rather than first finding the normal equations for the model. Perhaps the most common approach is the Gauss–Newton procedure (Myers, 1986; Neter et al., 1985).

This technique uses a first-order Taylor series expansion to approximate the nonlinear model with linear terms and then employs OLS to estimate the parameters in an iterative fashion. A Taylor series expansion of a function of x, $f(x)$, uses a polynomial in x, $p(x)$, to approximate the value of the function in some "neighborhood" of a given value, a. (A neighborhood of a is some small interval of numbers centered at $X = a$.) This is accomplished by forcing $p(x)$ and its first n derivatives to match the value of $f(x)$ and its first n derivatives at $X = a$ (Anton, 1984). In the current example, we require only the *first* partial derivatives of $Y = g_0 e^{g_1 X} + e$ with respect to g_0 and g_1. These are

$$\frac{\partial Y}{\partial g_0} = e^{g_1 X},$$

$$\frac{\partial Y}{\partial g_1} = g_0 X e^{g_1 X}.$$

We then choose some values for the parameter estimates as starting values for the iterative search. Call these values $g_0^{(0)}$ and $g_1^{(0)}$ One choice of starting values, followed in the current example, was to use the coefficient estimates from the $\log y$ model in Table 5.5. A first-order Taylor series expansion of $g_0 e^{g_1 X}$ about $g_0^{(0)}$ and $g_1^{(0)}$ is

$$g_0^{(0)} e^{g_1^{(0)} X} + e^{g_1^{(0)} X} (\gamma_0 - g_0^{(0)}) + g_0^{(0)} X e^{g_1^{(0)} X} (\gamma_1 - g_1^{(0)}),$$

where γ_0 and γ_1 are the parameter values that we are trying to estimate (i.e., they are the "variables" in this expression). A Taylor series approximation to the model for Y is, therefore,

$$Y \approx g_0^{(0)} e^{g_1^{(0)} X} + e^{g_1^{(0)} X} (\gamma_0 - g_0^{(0)}) + g_0^{(0)} X e^{g_1^{(0)} X} (\gamma_1 - g_1^{(0)}) + \varepsilon.$$

If we let $Y^{(0)} = Y - g_0^{(0)} e^{g_1^{(0)} X}$, then

$$Y^{(0)} \approx e^{g_1^{(0)} X} (\gamma_0 - g_0^{(0)}) + g_0^{(0)} X e^{g_1^{(0)} X} (\gamma_1 - g_1^{(0)}) + \varepsilon. \tag{5.16}$$

Myers (1986) refers to the left-hand-side of equation (5.16) as the "residual" for Y, where the parameters of the term $\gamma_0 e^{\gamma_1 X}$ are replaced by their starting values. Equation (5.16) is now a linear regression model with no intercept (i.e., a regression through the origin). The "parameters" to be estimated are $\gamma_0 - g_0^{(0)}$ and $\gamma_1 - g_1^{(0)}$, which enter the model linearly, and the independent variables are $e^{g_1^{(0)} X}$ and $g_0^{(0)} X e^{g_1^{(0)} X}$, which are two different transformations of X. OLS is then employed to estimate equation (5.16), giving us the parameter estimates

$$b_0^{(0)} = \hat{\gamma}_0 - g_0^0,$$
$$b_1^{(0)} = \hat{\gamma}_1 - g_1^0.$$

These are then used to obtain revised estimates of γ_0 and γ_1, since

$$\hat{\gamma}_0 = g_0^{(0)} + b_0^{(0)},$$
$$\hat{\gamma}_1 = g_1^{(0)} + b_1^{(0)}.$$

We then take $\hat{\gamma}_0$ and $\hat{\gamma}_1$ as our new starting values for equation (5.16) and reestimate the equation, again using OLS. We continue in this fashion, each time updating our old estimates, plugging the updates into equation (5.16), and reestimating the parameters until the difference between successive coefficient estimates and/or the difference in successive *SSE*'s becomes negligible. If the procedure is working correctly, *SSE* should continue to get smaller with each successive iteration. At the present time, many software programs for nonlinear regression (e.g., SAS) require the user to supply the expressions for the first partial derivatives of the model with respect to the parameters, as well as the starting values for the parameters, in order for the program to run. Assuming that the assumptions on the errors are valid, the resulting parameter estimates are approximately efficient, unbiased, and normally distributed in large samples. This means that for large n, the usual regression test statistics are applicable.

Estimates for the exponential model with additive errors, based on the Gauss–Newton procedure, are shown in the column "exponential model" in panel (a) of Table 5.5. They can be compared to those for the log y model by noting that the comparable parameter in the log y model to the intercept in the exponential model is $\exp(2.8678) = 17.5983$. Although the intercepts are slightly different, the coefficient for X (*diagnostic quiz* score) is virtually the same in each model. R^2 for the exponential model is calculated as the square of the correlation between its fitted values ($\hat{y}$) and Y. Again, this is virtually identical to the R^2 for the log y model. Panel (b) of the table shows the iteration history for the model. The initial estimates are in the row labeled "iteration 0" and are just the parameter estimates from the log y model. Convergence, in this case, was quite rapid, occurring in three iterations. Ratkowsky (1990) observes that convergence to the least squares estimates usually occurs fairly rapidly from reasonable starting values, especially for relatively simple models. In fact, even if one mistakenly uses 2.8678 instead of $\exp(2.8678)$ as the starting value for the intercept, convergence still occurs in five iterations. At any rate, all three models for *final exam* score in this example appear to produce approximately the same substantive conclusion regarding the impact of *diagnostic quiz* scores. In this particular instance there is no special advantage to employing the nonlinear model. However, in other cases, it may be the only suitable choice.

EXERCISES

5.1 Based on a probability sample of 680 married couples, Mirowsky (1985) examined the relationship between *depression* (*Y*), a continuous scale ranging from 0 ("no depression") to 112 ("maximum depression") and *marital power*

(*X*). *Power* was operationalized in terms of who usually made major decisions in the household, and was an approximately interval-level scale ranging from 1 ("wife always") to 5 ("husband always"), with a value of 3 indicating that decisions were made with both partners "having equal say." One hypothesis of the study was that *depression* would exhibit a U-shaped relationship with *power*, showing a minimum when power was equally shared. The equations for husbands and wives were:

Husbands: $\hat{y} = a + \mathbf{g'controls} - 12.343\ power + 1.944\ power^2$.

Wives: $\hat{y} = a + \mathbf{g'controls} - 11.567\ power + 2.537\ power^2$.

(a) Give the partial slope of each equation with respect to *power*.
(b) Show that each equation represents a U-shaped trend, using selected values of *power*.
(c) Using the fact that a U-shaped curve of the form $y = a + bx + cx^2$ achieves a minimum value when $x = -b/2c$, show that minimum depression occurs for each partner when neither partner dominates decision making.

5.2 Based on model 4 in Table 5.2, give the partial slope for the impact of *health* at 1 standard deviation below mean *age*, at mean *age*, and at 1 standard deviation above mean *age*, given that the standard deviation of age is 16.774.

5.3 Demonstrate with some selected values for γ_0, γ_1, x, and ε that for the multiplicative exponential model [i.e., model (5.13)], the variance of Y increases with increasing Y, whereas the variance of Y is constant with increasing Y in the nonlinear exponential model [i.e., model (5.12)]. (*Hint*: One choice of values is $\gamma_0 = 1.5$, $\gamma_1 = .25$, $x = 1, 2, 3, 4, 5$, and $\varepsilon = -.5, .5$).

5.4 An analysis of the 2320 respondents in the *GSS98*, examining the relationship between *sexual frequency* (*Y*), *age* (centered), and *education* (centered), found the following results for three different models:

Model 1: $\hat{y} = 3.103 - .04\ age - .001\ age^2 + .019\ education$; $R^2 = .2233$.
Model 2: $\hat{y} = 2.818 - .049\ age + .035\ education + .001\ age * education$; $R^2 = .1982$.
Model 3: $\hat{y} = 3.108 - .039\ age - .001\ age^2 + .029\ education + .001\ age * education - .00005\ age^2 * education$; $R^2 = .2238$.

(a) Perform global tests for the *age * education* interaction and for the curvilinearity of the relationship between *sexual frequency* and *age*.
(b) Given that the standard deviation of *education* is 2.894, give the partial slope of $\hat{y}$ in model 3 with respect to *age* at 1 standard deviation below mean *education*, at mean *education*, and 1 standard deviation above mean

education. Compare the nonlinear interaction effect here to that for the nonlinear interaction of *age* with *health* in Table 5.2.

5.5 For the data in Table 5.4, if the diagnostic quiz is divided up approximately into quartiles, an unconstrained model for *final exam* scores, using three dummies representing quartiles 2 to 4 produces $R^2 = .7937$. A model with just the quartile variable (coded 1 to 4) gives $R^2 = .7848$. Moreover, a model with the *centered*, continuous versions of *diagnostic quiz* and its square (the quadratic model) has $\hat{y} = 69.716 + 1.376 \ quiz + .012 \ quiz^2$.

(a) Test whether a linear model fits as well as the unconstrained model.

(b) Using the quadratic model, if the mean quiz score is 77.667 and the standard deviation is 18.642, give the predicted *final exam* scores for quiz scores of 60 and 92.

(c) For the quadratic model, give the partial slope for quiz score at 1 standard deviation below mean quiz scores, at mean quiz scores, and at 1 standard deviation above mean quiz scores.

5.6 For the following nonlinear models from Ratkowsky (1990), find $\partial y/\partial x$ and $\partial y/\partial \theta_p$ for each parameter θ_p (i.e., for each different parameter in the model):

(a) $y = \log(x - \alpha)$.

(b) $y = 1/(x + \alpha)$.

(c) $y = \log(\alpha + \beta x)$.

(d) $y = \alpha - \beta \gamma^x$.

5.7 Using the *couples dataset*, estimate the multiplicative exponential model for COITFREQ as a function of MALEAGE. In particular:

(a) Give the equation for $\log \hat{y}$.

(b) Interpret the effect of MALEAGE. What percent reduction in *coital frequency* is expected for a 10-year gain in male's age?

(c) Give the estimated expected value of Y (i.e., $\hat{y}$) for couples in which the male partner is 25, 35, and 55.

(d) Give the R^2 value for the model.

(e) Examine the residuals for plausibility of the assumptions on the errors.

5.8 Using the *couples dataset*, estimate the additive exponential model for COITFREQ as a function of MALEAGE (you'll need software that has a nonlinear regression procedure here). Use the sample coefficients from the multiplicative exponential model as starting values. In particular:

(a) Give the equation for $\hat{y}$.

(b) Interpret the effect of MALEAGE. What percent reduction in *coital frequency* is expected for a 10-year gain in male's age?

(c) Give the estimated expected value of Y (i.e., $\hat{y}$) for couples in which the male partner is 25, 35, and 55.

(d) Give the R^2 value for the model.

(e) Examine the residuals for plausibility of the assumptions on the errors.

5.9. Regard the following data for 30 cases:

X	Z	Y	X	Z	Y
0	−2.0845	3.56774	5	3.7676	.41187
0	.3866	.32872	5	6.5101	−.78784
0	.9488	1.03347	5	7.7405	−.97970
1	−.7751	1.43186	6	6.6440	−1.62030
1	−2.892	2.87661	6	7.9871	−.56195
1	−2.8461	3.24443	6	7.7051	−1.65471
2	5.2616	−.16583	7	5.0887	−.60902
2	5.9740	−1.60710	7	11.4823	−.94752
2	2.1966	1.22054	7	9.5552	−.41412
3	3.8658	−.72971	8	7.8112	−.98601
3	.3804	.47408	8	8.2402	−1.35958
3	3.4555	.12573	8	4.1219	−.90526
4	6.8522	−1.21795	9	4.9838	.16067
4	4.7659	−.95917	9	6.3705	−.51194
4	9.2228	−3.19808	9	11.9198	−.47474

(a) Estimate the following model for these data: $E(Y) = \alpha + \beta X^{1/3} + \gamma Z + \delta Z X^{1/3}$, and give the sample equation.

(b) Give $\partial \hat{y}/\partial X$ and $\partial \hat{y}/\partial Z$.

(c) Give the impact (i.e., partial slope) of X at the mean of Z and at 1 standard deviation below and above the mean.

(d) Give the impact of Z at the mean of X and at 1 standard deviation below and above the mean.

(e) Graph Y against X at the mean of Z and at 1 standard deviation below and above the mean.

5.10 Using the *kids dataset*, partition the variable PERMISIV approximately into sextiles using the following ranges: ≤ 8, (8,10], (10,12], (12,14], (14,16], and > 16.

(a) Plot the mean of ADVENTRE against sextiles of PERMISIV. Does the trend appear linear?

(b) Use the sextile-coded version of PERMISIV to perform the test of linearity in the regression of ADVENTRE on PERMISIV.

(c) Test the quadratic model in PERMISIV sextiles against the unconstrained model employed in part (b).

(d) Using the continuous and *centered* version of PERMISIV, estimate the quadratic model of ADVENTRE. Regardless of significance levels of the coefficients, give the slope of PERMISIV on ADVENTRE at the mean of PERMISIV and at 1 standard deviation below and above the mean. What type of curve is indicated? (Note that PERMISIV ranges from 4 to 20.)

5.11 Using the *couples dataset*, partition the variable DURYRS approximately into deciles, using the following ranges: ≤ 1.42, (1.42, 2.92], (2.92, 4.75], (4.75, 7],

(7, 10.92], (10.92, 15.92], (15.92, 20.58], (20.58, 29.75], (29.75, 40.83],
> 40.83.

(a) Plot the mean of DISAGMT against deciles of DURYRS. Does the trend appear linear?

(b) Use the decile-coded version of DURYRS to perform the test of linearity in the regression of DISAGMT on DURYRS.

(c) Test the quadratic model in DURYRS deciles against the unconstrained model employed in part (b).

(d) Using the continuous and *centered* version of DURYRS, estimate the quadratic model of DISAGMT. Regardless of the significance levels of the coefficients, give the slope of DURYRS on DISAGMT at the mean of DURYRS and at 1 standard deviation below and above the mean. What type of curve is indicated? (*Note*: DURYRS ranges from .17 to 60.58.)

5.12 Regard the following nonlinear model for $E(Y)$: $E(Y) = \alpha - \beta\gamma^x$. Letting $\alpha = 5$, $\beta = 2$, and $\gamma = .35$:

(a) Evaluate the slope (i.e., the first derivative of Y with respect to X) of X's impact on Y at $X = 0$, 2.5, and 5.

(b) Find $E(Y)$ at $X = 0$, 2.5, and 5.

(c) Graph the relationship between Y and X for X in the range $[0,5]$ assuming these parameter values.

5.13 Regard the following nonlinear model for $E(Y)$: $E(Y) = \log(\alpha + \beta X)$. Letting $\alpha = 5$, $\beta = 2$:

(a) Evaluate the slope of X's impact on Y at $X = 0$, 2.5, and 5.

(b) Find $E(Y)$ at $X = 0$, 2.5, and 5.

(c) Graph the relationship between Y and X for X in the range $[0,5]$ assuming these parameter values.

5.14 Regard the following nonlinear model for $E(Y)$:

$$E(Y) = \frac{\beta_1}{1 + \exp(\beta_2 + \beta_3 X)}.$$

Letting $\beta_1 = 5$, $\beta_2 = 2$, and $\beta_3 = .35$:

(a) Evaluate the slope of X's impact on Y at $X = 0$, 2.5, and 5.

(b) Find $E(Y)$ at $X = 0$, 2.5, and 5.

(c) Graph the relationship between Y and X for X in the range $[0,5]$ assuming these parameter values.

5.15 Identify the type of curve exemplified by each of the following equations for X in the range $[-10, 10]$.

(a) $\hat{y} = .5 + 2x + .5x^2$.

(b) $\hat{y} = 3 - .75x + .25x^2$.

(c) $\hat{y} = 5 - 3.25x - .15x^2$.

(d) $\hat{y} = 4 + 3x - 1.25x^2$.

5.16 Suppose that the model for $E(Y)$ is $E(Y) = 25 + .25x + .15x^2 + 3z$, where X and Z are centered and $s_z = 1.75$. Give the value of $E(Y)$ at the mean value of X and Z, at the mean of X but Z at $\bar{z} - 1$ s_z, and at the mean of X but Z at $\bar{z} + 1$ s_z.

5.17 Suppose that the model for $E(Y)$ is $E(Y) = -4 - 3x + .25x^2 - 1.5z + .15xz$, where X and Z are centered and $s_z = 1.75$. Give the value of $\partial[E(Y)]/\partial X$ at $\bar{z}$, $\bar{z} - 1$ s_z, and $\bar{z} + 1$ s_z.

5.18 Suppose that the model for $E(Y)$ is $E(Y) = 5 + 1.75x - 3z + .35xz$, where X and Z are centered and $s_z = 1.75$. Give the value of $\partial[E(Y)]/\partial X$ at $\bar{z}$, $\bar{z} - 1$ s_z, and $\bar{z} + 1$ s_z.

5.19 Suppose that the model for $E(Y)$ is $E(Y) = 15 - 3x + .25x^2 - 2z - .15xz + .25x^2z$, where X and Z are centered and $s_z = 1.75$. Give the value of $\partial[E(Y)]/\partial X$ at $\bar{z}$, $\bar{z} - 1$ s_z, and $\bar{z} + 1$ s_z.

5.20 Suppose that the model for $E(Y)$ is $E(Y) = -5.2 + 5x - 3z - .45x^2 + .25z^2 - .13xz - .07xz^2 + .15x^2z + .09x^2z^2$, where X and Z are centered and $s_z = 1.75$. Give the value of $\partial[E(Y)]/\partial X$ at $\bar{z}$, $\bar{z} - 1$ s_z, and $\bar{z} + 1$ s_z.

In Exercises 5.21 to 5.25, identify whether the equation for E(Y) is characterized by (a) a linear vs. nonlinear model, (b) a linear vs. nonlinear effect of X or Z (in the absence of any interaction effects only), and (c) a linear vs. nonlinear interaction in each of X and Z.

5.21 $E(Y) = \alpha + \beta \log x + \gamma z + \delta z \log x$.

5.22 $E(Y) = \alpha + \beta x + \gamma x^2 + \delta x^3 + \phi z^{1/2} + \lambda x z^{1/2}$.

5.23 $E(Y) = \log(\alpha + \beta x + \gamma z + \lambda x z)$.

5.24 $E(Y) = \alpha + \beta^x + \gamma^z + \lambda^{xz}$.

5.25 $E(Y) = \alpha + x^\beta + \gamma z + \lambda z^2 + \phi z x^\beta$.

CHAPTER 6

Advanced Issues in Multiple Regression

CHAPTER OVERVIEW

In this chapter I address a number of topics that are of a more advanced nature. Their complexity is due primarily to the necessity to resort to matrix algebra for much of their theoretical development. As matrix algebra may be foreign to many readers, this is arguably the most difficult chapter in the book. For this reason, the reader is strongly encouraged to read Section V of Appendix A before proceeding with this chapter. A familiarity with the notation and major concepts of matrix algebra will be extremely helpful for getting the most out of this material. On the other hand, those uncomfortable with the matrix developments can simply skip them and attend only to the "bottom line," as expressed in equations such as (6.5), (6.11), and (6.12) and accompanying discussions.

I begin by reviewing the matrix representation of the multiple regression model. I then take up the topic of heteroscedasticity and weighted least squares (WLS) estimation, the optimal estimation procedure when the homoscedasticity assumption fails. Along with this I discuss the use of WLS in testing slope homogeneity across groups when the assumption of equal error variance fails. I also consider the issue of using weighted regression on data from complex sampling schemes, employing WLS with sampling weights. This technique is referred to as *weighted ordinary least squares* (WOLS) (Winship and Radbill, 1994). I then return to the issue of omitted-variable bias, giving a formal development of the problem in the context of multiple regression. I also give an example showing how omitted-variable bias can affect interaction terms. The latter part of the chapter is devoted to regression diagnostics. In particular, I explain how to diagnose regression analyses for undue influence exerted by one or more observations. I also give a detailed explication of the problem

Regression with Social Data: Modeling Continuous and Limited Response Variables,
By Alfred DeMaris
ISBN 0-471-22337-9   Copyright © 2004 John Wiley & Sons, Inc.

of multicollinearity, its diagnosis, and possible remedies. The chapter ends by considering two techniques designed to improve on OLS estimates in the presence of severe collinearity: ridge regression and principal components regression.

MULTIPLE REGRESSION IN MATRIX NOTATION

The Model

In Section V of Appendix A, I outline the matrix representation of the multiple regression model. Let's review the basic concepts covered there. Recall that the matrix representation of the model for the ith observation is $y_i = \mathbf{x}^{i\prime}\boldsymbol{\beta} + \varepsilon_i$, where $\mathbf{x}^{i\prime}$ is a $1 \times p$ vector of scores on the p regressors in the model for the ith observation. Here, $p = K + 1$, and the first regressor score is a "1" that serves as the regressor for the intercept term. Further, $\boldsymbol{\beta}$ is a $p \times 1$ vector of the parameters in the model, with the first parameter being the intercept, β_0. Y_i and ε_i are the ith response score and the ith error term, as always. The matrix representation of the model for all n of the y scores is $\mathbf{y} = X\boldsymbol{\beta} + \boldsymbol{\varepsilon}$. Here, $\mathbf{y}$ is an $n \times 1$ vector of response scores, X is an $n \times p$ matrix of the regressor scores for all n observations, and $\boldsymbol{\varepsilon}$ is an $n \times 1$ vector of equation errors for the n observations. The ith row of X is, of course, $\mathbf{x}^{i\prime}$. As always, it is assumed that the errors have mean zero and constant variance σ^2 and are uncorrelated with each other. These assumptions are encapsulated in the notation $\boldsymbol{\varepsilon} \sim f(0, \sigma^2 I)$. This means that the errors have some density function, $f(\cdot)$ (typically assumed to be symmetric about zero, but not necessarily normal except for small samples) with zero mean and variance–covariance matrix $\sigma^2 I$. (Readers possibly used to the notation $\mathbf{x}_i\prime$ for the representation of the vector of regressor scores for the ith case may find the notation $\mathbf{x}^{i\prime}$ used in this book to be somewhat unusual. However, in that the ith case's regressor values are contained in the ith *row* of the $n \times p$ matrix of regressor values for all n observations, and as I use the superscript i to denote row vectors, the use of $\mathbf{x}^{i\prime}$ seems more appropriate. Note that the ith case's collection of regressor values written as a column vector is therefore denoted $\mathbf{x}^i$ throughout the book.)

OLS Estimates

The vector of OLS estimates of the model parameters is denoted $\mathbf{b}$, and as noted in Appendix A, its solution is $\mathbf{b} = (X'X)^{-1}X'\mathbf{y}$. In Chapter 2 I noted that b_1 in SLR was a weighted sum of the y_i and therefore normally distributed in large samples, due to the CLT. Similarly, each of the b_k in the multiple regression model is a weighted sum, or linear combination, of the y_i and is therefore also asymptotically normal. This is readily seen by denoting the $p \times n$ matrix $(X'X)^{-1}X'$ by the symbol G, and its kth row (where $k = 0, 1, \ldots, K$) as $\mathbf{g}^{k\prime}$. Then the kth regression estimate has the form $\mathbf{g}^{k\prime}\mathbf{y}$. Assuming that the X's are fixed over repeated sampling (the standard fixed-X assumption), this is nothing more than a weighted sum of the y's. The estimates are unbiased for their theoretical counterparts, since, as shown in Appendix A, $E(\mathbf{b}) = \boldsymbol{\beta}$. The variance–covariance matrix for $\mathbf{b}$, denoted $V(\mathbf{b})$, is $\sigma^2(X'X)^{-1}$. The variances of the parameter estimates lie on the diagonal of this matrix.

To illustrate the form of $\sigma^2(X'X)^{-1}$, let's derive the expressions for the variances of the SLR estimates, as given in Chapter 2, using matrix manipulations. In SLR, the X matrix can be written $[\mathbf{1} \quad \mathbf{x}]$, where $\mathbf{1}$ is a vector of ones and $\mathbf{x}$ is the column vector of scores on the independent variable. Therefore, $X'X$ is

$$X'X = \begin{bmatrix} \mathbf{1}' \\ \mathbf{x}' \end{bmatrix}[\mathbf{1} \quad \mathbf{x}] = \begin{bmatrix} \mathbf{1}'\mathbf{1} & \mathbf{1}'\mathbf{x} \\ \mathbf{x}'\mathbf{1} & \mathbf{x}'\mathbf{x} \end{bmatrix} = \begin{bmatrix} n & \sum x \\ \sum x & \sum x^2 \end{bmatrix}.$$

The determinant of $X'X$ is

$$n\sum x^2 - \left(\sum x\right)^2 = n\sum x^2 - n^2\bar{x}^2 = n\left(\sum x^2 - n\bar{x}^2\right) = nS_{xx},$$

where $S_{xx} = \sum(x - \bar{x})^2$ [see Appendix A, Section II.C(1)]. The inverse of $X'X$ is, therefore,

$$(X'X)^{-1} = \frac{1}{nS_{xx}}\begin{bmatrix} \sum x^2 & -\sum x \\ -\sum x & n \end{bmatrix} = \begin{bmatrix} \dfrac{\sum x}{nS_{xx}} & \dfrac{-\sum x}{nS_{xx}} \\ \dfrac{-\sum x}{nS_{xx}} & \dfrac{1}{S_{xx}} \end{bmatrix}$$

Finally, $\sigma^2(X'X)^{-1}$ is

$$\sigma^2(X'X)^{-1} = \begin{bmatrix} \dfrac{\sigma^2\sum x^2}{nS_{xx}} & \dfrac{-\sigma^2\sum x}{nS_{xx}} \\ \dfrac{-\sigma^2\sum x}{nS_{xx}} & \dfrac{\sigma^2}{S_{xx}} \end{bmatrix}.$$

As the reader can see, the expressions for $V(b_0)$ and $V(b_1)$ on the diagonal of $\sigma^2(X'X)^{-1}$ are the same as given in Chapter 2.

Hat Matrix. The fitted values in regression, denoted $\hat{y}_i$, are given by $\hat{y}_i = \mathbf{x}^{i\prime}\mathbf{b}$. The vector of fitted values is therefore given by $\hat{\mathbf{y}} = X\mathbf{b}$. Substituting for $\mathbf{b}$, we have $\hat{\mathbf{y}} = X(X'X)^{-1}X'\mathbf{y}$, or $\hat{\mathbf{y}} = H\mathbf{y}$. H, equal to $X(X'X)^{-1}X'$, is called the *hat matrix* (Belsley et al., 1980), since it converts $\mathbf{y}$ into $\hat{\mathbf{y}}$. This matrix plays a key role in the influence diagnostics discussed later in the chapter. Of particular interest are the diagonals of this matrix, denoted h_{ii}. These tap into the *leverage*, or potential for influence, exerted on the regression estimates by the ith observation. The matrix formula for h_{ii} is $h_{ii} = \mathbf{x}^{i\prime}(X'X)^{-1}\mathbf{x}^i$.

Regression Model in Standardized Form

Recall from Chapter 2 that the standardized slope in SLR results from the OLS regression of the standardized version of y on the standardized version of X. The same holds true in multiple regression. However, to understand the standardized representation of the MULR model, we must first examine the matrix representation of the standardized variable scores. Suppose that we denote by $\mathbf{y}_z$ the $n \times 1$ vector of standardized y-scores and by Z the $n \times K$ matrix of standardized $X=$ scores (the "1" disappears from this matrix in the standardization process; the standardized equation therefore has no intercept). Now, note that $(1/n)Z'Z = R_{xx}$, the correlation matrix for the X's, and $(1/n)Z'\mathbf{y}_z = \mathbf{r}_{xy}$, the vector of correlations between the X's and y. Why?

First, understand that the ith element in the kth column of Z is of the form

$$z_{ik} = \frac{x_{ik} - \bar{x}_k}{s_k},$$

where s_k is the standard deviation of the kth regressor. That is, the ikth element of Z is the kth variable minus its mean divided by its standard deviation for the ith case. Partitioning Z by its columns, $Z'Z$ is

$$Z'Z = \begin{bmatrix} \mathbf{z}_1' \\ \cdot \\ \cdot \\ \cdot \\ \mathbf{z}_K' \end{bmatrix} [\mathbf{z}_1 \cdots \mathbf{z}_K] = \begin{bmatrix} \mathbf{z}_1'\mathbf{z}_1 & \cdots & \mathbf{z}_1'\mathbf{z}_K \\ \cdot & \cdots & \cdot \\ \cdot & \cdots & \cdot \\ \cdot & \cdots & \cdot \\ \mathbf{z}_K'\mathbf{z}_1 & \cdots & \mathbf{z}_K'\mathbf{z}_K \end{bmatrix}.$$

This is a $K \times K$ matrix whose kth diagonal element is

$$\mathbf{z}_k'\mathbf{z}_k = \frac{\sum(x_{ik} - \bar{x}_k)^2}{s_k^2},$$

and whose off-diagonal elements are of the form

$$\mathbf{z}_k'\mathbf{z}_l = \frac{\sum(x_{ik} - \bar{x}_k)(x_{il} - \bar{x}_l)}{s_k s_l},$$

where k and l denote two different regressors in the model. Multiplying this matrix by $1/n$ results in the kth diagonal element being of the form

$$\frac{\sum(x_{ik} - \bar{x}_k)^2/n}{s_k^2} = 1,$$

and the off-diagonal elements being of the form

$$\frac{\sum(x_{ik} - \bar{x}_k)(x_{il} - \bar{x}_l)/n}{s_k s_l} = r_{kl}.$$

That is, the result is the correlation matrix for the X's in the model. [Technically, we should be multiplying by $1/(n-1)$ instead of $1/n$, but asymptotically these are equivalent, and $1/n$ simplifies the expression. It should be evident that the difference between $n-1$ and n is virtually nil in large samples.] A similar argument demonstrates that $(1/n)Z'\mathbf{y}_z = \mathbf{r}_{xy}$.

To write the model in standardized form, we shall find it convenient to transform the response and regressors as follows. Let $X^* = (1/\sqrt{n})Z$ and $\mathbf{y}^* = (1/\sqrt{n})\mathbf{y}_z$. Myers (1986, p. 76) refers to $\mathbf{y}^*$ and X^* as the vector and matrix, respectively, of *centered and scaled* variables. That is, the kth column of X^*, for example, represents a variable of the form

$$\frac{x_{ik}-\overline{x}_k}{\sqrt{\sum_{i=1}^{n}(x_{ik}-\overline{x}_k)^2}}.$$

The vector $\mathbf{y}^*$ represents Y in similar form. Then the standardized version of the model is $\mathbf{y}^* = X^*\boldsymbol{\beta}^s + \boldsymbol{\varepsilon}^*$, and the standardized estimates are obtained via the OLS solution:

$$\mathbf{b}^s = (X^{*\prime}X^*)^{-1}X^{*\prime}\mathbf{y}^* = \left[\left(\frac{1}{\sqrt{n}}Z\right)'\left(\frac{1}{\sqrt{n}}Z\right)\right]^{-1}\left(\frac{1}{\sqrt{n}}Z\right)'\frac{1}{\sqrt{n}}\mathbf{y}_z$$

$$= \left(\frac{1}{n}Z'Z\right)^{-1}\frac{1}{n}Z'\mathbf{y}_z = R_{xx}^{-1}\mathbf{r}_{xy}.$$

In other words, the standardized regression coefficients are the product of the inverse of the correlation matrix for the X's times the correlations of the X's with y. Further, letting σ_*^2 represent $V(\varepsilon^*)$, we then denote the variance–covariance matrix of the errors in the standardized equation by $\sigma_*^2 I$. Then the variance–covariance matrix of standardized parameter estimates is $V(\mathbf{b}^s) = \sigma_*^2(X^{*\prime}X^*)^{-1} = \sigma_*^2 R_{xx}^{-1}$. Having established the matrix representations for key elements in the MULR model, we are now ready to consider additional MULR topics.

HETEROSCEDASTICITY AND WEIGHTED LEAST SQUARES

Until now the standard assumption we have been operating under is that the equation errors are homoscedastic; that is, they have constant variance at each covariate pattern. In that case, $V(\varepsilon) = \sigma^2 I$. Suppose that this isn't the case. That is, suppose that the error variances vary across covariate patterns such that $V(\varepsilon_i) = \sigma_i^2$ for $i = 1, 2, \ldots, n$. Assuming that the errors are still uncorrelated, the form of $V(\varepsilon)$ is now

$$\begin{bmatrix} \sigma_1^2 & 0 & \cdots & 0 \\ 0 & \sigma_2^2 & \cdots & \cdot \\ \cdot & \cdots & \cdots & \cdot \\ \cdot & \cdots & \cdots & 0 \\ 0 & \cdots & 0 & \sigma_n^2 \end{bmatrix} = V.$$

Under this scenario the appropriate estimator is the *generalized least squares* (GLS) *estimator*, given by $\mathbf{b}_w = (X'V^{-1}X)^{-1}X'V^{-1}\mathbf{y}$ (Myers, 1986). If $V = \sigma^2 I$, $\mathbf{b}_w$ is simply $\mathbf{b}$, the OLS estimator. Because V is diagonal, and taking the inverse of a diagonal matrix is accomplished by simply inverting the diagonal elements, $\mathbf{b}_w$ has the form

$$\mathbf{b}_w = (X'D_{w_i}X)^{-1}X'D_{w_i}\mathbf{y},$$

where D_{w_i} indicates a diagonal matrix with diagonal entries $w_i = 1/\sigma_i^2$. Now, if we regard $\mathbf{b}_w$ more closely, we see that it can be expressed as

$$\mathbf{b}_w = [(D_{\sqrt{w_i}}X)'(D_{\sqrt{w_i}}X)]^{-1}(D_{\sqrt{w_i}}X)'D_{\sqrt{w_i}}\mathbf{y}.$$

Recall that premultiplying a matrix or vector by a diagonal matrix simply multiplies each row of that matrix or vector by the diagonal elements. Therefore, the design matrix now has the form

$$D_{\sqrt{w_i}}X = D_{\sqrt{w_i}}[\mathbf{1} \quad \mathbf{x}_1 \quad \cdots \quad \mathbf{x}_K] = \begin{bmatrix} \sqrt{w_1} & \sqrt{w_1}\,x_{11} & \cdots & \sqrt{w_1}\,x_{1K} \\ \sqrt{w_2} & \sqrt{w_2}\,x_{21} & \cdots & \sqrt{w_2}\,x_{2K} \\ \cdot & \cdot & \cdots & \cdot \\ \cdot & \cdot & \cdots & \cdot \\ \cdot & \cdot & \cdots & \cdot \\ \sqrt{w_n} & \sqrt{w_n}\,x_{n1} & \cdots & \sqrt{w_n}\,x_{nK} \end{bmatrix} = X_w,$$

and the response vector has the form

$$D_{\sqrt{w_i}}\mathbf{y} = \begin{bmatrix} \sqrt{w_1}\,y_1 \\ \sqrt{w_2}\,y_2 \\ \cdot \\ \cdot \\ \cdot \\ \sqrt{w_n}\,y_n \end{bmatrix} = \mathbf{y}_w.$$

If we regress $\mathbf{y}_w$ on X_w using OLS, the resulting estimator, $(X_w'X_w)^{-1}X_w'\mathbf{y}_w$, will be $\mathbf{b}_w$ (as the reader can verify by substituting for X_w and $\mathbf{y}_w$ in this expression). What this means is that $\mathbf{b}_w$ can be found by transforming the regressors and the response and then performing OLS on the transformed variables. The transformation involves multiplying the regressors and the response by the square root of the weight variable, where the weights are the reciprocals of the error variances. Hence this estimator is called the *weighted least squares* (WLS) *estimator*. Notice that the first column of the transformed design matrix is no longer a column of ones. Instead, it is a column of weights, which constitutes another variable. Consequently, the appropriate OLS regression analysis is a regression through the origin. The constant term for the WLS analysis is the coefficient for the weight variable resulting from this run (McClendon, 1994).

Properties of the WLS Estimator

If the proper weights, $1/\sigma_i^2$, are known, the WLS estimator $\mathbf{b}_w$ has the following properties: (1) it is unbiased for β; (2) it achieves the minimum variance of all linear unbiased estimators; and (3) it is the MLE for β if the errors are also normally distributed (Myers, 1986). Unfortunately, the true error variances are typically unavailable, so the proper weights are rarely known in practice. In this case, the error variances must first be estimated from the data available and then used in the weighting procedure. The resulting WLS estimator, also known as the *feasible generalized least squares* (FGLS) *estimator*, is no longer unbiased; however, it is still consistent and has a smaller sampling variance than that of the OLS estimator in large samples (Wooldridge, 2000).

Consequences of Heteroscedasticity

What are the consequences of using OLS in the presence of heteroscedasticity? First, the OLS estimators are inefficient. That is, there exist estimators with a smaller sampling variance: namely, the WLS (i.e., FGLS) estimator. This means that tests of significance will typically be more sensitive when using WLS than when using OLS. Perhaps more important, however, the estimated standard errors of the OLS coefficients obtained via the formula $\hat{\sigma}^2(X'X)^{-1}$ (the formula employed by all regression software) are no longer valid under heteroscedasticity. To see why, recall from Appendix A that for the OLS **b**,

$$V(\mathbf{b}) = V[(X'X)^{-1}X'\mathbf{y}] = (X'X)^{-1}X'V(\mathbf{y})X(X'X)^{-1}.$$

As long as $V(\mathbf{y}) = \sigma^2 I$, this matrix reduces to $\sigma^2(X'X)^{-1}$ and then $\hat{\sigma}^2(X'X)^{-1}$ becomes an unbiased estimator of $V(\mathbf{b})$. However, when heteroscedasticity prevails, $V(\mathbf{y}) = V$ (shown above). The variance of **b** is then

$$V(\mathbf{b}) = V[(X'X)^{-1}X'\mathbf{y}] = (X'X)^{-1}X'VX(X'X)^{-1}.$$

Hence, $\hat{\sigma}^2(X'X)^{-1}$ is no longer a valid estimator of this variance–covariance matrix.

White's Estimator of V(b). An alternative estimator of $V(\mathbf{b})$ which is robust to heteroscedasticity is the *White estimator* (White, 1980). This is based on the idea that the OLS **b** is a consistent estimator of $\boldsymbol{\beta}$, which implies that the OLS residuals are "pointwise consistent estimators" (Greene, 2003, p. 198) of the population ε_i. Moreover, the squared residuals would be consistent estimators of the squared ε_i, whose average values represent the variances of the ε_i, since $V(\varepsilon_i) = E(\varepsilon_i - E(\varepsilon_i))^2 = E(\varepsilon_i^2)$. Assuming at least one continuous predictor, the squared error for each case would typically be unique, so it represents its own average. Hence, the squared OLS residuals can be used to estimate the error variances in V, leading to the White heteroscedasticity-robust estimator,

$$V_w(\mathbf{b}) = (X'X)^{-1}X'\hat{V}X(X'X)^{-1},$$

where $\hat{V}$ is a diagonal matrix containing the squared OLS residuals (White, 1980).

Testing for Heteroscedasticity

In this section I discuss two tests for heteroscedasticity that are relatively easy to implement in any regression software that makes the residuals available for further manipulation. These tests are also available on request in some software (e.g., STATA, LIMDEP). The null hypothesis for either test is that the errors are homoscedastic. The first is White's test (White, 1980). For the substantive model $E(Y) = \beta_0 + \beta_1 X_1 + \beta_2 X_2 + \cdots + \beta_K X_K$, the test is accomplished by first estimating the model with OLS and then saving the residuals. One then regresses the squares of the residuals on all predictors in the model plus all nonredundant crossproducts among the predictors. For example, if the model is $E(Y) = \beta_0 + \beta_1 X_1 + \beta_2 X_2 + \beta_3 X_3$, one regresses the squared residuals from estimating this model on X_1, X_2, X_3, X_1^2, X_2^2, X_3^2, $X_1 X_2$, $X_1 X_3$, and $X_2 X_3$. The test is then nR^2 from this run, which, under the null hypothesis of homoscedasticity, is distributed as chi-squared with degrees of freedom equal to the number of regressors used to model the squared residuals (in this case, that would be 9). Be advised, however, that more than just homoscedasticity is being tested with this statistic. In fact, White (1980, p. 823) describes this test as follows: ". . . the null hypothesis maintains not only that the errors are homoskedastic, but also that they are independent of the regressors, *and* that the model is correctly specified . . . ." In other words, this is more of a general test for model misspecification.

The second test is the Breusch–Pagan test (Breusch and Pagan, 1979). This test is much more focused on heteroscedasticity than White's test, and, in fact, assumes that the error variance is related in some systematic fashion to the model predictors (Greene, 2003). There are various forms of the test, but the one I advocate here is that suggested by Wooldridge (2000). In this case, one regresses the squared OLS residuals from one's substantitve model on just the predictors in one's substantive model. Once again, the test is nR^2 from this regression, which is distributed as chi-squared with K degrees of freedom under the null hypothesis of homoscedasticity.

Example: Regression of *Coital Frequency*

Let's consider an example of a heteroscedastic model. Figure 6.1 presents a scatterplot of *coital frequency in the last month* against the *male partner's age* for the 416 couples in the *couples dataset*. (This is similar to Figure 2.7, which shows a plot of the residuals from this regression against *male partner's age*.) Nonconstant error variance is suggested by the wedge-shaped trend in the points in which the spread of points tapers down dramatically from left to right. This phenomenon makes substantive sense. One would expect that there would be considerable variability in coital frequency among young couples, since some are by choice more sexually active than others. However, age brings its own limitations to sexual activity, regardless of individual proclivities. We would therefore expect less variability in sexual frequency as couples enter middle and old age.

Table 6.1 presents regression results for these data. The first column shows the OLS results for the regression of *coital frequency* on *male's age*. The effect of *male's age* is negative and significant and suggests that each additional year of *age* reduces the couple's *coital frequency* by about .15 time per month. The effect is quite significant. The second column shows the standard errors estimated by White's heterscedasticity-robust technique. Compared to the OLS standard errors, the White standard errors are larger for the intercept but slightly smaller for the effect of *age*. No real substantive difference would result from using these standard errors in place of the OLS ones, however. The third column presents the model for White's test, which includes both *male age* and its

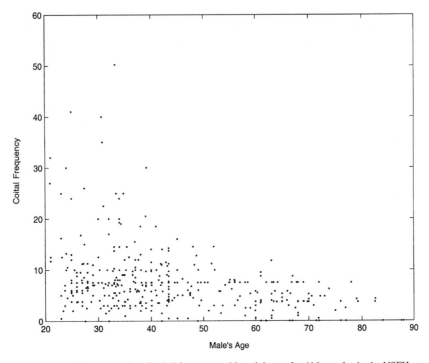

Figure 6.1 Scatterplot of coital frequency with male's age for 416 couples in the NSFH.

square. The dependent variable for this run is the squared residual from the model in the first column. White's test is therefore $416(.0423) = 17.597$. With 2 degrees of freedom, this is quite significant ($p = .00015$). For the Breusch–Pagan test, we regress the squared residuals from the substantive model on *male's age* (results not shown). The R^2 from this run is .0328. Hence, the Breusch–Pagan test is $416(.0328) = 13.645$,

Table 6.1   OLS and WLS Results for the Regression of Coital Frequency on Male Age for 416 Couples in the NSFH

Predictor	OLS: $b(\hat{\sigma}_b)$	White $\hat{\sigma}_b$	Model for White's Test[a]	WLS: $b(\hat{\sigma}_b)$
Intercept	14.028***		201.057***	13.522***
	(.874)	(.998)		(.992)
Male age	−.149***		−6.234**	−.143***
	(.019)	(.018)		(.016)
(Male age)^2			.049*	
R^2_{OLS}	.1295		.0423	.1570
R^2_{WLS}				.1280

[a] Response variable is the squared OLS residual.
* $p < .05$. ** $p < .01$. *** $p < .001$.

which with 1 degree of freedom is, again, highly significant ($p = .00022$). Apparently, both tests result in a sound rejection of the null hypothesis of constant error variance.

WLS in Practice: Two-Step Procedure

To estimate the regression for coital frequency via WLS, we must first estimate the error variances. One possibility is to regress the squared OLS residuals from the substantive model on the model's predictors and then to use the fitted values from this run as our estimates of σ_i^2. The justification for this is that the fitted squared residuals are consistent estimators of the expected squared residuals, which, as argued above, represent the error variances (McClendon, 1994). More generally, we regress the squared residuals on a set of explanatory variables which may or may not coincide with the substantive predictors, but which appear to determine the error variance (Greene, 2003). In the present case, I regressed the squared residuals on *male's age* and the square of *male's age*—the same model as used for White's test—and used the fitted values to create the weights. Failure to include the square of *male age* would result in several cases having negative weights and therefore being excluded from the regression. In this case, including the quadratic term is an easy way to prevent that. (Below I consider another technique for ensuring that the weights remain positive.) The weights are then simply the reciprocals of the fitted values. That is, I take $w_i = 1/\hat{e}_i^2$ as the weights for the WLS regression. (Most regression software has an option for running weighted regression; in SAS, for example, one simply includes a WEIGHT statement followed by the name of the weight variable.) The results are shown as the last column of Table 6.1. The magnitudes of both intercept and slope have been reduced slightly. The standard error of the slope has also been reduced, consistent with the WLS estimator exhibiting lower variance compared to OLS. Nonetheless, substantive conclusions are little affected by correcting for heteroscedasticity here.

R^2 for WLS. The R^2 for the OLS model is shown at the bottom of the first column of Table 6.1, suggesting that about 13% of the variation in *coital frequency* is accounted for by *male age*. Two R^2's are shown for the WLS run in the last column. The first, R_{OLS}^2, is the one reported by the software from the WLS analysis. It should appear suspicious to the reader. Because the OLS estimators result in the smallest *SSE* compared to any other estimators, they necessarily produce the highest R^2 in any given sample. Therefore, the WLS estimates *cannot* result in a higher R^2. The "catch" is that the number reported by software is for the transformed (by the weights) data, not the original data. To get the correct R^2 value, we need to "hand-calculate" (with the aid of a computer) the fitted values using the WLS estimates, then use these fitted values to construct the WLS residuals. We then sum the squared WLS residuals to get SSE_w, and then R_{WLS}^2 is calculated as

$$R_{WLS}^2 = 1 - \frac{SSE_w}{TSS}.$$

This value is also shown in the WLS column and is slightly smaller than R_{OLS}^2, as expected.

Ensuring Positive Weights. Estimating the error variances by performing an OLS regression on the squared residuals to get the fitted values will not always work. Very often, several of the fitted values will be negative and there is no way to model that problem away. A solution proposed by Wooldridge (2000) is to assume that the squared errors are related to the model predictors via an exponential function. In particular, we suppose that the squared errors are related to the predictors via the following equation:

$$\varepsilon^2 = \sigma^2 \exp(\delta_0 + \delta_1 X_1 + \delta_2 X_2 + \cdots + \delta_K X_K)u,$$

where u, the error term in this model, has mean 1 and is orthogonal to the predictors. This is really no more arbitrary than the assumption that the squared error is linearly related to the predictor set but has the additional advantage of ensuring that ε^2 is always positive. The equation is then transformed so that it can be estimated via OLS:

$$\log \varepsilon^2 = \alpha_0 + \delta_1 X_1 + \delta_2 X_2 + \cdots + \delta_K X_K + \upsilon, \tag{6.1}$$

where $\alpha_0 = \log \sigma^2 + \delta_0$ and $\upsilon = \log u$. According to Wooldridge, (6.1) now satisfies the classic regression assumptions and we can therefore use OLS to get unbiased estimates of its parameters. Using these, we obtain the fitted values and then transform them using the exponential function to get the estimated squared residuals for the weights. In other words, the procedure is:

1. Run the regression of Y on $X_1, X_2, \ldots, X_K$ and save the residuals, e_i.
2. Create the variable $\log e_i^2$.
3. Regress $\log e_i^2$ on $X_1, X_2, \ldots, X_K$ and get the fitted values, $\log \hat{e}_i^2$.
4. Exponentiate $\log \hat{e}_i^2$ to recover $\hat{e}_i^2$.
5. Regress Y on $X_1, X_2, \ldots, X_K$ via WLS using as weights $w_i = 1/\hat{e}_i^2$.

Coital Frequency *Revisited*. Table 6.2 presents results for the regression of *coital frequency* on *male age* (in years), *female age* (in years), *union duration* (duration of the union in years as of wave 1), *couple modernism*, and *couple disagreement* (interval variable ranging from 1 = "minimal disagreement" to 6 = "maximum disagreement") for our 416 couples. The first column is, again, the result of OLS estimation. It appears that in addition to *male age*, *couple modernism* has a negative impact on sexual activity. This is probably due to more modern couples being more gender egalitarian. Such couples are likely to accord greater weight to the woman's desires regarding the frequency of sex, and women typically desire sex less often than men. Net of other factors, the female's age is also negatively related to coital activity, but the effect is not quite significant.

As in the SLR model, I again suspect heteroscedasticity. This appears to be confirmed in Figure 6.2, which is a plot of the residuals against the fitted values. The figure reveals that the error variance appears to increase dramatically with increasing values of predicted *coital frequency*. The White standard errors for the OLS run

Table 6.2   OLS and WLS Results for the Multiple Regression of Coital Frequency on Selected Predictors for 416 Couples in the NSFH

Predictor	OLS: $b(\hat{\sigma}_b)$	White $\hat{\sigma}_b$	Model for B-P Test[a]	WLS: $b(\hat{\sigma}_b)$
Intercept	24.158***		155.102*	19.433***
	(3.723)	(3.972)		(3.147)
Male age	−.092*		−.815	−.093**
	(.040)	(.028)		(.030)
Female age	−.065		−.405	−.047
	(.044)	(.031)		(.032)
Union duration	−.033		−.397	−.026
	(.033)	(.024)		(.022)
Modernism	−.287*		.2.482	−.156
	(.114)	(.120)		(.095)
Disagreement	−.958		.738	−.791
	(.513)	(.528)		(.441)
R^2_{OLS}	.1577		.0373	.2431
R^2_{WLS}				.1530

[a] Response variable is the squared OLS residual.
* $p < .05$. ** $p < .01$. *** $p < .001$.

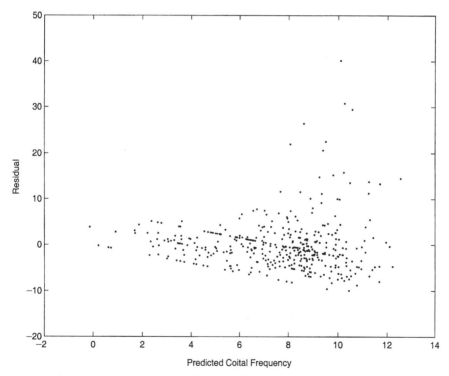

Figure 6.2   Scatterplot of residuals against fitted values from the multiple regression of coital frequency on selected predictors for 416 couples in the NSFH.

are shown in the second column of Table 6.2. About half are larger than the OLS standard errors and about half are smaller. Using the White standard errors, however, the main substantive change would be that *female age* would become significant at $p < .04$. The model for the Breusch–Pagan test is shown in the third column of the table. The test statistic value is 15.517 and is chi-squared with 5 degrees of freedom under the null. With a p-value of .0084, we would reject homoscedasticity. White's test statistic (not shown), on the other hand, is 27.997 and has 20 degrees of freedom, a nonsignificant result $(p > .1)$. Nonetheless, as the Breusch–Pagan test concentrates its power specifically on heteroscedasticity, it will be more trustworthy here. The last column of the table shows the WLS analysis based on using equation (6.1) to estimate the weights. Notice that all of the standard errors are smaller compared to those from OLS, evidence again of the greater efficiency of WLS. However, the only significant predictor of *coital frequency* in the model is *male age*.

Testing Slope Homogeneity with WLS

In Chapter 3 I presented salary models for male and female faculty at BGSU and tested whether they were the same. Although the test was significant, suggesting that the effects of predictors on salary were different across gender, a critical assumption for this test—equal error variances across groups—was found in Chapter 4 to be violated. As I indicated in Chapter 4, when the equal error variance assumption is violated, there is a WLS procedure that still allows us to test regression slope homogeneity—or equality of predictor effects—across groups. This approach is, however, restricted to having only two groups. As outlined by Overton (2001), the idea is as follows. Suppose that we have two groups in which the following models hold:

$$Y_1 = \sum \beta_k^1 X_k + \varepsilon_1, \; V(\varepsilon_1) = \sigma_1^2,$$
$$Y_2 = \sum \beta_k^2 X_k + \varepsilon_2, \; V(\varepsilon_2) = \sigma_2^2,$$

where the subscripts/superscripts 1 and 2 stand for groups 1 and 2, and $\sigma_1^2 \neq \sigma_2^2$. Suppose further that each group's data are weighted by the reciprocals of their error standard deviations. Then we have

$$\frac{1}{\sigma_1} Y_1 = \frac{1}{\sigma_1} \sum \beta_k^1 X_k + \frac{1}{\sigma_1} \varepsilon_1$$

and

$$\frac{1}{\sigma_2} Y_2 = \frac{1}{\sigma_2} \sum \beta_k^2 X_k + \frac{1}{\sigma_2} \varepsilon_2.$$

In this case, the new error variances are

$$V\left(\frac{1}{\sigma_1} \varepsilon_1\right) = \frac{1}{\sigma_1^2} V(\varepsilon_1) = \frac{\sigma_1^2}{\sigma_1^2} = 1 \quad \text{and} \quad V\left(\frac{1}{\sigma_2} \varepsilon_2\right) = \frac{1}{\sigma_2^2} V(\varepsilon_2) = \frac{\sigma_2^2}{\sigma_2^2} = 1.$$

That is, the weighting procedure restores error-variance homogeneity across groups, rendering tests for group-covariate interactions valid (Overton, 2001). It should be noted that the case of unequal error variances across groups is not the same as heteroscedasticity, which is inequality of error variances across cases *within* groups. It is still assumed in this procedure that the errors are homoscedastic within each group. WLS in this case is used to correct only for unequal error variance *across* groups (Overton, 2001).

To employ WLS to test slope homogeneity across two groups, we proceed as follows:

1. Regress Y on the model's main effects separately in each of the two groups.
2. Compute $MSE^* = SSE/(df_E - 2)$ separately for each group, based on each regression. The adjustment to the error degrees of freedom is necessary to correct for the bias in the reciprocals of the error variances when used as weights (see Overton, 2001, pp. 221–222).
3. Invert each MSE^* to create each group's weight.
4. Run WLS using the group weights in the combined sample, along with group-covariate cross-product terms, to examine differences in predictor effects across groups.

Gender Differences in Salary Models, Revisited

Let's tackle the issue of potential gender differences in the models for faculty salary one more time. This time, having found that the error variances are unequal, I use the WLS procedure outlined above. I employ the same model as in Table 3.5 except that this time I omit the interaction of *years at the university* with *years in rank*. The basic model is a regression of faculty salary on centered versions of *prior experience, years in rank, years at BG*, and *marketability*. First I estimate the model separately for each gender (results not shown). A test for error-variance homogeneity (just to be sure that omitting the interaction term doesn't change things) results in an F of 1.527, which with 506 and 209 degrees of freedom, is again highly significant ($p < .001$). I then create the weights for each gender. For males, $SSE/(df_E - 2) = 41286268592/504 = 81917199.587$. The reciprocal of this is the males' weight. For females $SSE/(df_E - 2) = 11166831326/207 = 53946045.053$. Its reciprocal is the females' weight. I then reestimate the basic model in the combined sample, weighting the cases differentially, according to gender. The results of the WLS analysis are shown in Table 6.3.

Model 1 is the main effects model. We see that the continuous covariates are all significant with the exception of *years in rank*. The dummy for being female is also quite significant and suggests that female salary is, on average, $4722.43 lower than male salary. Model 2 adds the cross-products of *female gender* with the continuous covariates. A nested F test, based on the R^2's from models 1 and 2 and reported at the bottom of the model 2 column, suggests that the block of interaction terms is significant. This essentially agrees with the results of the Chow test reported in Chapter 3. Apparently this result is robust to correction for error variance heterogeneity. Two of the individual cross-product terms are significant. The more significant interaction effect is that of *female gender* with *prior experience*. The effect suggests that, for males, each additional year of prior experience is worth a $1042.845 increment in

Table 6.3   WLS Results for Testing Slope Homogeneity for Male vs. Female Faculty in Effects of Continuous Covariates on Salary for 725 Faculty Members

Predictor	Model 1	Model 2
Intercept	49195.000***	49036.000***
Female	−4722.427***	−5213.543***
Prior experience[a]	820.422***	1042.845***
Years in rank[a]	−115.306	−47.809
Years at BG[a]	988.633***	1009.120***
Marketability[a]	31417.000***	33601.000***
Female × prior experience[a]		−662.797***
Female × years in rank[a]		−447.862*
Female × years at BG[a]		115.478
Female × marketability[a]		−7618.157
ΔF		7.654***

[a] Centered variable.
* $p < .05$. ** $p < .01$. *** $p < .001$.

academic-year salary. For females, each additional year is worth only 1042.845 − 662.797 = \$380.048 in additional salary.

Overton (2001) notes that an advantage of the WLS procedure employed here is that correct follow-up tests can be pursued to further explore the nature of the interaction effect. For example, given that the strongest interaction appears to be that between *female gender* and *prior experience*, suppose that I wish to test whether the gender difference in salary is significant at different levels of prior experience. A useful technique is to use the *uncentered* version of *prior experience* and rescale it as necessary to isolate the difference coefficient. (In Chapter 8 I refer to this scaling technique as *targeted centering* and discuss it in more detail there.) In other words, the equation is: $\hat{y} = b_0 + d\,female + b_1\,PE + b_2\,years\ in\ rank + b_3\,years\ at\ BG + b_4\,marketability + g_1\,female * PE$. To begin, I let $PE = prior\ experience$. Then for those with no prior experience, $PE = 0$. For males, estimated mean salary is $\hat{y} = b_0 + b_2\,years\ in\ rank + b_3\,years\ at\ BG + b_4\,marketability$. For females, it's $\hat{y} = b_0 + d + b_2\,years\ in\ rank + b_3\,years\ at\ BG + b_4\,marketability$. The estimated difference in mean salary is d; therefore, a test for the significance of d is a test for gender differences in mean salary for those with no experience. Okay, you already knew that. But suppose that we want to test gender differences at two years of experience. Then we let $PE = prior\ experience − 2$, and also employ it for the cross-product term. For those with two years of experience, $PE = 0$ again, and again, a test for d is a test for the gender difference in mean salary for *those with two years of experience*. In general, to conduct the test of gender difference at c years of experience, we estimate the model with $PE = prior\ experience − c$, form the cross-product using this version of PE, and use the t test for d. Running this procedure with c set alternately to 0, 2, and 10 years of *prior experience* produces the following mean differences in female minus male average salary: −2885.81 (0 years), −4119.68 (2 years), and −9055.18 (10 years). All differences are highly significant.

WLS with Sampling Weights: WOLS

A final application of WLS that should be mentioned is the use of WLS for weighting regression analyses with sampling weights. Sampling weights are typically provided with secondary data gathered from complex sampling designs. In such designs, different individuals have different probabilities of selection into the sample, and the weights are used to make the sample distributions on one's variables resemble their population counterparts. This is important for the estimation of univariate parameters such as means and proportions. It is often thought that regression analyses should also be weighted with sampling weights in order to achieve correct inferences. Winship and Radbill (1994), however, explain why this is not an advisable practice most of the time.

First, if the unweighted data are homoscedastic, sampling weights will make them heteroscedastic. Why? Suppose that the correct model for the data is

$$Y_i = \sum \beta_k X_{ik} + \varepsilon_i,$$

where ε_i is normal with zero mean and variance σ^2. However, we weight the data before analysis, so that the data are actually

$$\sqrt{w_i} Y_i = \sqrt{w_i} \sum \beta_k X_{ik} + \sqrt{w_i} \varepsilon_i.$$

Now the variance of the errors is $V(\sqrt{w_i}\varepsilon_i) = w_i\sigma^2$, which of course implies heteroscedasticity. The immediate consequence of this is that as we have seen, standard errors produced by regression software are no longer valid. Weighted estimators are optimal and produce correct estimates of standard errors only when the weights are a function of the error variances (Myers, 1986).

Second, if sampling weights are only a function of the independent variables included in the model being estimated, unweighted OLS is really the appropriate procedure. Using versus not using weights would be equivalent to drawing samples exhibiting different distributions on the X's. Yet none of the classic regression assumptions requires that the distributions of the X's in the sample mirror those in the population. In fact, if the model is specified correctly, samples exhibiting different distributions on the X's should produce the same OLS estimates (Winship and Radbill, 1994). But if the model is misspecified, samples with different X distributions may very well produce different regression estimates. So if the parameter estimates from weighted and unweighted analyses differ, this suggests either that the model is misspecified or that the weights are a function of the dependent variable (Winship and Radbill, 1994).

To assess whether weighted and unweighted analyses produce different estimates, Winship and Radbill (1994) recommend employing the test devised by DuMouchel and Duncan (1983). Letting W represent the weight variable, the test is as follows. First, we estimate the substantive model:

$$E(Y) = \beta_0 + \sum \beta_k X_k. \tag{6.2}$$

Then we add the weight variable plus all interactions of the weight variable with the model's predictors:

$$E(Y) = \beta_0 + \sum \beta_k X_k + \delta W + \sum \gamma_k W X_k. \tag{6.3}$$

Both analyses are done using OLS. The test is a nested F test of (6.2) versus (6.3). If it is nonsignificant, weighted and unweighted analyses do not differ and the analyst should proceed with OLS. If the test is significant, model (6.3) should be examined more closely. If the weight variable itself has a significant effect, this suggests that the weights are a function of X's omitted from the model. If one or more WX interaction terms is signficant, there may be imporant interaction terms missing from the model. The analyst may then be able to respecify the model and perform the test again. If the difference between weighted and unweighted analyses is still significant, the analyst should then employ weighted regression using the sampling weights. In that case, however, Winship and Radbill (1994) suggest employing White's heteroscedasticity-robust estimator of $V(\mathbf{b})$ to obtain the standard errors of coefficients.

Example. Table 6.4 presents weighted and unweighted analyses of data from wave one of the NSFH on 7273 intimate couples (married as well as cohabiting unmarried).

Table 6.4   OLS and WOLS Results for the Regression of Couple Disagreement on Demographic Predictors for 7273 Couples in the NSFH

Predictor	Model 1	Model 2	Model 3
Intercept	11.122***	11.044***	11.171***
Cohabiting couple	.400*	−.094	.517*
Union duration	−.066***	−.066***	−.066***
Biological children	1.894***	1.951***	1.873***
Stepchildren	1.194***	1.030***	1.260***
Minority couple	.109	.584**	−.043
Male education	.001	−.018	.006
Female education	.007	.018	.003
Income-to-needs ratio	−.010	−.002	−.012
Case weight		.063	
Weight × cohabiting couple		.631	
Weight × union duration		−.000[a]	
Weight × biological children		−.060	
Weight × stepchildren		.215	
Weight × minority couple		−.462**	
Weight × male education		.017	
Weight × female education		−.009	
Weight × income-to-needs ratio		−.007	
RSS	17597.692	17773.178	

[a] Actual value is −.0003.
* $p < .05$. ** $p < .01$. *** $p < .001$.

The dependent variable is *couple disagreement*: the average of male and female partners' reports of how often the couple has a serious disagreement (interval variable ranging from 6 = "minimal disagreement" to 36 = "maximum disagreement"). The NSFH data are from a complex sampling design in which certain groups were oversampled: cohabitors, recently married couples, minorities, stepparent families, and one-parent families. Model 1 presents the results from the unweighted regression. Among the predictors are male and female *education* (in years of schooling completed) as well as the *income-to-needs ratio* (the ratio of household income to the poverty level for that type of household). As the model also includes dummies for *cohabitation*, being a *minority couple*, having only *biological children*, and *having stepchildren* (the contrast is being childless), as well as the continuous covariate *union duration* (in years), the sampling weights are functions of model predictors. We would therefore expect that weighting would make no difference in the parameter estimates. *MSE* for model 2 is 16.509. The last row of the table shows the *RSS* values necessary to compute DuMouchel and Duncan's (1983) test statistic, which is

$$F = \frac{(17773.178 - 17597.692)/9}{16.509} = 1.181.$$

The *p*-value for this test statistic is greater than .3, suggesting that weighted and unweighted analyses do not differ. Model 3 presents the weighted estimates for comparison purposes. Clearly, there is no substantive difference between model 3 and the OLS results. Although the nested *F* is nonsignificant, we notice that there is one significant interaction in model 2 between the weight variable and the dummy *minority couple*. If the DuMouchel and Duncan test had been significant, this term might suggest an omitted interaction. In previous analyses of these data, I did find an interaction between *minority status* and the *income-to-needs ratio* in their effects on *couple violence* (DeMaris, 2003). I therefore checked to see if the same interaction effect was significant for *couple disagreement* (results not shown), but it was not. In sum, the estimates for model 1 would seem to represent the optimal estimates for these data.

OMITTED-VARIABLE BIAS IN A MULTIVARIABLE FRAMEWORK

In Chapter 3 I presented a relatively simplified explication of omitted-variable bias. In this section I present a more general framework from which to understand this issue. In the process I show that omitted variables can also confound, suppress, or mediate the effects of higher-order terms such as cross-products and quadratic effects. First, recall that an interaction effect is of the form $\beta_k X_k + \gamma_k X_j X_k$, where the subscripts j and k refer to two different predictors in the model. The effect of X_k is $\beta_k + \gamma_k X_j$, which shows that the effect of X_k depends on the level of X_j. If γ_k is zero, there is no interaction and the effect of X_k is constant over levels of X_j. Similarly, a quadratic effect is of the form $\beta_k X_k + \gamma_k X_k^2$. The effect (partial derivative) of X_k is $\beta_k + 2\gamma_k X_k$. If γ_k is again zero, there is no quadratic effect and the relationship between Y and X_k is linear. The point is that the coefficient of the cross-product term or the quadratic term represents the departure from additivity or linearity, respectively. That is, the higher-order term is what "carries" the effect in question. Hence, it is the association of this term with omitted variables that may lead to bias.

Mathematics of Omitted-Variable Bias

To understand the effect of variable omission, let's suppose that the true model for Y is $y = X\beta + \varepsilon$, with ε being normal with zero mean and variance $\sigma^2 I$. Let's further partition the X matrix as $[X_1 \quad X_2]$, where the matrix X_1 consists of the variables that the researcher includes in his or her model, and the matrix X_2 consists of variables left out of the researcher's model. The corresponding partitioning of β is $\beta' = [\beta_1' \quad \beta_2']$, where β_1 contains the parameters associated with X_1 and β_2 those associated with X_2. We further stipulate that X_1 has K variables, the first being a column of ones, and X_2 has Q variables. Similarly, β_1 has K parameters, including the intercept, and β_2 has Q parameters. Without loss of generality I also assume that the Kth variable in X_1 is a cross-product term of the form $X_j X_k$. Then $y = X\beta + \varepsilon$ becomes $y = X_1\beta_1 + X_2\beta_2 + \varepsilon$ (in multiplying these partitioned entities we treat X like a row vector). But instead, the researcher estimates $y = X_1\beta_1 + \upsilon$, where $\upsilon = X_2\beta_2 + \varepsilon$. What is the result?

The OLS estimator for the researcher's model is

$$b_1 = (X_1'X_1)^{-1}X_1'y$$

which, in actuality, is

$$\begin{aligned} b_1 &= (X_1'X_1)^{-1}X_1'[X_1\beta_1 + X_2\beta_2 + \varepsilon] \\ &= (X_1'X_1)^{-1}X_1'X_1\beta_1 + (X_1'X_1)^{-1}X_1'X_2\beta_2 + (X_1'X_1)^{-1}X_1'\varepsilon \\ &= \beta_1 + (X_1'X_1)^{-1}X_1'X_2\beta_2 + (X_1'X_1)^{-1}X_1'\varepsilon \end{aligned}$$

which means that

$$E(b_1) = \beta_1 + (X_1'X_1)^{-1}X_1'X_2\beta_2 + (X_1'X_1)^{-1}X_1'E(\varepsilon) = \beta_1 + (X_1'X_1)^{-1}X_1'X_2\beta_2.$$

Hence, the term $(X_1'X_1)^{-1}X_1'X_2\beta_2$ represents the bias in b_1 due to the excluded regressors (since, formally, the bias in b_1 is defined as $E(b_1 - \beta_1) = E(b_1) - \beta_1$).

It is worthwhile to consider this bias in greater detail. The matrix $(X_1'X_1)^{-1}X_1'X_2$, referred to as the *alias matrix* (Myers, 1986), represents the matrix of least squares solutions for the regression of each of the predictors in X_2 on the set of regressors in X_1. Why? Let's partition X_2 as $[x_{21} \quad x_{22} \quad \cdots \quad x_{2Q}]$, adding a "2" subscript to the column vectors to denote that these are the columns of X_2. Then

$$\begin{aligned} (X_1'X_1)^{-1}X_1'X_2 &= (X_1'X_1)^{-1}X_1'[x_{21} \quad x_{22} \quad \cdots \quad x_{2Q}] \\ &= [(X_1'X_1)^{-1}X_1'x_{21} \quad (X_1'X_1)^{-1}X_1'x_{22} \cdots (X_1'X_1)^{-1}X_1' \ x_{2Q}]. \end{aligned} \quad (6.4)$$

It should be clear from (6.4) that each column of the $K \times Q$ alias matrix is a vector of OLS coefficients for the regression of a given column of X_2 (i.e., a given regressor in X_2) on all of the regressors in X_1. Suppose that we denote the alias matrix as G. Then it has the form

$$
G = \begin{bmatrix} \mathbf{g}^{1'} \\ \mathbf{g}^{2'} \\ . \\ . \\ . \\ \mathbf{g}^{K'} \end{bmatrix} = \begin{bmatrix} g_{11} & g_{12} & \cdots & g_{1Q} \\ g_{21} & g_{22} & \cdots & g_{2Q} \\ . & . & \cdots & . \\ . & . & \cdots & . \\ . & . & \cdots & . \\ g_{K1} & g_{K2} & \cdots & g_{KQ} \end{bmatrix},
$$

where each g_{ki} is the partial regression coefficient for the impact of the kth regressor in X_1 on the ith regressor in X_2. Hence, g_{11} is the partial effect of regressor 1 in X_1 on regressor 1 in X_2, g_{21} is the partial effect of regressor 2 in X_1 on regressor 1 in $X_2, \ldots, g_{K1}$ is the partial effect of regressor K in X_1 (in this case, the cross-product term) on regressor 1 in X_2, and so on. The bias in $\mathbf{b}_1$ as an estimator of $\boldsymbol{\beta}_1$ is then

$$
(X_1'X_1)^{-1}X_1'X_2\boldsymbol{\beta}_2 = \begin{bmatrix} g_{11} & g_{12} & \cdots & g_{1Q} \\ g_{21} & g_{22} & \cdots & g_{2Q} \\ . & . & \cdots & . \\ . & . & \cdots & . \\ . & . & \cdots & . \\ g_{K1} & g_{K2} & \cdots & g_{KQ} \end{bmatrix} \begin{bmatrix} \beta_{1y} \\ \beta_{2y} \\ . \\ . \\ . \\ \beta_{Qy} \end{bmatrix},
$$

where β_{qy} in the rightmost vector is the partial effect of the qth variable in X_2 on Y. The bias in the kth coefficient in X_1 is therefore

$$
\mathbf{g}^{k'}\boldsymbol{\beta}_2 = \sum_{q=1}^{Q} g_{kq}\beta_{qy}, \tag{6.5}
$$

which is the sum of the products of the elements in the kth row of G with the elements in the vector $\boldsymbol{\beta}_2$. That is, there is bias in the kth coefficient whenever the kth regressor in X_1 is associated with the qth regressor in X_2, net of the other regressors in X_1 (i.e., g_{kq} is nonzero), and the qth regressor in X_2 is a predictor of Y, net of the other regressors in X_2 (i.e., β_{qy} is nonzero).

Bias in the Cross-Product Term

The bias in the Kth term in X_1, the cross-product term, takes the same form as equation (6.5): namely,

$$
\mathbf{g}^{k'}\boldsymbol{\beta}_2 = \sum_{q=1}^{Q} g_{Kq}\beta_{qy}.
$$

This suggests that *the cross-product term is biased whenever X_k and X_j interact in their effects on some excluded regressor that also affects Y.* (The same applies to quadratic effects: They are biased whenever X_k has a quadratic effect on some excluded regressor that also affects Y.) Although researchers are usually very careful to include controls to preclude omitted-variable bias in the main effects of focus variables, they typically treat interaction terms as though they are impervious to the same phenome-non. The foregoing shows that they are not. This suggests that analysts should also be aware of potential confounds or suppressors when they are investigating interaction (or quadratic) effects. Equally important, they may want to consider covariates that could mediate these effects. That is, certain covariates may be the mechanisms that are responsible for bringing about the interaction (or nonlinearity).

Example: Bias in Models for Faculty Salary

Table 6.5 presents the results of several models that examine the main effect of gen-der and the interaction of gender with college in their effects on faculty salary for our 725 faculty members in the *faculty salary dataset*. Model 1 contains the main effects of *gender* and *college*, both factors represented by dummy variables. Here we see that the "firelands college" and "other departments" are significantly lower, and the "business college" is significantly higher, in average salary than "arts and sciences," controlling for *gender*. We see also that holding *college* constant, female faculty members' salaries are, on average, $11,799 lower than males'. Model 2 adds the block of four cross-product terms representing the interaction of gender with col-lege. The test for the significance of the interaction is

$$F_{(4,715)} = \frac{(.2372 - .2130)/4}{(1 - .2372)/715} = 5.671.$$

Table 6.5 Omitted-Variable Bias in the Interaction of Gender and College in Their Effects on Salary for 725 Faculty Members

Predictor	Model 1	Model 2	Model 3	Model 4
Intercept	51793.000***	52420.000***	−2313.589	−1803.918
Female	−11799.000***	−14641.000***	−690.883	−927.114
Firelands	−8552.429***	−9934.029***	60.204	−561.723
Business	5179.857***	6156.490***	5309.482***	6095.228***
Education	−1057.819	−2603.822	−1173.765	−1295.265
Other departments	−3840.726**	−7588.529***	−960.892	−1724.595*
Female × Firelands		6142.285		2681.444
Female × Business		−3974.483		−3230.073
Female × Education		4529.827		300.709
Female × other departments		10704.000***		1908.234
Prior experience			348.234***	335.251***
Years in rank			720.781***	723.296***
Years at BG			−146.391*	−158.452*
Marketability			19638.000***	19294.000***
Rank			9530.083***	9546.362***
R^2	.213	.237	.796	.798

$* p < .05. ** p < .01. *** p < .001.$

With a p-value of .00017, this is quite significant. Tests of the individual terms indicate that the effect is driven largely by the interaction of gender with "other departments." Ignoring the other interaction terms, the gender gap in salary can be written $-14641 + 10704$ *other departments*. This suggests that the gender gap is -14641 in "arts and sciences" compared to -3937 in "other departments."

Missing from these models, however, are several key factors in salary determination that are also related to both *gender* and *college*, as well as to their interaction. Model 3 adds the covariates *prior experience, years in rank, years at BG, marketability*, and *rank* to model 1. *Rank* is an ordinal variable (being treated as continuous) with four values: 1 ("instructor/lecturer"), 2 ("assistant professor"), 3 ("associate professor"), and 4 ("professor"). Among other associations in the data, it is well known that *gender* is strongly related to *rank*, with females generally occupying lower ranks than males (see, e.g., Balzer et al., 1996; Boudreau et al., 1997). For this reason, we would expect the *gender* effect to be at least somewhat diminished when *rank* is held constant. In fact, it is diminished substantially when these additional covariates are added in model 3 and is no longer significant. The effect of *college* is also reduced when adding the covariates, as was seen in Chapter 4. Now only "business" is significantly different from "arts and sciences" in average salary. More important, when the interaction terms are reentered, in model 4, the block is no longer significant ($F = 2.112, p = .077$). The interaction is apparently accounted for by the covariates. But which ones?

Subsequent analyses (not shown) reveal that both *years at BG* and *years in rank* exhibit significant *gender * college* interaction effects. The pattern of relationships is exhibited in Figure 6.3. The *gender * college* cross-product term has a positive effect on salary net of the covariates and has positive effects on both *years at BG* and *years in rank*. *Years at BG* has a negative effect on salary, while *years in rank* has a positive effect. Of the two covariates, *years in rank* has a much stronger effect on salary, 723.296 versus -158.452 in model 4 (as these covariates are in the same metric, the magnitudes of the unstandardized coefficients can be directly compared). What this means is that

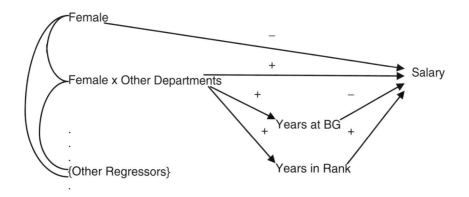

Figure 6.3 Model illustrating omitted-variable bias in the interaction of gender with membership in other departments in their effects on faculty salary.

when these covariates are ignored, the positive indirect path from the cross-product to salary through *years in rank* overrides the negative indirect path through *years at BG* to combine with the positive direct path from the cross-product to salary. The result is a much larger positive "direct" path that manifests itself as a significant interaction. In the current case, since *gender* is causally antecedent to other variables, we would say that the covariates mediate, or account for, the interaction effect. Specifically, the gender gap in *years in rank* is virtually nil in "other departments" vis-à-vis "arts and sciences," and more years in rank translate into higher salaries. Therefore, the smaller gender gap in salary in "other departments" versus "arts and sciences" (the interaction effect) is due primarily to the smaller gender gap in *years in rank* in "other departments," in combination with *years in rank*'s positive effect on salary.

REGRESSION DIAGNOSTICS I: INFLUENTIAL OBSERVATIONS

In the final sections of the chapter I take up the issue of regression diagnostics and begin by outlining the discovery and treatment of influential observations. Influential observations are cases that exert an "undue" amount of influence on the estimated regression model. That is, these are one or more cases that are essentially "driving" the results, in the sense that the estimated model might be substantially different were these cases deleted. Interest in the discovery of such cases was stimulated largely by the work of Belsley et al. (1980). Today most regression software offers several tools that allow the analyst to comb the data for such observations.

Why is it important to know which cases are especially influential? There are a couple of reasons. First, it is possible that an influential data point is a coding error, a bogus response, or some other such anomaly that was not caught in the data-cleaning process. We certainly don't want this type of case to remain in the analysis as is. As an example, in DeMaris (1997) I discovered a highly influential case in my investigation of heightened sexual activity in violent intimate relationships, using the NSFH data from wave 1. A key hypothesis in that work was that greater sexual activity in violent romantic partnerships was due partially to the climate of fear created by male violence. I argued that females in these liaisons were likely to accede to sexual activity more often than they preferred, out of fear of the consequences of displeasing the partner. Given that this sexual activity would in a sense be coerced, due to women's fear of violence by a displeased partner, I expected an increase in depression among such women. To marshal evidence for this, I tested whether sex in the presence of male violence was associated with an increase in the female's depressive symptomatology. Although this finding was supported, I also found that the most influential case in the data appeared to have a bogus value of 93 for the number of times the couple had "had sex" in the past month. However, deleting this observation only strengthened the original finding. In this case the findings appeared robust to the influence of the "rogue" observation. Such may not always be the case, however.

Another possibility is that the existence of a few very influential observations reveals a flaw in the model specification that the researcher may be able to correct. A handy example of this can be found by examining influence diagnostics for model 3 in

Table 6.5. One will find that the three most influential faculty members in the dataset have one element in common: They are all "eminent scholars." These are professors with particularly luminous reputations in their fields who had been recently hired under a statewide program to enhance the quality of scholarship at Ohio universities. Their salaries were, as a consequence, considerably higher than those of other faculty with comparable rank and experience at BGSU. In this case, however, it is easy to model this phenomenon by including in the model a dummy variable that identifies these types of scholars (see, e.g., Balzer et al., 1996). Once the dummy is included, these cases no longer exert such influence, because they are better fitted by the model.

If influential observations represent neither "bad" data nor flaws in model specification, nothing further can be done about them. There is no rationale for deleting legitimate data from the analysis, regardless of their influence. It may nevertheless be fruitful to know that the results are largely due to perhaps only a few "interesting" cases in the data. The analyst may then wish to be more cautious in attempting to generalize beyond the current sample until the findings can be replicated in other datasets. At any rate, let's consider some tools to use in the exploration of such cases.

Building Blocks of Influence: Outliers and Leverage

The degree to which a case has the ability to influence the regression analysis depends on two characteristics: the extent to which it is an outlier, and the extent to which it has leverage. An *outlier* is a case that is far from the regression trend exhibited by the other data points. That is, if the regression of Y on the X's is represented by a "swarm" of points in p-dimensional space, an outlier is a point that is at a noticeable distance from this swarm, in the Y direction. An outlier is typically identified by having a comparatively large residual, indicating that its actual Y value is nowhere near where it is "supposed to be" [i.e., the regression line (plane) that runs through the swarm]. An example of an outlier can be seen in Figure 2.1, discussed earlier, showing the regression of the first exam score on the math diagnostic for 213 students. It is the lowest point in this swarm of points and occurs in the middle of the plot, the point with an X score of 37 and a Y score of 17. It is also the point having the largest residual in the data (in absolute value), or the largest standardized residual (see also Figures 2.6 and 2.8).

However, being an outlier by itself does not give an observation the power to affect the regression. The location of this particular outlier in the middle of the data does not allow it to exert much "pull" on the regression line. We say that it lacks *leverage* to affect the regression. An observation has leverage to the extent that its covariate pattern is atypical. That is, its combination of X scores is far from the *centroid*, or vector of means, of the X's. For the data in Figure 2.1, in which there is only one X, the centroid is 40.925, the mean on the *math diagnostic*. Since a score of 37 is relatively close to this, the outlier has little leverage. The two observations with *math diagnostic* scores of 28 (at the far left in the figure), on the other hand, have considerable leverage. However, their location in the Y direction is consistent with the trend in the swarm of points—their Y scores are 37 and 45, respectively. Hence one can imagine that the regression line would not change much were they omitted from the analysis. In other words, the two cases with diagnostic scores of 28 have leverage but are not outliers and thus have little influence, too. In sum, the only types of cases with real power to influence the analysis are outliers with leverage.

Measuring Influence

There are several measures available that can help the analyst identify potentially influential observations. First, let's consider simply how to find such observations. Then we'll take up the issue of assessing how much influence they exert.

Externally Studentized Residuals. A handy diagnostic for flagging influential observations is based on the residual for a given case. Now, the residual by itself is not always effective in signaling influence. The reason for this is that an outlier with lots of leverage will typically cause the fitted line to be shifted toward it, thereby reducing the magnitude of the residual. What is needed is a residual that is sensitive both to an observation's distance from the fitted line and to its degree of leverage. The ideal measure is called the *externally studentized residual* (Myers, 1986) and symbolized by t_i. Its formula is

$$t_i = \frac{y_i - \hat{y}_i}{s_{-i}\sqrt{1 - h_{ii}}}. \tag{6.6}$$

The numerator of this measure is e_i, the ordinary residual. The first product in the denominator, s_{-i}, is the *standard error of estimate* (i.e., the square root of *MSE*) from a regression that leaves out the ith observation. If the ith observation reflects a model misspecification, s_{-i} is a better estimate of σ than is the square root of *MSE* using all of the cases (Myers, 1986). The second term involves the hat diagonal, h_{ii}, discussed earlier. As mentioned, h_{ii} is a measure of the leverage exerted by the ith case. For models with an intercept, $1/n \le h_{ii} \le 1$, and h_{ii} represents the standardized squared distance from the ith case's covariate pattern to the centroid of the X's (Myers, 1986). In SLR, the formula for h_{ii} is

$$h_{ii} = \frac{1}{n} + \frac{(x_i - \bar{x})^2}{s_{xx}}.$$

As is evident, h_{ii} reaches its minimum in SLR when X is at its mean. Similarly, h_{ii} reaches its minimum in a MULR model when $\mathbf{x}^i$ is equal to $\bar{\mathbf{x}}$, the vector of covariate means. A rough guideline for evaluating the amount of leverage exerted by an observation is that h_{ii}'s greater than $2p/n$ represent cases with noticeable leverage.

In sum, t_i is a kind of standardized residual adjusted for a case's leverage. Large values of t_i will be observed whenever cases have large residuals and/or high leverage values, and t_i will be especially high for cases with both properties. This statistic also represents a formal test for outlier status. Under the null hypothesis that the ith case is not an outlier, and given standard assumptions on the ε_i, including normality, t_i follows a t distribution with $n - p - 1$ degrees of freedom, where $p = K + 1$ is the number of model parameters (Myers, 1986). This suggests that t_i values of about 2 or higher signal outliers. (Note that t_i is usually referred to as "rstudent" in software packages such as SAS.)

Dffits$_i$. Several measures are available to tap the degree of influence exerted by the ith case on various regression results. The influence on the fitted values is best tapped by a measure called *Dffits$_i$*:

$$Dffits_i = \frac{\hat{y}_i - \hat{y}_{i,-i}}{s_{-i}\sqrt{h_{ii}}}. \tag{6.7}$$

The numerator of (6.7) is the difference between the fitted value for the ith case with and without the ith case in the analysis (i.e., $\hat{y}_{i,-i}$ is the fitted value for the ith case based on a regression that omits the ith case). What about the denominator? The standard error of prediction (i.e., the standard error of the ith fitted value) is $\sigma\sqrt{x^{i\prime}(X'X)^{-1}x^i}$, and recall that $h_{ii} = x^{i\prime}(X'X)^{-1}x^i$. Therefore, the denominator is the estimated standard error of the ith fitted value. *Dffits$_i$* can therefore be intepreted as the estimated number of standard errors the fitted value changes when the ith case is omitted from the regression. A size-adjusted cutoff for *Dffits$_i$*, above which the value suggests noticeable influence, is $2\sqrt{p/n}$ (Belsley et al., 1980).

Dfbetas$_{ji}$. The influence of the ith case on the jth parameter estimate is tapped by *Dfbetas$_{ji}$*:

$$Dfbetas_{ji} = \frac{b_j - b_{j,-i}}{s_{-i}\sqrt{c_{jj}}}. \tag{6.8}$$

The numerator of (6.8) is the difference in the jth parameter estimate with and without the ith case in the regression. The denominator is the estimated standard error of the jth coefficient, since c_{jj} is the jth diagonal element of $(X'X)^{-1}$. *Dfbetas$_{ji}$* is interpreted as the number of standard errors the jth parameter estimate changes when the ith case is omitted from the regression. A recommended cutoff for *Dfbetas$_{ji}$* is $2/\sqrt{n}$.

Cook's D$_i$. When there are many regressors in the model and n is large, it may be very tedious to scrutinize all of the *Dfbetas$_{ji}$*, particularly when no single parameter estimate is of paramount importance. With this in mind, a convenient summary measure of the influence of the ith case on the collection of parameter estimates is Cook's distance measure, D_i:

$$D_i = \frac{(\mathbf{b} - \mathbf{b}_{-i})'(X'X)(\mathbf{b} - \mathbf{b}_{-i})}{pMSE}. \tag{6.9}$$

As $\mathbf{b}_{-i}$ represents the vector of parameter estimates based on a regression that omits the ith case, (6.9) is the standardized distance between the vector of parameter estimates with and without the ith case in the regression. It may be easier to get an intuitive feeling for D_i by regarding its expression in the SLR model:

$$D_{i,\text{SLR}} = \frac{n(b_0 - b_{0,-i})^2 + 2n\bar{x}(b_0 - b_{0,-i})(b_1 - b_{1,-i}) + (b_1 - b_{1,-i})^2\sum x^2}{2MSE}.$$

The larger the difference in parameter estimates from regressions with and without the ith case, the larger is D_i. In the extreme opposite case in which there is no change in the parameter estimates after deleting the ith case, D_i is clearly zero. Cook's D_i has been described as an "F-like statistic" with degrees of freedom p and $n - p$ (Myers, 1986). However, the usual critical values of F used for hypothesis testing are not applicable here. Neter et al. (1985) suggest instead that the 50th (rather than, say, the 95th) percentile of the F distribution should be the cutoff for declaring a case influential. In practice, it may be prudent to investigate any case more closely if its D_i value is markedly larger than all others (see, e.g., DeMaris, 1997).

Covratio$_i$. Finally, a measure that assesses the influence of the ith case on the estimated variance–covariance matrix of parameter estimates is *Covratio$_i$*:

$$Covratio_i = \frac{|s^2_{-i}(X'_{-i}X_{-i})^{-1}|}{|s^2(X'X)^{-1}|}. \tag{6.10}$$

Recalling that s^2 is *MSE*, the estimate of σ^2, this represents the ratio of the determinants of the estimated variance–covariance matrix without, versus with, the ith case in the regression. The rationale for this measure is that the determinant of the variance–covariance matrix is a scalar measure of the generalized variance of the regression coefficients (Graybill, 1976; Myers, 1986). All else equal, a smaller generalized variance implies regression coefficients that have greater precision. A value of *Covratio$_i$* greater than 1 suggests that the ith data point brings about a reduction in the generalized variance (the determinant is smaller with the ith case in the analysis), while a value less than 1 implies that the ith case increases the generalized variance. Belsley et al. (1980) suggest that influential observations will be indicated by *Covratio$_i$* being either greater than $1 + 3p/n$ or less than $1 - 3p/n$. [It should be mentioned here that although the diagnostic measures are all based on "omitting the ith case from the regression," all measures are computed in one pass through the data. In other words, all measures can be calculated from one regression run. There is no need to actually run the regression n times, each time omitting the ith case. See Myers (1986) for details.]

Illustration of Influence Diagnosis

As an example of the evaluation of influence, I examine the regression of the *first exam score* on the *math diagnostic score*, *college GPA*, and *attitude toward statistics* for 214 students in my introductory statistics classes. The results are shown in Table 6.6. Panel A presents regression results, and panel B presents diagnostics for the four most influential observations in the dataset. Cutoffs for all of the diagnostic measures are shown in parentheses. For example, since p is 4, the cutoff for leverage is $2(4)/214 = .037$. The cutoff for *Dffits$_i$* is $2\sqrt{4/214} = .273$, the cutoff for *Dfbetas$_i$* is $2/\sqrt{214} = .137$, and so on.

The single most influential observation is case number 214. Its t_i value is over 4, and its *Dffits*, *Covratio*, and *D* values are all the most extreme of any of the cases. With the exception of the coefficient for *college GPA*, case 214's influence on the fitted values and the parameter estimates is especially noteworthy. This particular student, the same

Table 6.6 Effect of Influential Observations on the OLS Regression for Score on the First Exam for 214 Students in Introductory Statistics

Panel A:

Predictor	b_{all}	b_{-214}
Intercept	−52.892***	−47.879***
Math diagnostic	2.151***	2.031***
College GPA	12.929***	12.872***
Attitude toward statistics	.374*	.441*
R^2	.432	.446

Panel B:	Influential Observations			
Diagnostic (Benchmark)	No.58	No.92	No.158	No.214
t_i (> 2 in absolute value)	−2.936	−2.649	−2.220	−4.287
h_{ii} (> .037)	.017	.025	.050	.021
$Dffits_i$ (> .273 in absolute value)	−.381	−.425	−.511	−.628
$Dfbetas_{0i}$ (> .137 in absolute value)	−.275	.191	−.461	−.437
$Dfbetas_{1i}$ (> .137 in absolute value)	.179	−.143	.477	.433
$Dfbetas_{2i}$ (> .137 in absolute value)	.159	−.093	−.044	.032
$Dfbetas_{3i}$ (> .137 in absolute value)	.105	−.273	−.019	−.417
$Covratio_i$ (< .944 or > 1.056)	.882	.916	.978	.743
Cook's D_i (> .84)	.035	.044	.064	.091

* $p < .05$. ** $p < .01$. *** $p < .001$.

one responsible for the outlier in Figure 2.1, represents somewhat of an atypical data point. He or she is fairly strong on the regressor values—with *math diagnostic, college GPA*, and *attitude* scores of 37, 3.00, and 11, respectively—but surprisingly weak on the exam, with a score of 17. On the other hand, his or her leverage is below the cutoff, with a value of only .021. Nevertheless, the combination of what leverage there is plus the case's outlier status adds up to a fair amount of influence.

Panel A shows the regression estimates with (b_{all}) and without (b_{-214}) this case in the analysis. To get an intuitive feeling for the coefficient influence measures, regard the $Dfbetas_{ji}$ for, say, the coefficient for the *math diagnostic score*. The value of .433 suggests that the coefficient should drop about four-tenths of a standard error when case 214 is omitted from the regression. The estimated standard error of the coefficient is .279 without this case in the analysis. Thus, the actual drop is (2.151 − 2.031)/.279 = .430, which is about what was predicted. Even deleting this most influential case, however, there is little substantive change in the model. All regressors have significant, positive effects on the *first exam score*. In fact, as the diagnostics indicate, no case has the ability to alter any of the coefficients by more than one-half of a standard error. In this example, then, the degree of influence exerted by any given case is less than dramatic. For case number 214, even though he or she exhibits a relatively unusual pattern of data values, there would be no compelling reason to delete his or her data from the analysis.

REGRESSION DIAGNOSTICS II: MULTICOLLINEARITY

Multicollinearity was touched on briefly in Chapter 3, in which I defined the problem and suggested some easy remedies. In this section I consider the topic in much greater technical detail and also discuss alternatives to OLS when the problem cannot easily be resolved. Recall that multicollinearity is a condition in which one or more of the independent variables is almost exactly determined by the other regressors. Another way of saying this is that there are one or more "near linear dependencies" among the columns of X, that is, among the regressors in the X matrix (Myers, 1986, p. 76). A complete understanding of the problem requires that we dissect the X matrix—in a manner of speaking. However, let's work with the centered and scaled matrix X* (discussed earlier), since this is the basis for the standardized regression coefficients, from which the unstandardized versions are easily recovered. We will refer to X* as the *design matrix*. Remember that X*'X* is R_{xx}, the correlation matrix for the X's.

Linear Dependencies in the Design Matrix

I begin by reminding the reader of the definition of linear dependence with respect to the columns of a matrix. In Section V.G of Appendix A I defined the linear dependence of the columns of a matrix A as follows: *If there is a nonnull vector x such that Ax = 0, then provided that no column of A is null, the columns of A are linearly dependent.* Let's see how this principle applies here. I will show that a linear dependence in the design matrix is associated with a zero eigenvalue of the matrix.

Since the $K \times K$ matrix X*'X* is symmetric, it can be spectrally decomposed (See Section V.H of Appendix A for an explanation of spectral decomposition). Recall that the spectral decomposition of a symmetric matrix A allows us to write the matrix as

$$A = \sum \lambda_j \mathbf{u}_j \mathbf{u}_j',$$

where λ_j is the jth eigenvalue of A and $\mathbf{u}_j$ is the jth eigenvector for $j = 1, 2, \ldots, K$. This is equivalent to writing $A = UD_\lambda U'$, where U is the matrix whose columns are the normalized eigenvectors of A and D_λ is the diagonal matrix of eigenvalues of A (verification that these expressions for A are equivalent is left as an exercise). Pre- and postmultiplication of A by U' and U, respectively, thereby produces $U'AU = D_\lambda$ (since U, being an orthogonal matrix, has the property that $U'U = UU' = I$). This is called the *eigenvalue decomposition* of A (Myers, 1986). Now substitute X*'X* for A, and we have that $U'(X*'X*)U = D_\lambda$, where λ now refers to the eigenvalues of X*'X*, and U to the matrix of its normalized eigenvectors.

Recall from Appendix A that a matrix has an inverse only if its columns are linearly independent, in which case its determinant is not zero. Moreover, its determinant is the product of its eigenvalues. So one or more zero eigenvalues imply that the determinant is zero and are therefore indicative of exact linear dependence among the columns of the matrix. Similarly, eigenvalues that are *near* zero reflect *near* linear dependence. Hence, a near-zero eigenvalue of $X^{*\prime}X^*$ is indicative of near linear dependence among the regressors. This is easier to see if we take advantage of the eigenvalue decomposition of $X^{*\prime}X^*$:

$$U'(X^{*\prime}X^*)U = \begin{bmatrix} \mathbf{u}_1' \\ \mathbf{u}_2' \\ \cdot \\ \cdot \\ \cdot \\ \mathbf{u}_K' \end{bmatrix} (X^{*\prime}X^*)[\mathbf{u}_1 \quad \mathbf{u}_2 \quad \cdots \quad \mathbf{u}_K]$$

$$= \begin{bmatrix} \mathbf{u}_1'(X^{*\prime}X^*)\mathbf{u}_1 & \mathbf{u}_1'(X^{*\prime}X^*)\mathbf{u}_2 & \cdots & \mathbf{u}_1'(X^{*\prime}X^*)\mathbf{u}_K \\ \mathbf{u}_2'(X^{*\prime}X^*)\mathbf{u}_1 & \mathbf{u}_2'(X^{*\prime}X^*)\mathbf{u}_2 & \cdots & \cdot \\ \cdot & \cdot & \cdots & \cdot \\ \cdot & \cdot & \cdots & \cdot \\ \cdot & \cdot & \cdots & \cdot \\ \mathbf{u}_K'(X^{*\prime}X^*)\mathbf{u}_1 & \mathbf{u}_K'(X^{*\prime}X^*)\mathbf{u}_1 & \cdots & \mathbf{u}_K'(X^{*\prime}X^*)\mathbf{u}_K \end{bmatrix}$$

$$= \begin{bmatrix} \lambda_1 & 0 & \cdots & & 0 \\ 0 & \lambda_2 & \cdots & & \cdot \\ \cdot & \cdot & \cdots & & \cdot \\ \cdot & \cdot & \cdots & & \cdot \\ \cdot & \cdot & \cdots & 0 & \\ 0 & \cdot & \cdot & 0 & \lambda_K \end{bmatrix}.$$

Now, if one of the eigenvalues, say λ_K, is near zero, we have that

$$\mathbf{u}_K'(X^{*\prime}X^*)\mathbf{u}_K = (X^*\mathbf{u}_K)'(X^*\mathbf{u}_K) \approx 0.$$

Notice that since $X^*\mathbf{u}_K$ is a vector (its dimensions are $K \times 1$), $(X^*\mathbf{u}_K)'(X^*\mathbf{u}_K)$ is essentially the sum of squares of all elements of $X^*\mathbf{u}_K$. And the sum of squares of all elements can be near zero only if the vector itself is approximately the zero vector. In that $X^*\mathbf{u}_K$ is a linear combination of the columns of X^*, this implies that the columns of X^* are approximately linearly dependent, according to the definition of linear dependence given above. Moreover, the elements of $\mathbf{u}_K$ reveal the nature of the dependency, since they indicate the weights for the linear combination of the columns of X^* that is approximately zero.

For example, Dunteman (1989) presents a regression, for 58 countries, of *educational expenditures as a percent of the gross national product* (Y) on six characteristics of countries: *population size* (X_1), *population density* (X_2), *literacy rate* (X_3), *energy consumption per capita* (X_4), *gross national product per capita* (X_5), and *electoral irregularity score* (X_6). The spectral decomposition of the design matrix reveals one relatively small eigenvalue of .047, with the following associated eigenvector:

$\mathbf{u}' = (.005 \quad .097 \quad -.065 \quad -.660 \quad .729 \quad -.152)$. The largest weights are for X_4 and X_5, suggesting that these two variables are somewhat linearly dependent. Ignoring the other elements of this vector, we have $-.660X_4 + .729X_5 \approx 0$, or $X_4 \approx 1.105X_5$. In fact, the correlation between these two regressors is .93.

Consequences of Collinearity

In Chapter 3 I suggested that two major consequences of multicollinearity were an inflation in the variances of OLS estimates and an inflation in the magnitudes of the coefficients themselves. To understand how collinearity causes these problems, we rely once again on the spectral decomposition of $X^{*\prime}X^*$. First, consider the matrix expression for the sum of the variances of the standardized coefficients. We begin with the expression $(\mathbf{b}^s - \boldsymbol{\beta}^s)'(\mathbf{b}^s - \boldsymbol{\beta}^s)$:

$$(\mathbf{b}^s - \boldsymbol{\beta}^s)'(\mathbf{b}^s - \boldsymbol{\beta}^s) = [b_1^s - \beta_1^s \quad b_2^s - \beta_2^s \quad \cdots \quad b_K^s - \beta_K^s] \begin{bmatrix} b_1^s - \beta_1^s \\ b_2^s - \beta_2^s \\ \cdot \\ \cdot \\ \cdot \\ b_K^s - \beta_K^s \end{bmatrix} = \sum \left(b_k^s - \beta_k^s \right)^2,$$

and therefore,

$$E[(\mathbf{b}^s - \boldsymbol{\beta}^s)'(\mathbf{b}^s - \boldsymbol{\beta}^s)] = \sum E(b_k^s - \beta_k^s)^2 = \sum V(b_k^s),$$

this last term being the sum of the variances of the standardized coefficients. Now the variances of the coefficients are on the diagonal of the matrix $\sigma_*^2(X^{*\prime}X^*)^{-1}$. The sum of the diagonal elements of this matrix is, of course, its trace, and the trace of a square matrix is the sum of its eigenvalues. Moreover, the eigenvalues of the inverse of a matrix are simply the reciprocals of the eigenvalues of the matrix itself (the proof of this is left as an exercise for the reader). Therefore, the eigenvalues of $(X^{*\prime}X^*)^{-1}$ are simply of the form $1/\lambda_k$. Thus,

$$\sum V(b_k^s) = \text{tr}[\sigma_*^2(X^{*\prime}X^*)^{-1}] = \sigma_*^2 \, \text{tr}[(X^{*\prime}X^*)^{-1}] = \sum_{i=1}^{K} \frac{\sigma_*^2}{\lambda_i}, \qquad (6.11)$$

and since

$$E[(\mathbf{b}^s - \boldsymbol{\beta}^s)'(\mathbf{b}^s - \boldsymbol{\beta}^s)] = \sum V(b_k^s) = \sum_{i=1}^{K} \frac{\sigma_*^2}{\lambda_i},$$

we have that

$$E[\mathbf{b}^{s\prime}\mathbf{b}^s - \mathbf{b}^{s\prime}\boldsymbol{\beta}^s - \boldsymbol{\beta}^{s\prime}\mathbf{b}^s + \boldsymbol{\beta}^{s\prime}\boldsymbol{\beta}^s] = \sum_{i=1}^{K} \frac{\sigma_*^2}{\lambda_i},$$

or

$$E(\mathbf{b}^{s\prime}\mathbf{b}^s) - E(\mathbf{b}^{s\prime})\boldsymbol{\beta}^s - \boldsymbol{\beta}^{s\prime}E(\mathbf{b}^s) + \boldsymbol{\beta}^{s\prime}\boldsymbol{\beta}^s = \sum_{i=1}^{K} \frac{\sigma_*^2}{\lambda_i},$$

or

$$E(\mathbf{b}^{s\prime}\mathbf{b}^s) = \boldsymbol{\beta}^{s\prime}\boldsymbol{\beta}^s + \sum_{i=1}^{K} \frac{\sigma_*^2}{\lambda_i}. \tag{6.12}$$

Equation (6.12) shows that the sums of squares of the coefficient estimates are heavily upwardly biased when there is multicollinearity, as indexed by one or more small eigenvalues. This means that the coefficients will have a tendency to be too large in magnitude if the regressors are collinear (Myers, 1986).

Why are the coefficient variances inflated? The spectral decomposition of $(X^{*\prime}X^*)^{-1}$ is

$$(X^{*\prime}X^*)^{-1} = \sum_{i=1}^{K} \frac{1}{\lambda_i}\mathbf{u}_i\mathbf{u}_i' = \frac{1}{\lambda_1}\mathbf{u}_1\mathbf{u}_1' + \frac{1}{\lambda_2}\mathbf{u}_2\mathbf{u}_2' + \cdots + \frac{1}{\lambda_K}\mathbf{u}_K\mathbf{u}_K'. \tag{6.13}$$

Now, to simplify things somewhat, let's consider what this matrix looks like for three regressors, focusing only on the diagonal elements. In this case $(X^{*\prime}X^*)^{-1}$ is

$$\frac{1}{\lambda_1}\begin{bmatrix} u_{11} \\ u_{21} \\ u_{31} \end{bmatrix}\begin{bmatrix} u_{11} & u_{21} & u_{31} \end{bmatrix} + \frac{1}{\lambda_2}\begin{bmatrix} u_{12} \\ u_{22} \\ u_{32} \end{bmatrix}\begin{bmatrix} u_{12} & u_{22} & u_{32} \end{bmatrix} + \frac{1}{\lambda_3}\begin{bmatrix} u_{13} \\ u_{23} \\ u_{33} \end{bmatrix}\begin{bmatrix} u_{13} & u_{23} & u_{33} \end{bmatrix}$$

$$= \begin{bmatrix} \frac{u_{11}^2}{\lambda_1} & \cdot & \cdot \\ \cdot & \frac{u_{21}^2}{\lambda_1} & \cdot \\ \cdot & \cdot & \frac{u_{31}^2}{\lambda_1} \end{bmatrix} + \begin{bmatrix} \frac{u_{12}^2}{\lambda_2} & \cdot & \cdot \\ \cdot & \frac{u_{22}^2}{\lambda_2} & \cdot \\ \cdot & \cdot & \frac{u_{32}^2}{\lambda_2} \end{bmatrix} + \begin{bmatrix} \frac{u_{13}^2}{\lambda_3} & \cdot & \cdot \\ \cdot & \frac{u_{23}^2}{\lambda_3} & \cdot \\ \cdot & \cdot & \frac{u_{33}^2}{\lambda_3} \end{bmatrix}$$

$$= \begin{bmatrix} \frac{u_{11}^2}{\lambda_1} + \frac{u_{12}^2}{\lambda_2} + \frac{u_{13}^2}{\lambda_3} & \cdots & \cdots \\ \cdot & \frac{u_{21}^2}{\lambda_1} + \frac{u_{22}^2}{\lambda_2} + \frac{u_{23}^2}{\lambda_3} & \cdots \\ \cdot & \cdots & \frac{u_{31}^2}{\lambda_1} + \frac{u_{32}^2}{\lambda_2} + \frac{u_{33}^2}{\lambda_3} \end{bmatrix}.$$

The diagonals of $\sigma_*^2(X^{*\prime}X^*)^{-1}$ represent the coefficient variances. So, for example, the variance of b_1^s is then equal to

$$\sigma_*^2\left(\frac{u_{11}^2}{\lambda_1} + \frac{u_{12}^2}{\lambda_2} + \frac{u_{13}^2}{\lambda_3}\right). \tag{6.14}$$

This expression should make it clear that collinearity, in the form of a small eigenvalue, will tend to inflate the variance of a given coefficient. In fact, a given near linear dependency has the potential to affect the variances of all of the coefficients; but there is a "catch." Suppose that X_2 and X_3 are highly correlated but both are close to orthogonal to X_1. Then most likely λ_3 will be fairly small, but so will u_{13}, since the high-magnitude weights will be u_{23} and u_{33}. In this case, the contribution to $V(b_1^s)$ of the last term inside the parentheses will be negligible. This suggests that the variables responsible for the near linear dependencies in the data are the ones whose variances are primarily affected by collinearity.

Diagnosing Collinearity

There are three major tools in the diagnosis of collinearity. Expression (6.14) is associated with two of them. The first is the *VIF*, introduced in Chapter 3. The *VIF* tells us *how many times the variance of a coefficient is magnified as a result of the collinearity compared to the ideal case of perfectly orthogonal regressors* (Myers, 1986). It turns out that the *VIF*'s are the diagonal elements of R_{xx}^{-1} (Neter et al., 1985); hence, the term inside the parentheses in (6.14) is the *VIF* for b_1^s (or for b_1, since the *VIF* is the same for the unstandardized coefficient as it is for the standardized one). As mentioned previously, *VIF*'s of about 10 or higher indicate collinearity problems. The second diagnostic allows us to discern which variables are, in fact, approximately linearly dependent. This is the variance proportion, or p_{ji}:

$$p_{ji} = \frac{\sigma_*^2 u_{ij}^2 / \lambda_j}{V(b_i^s)}.$$

This is interpreted as the *proportion of the variance of b_i attributable to the collinearity characterized by λ_j*. For example, from (6.14) we have

$$p_{31} = \frac{\sigma_*^2 u_{13}^2 / \lambda_3}{V(b_1^s)}.$$

Typically, high variance proportions associated with the same eigenvalue for two or more regressors indicate that those regressors are approximately linearly dependent. The third useful diagnostic is called the *condition number* of the $(X^{*\prime}X^*)^{-1}$ matrix. It is the ratio of the largest to the smallest eigenvalue and is typically symbolized by ϕ. Condition numbers greater than 1000 are indicative of collinearity problems in the matrix. Often, what is reported as the condition number by software (e.g., SAS) is the square root of ϕ, in which case the cutoff is about 32. According to Myers (1986), a complete collinearity diagnosis uses the condition number to assess the seriousness of linear dependencies in the design matrix, the variance proportions to identify which variables are involved, and the *VIF*'s to determine the amount of "damage" to individual coefficients.

Illustration

Table 6.7 presents collinearity diagnostics for data reported in Neter et al. (1985) on 20 healthy females aged 25–34. The response variable is *body fat*, while the regressors are *triceps skinfold thickness* (X_1), *thigh circumference* (X_2), and *midarm circumference* (X_3). Although not exactly typical "social" data, this example was chosen

Table 6.7   OLS, Ridge, and Principal Components Regression Results, and Collinearity Diagnostics for Body Fat Data from 20 Women

Panel A: Regression Results

Predictor	OLS Estimator	b^s	VIF	Ridge Estimator	PC Estimator
Intercept	117.085			− 7.403	− 12.205
Triceps skinfold thickness	4.334	4.264	708.843	.555	.422
Thigh circumference	− 2.857	− 2.929	564.343	.368	.492
Midarm circumference	− 2.186	− 1.561	104.606	− .192	− .125
R^2	.801***				

Panel B: Collinearity Diagnostics

λ No.	λ Value	$\sqrt{\phi}$	Variance Proportions p_{j0}	p_{j1}	p_{j2}	p_{j3}
1	3.968	1.000	.0000	.0000	.0000	.0000
2	.021	13.905	.0004	.0013	.0000	.0014
3	.012	18.566	.0006	.0002	.0003	.0069
4	.000009	677.372	.9990	.9985	.9996	.9917

*** $p < .001$.

primarily because it illustrates the symptoms and warning signs of extremely collinear data. The OLS regression of *body fat* on the three independent variables is shown in the first column of panel A of the table. A couple of symptoms of collinearity are already evident here. To begin, the sign of the coefficient for *thigh circumference* is counterintuitive: We would expect a greater thigh circumference to predict more, not less, body fat. Also, the standardized coefficients are outside the range $[-1, 1]$. The *VIF*'s confirm that there is a collinearity problem that affects all of the coefficients: All *VIF*'s are well over 10—by a factor of at least 10!

Panel B shows the eigenvalues of the $X'X$ matrix as well as its condition numbers (i.e., $\sqrt{\phi}$ for the ratio of the largest eigenvalue to each successively smaller one) and the variance proportions. In this case, we are using the decomposition of $X'X$ instead of $X^{*'}X^*$. Either matrix is useful for diagnosing collinearity, although Myers (1986) recommends using $X^{*'}X^*$ if the intercept is not of special interest. At any rate, the smallest eigenvalue is associated with a condition number of 677.372. Bearing in mind that 32 is the cutoff, this is quite large indeed. This suggests that there is a serious linear dependency in the matrix. The variance proportions indicate that all three regressors, as well as the intercept, are tied together in a near linear dependency.

A somewhat more realistic example, at least by social science criteria, is shown in Table 6.8, which presents collinearity diagnostics for a regression using the NSFH data employed for Table 6.4. This time I regress *couple disagreement* on several controls plus the variables current *male-* and *female age*, *male-* and *female age at the beginning of the union*, (shown simply as male- and female "age at union" in the

Table 6.8   OLS, Ridge, and Principal Component Regression Results, and Collinearity Diagnostics for Regression of Couple Disagreement on Demographic Predictors for 7273 Couples in the NSFH

Panel A: Regression Results

Predictor	OLS Estimator	b^s	VIF	Ridge Estimator	PC Estimator
Intercept	15.077***			14.995	14.994
Cohabiting couple	.083	.005	1.187	.109	.132
Biological children	1.306***	.147	1.575	1.242	1.332
Stepchildren	.975***	.079	1.303	.923	1.007
First union	−.552***	−.063	1.605	−.484	−.510
Minority couple	.022	.002	1.100	.041	.027
Male age (b_6)	−.011	−.036	10.184	−.024	−.028
Female age (b_7)	.003	.009	16.863	−.021	−.033
Male age at union (b_8)	−.048	−.093	5.245	−.032	−.029
Female age at union (b_9)	−.059	−.107	6.615	−.034	−.022
Union duration (b_{10})	−.075	−.258	26.583	−.040	−.025
R^2	.155***				

Panel B: Partial Collinearity Diagnostics

			Variance Proportions				
λ No.	λ Value	$\sqrt{\phi}$	p_{j6}	p_{j7}	p_{j8}	p_{j9}	p_{j10}
10	.007	31.197	.734	.121	.653	.128	.051
11	.004	41.329	.186	.836	.141	.685	.914

*** $p < .001$.

table), and *union duration*. In a complete dataset with no missing values, these variables would be exactly linearly dependent since *current age = age at union + union duration*. However, it is typical in survey data that answers are not recorded for a number of cases. In this instance, I used mean substitution for the missing data. Filling in the missing data with imputed values nullifies the exact linear dependency, making it possible to estimate a regression. Again, the first column of panel A shows the OLS results. This time none of the standardized coefficients is outside the range $[-1, 1]$. However, although the coefficient is not significant, the sign of the effect of *female age* is somewhat counterintuitive, since older couples typically have fewer disagreements. The *VIF*'s suggest that at least three of the coefficient variances are affected by collinearity: those for *male age, female age*, and *union duration*. The coefficients for *male-* and *female age at the beginning of the union* have variances that are somewhat inflated, but not by enough to cause concern.

Panel B again shows condition numbers and variance proportions. However, this time I have only shown diagnostics connected with the two smallest eigenvalues. The smallest eigenvalues are associated with condition numbers of 31.197 and 41.329, which are just large enough to signal that there are some approximate linear dependencies among the regressors. The variance proportions associated with eigenvalue number 10 indicate a correlation between *male age* and *male age at union*. Those associated with eigenvalue number 11 suggest that *female age*, *female age at union*, and *union duration* are somewhat linearly dependent. These are all precisely the variables that would be exactly collinear were it not for missing data.

Alternatives to OLS When Regressors Are Collinear

Several simple remedies for collinearity problems were discussed in Chapter 3, including dropping redundant variables, incorporating variables into a scale, employing nonlinear transformations, and centering (for collinearity arising from cross-product terms). However, there are times when none of these solutions are satisfactory. For example, in the body fat data, I may want to know the effect of, say, *triceps skinfold thickness* on *body fat*, net of (i.e., controlling for) the effects of *thigh circumference* and *midarm circumference*. Or, in the NSFH example, I may want to tease out the separate effects of *male-* and *female age*, *male-* and *female age at the beginning of the union*, and *union duration*, on *couple disagreement*. None of the simple remedies are useful in these situations. With this in mind, I will discuss two alternatives to OLS: *ridge regression* and *principal components regression*. These techniques are somewhat controversial (see, e.g., Draper and Smith, 1998; Hadi and Ling, 1998). Nevertheless, they may offer an improvement in the estimates of regressor effects when collinearity is severe.

First we need to consider a key tool in the evaluation of parameter estimators: the *mean squared error* of the estimator, denoted *MSQE* (to avoid confusion with the *MSE* in regression). Let θ be any parameter and $\hat{\theta}$ its sample estimator. Then $MSQE(\hat{\theta}) = E_{\hat{\theta}}(\hat{\theta} - \theta)^2$. That is, $MSQE(\hat{\theta})$ is the average, over the sampling distribution of $\hat{\theta}$, of the squared distance of $\hat{\theta}$ from θ. All else equal, estimators with a small *MSQE* are preferred, since they are by definition closer, on average, to the true value of the parameter, compared to other estimators. It can be shown that $MSQE(\hat{\theta}) = V(\hat{\theta}) + [B(\hat{\theta})]^2$, where $B(\hat{\theta})$ is the bias of $\hat{\theta}$ (defined in Chapter 1). Both ridge and principal components regression employ biased estimators. However, both techniques offer a trade-off of a small amount of bias in the estimator for a large reduction in its sampling variance. Ideally, this means that these techniques bring about a substantial reduction in the *MSQE* of the regression coefficients compared to OLS.

At the same time, both techniques have a major drawback, particularly in the social sciences, where hypothesis testing is so important: The extent of bias in the regression coefficients is unknown. Therefore, significance tests are not possible. To understand why, consider the test statistic for the null hypothesis that $\beta_k = 0$. For simplicity, suppose that the true variance of b_k is known and that n is large, so that t tests and z tests are equivalent. The test relies on the fact that b_k is unbiased for β_k. Now the test statistic is

$$z = \frac{b_k - \beta_{k,0}}{\sigma_{b_k}},$$

where $\beta_{k,0}$ is the null-hypothesized value of β_k, which in this case is zero. Therefore, if β_k is truly zero, that is, the null hypothesis is true, then

$$E(z) = E\left(\frac{b_k - \beta_{k,0}}{\sigma_{b_k}}\right) = \frac{1}{\sigma_{b_k}}[E(b_k) - \beta_{k,0}] = \frac{1}{\sigma_{b_k}}(0 - 0) = 0 \qquad (6.15)$$

and

$$V\left(\frac{b_k - \beta_{k,0}}{\sigma_{b_k}}\right) = \frac{1}{\sigma_{b_k}^2}V(b_k) = \frac{\sigma_{b_k}^2}{\sigma_{b_k}^2} = 1. \qquad (6.16)$$

That is, the test statistic has the standard normal distribution under the null hypothesis. Therefore, a value of 2 implies a sample coefficient that is 2 standard deviations above its expected value under the null, and the probability of this is less than .05. But suppose that b_k is biased and its expected value is unknown. Then z would still be normally distributed and (6.16) would still hold under the null hypothesis. But (6.15) is no longer necessarily valid, since the expected value of the estimator is no longer necessarily zero under the null. This means that the exact distribution of the test statistic under the null hypothesis is no longer known, and the probability of getting, say, a value of 2 cannot be determined. Given this limitation, these techniques are useful only to the extent that the values of the coefficients themselves, rather than whether they are "significant," are of primary importance.

Ridge Regression. In ridge regression, we add a small value, called the *ridge constant*, to the diagonals of the design matrix prior to computing $R_{xx}^{-1}r_{xy}$. What does this do for us? To answer this, first consider the design matrix when there is no collinearity. Suppose that we have a simple model with only two regressors and the correlation between them (r_{12}) is .5. Suppose further that the true model for the standardized variables is $Y_z = .25Z_1 + .15Z_2 + \varepsilon^*$, where ε^* is uncorrelated with the regressors. This implies that the correlation of Y with Z_1 is .325 and with Z_2 is .275 (as the reader can verify using covariance algebra). If we have sample data that reflect population values perfectly, the standardized coefficient estimates are

$$\mathbf{b}^s = R_{xx}^{-1}\mathbf{r}_{xy} = \begin{bmatrix} 1 & .5 \\ .5 & 1 \end{bmatrix}^{-1}\begin{bmatrix} .325 \\ .275 \end{bmatrix} = \begin{bmatrix} 1.3333 & -.6667 \\ -.6667 & 1.3333 \end{bmatrix}\begin{bmatrix} .325 \\ .275 \end{bmatrix} = \begin{bmatrix} .25 \\ .15 \end{bmatrix}.$$

The OLS estimator gives us the true coefficients. Now suppose that we perturb $\mathbf{r}_{xy}$ slightly to simulate sampling variability. Arbitrarily, I add .01 to the first correlation and subtract .01 from the second. Now we have

$$\mathbf{b}^s = R_{xx}^{-1}\mathbf{r}_{xy} = \begin{bmatrix} 1.3333 & -.6667 \\ -.6667 & 1.3333 \end{bmatrix}\begin{bmatrix} .335 \\ .265 \end{bmatrix} = \begin{bmatrix} .27 \\ .13 \end{bmatrix}.$$

The OLS estimator is slightly off but still pretty close to the true coefficients.

On the other hand, let's say the correlation between the regressors is .995. Using the same true regression parameters, the correlation of Y with Z_1 is now .39925 and with Z_2 is .39875. As before, with sample data that exactly reflect population values, we have

$$\mathbf{b}^s = \mathbf{R}_{xx}^{-1}\mathbf{r}_{xy} = \begin{bmatrix} 1 & .995 \\ .995 & 1 \end{bmatrix}^{-1}\begin{bmatrix} .39925 \\ .39875 \end{bmatrix}$$

$$= \begin{bmatrix} 100.2506 & -99.7494 \\ -99.7494 & 100.2506 \end{bmatrix}\begin{bmatrix} .39925 \\ .39875 \end{bmatrix} = \begin{bmatrix} .25 \\ .15 \end{bmatrix}.$$

Once again, the OLS estimator gives us the true coefficients. However, now when we perturb $\mathbf{r}_{xy}$ in the same fashion—adding and subtracting .01—we get

$$\mathbf{b}^s = \mathbf{R}_{xx}^{-1}\mathbf{r}_{xy} = \begin{bmatrix} 100.2506 & -99.7494 \\ -99.7494 & 100.2506 \end{bmatrix}\begin{bmatrix} .40925 \\ .38875 \end{bmatrix} = \begin{bmatrix} 2.25 \\ -1.85 \end{bmatrix}.$$

In this case, both standardized coefficients are not only way off but also outside their conventional range of $[-1, 1]$. Moreover, the coefficient for Z_2 is of the wrong sign. The problem is that the diagonal elements of $\mathbf{R}_{xx}^{-1}$ no longer dominate the matrix as they did when r_{12} was .5. In the former case the ratio of the absolute value of the diagonal to the off-diagonal element was 2. In the current case it is 1.005. So the ridge-regression approach adds a constant, δ, where $0 \le \delta \le 1$, to the diagonal entries of $\mathbf{R}_{xx}$ in order to make the diagonal of $\mathbf{R}_{xx}^{-1}$ more dominant. After experimenting with different constants (more on choosing δ below), I decided to use a value of .15. The estimator is now

$$\mathbf{b}_{rr}^s = \mathbf{R}_{xx}^{-1}\mathbf{r}_{xy} = \begin{bmatrix} 1.15 & .995 \\ .995 & 1.15 \end{bmatrix}^{-1}\begin{bmatrix} .40925 \\ .38875 \end{bmatrix}$$

$$= \begin{bmatrix} 3.4589 & -2.9927 \\ -2.9927 & 3.4589 \end{bmatrix}\begin{bmatrix} .40925 \\ .39975 \end{bmatrix} = \begin{bmatrix} .2521 \\ .1199 \end{bmatrix},$$

where "rr" stands for "ridge regression." Notice, first, that the ratio of diagonal to off-diagonal elements of $\mathbf{R}_{xx}^{-1}$ has increased to 1.156. More important, the coefficient estimates are now fairly close to the true parameter values.

In general, ridge regression calls for replacing $(\mathbf{X}^{*\prime}\mathbf{X}^*)$ with $(\mathbf{X}^{*\prime}\mathbf{X}^* + \delta\mathbf{I})$. The estimator is then $\mathbf{b}_{rr}^s = (\mathbf{X}^{*\prime}\mathbf{X}^* + \delta\mathbf{I})^{-1}\mathbf{r}_{xy}$. The ridge estimator has substantially smaller variance than the OLS estimator, since the sum of the variances of the ridge coefficients is (Myers, 1986)

$$\sum V(b_{k,rr}^s) = \sum_{i=1}^{K} \frac{\sigma_*^2\lambda_i}{(\lambda_i + \delta)^2},$$

which can be compared to the comparable expression for OLS shown in equation (6.11). On the other hand, the bias of the estimator is easily seen. Because the OLS estimator can be written

$$\mathbf{b}^s = \mathbf{R}_{xx}^{-1}\frac{1}{n}\mathbf{Z}'\mathbf{y}_z = \mathbf{R}_{xx}^{-1}\frac{1}{n}\mathbf{Z}'(\mathbf{Z}\boldsymbol{\beta}^s + \boldsymbol{\varepsilon}^*) = \mathbf{R}_{xx}^{-1}\frac{1}{n}\mathbf{Z}'\mathbf{Z}\boldsymbol{\beta}^s + \mathbf{R}_{xx}^{-1}\frac{1}{n}\mathbf{Z}'\boldsymbol{\varepsilon}^*,$$

its expected value is

$$E(\mathbf{b}^s) = \mathbf{R}_{xx}^{-1}\frac{1}{n}\mathbf{Z}'\mathbf{Z}\boldsymbol{\beta}^s + \mathbf{R}_{xx}^{-1}\frac{1}{n}\mathbf{Z}'E(\boldsymbol{\varepsilon}^*) = \mathbf{R}_{xx}^{-1}\mathbf{R}_{xx}\boldsymbol{\beta}^s = \boldsymbol{\beta}^s,$$

showing that the OLS estimator is unbiased. Comparable operations for the ridge estimator give us

$$E(\mathbf{b}_{rr}^s) = (R_{xx} + \delta I)^{-1} R_{xx} \boldsymbol{\beta}^s \neq \boldsymbol{\beta}^s.$$

Choosing δ. Choosing the appropriate value of the ridge constant is more art than science. There are a number of criteria to use [see Myers (1986) for a description of several], but all revolve around plotting some criterion value against a succession of values for δ and choosing the δ that produces the "best" criterion. One relatively straightforward method is to use the *ridge trace* for each parameter estimate. This is a plot of the estimate against a succession of values for δ ranging from 0 to 1. Recall that the coefficients tend to be wildly inflated in magnitude and perhaps of the wrong sign under collinearity. The ridge traces for the different coefficients reveal how the coefficients shrink toward more tenable values as δ is increased until, after some value of δ, call this δ^*, there is little additional change in the coefficients. The value of δ^* is then chosen as the best ridge constant to use.

Ridge Regression with the Body Fat and NSFH Data. Figures 6.4 through 6.7 present the ridge traces for the intercept and the three regressors in the body fat data.

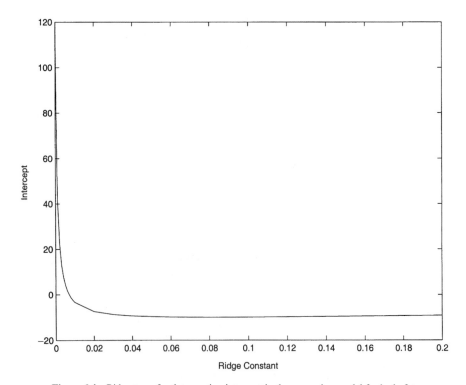

Figure 6.4 Ridge trace for the equation intercept in the regression model for body fat.

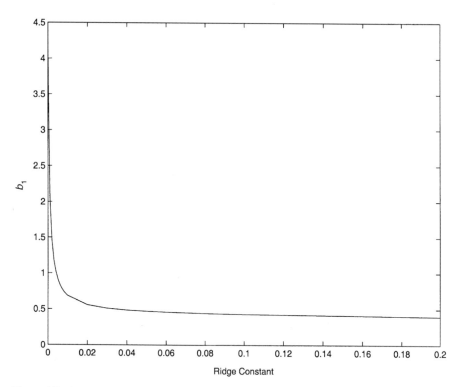

Figure 6.5 Ridge trace for the coefficient of triceps skinfold thickness (b_1) in the regression model for body fat.

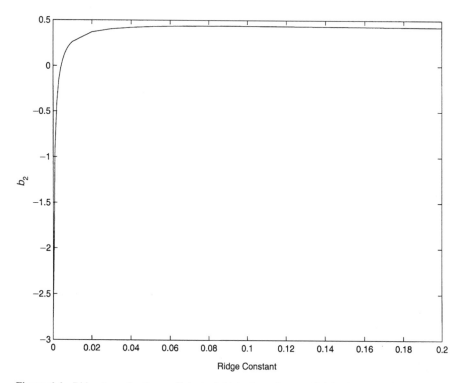

Figure 6.6 Ridge trace for the coefficient of thigh circumference (b_2) in the regression model for body fat.

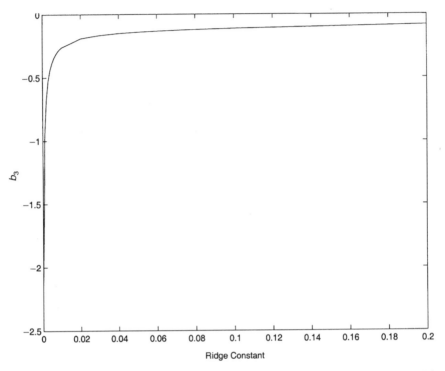

Figure 6.7 Ridge trace for the coefficient of midarm circumference (b_3) in the regression model for body fat.

Each of the plots suggests that the coefficients stabilize right around $\delta = .02$, and hence, this is the value chosen for the ridge constant. Notice in Figure 6.6, in particular, that the coefficient for *thigh circumference* changes sign from negative to positive very quickly as δ increases away from zero, and then remains positive. As expected, this suggests that a greater *thigh circumference* is associated with more, rather than less, body fat. The ridge regression estimates are shown in the "ridge estimator" column of panel A of Table 6.7. These are the unstandardized estimates and so can be compared to the OLS estimates in the first column. All have been reduced markedly in magnitude, consistent with the expected reduction in the coefficients once the multicollinearity problem has been addressed.

The ridge traces for *male-* and *female age, male-* and *female age at the beginning of the union*, and *union duration* for the NSFH data are shown in Figures 6.8 to 6.12. As these are the most problematic variables in the analysis, I choose a value of δ based on an attempt to stabilize their effects. In this case, all of the graphs point toward the value of .05 for δ, so this is the ridge constant that was chosen. Notice that the coefficient for *female age* changes from positive to negative as δ moves away from zero. Again, a negative effect was expected theoretically. Of the five plots, Figure 6.12, showing the trend in the coefficient for *union duration*, appears to evince the most dramatic change. The ridge estimates are shown in the "ridge estimator" column

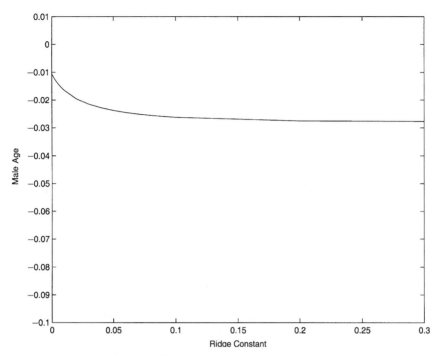

Figure 6.8 Ridge trace for the coefficient of male age in the regression model for couple disagreement.

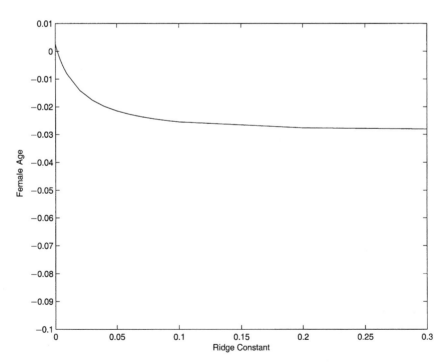

Figure 6.9 Ridge trace for the coefficient of female age in the regression model for couple disagreement.

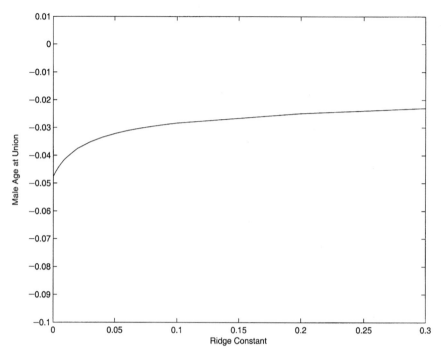

Figure 6.10 Ridge trace for the coefficient of male age at union in the regression model for couple disagreement.

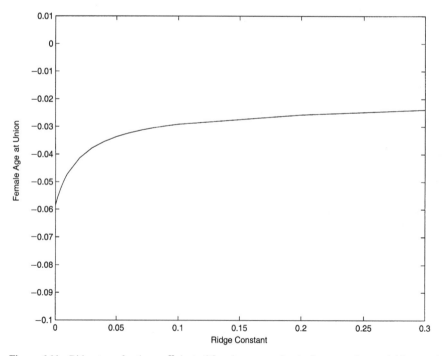

Figure 6.11 Ridge trace for the coefficient of female age at union in the regression model for couple disagreement.

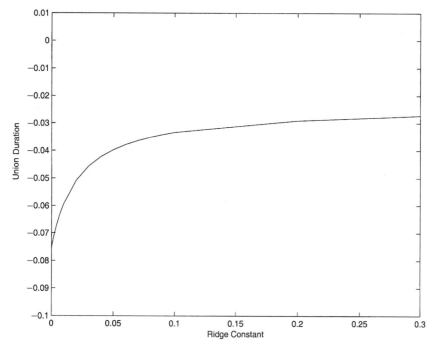

Figure 6.12 Ridge trace for the coefficient of union duration in the regression model for couple disagreement.

of panel A of Table 6.8. Again, these are the unstandardized coefficients. In this case, there is some shrinkage in all of the coefficients, compared to the OLS estimates, but the changes are nowhere near as pronounced as in Table 6.7. The most notable change is the aforesaid reversal of sign for the effect of *female age*.

Principal Components Regression. *Principal components regression*, also called *regression on principal components*, gets its name from the fact that the n scores on the K independent variables can be linearly transformed into a comparable set of scores on K principal components. However, the principal components have the property that they are orthogonal to each other. Each principal component is a weighted sum of all K of the original variables. Moreover, the principal components contain all of the variance of the original variables, but the first $J < K$ principal components typically account for the bulk of that variance. The variance of the jth component is equal to the jth eigenvalue of R_{xx}. So components associated with small eigenvalues contribute very little to the data and as a result, can be omitted from the analysis. As the small eigenvalues are associated with linear dependencies, this omission also greatly reduces the impact of those linear dependencies on the design matrix. Standard treatments of principal components regression (e.g., Jolliffe, 1986; Myers, 1986) usually develop the technique by writing the regression model in terms of the principal components.

Instead, I employ an alternative, but equivalent development that is more consistent with the notion of altering the correlation matrix, as discussed above with respect to ridge regression. Once again, we rely on the spectral decomposition of the design matrix to understand the procedure. Recall that the design matrix can be decomposed as $R_{xx} = \sum \lambda_j u_j u_j'$, which is the sum of K rank 1 matrices, each of the form $\lambda_j u_j u_j'$ (that these matrices only have rank 1 is left as a proof for the reader). However, if λ_j is small, the contribution of $\lambda_j u_j u_j'$ to R_{xx} is relatively insignificant. But it is at the same time the contribution that is associated with a linear dependency, so it is also troublesome to include (as we see below). In principal components regression we simply leave this contribution out of R_{xx} (and its inverse, R_{xx}^{-1}) when calculating the regression estimates.

To see the benefits of this, let's return to our simple two-regressor example used earlier. For the design matrix in which X_1 and X_2 are correlated .995, the eigenvalues are 1.995 and .005, with corresponding eigenvectors $u_1' = [.7071 \quad .7071]$ and $u_2' = [-.7071 \quad .7071]$. The spectral decomposition of R_{xx} shows that it is the sum of two parts:

$$\lambda_1 u_1 u_1' = 1.995 \begin{bmatrix} .7071 \\ .7071 \end{bmatrix} [.7071 \quad .7071] = \begin{bmatrix} .9975 & .9975 \\ .9975 & .9975 \end{bmatrix} \qquad (6.17)$$

and

$$\lambda_2 u_2 u_2' = .005 \begin{bmatrix} -.7071 \\ .7071 \end{bmatrix} [-.7071 \quad .7071] = \begin{bmatrix} .0025 & -.0025 \\ -.0025 & .0025 \end{bmatrix}. \qquad (6.18)$$

The reader can easily verify that the sum of the rightmost matrices in (6.17) and (6.18) is R_{xx}. But notice that the matrix in (6.17), which I refer to as the *reduced correlation matrix* and denote as R_{red}, almost perfectly reproduces R_{xx}, while the contribution of the matrix in (6.18) is almost negligible. However, the decomposition of R_{xx}^{-1}, following equation (6.13), shows that its two parts are

$$\frac{1}{\lambda_1} u_1 u_1' = .5013 \begin{bmatrix} .7071 \\ .7071 \end{bmatrix} [.7071 \quad .7071] = \begin{bmatrix} .2506 & .2506 \\ .2506 & .2506 \end{bmatrix} \qquad (6.19)$$

and

$$\frac{1}{\lambda_2} u_2 u_2' = 200 \begin{bmatrix} -.7071 \\ .7071 \end{bmatrix} [-.7071 \quad .7071] = \begin{bmatrix} 100 & -100 \\ -100 & 100 \end{bmatrix}. \qquad (6.20)$$

Adding the rightmost expressions in (6.19) and (6.20) gives us R_{xx}^{-1}, as shown above. Here it is clear that the contribution of the second eigenvalue, in the form of its inverse, and associated eigenvector are what "blow up" R_{xx}^{-1}. Therefore, in constructing "R_{xx}^{-1}," we simply omit the matrix in (6.20). I'm using quotation marks here since the reduced form of R_{xx}^{-1}, which I now denote R_{red}^-, is not actually the inverse of R_{red} (which can easily be verified by noting that $R_{red}^- R_{red} \neq I$). In fact, R_{red} is not invertible, since it is a 2×2 matrix with rank 1, as is immediately evident—there is only one independent vector present. In general, R_{red} is not invertible because it is $K \times K$ but is the sum of fewer than K rank 1 matrices. As the rank of a summed matrix is no greater than the sum of the ranks of its component matrices (Searle, 1982), the rank of R_{red} is always less than K, and therefore R_{red} cannot have an inverse. Consequently, I refer to R_{red}^- in (6.19) as the *pseudoinverse* of R_{red}. The standardized principal components coefficients are therefore

$$\mathbf{b}_{pc}^s = R_{red}^- \mathbf{r}_{xy} = \begin{bmatrix} .2506 & .2506 \\ .2506 & .2506 \end{bmatrix} \begin{bmatrix} .40925 \\ .38875 \end{bmatrix} = \begin{bmatrix} .2 \\ .2 \end{bmatrix}.$$

Although these are not quite as close to the true values of .25 and .15 as are the ridge estimates, they represent a vast improvement over the OLS estimates.

In general, then, the principal components estimator of $\boldsymbol{\beta}^s$ is of the form $\mathbf{b}_{pc}^s = R_{red}^- \mathbf{r}_{xy}$, where

$$R_{red}^- = \sum_{i=1}^{J} \frac{1}{\lambda_i} \mathbf{u}_i \mathbf{u}_i'$$

and $J < K$. Some authors recommend using the percentage of variance accounted for by the J retained components as a guide to how many components to omit (e.g., Hadi and Ling, 1998). However, typically, dropping the last component, which is associated with the smallest eigenvalue, will be sufficient. As with ridge regression, the bias of the principal components estimator is easy to see, since

$$E(\mathbf{b}_{pc}^s) = E(R_{red}^- \mathbf{r}_{xy}) = R_{red}^- \frac{1}{n} Z' E(\mathbf{y}_z) = R_{red}^- \frac{1}{n} Z'Z\boldsymbol{\beta}^s = R_{red}^- R_{xx}\boldsymbol{\beta}^s \neq \boldsymbol{\beta}^s.$$

Body Fat and NSFH Data, Revisited. The last column in panel A of Tables 6.7 and 6.8 presents the unstandardized principal components estimates for the body fat and NSFH data, respectively. Notice that the principal components estimates are quite close to the ridge regression estimates, and both are substantially different from the OLS estimates for the body fat data. For the NSFH data, all of the estimates, whether OLS, ridge, or principal components, are fairly similar. Perhaps the key substantive difference between OLS and the other estimators are that the latter have more intuitive signs for the effect of *thigh circumference* in the body fat data and for the effect of *female age* in the NSFH data. Again, the primary limitation with the latter estimators is that inferences to the population parameters cannot be made. On the other hand, the ridge and principal components estimators are probably closer than the OLS coefficients to the true values of the parameters. It should be mentioned that not all software makes these two techniques available to the analyst. SAS offers both estimators as options to the OLS regression procedure, invoked using the keywords RIDGE (for ridge regression) and PCOMIT (for principal components regression).

Concluding Comments. Although I confine my discussion of influential observations as well as collinearity problems and remedies to this chapter, these issues apply to all generalized linear models. Influence diagnostics have been devised for techniques such as logistic regression (Pregibon, 1981) and are included in such software packages as SAS. Collinearity problems can plague any model that employs multiple regressors; however, not all procedures offer collinearity diagnostics. On the other hand, multicollinearity is strictly a problem in the design matrix and does not depend on the nature of the link to the response. Therefore, it can always be diagnosed with an OLS procedure that provides collinearity diagnostics, which most of them do. One simply needs to code the dependent variable in some manner that is consistent with the use of OLS. Versions of ridge and principal components regression have been developed for logistic regression (see, e.g., Barker and Brown, 2001; Schaefer, 1986), suggesting that such techniques should become more widely available for other generalized linear models in the future.

EXERCISES

6.1 Suppose that we have a population of $N = 4$ cases. Let X be a matrix with columns $\mathbf{x}_1' = [1 \quad 2 \quad 4 \quad 8]$ and $\mathbf{x}_2' = [2 \quad 1 \quad 3 \quad 4]$. Suppose further that $\sigma_{x_1} = 2.681$ and $\sigma_{x_2} = 1.118$. Also, let C be the centering matrix with dimensions $C_{44} = I_{44} - \frac{1}{4}J_{44}$.

    (a) Show that $Z = CXD^{-1/2}$ is the matrix of standardized variable scores for these four cases, where $D^{-1/2}$ is a diagonal matrix with elements $1/\sigma_{x_i}$.

    (b) Calculate the correlation matrix using $R_{xx} = (1/n)Z'Z$.

6.2 Prove that the inverse of D_{c_i} is D_{1/c_i} for $i = 1, 2, \ldots, n$.

6.3 Show that $A = \sum \lambda_j \mathbf{u}_j \mathbf{u}_j' = UD_\lambda U'$.

6.4 Prove that if λ_j are the eigenvalues of A, $1/\lambda_j$ are the eigenvalues of A^{-1}. (*Hint*: Start with the spectral decomposition of A and then take the inverse of both sides of the equation.)

6.5 Let $r_{x_1 x_2} = .675$ and show that VIF_1 and VIF_2 are the diagonal elements of R_{xx}^{-1}.

6.6 Prove that $MSQE(\hat{\theta}) = V(\hat{\theta}) + [B(\hat{\theta})]^2$. (*Hint*: Start with the definition of $MSQE$; let $E\hat{\theta} = E(\hat{\theta})$ for economy of notation, then subtract and add this term from the expression inside parentheses and expand the expression.)

6.7 Prove that the matrix $\lambda \mathbf{u}\mathbf{u}'$ has only rank 1, for any scalar λ and any vector $\mathbf{u}$.

6.8 Verify that, in general, for any square matrix A, $tr(cA) = c\,tr(A)$, for any scalar c.

6.9 For the following data:

Case	X	Y	Case	X	Y
1	−2	2	6	6	12
2	3	3	7	9	5.5
3	3	6	8	9	8
4	6	4.5	9	15	2
5	6	7			

Do the following using a calculator:

(a) Calculate h_{ii} for all nine cases.

(b) For case 9, calculate t_i, $dffits_i$, $dfbetas_{ji}$ for the slope of the SLR and Cook's D_i. Note that another formula for Cook's D is (Neter et al., 1985)

$$D_i = \frac{e_i^2}{pMSE} \frac{h_{ii}}{(1 - h_{ii})^2}.$$

Also, for $dfbetas_{ji}$, use $\hat{\sigma}_{b_1}$ from the regression that omits case 9.

6.10 Let $Z_1 = \lfloor z_{11} \quad z_{11}^2 \rfloor$ and let $Z_2 = [z_{21} \quad z_{22}]$ represent matrices of standardized variable scores (i.e., assume that Z_{11}^2 is standardized *after* being created from the square of Z_{11}) and suppose that the following model (in standardized coefficients) characterizes the data:

$$Y_z = .25Z_{11} + .01Z_{11}^2 + .3Z_{21} + .25Z_{22} + \varepsilon_y,$$

$$Z_{21} = .2Z_{11}^2 + \varepsilon_{21},$$

$$Z_{22} = .15Z_{11}^2 + \varepsilon_{22},$$

$$\text{Cov}(Z_{11}, Z_{11}^2) = .8,$$

$$\text{Cov}(Z_{11}, \varepsilon_y) = \text{Cov}(Z_{11}, \varepsilon_{21}) = \text{Cov}(Z_{11}, \varepsilon_{22})$$

$$= \text{Cov}(Z_{11}^2, \varepsilon_y) = \text{Cov}(Z_{11}^2, \varepsilon_{21}) = \text{Cov}(Z_{11}^2, \varepsilon_{22}) = 0.$$

(a) If the analyst, instead, estimates $y_z = Z_1 \beta_1^s + \upsilon$, where $\upsilon = Z_2 \beta_2^s + \varepsilon_y$, give the value of $E(b_1^s)$. [*Hint:* First, note that $E(b_1^s) = \beta_1^s + r_{11}^{-1} r_{12} \beta_2^s$, where $r_{11}^{-1} r_{12} \beta_2^s$ is the bias in b_1^s and

$$r_{11} = \begin{bmatrix} 1 & \text{Cov}(Z_{11}, Z_{11}^2) \\ \text{Cov}(Z_{11}, Z_{11}^2) & 1 \end{bmatrix} \quad \text{and}$$

$$r_{12} = \begin{bmatrix} \text{Cov}(Z_{11}, Z_{21}) & \text{Cov}(Z_{11}, Z_{22}) \\ \text{Cov}(Z_{11}^2, Z_{21}) & \text{Cov}(Z_{11}^2, Z_{22}) \end{bmatrix}.$$

Then use covariance algebra to derive all unknown correlations (recall that the covariance between standardized variables is their correlation), and do the appropriate matrix operations.]

(b) Interpret the nature of the bias in the sample quadratic term.

6.11 Let $Z_1 = [z_{11} \quad z_{12} \quad cp]$, where $CP = Z_{11}Z_{12}$, and let $Z_2 = [z_{21} \quad z_{22}]$. Suppose further that all variables, including CP, are standardized. (*Note:* Normally, we don't standardize quadratic or cross-product terms; rather, they are formed as products of standardized variables; see Aiken and West, 1991. However, we do so here to simplify the covariance algebra.) Suppose that the following model (in standardized coefficients) characterizes the data:

$$Y_z = .25Z_{11} + .15Z_{12} + .10CP + .35Z_{21} + .3Z_{22} + \varepsilon_y,$$

$$Z_{21} = -.2CP + \varepsilon_{21},$$

$$Z_{22} = -.1CP + \varepsilon_{22},$$

$$\text{Cov}(Z_{11}, Z_{12}) = .45,$$

$$\text{Cov}(Z_{11}, CP) = .85,$$

$$\text{Cov}(Z_{12}, CP) = .90,$$

and the covariances of Z_{11}, Z_{12}, and CP, with all equation errors equal zero.

(a) If the analyst, instead, estimates $Y_z = Z_1\boldsymbol{\beta}_1^s + \upsilon$, where $\upsilon = Z_2\boldsymbol{\beta}_2^s + \varepsilon_y$, give the value of $E(\mathbf{b}_1^s)$. (*Hint*: Follow the hint for Exercise 6.10.)

(b) Interpret the nature of the bias in the sample estimate of the coefficient of the cross-product term (i.e., CP).

The following information is for Exercises 6.12 and 6.13: Suppose that the true model for Y is $\beta_1 Z_1 + \beta_2 Z_2 + \varepsilon$, where all observed variables are standardized. Suppose, further, that the sample correlation matrix is $R_{xx} = \begin{bmatrix} 1 & .997 \\ .997 & 1 \end{bmatrix}$ *and the sample vector of correlations of the regressors with Y is* $\mathbf{r}_{xy} = \begin{bmatrix} .7192 \\ .6786 \end{bmatrix}$. *Note also that the eigenvalues of R_{xx} are $\lambda_1 = 1.997$ and $\lambda_2 = .003$, while the eigenvectors are* $\mathbf{u}_1 = \begin{bmatrix} .7071 \\ .7071 \end{bmatrix}$ *and* $\mathbf{u}_2 = \begin{bmatrix} -.7071 \\ .7071 \end{bmatrix}$.

6.12 Estimate $\boldsymbol{\beta}^s = \begin{bmatrix} \beta_1^s \\ \beta_2^s \end{bmatrix}$ using OLS. Then perform a ridge regression using the following values for the ridge constant: .05, .10, .15, .20, and .25. You will have five different pairs of coefficient estimates. What appears to be the best ridge constant based on the changes in the coefficients?

6.13 Conduct a principal components regression for the data in Exercise 6.12.

The following information is for Exercises 6.14 and 6.15: In the model $y* = X*\boldsymbol{\beta}^s + \varepsilon^*$, the spectral decomposition of the 4×4 correlation matrix, R_{xx}, results in the following:

$$D_\lambda = \begin{bmatrix} 2.9305 & 0 & \cdot & 0 \\ 0 & .9996 & \cdot & \cdot \\ \cdot & \cdots & .0689 & 0 \\ 0 & \cdots & 0 & .00091806 \end{bmatrix},$$

$$U = \begin{bmatrix} .5725 & .0061 & .7549 & -.3199 \\ .5837 & .0093 & -.1015 & .8055 \\ .5756 & .0082 & -.6480 & -.4988 \\ .0137 & -.9999 & -.0017 & .0014 \end{bmatrix}.$$

6.14 (a) Give ϕ for R_{xx}.

(b) Give R_{xx}.

(c) Give R_{xx}^{-1}. [*Hint*: for parts (b) and (c) you may want to use a matrix program such as SAS IML or MATLAB to perform the necessary calculations.]

6.15 (a) Show the decomposition of $V(b_4^s)$ after the fashion of equation (6.14), assuming that $V(\varepsilon^*) = \sigma_*^2$.

(b) Give the variance proportions p_{13}, p_{23}, p_{33}, and p_{43}.

6.16 For the following data:

Case	X	Y	Case	X	Y
1	0	0	4	10	12.5
2	0	5	5	15	7.5
3	10	2.75	6	15	27.5

Assume that the model for the squared errors is $\varepsilon^2 = \sigma^2 \exp(\delta_0 + \delta_1 X)u$, and using a calculator, find the weights to be used in a WLS regression of Y on X.

6.17 Using the *couples dataset*, perform a WLS regression of COITFREQ on MALEAGE via an OLS regression of $\sqrt{w}$ COITFREQ on $\sqrt{w}$ and $\sqrt{w}$ MALEAGE, as discussed in the chapter (remember to specify a regression through the origin). Confirm that the resulting coefficients for $\sqrt{w}$ and for MALEAGE are the intercept and slope in the last column of Table 6.1. (*Note:* The solution is also in this column.)

6.18 Use the *kids dataset* to test for heteroscedasticity in the regression of PERMISIV on ADVENTRE, FSTYLE1, MSEXATT, and FSEXATT. Do both White's test and the Breusch–Pagan test. Also, obtain the White heteroscedasticity-robust estimates of the OLS standard errors (using appropriate software).

6.19 Use the *kids dataset* to estimate a WLS regression for the model in Exercise 6.18, employing an exponential function of the predictors to model the error variance (as in Exercise 6.16). Be sure to provide the correct R^2 for the WLS analysis.

6.20 Using the *students dataset*, test whether the error variance for the model EXAM1 = $\beta_0 + \beta_1$ SCORE + β_2 COLGPA + β_3 STATMOOD + ε is different for males and females. Then test for gender differences in the model's coefficients using both OLS and WLS approaches. Missing imputation: Substitute the parenthetical values for missing data on each variable indicated: SCORE (40.9358974) and COLGPA (3.0827835).

6.21 Using the *students dataset*, and the results from Exercise 6.20, test for gender differences in EXAM1 scores at college GPAs of 2.5, 3.0, and 3.5 employing the WLS approach, in which the weights are functions of each gender's error variance.

For Exercises 6.22 to 6.25, use the faculty salary dataset and note that "the model" refers to the regression of AYSALARY on YRDG, YRBG, PRIOREXP, R1, R2, R4, YRRANK, SALFAC, GRAD, ADMIN, and EMINENT. This model should be regarded as having heuristic rather than substantive value. First, YRDG, YRBG, and PRIOR-EXP in ordinary circumstances would be exactly linearly dependent, since PRIOR-EXP = YRDG − YRBG. However, in this case, due to inaccuracies in either recall or recording of dates, some values of PRIOREXP were negative. These were simply recoded to zero, a procedure which, as in the case of missing imputation, nullifies the exact linear dependency. Second, it is probably not realistic to attempt to estimate the impact of each of YRDG, YRBG, and PRIOREXP while holding the other two variables constant, given how much information these variables have in common. Nevertheless, teasing out the separate impacts of each of these factors might be of interest.

6.22 Examine the influence diagnostics for the model. Describe the characteristics of the most influential observation(s) in the data. Is there any justification for dropping any of the cases from the analysis?

6.23 Examine collinearity diagnostics for the model. Which variables appear to be tied together by near-linear dependencies? How severe is the collinearity? Which coefficients are affected, and by how much?

6.24 Using appropriate software, perform a ridge regression for the model. Use the ridge traces of YRDG, YRBG, and PRIOREXP to choose the best ridge constant.

6.25 Using appropriate software, perform a principal components regression for the model, leaving out the last component of R_{xx}^{-1}. Which set of coefficients is more appealing: those from the ridge regression or those from the principal components regression? Why?

CHAPTER 7

Regression with a Binary Response

CHAPTER OVERVIEW

In the social sciences some of the more interesting response variables are binary, or dichotomous. In certain instances, such variables arise due to the crudeness of measurement. For example, the NSFH asks respondents involved in marriages or cohabiting relationships whether they and their partner experienced any "physical arguments" in the past year, with possible responses of "yes" and "no." Although *physical aggression* might be treated as a continuous variable if it were measured more precisely, we are limited to a binary variable in this case, due to the measurement strategy. Other variables are naturally binary, such as whether someone voted in the last presidential election, or whether a woman experienced a pregnancy prior to age 18. In either case, employing a linear regression model is not the optimum strategy with a binary response. Hence, in this chapter we consider some alternative, nonlinear regression models that are especially suited to dichotomous dependent variables. I begin by considering the problems encountered in using linear regression for such situations. I then introduce the two most popular alternatives, logistic and probit regression, along with the theoretical rationale for these techniques. Details of interpretation, estimation, and inference are covered, with analogies to counterparts in linear regression. I then introduce two other, less known variants on binary response models, the scobit and complementary log-log models. Although not as easily interpreted as the logit model, these techniques overcome some of the limitations of logit and probit. Because of its interpretational advantages, primary emphasis is given in this chapter to the logistic regression model. The chapter closes with a discussion of assessing both discriminatory power and empirical consistency for this model.

Regression with Social Data: Modeling Continuous and Limited Response Variables,
By Alfred DeMaris
ISBN 0-471-22337-9   Copyright © 2004 John Wiley & Sons, Inc.

LINEAR PROBABILITY MODEL

Suppose that we have a binary response, Y_i, coded 1 if the ith case is in the category of interest, and 0 otherwise. (The coding of a binary response is actually arbitrary, but dummy coding is especially convenient, as will become apparent.) Recall that the linear regression model for the conditional mean of Y, given $\mathbf{x}$, is

$$E(Y_i) = \beta_0 + \beta_1 X_{i1} + \beta_2 X_{i2} + \cdots + \beta_K X_{iK}.$$

However, the mean of a dummy coded variable is the proportion of people in the category of interest, or equivalently, the probability of being in the interest category, denoted π. Letting π_i be the probability of being in the interest category given the ith covariate pattern, the linear regression model for the conditional mean of a binary response is

$$\pi_i = \beta_0 + \beta_1 X_{i1} + \beta_2 X_{i2} + \cdots + \beta_K X_{iK}. \tag{7.1}$$

Because the probability is being modeled as a linear function of the parameters, this is referred to as the *linear probability model* (LPM) (Aldrich and Nelson, 1984; Long, 1997). The regression coefficients are interpreted in terms of the probability of being in the interest category on Y. Hence, β_1 represents the change in the *probability* for each unit increase in X_1, net of the other covariates, and so on.

Example

Employing questions from the NSFH on partners' violence toward each other, I categorized 4095 married and cohabiting couples surveyed between 1987 and 1994 according to whether or not either partner had been violent toward the other during that period. These data are in the *violence dataset*. A total of 555 couples, or 13.55%, had experienced intimate violence. Of interest here is the extent to which *couple violence* is a function of several couple characteristics, including whether they were cohabiting, as opposed to legally married (*cohabiting*); the duration of the relationship, in years, as of the initial survey (*relationship duration*); whether either partner in the couple was a minority (*minority couple*); the female's age at the start of the union (*female's age at union*); the degree to which the male was socially or emotionally isolated from his or his partner's immediate kin (*male's isolation*); the degree of economic disadvantage exhibited by the couple's neighborhood of residence at the time of the initial survey (*economic disadvantage*); and whether either partner had a problem with alcohol or drugs (*alcohol/drug problem*). The continuous variables *relationship duration, female's age at union, male's isolation,* and *economic disadvantage* are all centered. Although the couple's *violence profile* is, in actuality, a three-category variable, here I simply distinguish the violent from the nonviolent. In Chapter 8 I consider the three-level response in greater detail. To keep things relatively simple, issues of sample selectivity and other critical explanatory variables are omitted in what

Table 7.1   OLS, Logit, and Probit Estimates for the Regression of Violence on Couple Characteristics

Predictor	Estimate	Model		
		OLS	Logit	Probit
Intercept	b	.117***	−2.151***	−1.247***
	$\hat{\sigma}_b$	.007	.066	.034
	t or z	17.750	−32.854	−37.058
Relationship duration	b	−.004***	−.046***	−.023***
	$\hat{\sigma}_b$	.0004	.005	.002
	t or z	−9.460	−9.299	−9.352
Cohabiting	b	.161***	.810***	.485***
	$\hat{\sigma}_b$	.037	.241	.145
	t or z	4.360	3.363	3.340
Minority couple	b	.023	.221*	.122*
	$\hat{\sigma}_b$	.012	.108	.059
	t or z	1.850	2.042	2.082
Female's age at union	b	−.003***	−.027***	−.014***
	$\hat{\sigma}_b$	.001	.007	.004
	t or z	−3.520	−3.588	−3.538
Male's isolation	b	.002*	.020*	.011*
	$\hat{\sigma}_b$	.001	.008	.004
	t or z	2.440	2.541	2.492
Economic disadvantage	b	.003**	.023*	.012*
	$\hat{\sigma}_b$	.001	.009	.005
	t or z	2.590	2.483	2.329
Alcohol/drug problem	b	.158***	1.029***	.589***
	$\hat{\sigma}_b$	.022	.156	.092
	t or z	7.060	6.611	6.429
F or model χ^2 (7 df)		29.130***	195.394***	191.860***
R^2		.048		
R_L^2			.060	.059
R_G^2			.047	.049
R_{GSC}^2			.085	.084
R_{MZ}^2			.127	
$\hat{\Delta}$			.051	

* $p < .05$. ** $p < .01$. *** $p < .001$.

follows. For the larger project on which this example is based, the reader should consult DeMaris et al. (2003).

The OLS column in Table 7.1 presents the results of regressing *violence* on couple characteristics using the LPM. The F test suggests that the model is significant as a whole, and the R^2 indicates that it accounts for about 5% of the variation in violence. Several factors have significant effects in expected directions. Consistent with past research (Stets, 1991), cohabitors are more likely than marrieds to be violent. The coefficient of .161 means that, net of other effects, cohabitors' probability of being violent is estimated to be .161 higher than for married people. Each additional year that the couple has been together is estimated to reduce the probability of violence by .004. The other effects are interpreted similarly.

Problems with the LPM

Although OLS is adequate for a quick exploration of the nature and significance of predictor effects, there are several difficulties with the LPM. To begin, the assumption of a model that is linear in the parameters is problematic. The conditional mean is a probability and is therefore confined to the [0,1] interval. The linear predictor, on the other hand, has no such constraints. For the linear predictor, $\sum_k \beta_k X_{ik}$, to be bounded by 0 and 1, it is necessary a priori to constrain the β's to be very small, particularly when the X's exhibit considerable variability [see Aldrich and Nelson (1984) for specific examples of this]. Without such constraints, probabilities would fall outside their logical range. This is evident in the OLS estimates in Table 7.1. If we calculate the estimated probability of violence for a "low-risk" couple, that is, a married nonminority couple without alcohol or drug problems that is 2 standard deviations above the mean on *relationship duration* ($SD = 12.823$) and *female's age at union* ($SD = 7.081$), 2 standard deviations below the mean on *male's isolation* ($SD = 6.296$), and 1 standard deviation below the mean on *economic disadvantage* ($SD = 5.128$), we have $\hat{\pi} = .117 - .004(2)(12.823) - .003(2)(7.081) + .002(2)$ $(-6.296) + .003(-5.128) = -.069$, which is clearly an impossible value.

On the other hand, for the sake of argument, let's suppose that the linear model in equation (7.1) *is* the true model for the probability of interest. First, as long as the mean structure—in this case, the structural model for π_i—is specified correctly, the parameter estimates will be consistent (Cameron and Trivedi, 1998). Nonetheless, recall that estimation via OLS assumes that the errors have zero mean and constant variance across all covariate patterns. Now, the model for an individual observation is $Y_i = \pi_i + \varepsilon_i$. The errors can take on only two possible values: $\varepsilon_i = 1 - \pi_i$ for cases in the interest category and $\varepsilon_i = -\pi_i$ for cases in the other category. As long as the mean of the errors at each covariate pattern is zero, the OLS estimates will be unbiased. To see that this holds here, note that the probability function f(ε) for the errors is as follows: $\varepsilon = 1 - \pi$ with probability π (i.e., the probability that $Y = 1$) and $\varepsilon = -\pi$ with probability $1 - \pi$ (the probability that $Y = 0$). Thus, the mean of the errors is

$$E(\varepsilon) = \sum_\varepsilon \varepsilon f(\varepsilon) = (1 - \pi)\pi + (-\pi)(1 - \pi) = 0.$$

What about the constant-variance assumption? The variance of the errors is

$$V(\varepsilon) = E(\varepsilon - E(\varepsilon))^2 = \sum_\varepsilon [\varepsilon - E(\varepsilon)]^2 f(\varepsilon) = \sum_\varepsilon \varepsilon^2 f(\varepsilon)$$

$$= (1 - \pi)^2 \pi + (-\pi)^2 (1 - \pi) = \pi (1 - \pi)[(1 - \pi) + \pi]$$

$$= \pi(1 - \pi). \tag{7.2}$$

In that π is a function of the explanatory variables, as shown by equation (7.1), the implication of expression (7.2) is that the errors vary with the X's and are therefore inherently heteroscedastic. The consequences of this, of course, are that the OLS estimators are no longer those with the smallest sampling variance and that OLS estimates of the standard errors of coefficient estimates are biased. We could correct these particular problems by using WLS estimation (Aldrich and Nelson, 1984). However, the preferred approach is to employ a nonlinear probability model, which, as we will see, makes more theoretical sense.

NONLINEAR PROBABILITY MODELS

Theoretically, the idea that probabilities are linear in explanatory variables is not particularly reasonable. Suppose that we are interested in modeling the effect of salary offers on the probability of changing jobs. For someone who is "on the fence" on the subject (i.e. someone with a 50–50 chance of changing jobs), a small increment in salary might be enough to entice one to switch to the new job. On the other hand, for those with either very low or very high probabilities of changing jobs, small differences in salary between the current and alternative positions probably will have little effect on their intentions. In other words, the effects of explanatory variables on the probability of an event should be stronger among those with probabilities in the middle range, but weaker when probabilities are near 0 or 1. Linear models, in contrast, assume that a predictor has the same effect on the response, regardless of the initial level of the response.

Latent-Variable Motivation of Probit and Logistic Regression

This idea can be developed more formally by assuming that our binary response is a proxy for a latent continuous variable, or *latent scale*, which follows a linear regression model. That is, assume that there is an unobserved, continuous variable, Y_i^*, such that

$$Y_i^* = \sum \beta_k X_{ik} + \varepsilon_i,$$

and assume further that the ε_i are independent random variables having a symmetric distribution with zero mean and constant variance across values of $\mathbf{x}$. What we observe, on the other hand, is a binary indicator, Y_i, that takes on the value 1 whenever Y_i^* is greater than some threshold value, c, and that takes on the value 0 otherwise. For convenience, it is usually assumed that c is 0. Then the probability that Y_i is 1 is

$$P(Y_i = 1) = P(Y_i^* > 0) = P\left(\sum \beta_k X_{ik} + \varepsilon_i > 0\right)$$

$$= P\left(\varepsilon_i > -\sum \beta_k X_{ik}\right) \qquad (7.3)$$

$$= P\left(\varepsilon_i < \sum \beta_k X_{ik}\right) \tag{7.4}$$

$$= F\left(\sum \beta_k X_{ik}\right). \tag{7.5}$$

The terms (7.3) and (7.4) are equal because the distribution of ε is assumed to be symmetric. (For example, the probability that a standard normal variable is greater than $-z$ is the same as the probability that it is less than z.)

Expressions (7.4) and (7.5) imply that the probability of Y being 1 can be modeled as a distribution function of ε, denoted $F(\varepsilon)$. The two distributions most often employed are the standard normal distribution and the standard logistic distribution. Figure 7.1 depicts the standard normal (probit) and standard logistic (logit) *densities*. They are both bell-shaped curves, but the logistic density has greater spread. The variances of these densities are 1 for the standard normal and $\pi^2/3$ (approximately equal to 3.29) for the standard logistic. The formulas for these functions are as follows. The standard normal density function (which uses the special symbol, ϕ) for a variable X, is

$$\phi(x) = \frac{1}{\sqrt{2\pi}} \exp\left(-\frac{1}{2}x^2\right), \tag{7.6}$$

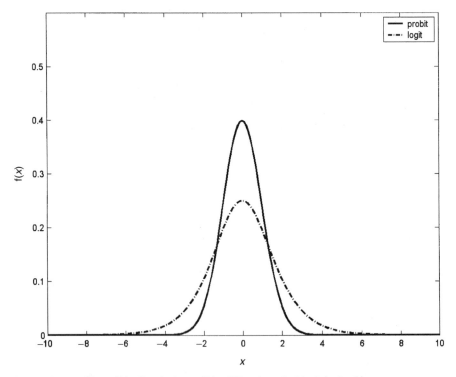

Figure 7.1 Standard normal (probit) and standard logistic densities.

whereas the standard logistic density (which uses the symbol λ) is

$$\lambda(x) = \frac{\exp(x)}{[1 + \exp(x)]^2}. \tag{7.7}$$

Recall that the distribution function for a variable X, $F(x)$, is the probability that $X \leq x$. This is just the area under the curve to the left of x. Hence, the standard normal distribution function, represented by the special symbol, Φ, is

$$\Phi(x) = \int_{-\infty}^{x} \frac{1}{\sqrt{2\pi}} \exp\left(-\frac{1}{2}u^2\right) du,$$

where the integration symbol ($\int_{-\infty}^{x}$) indicates the area under the stated function between negative infinity and x. This integral has no closed-form solution and must be obtained by actually summing areas under the standard normal curve. However, tables of areas under the standard normal curve can easily be used to find $F(x)$ given any value of x. The standard logistic distribution function, denoted with the symbol Λ, is

$$\Lambda(x) = \frac{\exp(x)}{1 + \exp(x)},$$

which, unlike the standard normal, does have a closed form.

Figure 7.2 depicts the standard normal and standard logistic distribution functions. It is evident that both are S-shaped curves that range from 0 to 1, as is logical for a probability. The curves are quite similar, except that the logistic curve approaches both 0 and 1 more gradually than does the normal. Indeed, it is not necessary to pose a latent variable as underlying the model. We could simply choose a distribution function as our model for probability due to its property of remaining within the [0,1] interval for any value of x. Notice also that for both curves, the rate at which the probability changes with x is minimal when the probability is either very small or very large. As mentioned above, this is also a theoretically desirable property of a probability function.

At this point, to simplify matters, let $z_i = \sum \beta_k X_{ik}$. Then the probit model for $P(Y_i = 1) = \pi_i$, following expression (7.5), is

$$\pi_i = \Phi(z_i) = \int_{-\infty}^{z_i} \frac{1}{\sqrt{2\pi}} \exp\left(-\frac{1}{2}u^2\right) du, \tag{7.8}$$

and the logistic regression model for π_i is

$$\pi_i = \Lambda(z_i) = \frac{\exp(z_i)}{1 + \exp(z_i)}. \tag{7.9}$$

It should be clear that these are both nonlinear functions of the parameters (i.e., the betas). Nevertheless, both models can be linearized, using the appropriate link function. Recall from Chapter 1 that in generalized linear models, which encompass both probit and logit, the link function is a transformation of the mean of Y (in this case, π) that equals the linear predictor, $\sum \beta_k X_{ik}$. The linear form of the probit model in equation (7.8) is, therefore,

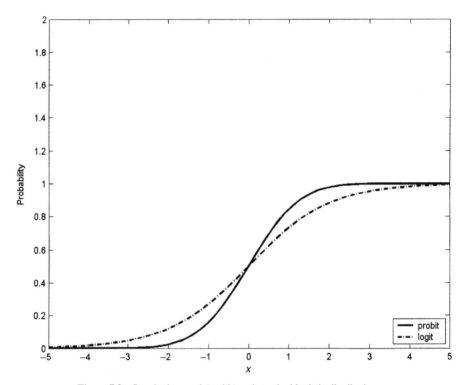

Figure 7.2   Standard normal (probit) and standard logistic distributions.

$$\Phi^{-1}(\pi_i) = \sum \beta_k X_{ik}, \tag{7.10}$$

where $\Phi^{-1}(\pi_i)$, called the *probit*, or *normit*, link, is the z_i such that $F(z_i) = \pi_i$. [For example, $\Phi^{-1}(.025) = -1.96$, since $\Phi(-1.96) = .025$.] The linear form of the logistic regression model in equation (7.9) uses the logit link (as derived in Chapter 1)

$$\ln\frac{\pi_i}{1 - \pi_i} = \sum \beta_k X_{ik}. \tag{7.11}$$

Equations (7.10) and (7.11) now resemble ordinary regression models in that the response variables can take on any value in the real numbers, just as can the linear predictor on the right-hand side. The coefficients, rather than indicating the effects of explanatory variables on Y_i, indicate the effects on the probit or logit link. Below I consider more intuitive interpretations of the β's, especially in the logit model.

Estimation

Both probit and logit models are constructed by assuming that a particular density underlies the data. Hence, these models are typically estimated using maximum likelihood rather than least squares. We proceed as follows. If π_i is the probability that Y_i equals 1, and $1 - \pi_i$ is the probability that $Y_i = 0$, we can write the discrete density function for the Y_i as

$$f(y_i) = \pi_i^{y_i}(1 - \pi_i)^{1-y_i}.$$

This function gives us the probability that Y takes on either value in its range $f(1) = \pi_i^1(1 - \pi_i)^{1-1} = \pi_i$ and $f(0) = \pi_i^0(1 - \pi_i)^{1-0} = 1 - \pi_i$. Now, the probability of observing $\mathbf{y}$, a particular collection of ones and zeros, is

$$f(\mathbf{y}) = \prod_{i=1}^{n} \pi_i^{y_i}(1 - \pi_i)^{1-y_i}. \qquad (7.12)$$

Substituting (7.8) for π_i in (7.12) results in the likelihood function for the probit model:

$$L(\boldsymbol{\beta}\,|\,\mathbf{y},\mathbf{x}) = \prod_{i=1}^{n} \Phi(z_i)^{y_i}[1 - \Phi(z_i)]^{1-y_i}.$$

Substituting (7.9) into (7.12) gives us the likelihood for the logit model:

$$L(\boldsymbol{\beta}\,|\,\mathbf{y},\mathbf{x}) = \prod_{i=1}^{n} \Lambda(z_i)^{y_i}[1 - \Lambda(z_i)]^{1-y_i}.$$

The idea behind maximum likelihood estimation is to find the β values that maximize the likelihood function, or equivalently, the log of the likelihood function. For example, the log of the likelihood function for logistic regression is

$$\ell(\boldsymbol{\beta}\,|\,\mathbf{y},\mathbf{x}) = \sum_{i=1}^{n} \left[y_i \ln \frac{\exp(\mathbf{x}^{i\prime}\boldsymbol{\beta})}{1+\exp(\mathbf{x}^{i\prime}\boldsymbol{\beta})} + (1 - y_i)\ln \frac{1}{1+\exp(\mathbf{x}^{i\prime}\boldsymbol{\beta})} \right], \qquad (7.13)$$

where $\mathbf{x}^{i\prime}\boldsymbol{\beta}$ represents $\sum \beta_k X_{ik}$ (see Appendix A, Section V.J). To find the coefficient estimates for the logit model, one takes the first partial derivatives of (7.13) with respect to each of the β's, sets them to zero, and then solves the set of simultaneous equations for $\mathbf{b}$. (The same idea applies to the probit model.) As this is a system of nonlinear equations, an iterative procedure is required [see Long (1997) for details]. The matrix of second derivatives of (7.13) with respect to the parameters is called the *Hessian* (the same applies to probit), and the inverse of the negative of the expected value of the Hessian is the variance–covariance matrix for the parameter estimates (Long, 1997). The estimate of this matrix provides the estimated standard errors of the coefficients. With large samples, as noted in Chapter 1, MLEs tend to be unbiased, consistent, efficient, and normally distributed.

Inferences in Logit and Probit

There are several inferential tests of interest in probit and logit that are analogous to those in linear regression. First, the logit/probit counterpart to the global F test in linear regression is the *likelihood-ratio chi-squared test*, also called the *model chi-squared*. Let L_0 denote the likelihood function evaluated at the MLE for an intercept-only model, and L_1 denote the likelihood function evaluated at the MLEs for the hypothesized model. Then the model χ^2 is

$$\text{model} \chi^2 = -2 \ln \frac{L_0}{L_1}$$

$$= -2 \ln L_0 - (-2 \ln L_1).$$

The log of L_1 can be computed by plugging the coefficient estimates into the probit or logit likelihood and then logging the result. In particular, $\ln L_1$ in logistic regression would be computed by evaluating (7.13) with $\hat{\beta}$ substituted in place of β. For the logistic regression of *violence* on the couple characteristics shown in Table 1, $-2 \ln L_1$ is 3054.132.

In binary response models, the likelihood evaluated under the MLE for an intercept-only model has an especially simple form. In the absence of any covariates in the model, the MLE for π_i is just p, the sample proportion in the interest category on Y. L_0 is, therefore,

$$L_0 = \prod_{i=1}^{n} p^{y_i}(1-p)^{1-y_i} = p^{n_1}(1-p)^{n_0},$$

where n_1 is the number of cases with Y equal to 1 and n_0 is the number of cases with Y equal to 0. For the violence example, the null likelihood is

$$L_0 = (.1355)^{555}(1 - .1355)^{3540},$$

and thus the log-likelihood is

$$\ln L_0 = 555 \ln(.1355) + 3540 \ln(1 - .1355) = -1624.763,$$

implying that

$$-2 \ln L_0 = -2(-1624.763) = 3249.526.$$

The model χ^2 for the logistic regression of *violence* is, therefore, $3249.526 - 3054.132 = 195.394$. Under the null hypothesis that the β's for all of the explanatory variables equal zero, the model χ^2 is distributed as chi-squared with K degrees of freedom, where K, as always, indicates the number of regressors in the model. In this case, the *df* is 7, and the result is very significant, suggesting that at least one of the regression coefficients is not zero.

Tests for nested models are accomplished with a *nested* χ^2 *test*, analogous to the nested F test in OLS. If model B is nested inside model A, a test for the validity of the constraints on A that lead to B is

$$\text{nested} \chi^2 = \text{model} \chi^2 \text{ for A} - \text{model} \chi^2 \text{ for B},$$

which under the null hypothesis that the constraints are valid is chi-squared with degrees of freedom equal to the number of constraints imposed (e.g., the number of parameters set to zero).

As the coefficient estimates are normally distributed for large n, we use a z test to test H_0: $\beta_k = 0$. The test is of the form

$$z = \frac{b_k}{\hat{\sigma}_{b_k}},$$

where $\hat{\sigma}_{b_k}$ is the estimated standard error of b_k. The square of z is what is actually reported in some software (e.g., SAS), and z^2 is referred to as the *Wald chi-squared* since it has a chi-squared distribution with 1 degree of freedom under H_0. This test is asymptotically equivalent to the nested χ^2 that would be found from comparing models with and without the predictor in question. However, the reader should be cautioned that Wald's test can behave in an aberrant manner when an effect is too large. In particular, the Wald statistic shrinks toward zero as the absolute value of the parameter estimate increases without bound (Hauck and Donner, 1977). Therefore, when in doubt, the nested χ^2 is to be preferred over the Wald test for testing individual coefficients.

Confidence intervals for logit or probit coefficients are also based on the asymptotic normality of the coefficient estimates. Thus, a 95% confidence interval for β_k in either type of model takes the form $b_k \pm 1.96 \hat{\sigma}_{b_k}$. This formula applies generically to any coefficient estimates that are based on maximum likelihood estimation and is relevant to all the models discussed from this point on in the book. I therefore omit coverage of confidence intervals in subsequent chapters.

More about the Likelihood. As the likelihood function is liable to be relatively unfamiliar to many readers, it is worth discussing in a bit more detail. It turns out that this function taps the indeterminacy in Y under a given model, much like the total and residual sums of squares do, in linear regression. By indeterminacy, I mean the uncertainty of prediction of Y under a particular model. For example, in OLS, if the "model" for Y is a constant, μ, estimated in the sample by $\bar{y}$, the indeterminacy in Y with respect to this model is TSS = the sum of squares around $\bar{y}$. This, of course, is the naive model, which posits that Y is unrelated to the explanatory variables. TSS measures the total amount of indeterminacy in Y that is potentially "explainable" by the regression. On the other hand, SSE, which equals the sum of squares around $\hat{y}$, is the indeterminacy in Y with respect to the hypothesized model. If the model accounts perfectly for Y, then $Y = \hat{y}$ for all cases, and SSE is zero.

In linear regression, we rely on the squared deviation of Y from its predicted value under a given model to tap uncertainty. The counterpart in MLE is the *likelihood of Y* under a particular model. The greater the likelihood of the data, given the parameters, the more confident we are that the process that generated Y has been identified correctly. Under the naive model, the process that generated Y is captured by p, and $-2 \ln L_0$ reflects the total uncertainty in Y that remains to be explained. The indeterminacy under the hypothesized model is $-2 \ln L_1$. What happens if Y is predicted perfectly by the model? Let's rewrite the likelihood function as a general expression, with $\hat{\pi}_i$ denoting the predicted probability that $Y = 1$ for the ith case. We begin with (7.12), but substitute $\hat{\pi}_i$, and write it as follows. Let 1 represent the set of cases with Y equal to 1, and 0 the set of cases with Y equal to zero. Then we have

$$L(\boldsymbol{\beta}|\mathbf{y},\mathbf{x}) = \prod_{y\in 1} \hat{\pi}_i \prod_{y\in 1} (1 - \hat{\pi}_i).$$

Now if Y is perfectly predicted, $\hat{\pi}_i = 1$ when $Y = 1$ and $\hat{\pi}_i = 0$ when $Y = 0$. We then have

$$L(\boldsymbol{\beta}|\mathbf{y},\mathbf{x}) = \prod_{y\in 1} 1 \prod_{y\in 1} (1 - 0) = 1,$$

in which case $-2\ln(1) = 0$. In other words, the closer to zero $-2\ln L$ is, the less uncertainty there is about Y under the model. The larger $-2\ln L$ is, the poorer the model is in accounting for the data. The model χ^2 simply tells us how much the original level of indeterminacy in Y—from using the naive model—is reduced under the hypothesized model. We will see below that this reasoning leads to one of the R^2 analogs used in models employing MLE.

Logit and Probit Analyses of Violence

Logit and probit estimates for the regression of violence on couple characteristics are shown in the "logit" and "probit" columns of Table 7.1. Model χ^2 values suggest that both models are significant. Substantively, the logit and probit results tend to agree with the OLS ones: *cohabiting* instead of being married, being a *minority couple*, *male isolation*, *economic disadvantage*, and having an *alcohol or drug problem* all elevate the probability of violence, while longer *relationship durations* and older *ages at inception of the union* lower it. Z ratios for tests of logit and probit coefficients, with the exception of that for the intercept, are roughly comparable to the OLS t ratios. The effect of being a minority couple is just significant in logit and probit, but just misses being significant in OLS. Otherwise, all regressor effects are significant, across all models.

However, predicted probabilities generated by logit and probit, particularly near the extremes of 0 or 1, depart from those of OLS. For example, suppose that we reestimate the probability of violence for our low-risk couple described above. We saw above that it was $-.069$ for OLS. For the logistic regression model, we first calculate the predicted logit:

$$\ln\frac{\hat{\pi}}{1-\hat{\pi}} = -2.151 - .046(2)(12.823) - .027(2)(7.081)$$
$$+ .020(2)(-6.296) + .023(-5.128) = -4.083.$$

Then the predicted probability is

$$\hat{\pi} = \frac{\exp(-4.083)}{1 + \exp(-4.083)} = .017.$$

In probit, we estimate

$$\Phi^{-1}(\hat{\pi}) = -1.247 - .023(2)(12.823) - .014(2)(7.081)$$

$$+ .011(2)(-6.296) + .012(-5.128) = -2.235.$$

Then the estimated probability is $\Phi(-2.235) = .013$.

Interpreting the Coefficients. One way to interpret logit and probit coefficients is in terms of their effects on Y^* in the regression model for the latent scale, as the logit/probit coefficients are estimates of the β_k in this regression. However, presuming that Y is a binary proxy for an underlying continuous variable may not always make sense. Instead, the β_k can be interpreted in terms of the probability of being in the interest category on Y—but there's a catch. In OLS, the partial derivative of $E(Y)$ with respect to X_k is a constant value, β_k, regardless of the levels of the regressors. This is no longer true in nonlinear models, as we saw in Chapter 5. Recalling that both the logit and the probit model for π_i are distribution functions, we can derive a general expression for the partial derivative of a distribution function of $\mathbf{x}^{i\prime}\boldsymbol{\beta}$, with respect to X_k. Employing the chain rule (Appendix A, Section IV.B) yields

$$\frac{\partial}{\partial X_k}F(\mathbf{x}^{i\prime}\boldsymbol{\beta}) = \frac{\partial}{\partial \mathbf{x}^{i\prime}\boldsymbol{\beta}}F(\mathbf{x}^{i\prime}\boldsymbol{\beta})\frac{\partial}{\partial X_k}(\mathbf{x}^{i\prime}\boldsymbol{\beta})$$

$$= f(\mathbf{x}^{i\prime}\boldsymbol{\beta})\beta_k. \tag{7.14}$$

Applying (7.14) to the logit and probit models, the partial derivatives are as follows. For logit:

$$\frac{\partial}{\partial X_k}\pi_i = \frac{\partial}{\partial X_k}\Lambda(\mathbf{x}^{i\prime}\boldsymbol{\beta}) = \lambda(\mathbf{x}^{i\prime}\boldsymbol{\beta})\beta_k$$

$$= \frac{\exp(\mathbf{x}^{i\prime}\boldsymbol{\beta})}{1+\exp(\mathbf{x}^{i\prime}\boldsymbol{\beta})}\frac{1}{1+\exp(\mathbf{x}^{i\prime}\boldsymbol{\beta})}\beta_k$$

$$= \pi_i(1-\pi_i)\beta_k. \tag{7.15}$$

For the probit model we have

$$\frac{\partial}{\partial X_k}\pi_i = \frac{\partial}{\partial X_k}\Phi(\mathbf{x}^{i\prime}\boldsymbol{\beta}) = \phi(\mathbf{x}^{i\prime}\boldsymbol{\beta})\beta_k$$

$$= \left\{\frac{1}{\sqrt{2\pi}}\exp\left[-\frac{1}{2}\left(\mathbf{x}^{i\prime}\boldsymbol{\beta}\right)^2\right]\right\}\beta_k. \tag{7.16}$$

Expressions (7.15) and (7.16) make it clear that the partial slope of X_k with respect to π_i in logit and probit is not a constant but is rather dependent on a particular value of the linear predictor, $\mathbf{x}^{i\prime}\boldsymbol{\beta}$, which, in turn, varies with the regressors. For example, in logistic regression the estimated partial slope for the probability of violence with respect to relationship duration, for our low-risk couple, is

$$\frac{\partial}{\partial X_k}\hat{\pi} = (.017)(1 - .017)(-.046) = -.00077.$$

For the probit model we have

$$\frac{\partial}{\partial X_k}\hat{\pi} = \left\{\frac{1}{\sqrt{2\pi}}\exp\left[-\frac{1}{2}(-2.235)^2\right]\right\}(-.023) = -.00075,$$

which, despite differences in b_k, is essentially the same as for logit. In that the correction factors in the partial slope, $\phi(\mathbf{x}^{i\prime}\boldsymbol{\beta})$ for probit and $\lambda(\mathbf{x}^{i\prime}\boldsymbol{\beta})$ for logit, are always positive and less than 1, the b_k can be interpreted as the effects on the probability apart from an attenuation factor. Hence positive coefficients indicate regressors with positive effects on the probability, and negative regressors indicate regressors with negative effects. Beyond this, the coefficients do not have an intuitively simple interpretation. (We will see below, however, that the logit coefficients, when exponentiated, have a particularly appealing interpretation.)

Alternative Models. An artifact of both logit and probit models is that the effect of a regressor on the probability, as indicated by the partial derivative, is always at its maximum when π equals .5. In this case, the attenuation factor for logit reaches its maximum value of $(.5)(1 - .5) = .25$. In the probit model, when the linear predictor is zero, $\pi = \Phi(0) = .5$, and the attenuation factor reaches its maximum value of $1/\sqrt{2\pi} = .399$. Nagler (1994) pointed out that this is a limitation of logit and probit models. People with an initial probability of an event of .5 will be indicated to be most susceptible to regressor effects (i.e., the partial effect of each X_k will reach its maximum value) due to a phenomenon that is imposed by the model specification. He therefore proposed using an alternative model when sensitivity of people's probabilities of events to regressor effects was important in one's investigation. The model, called *scobit* for "skewed logit" (Nagler, 1994, p. 235), allows the probability at which people are most susceptible to regressors to be estimated from the data.

The scobit model is based on the Burr-10 distribution, which unlike logit and probit, is an *asymmetric* distribution. The formula for this distribution function is

$$B_{10}(x) = \left[\frac{\exp(x)}{1 + \exp(x)}\right]^{\alpha}.$$

When $\alpha = 1$, this reduces to the logit distribution. Hence, the logit model is nested inside the scobit model, and a nested chi-squared test can be used to determine whether scobit is an improvement over logit. The scobit model is developed with the same latent-scale formulation as logit and probit except that the Burr-10 distribution replaces the standard normal or standard logistic distribution. Once again, we assume a latent scale, Y^*, such that $Y_i^* = \sum \beta_k X_{ik} + \varepsilon_i$. This time, we assume that ε_i follows the

Burr-10 distribution. Then

$$P(Y_i = 1) = P(Y_i^* > 0) = P\left(\sum \beta_k X_{ik} + \varepsilon_i > 0\right)$$

$$= P\left(\varepsilon_i > -\sum \beta_k X_{ik}\right).$$

However, because the Burr-10 distribution is not symmetric, this does *not* equal expression (7.4). Instead, because of the principle that $P(x > c) = 1 - P(x < c)$, we have that

$$P(Y_i = 1) = 1 - P\left(\varepsilon_i < -\sum \beta_k X_{ik}\right)$$

$$= 1 - \left[\frac{\exp(-\sum \beta_k X_{ik})}{1 + \exp(-\sum \beta_k X_{ik})}\right]^\alpha, \qquad (7.17)$$

which we denote as $G(\mathbf{x}^{i\prime}\boldsymbol{\beta})$. Again, it is easy to show that if $\alpha = 1$, $G(\mathbf{x}^{i\prime}\boldsymbol{\beta}) = \Lambda(\mathbf{x}^{i\prime}\boldsymbol{\beta})$. Unlike the case with logit and probit models, there is no link function that linearizes (7.17) when α is not equal to 1. The probability, π^*, at which people are most susceptible to effects of explanatory variables is given by

$$\pi^* = 1 - \left(\frac{\alpha}{1 + \alpha}\right)^\alpha$$

(Nagler, 1994). As is evident, if $\alpha = 1$, $\pi^* = .5$, the π^* for the logit model. Also, π^* converges to zero as α converges to zero; as α tends to infinity, however, π^* reaches a maximum value of about .632 (Nagler, 1994).

Another asymmetric distribution that can be employed as a probability model is the *complementary log-log model* (Agresti, 2002; Long, 1997):

$$\pi_i = H(\mathbf{x}^{i\prime}\boldsymbol{\beta}) = 1 - \exp[-\exp(\mathbf{x}^{i\prime}\boldsymbol{\beta})]. \qquad (7.18)$$

The link function that linearizes (7.18) is

$$\ln[-\ln(1 - \pi_i)] = \sum \beta_k X_{ik}.$$

The likelihood function for this model becomes important in Chapter 10; therefore, I give it here:

$$L(\boldsymbol{\beta} \mid \mathbf{y}, \mathbf{x}) = \prod_{i=1}^{n} \pi_i^{y_i}(1 - \pi_i)^{1-y_i}$$

$$= \prod_{i=1}^{n} [1 - \exp(-\exp(\mathbf{x}^{i\prime}\boldsymbol{\beta}))]^{y_i}[\exp(-\exp(\mathbf{x}^{i\prime}\boldsymbol{\beta}))]^{1-y_i}$$

$$= \prod_{y \in 0} [\exp(-\exp(\mathbf{x}^{i\prime}\boldsymbol{\beta}))] \prod_{y \in 1} [1 - \exp(-\exp(\mathbf{x}^{i\prime}\boldsymbol{\beta}))].$$

Figure 7.3 shows the density functions, and Figure 7.4 shows the distribution functions for the scobit and complementary log-log models. The asymmetery of the

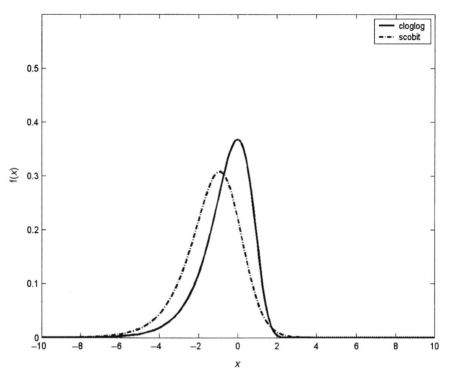

Figure 7.3 Scobit and complementary log-log densities.

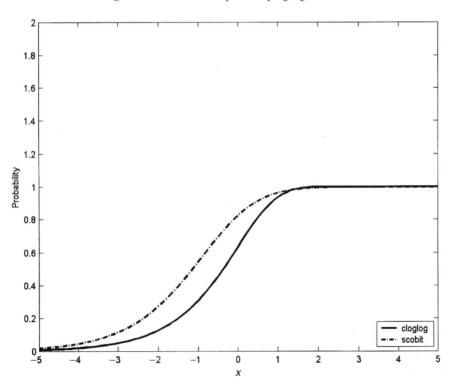

Figure 7.4 Scobit and complementary log-log distributions.

densities is evident in Figure 7.3, which reveals both densities to be skewed to the left. In contrast to the logit and probit distributions, the functions in Figure 7.4 both approach 1 more rapidly than they approach 0; this is particularly true for the complementary log-log model. In fact, Agresti (2002) remarks that logit and probit models are not appropriate when π increases from 0 fairly slowly but approaches 1 quite suddenly (see his monograph for an application in which this is the case). In that situation, the complementary log-log model would be most appropriate. Table 7.2 presents the scobit and complementary log-log estimates for the regression of violence. Although the intercepts diverge, the regressor coefficients from each model are very similar in value. Apart from an attenuation factor, the coefficients indicate effects on π and are in substantive agreement with the logit and probit effects. The nested test for scobit versus logit is shown as the "χ^2 for alpha" in the table and is a test for H_0: $\alpha = 1$. Because the test is nonsignificant, scobit appears to offer no improvement over the logit model. (The scobit model is available in STATA, whereas the complementary log-log model is available in SAS.)

Interpreting the Partial Derivative, Revisited. It is tempting to interpret the partial derivative in nonlinear probability models as the change in the probability for a unit increase in a given predictor, net of other regressors (e.g., Cleary and Angel, 1984). This is, of course, the unit-impact interpretation appropriate in the LPM. Although the partial derivative is, at times, a very close approximation to such a change, this interpretation is not technically correct, as has been observed elsewhere (DeMaris, 1993, 2003; Petersen, 1985). The reason is illustrated in Figure 7.5, which shows how closely the partial derivative approximates the change in $P(x) = \exp(x)/[1 + \exp(x)]$ from $x = 1$ to $x = 2$, a unit change. Notice that this is just a logistic regression model with $\beta_0 = 0$ and $\beta_1 = 1$. The partial derivative at $x = 1$ is $P'(1) = P(1)[1 - P(1)]\beta_1 = (.731)(1 - .731)(1) = .1966$. This is the slope of the line tangent to $P(x)$ at $x = 1$, as shown in the figure. Now the slope of that line indicates change *along the line* for each unit increase in x, but not change along the *function*, as it is clear that the line does not follow the function very closely. Change along the function for a unit increase in x at $x = 1$ is given by

$$P(2) - P(1) = \frac{e^2}{1 + e^2} - \frac{e}{1 + e}$$

$$= .8808 - .73106 = .1497.$$

As .1966 is not very close to .1497, $P'(1)$ is not a good approximation to the unit impact. On the other hand, if a unit change represents a very small change in a predictor, the partial derivative will be a close approximation to the unit impact. But in general, the accurate way to assess the change in probability for a unit increase in X_k, net of other predictors, is to evaluate $F(x_k + 1 | \mathbf{x}_{-k}) - F(x_k | \mathbf{x}_{-k})$, where F represents the probability model of interest (e.g., logit, probit, scobit, complementary log-log).

Table 7.2 Scobit and Complementary Log-Log Estimates for the Regression of Violence on Couple Characteristics

Predictor	Estimate	Model	
		Scobit	Cloglog
Intercept	b	−14.922	−2.207***
	$\hat{\sigma}_b$	728.360	.061
	t or z	−.020	−36.064
Relationship duration	b	−.044***	−.044***
	$\hat{\sigma}_b$	.005	.005
	t or z	−9.311	−9.330
Cohabiting	b	.685***	.685***
	$\hat{\sigma}_b$	.195	.194
	t or z	3.519	3.530
Minority couple	b	.192*	.192*
	$\hat{\sigma}_b$	.098	.098
	t or z	1.959	1.959
Female's age at union	b	−.025***	−.025***
	$\hat{\sigma}_b$	.007	.007
	t or z	−3.634	−3.624
Male's isolation	b	.018*	.018*
	$\hat{\sigma}_b$	.007	.007
	t or z	2.528	2.515
Economic disadvantage	b	.022**	.022**
	$\hat{\sigma}_b$	.008	.008
	t or z	2.705	2.710
Alcohol/drug problem	b	.902***	.902***
	$\hat{\sigma}_b$	.130	.130
	t or z	6.944	6.945
α		332675.100	
Model χ^2 (7 df)		196.686***	196.685***
χ^2 (1 df) for α		1.290	
R_L^2		.061	.061
R_G^2			.047
R_{GSC}^2			.086

* $p < .05$. ** $p < .01$. *** $p < .001$.

Interpeting Logit Models: Odds and Odds Ratios. Logit models have an advantage over other models in interpretability, because $\exp(\beta_k)$ can be interpreted as the *multiplicative impact on the odds of an event for a unit increase in X_k, net of the other covariates.* Indeed, $\exp(\beta_k)$ is the multiplicative analog of β_k in the linear regression model. To understand why this is so, we exponentiate both sides of equation (7.11) to express the logit model as

$$\exp\left(\ln\frac{\pi_i}{1 - \pi_i}\right) = \exp\left(\sum \beta_k X_{ik}\right)$$

or

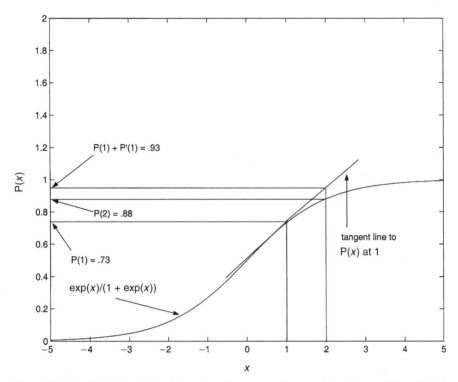

Figure 7.5   First derivative of the logit function with respect to X, versus the unit impact of X on the logit function.

$$\frac{\pi_i}{1 - \pi_i} = \exp(\beta_0) \exp(\beta_1 X_{i1}) \exp(\beta_2 X_{i2}) \cdots \exp(\beta_K X_{iK}). \qquad (7.19)$$

The left-hand side of (7.19) is called the *odds* of event occurrence for the ith case and is denoted O_i. The odds is a ratio of probabilities. In particular, it is the ratio of the probability of the event to the probability of not experiencing the event. If the odds is, say, 2, the event is twice as likely to occur as not to occur. If the odds is .5, the event is only one-half as likely to occur as not to occur, and so on. Now suppose that we have two people, with all covariates the same, except that the first has X_k equal to $x_k + 1$, and the second has X_k equal to x_k. The ratio of their odds, or their *odds ratio*, denoted ψ_{x_k+1}, is

$$\psi_{x_k+1} = \frac{O_{x_k+1}}{O_{x_k}} = \frac{\exp(\beta_0) \exp(\beta_1 X_1) \exp(\beta_2 X_2) \cdots \exp(\beta_k(x_k + 1)) \cdots \exp(\beta_K X_K)}{\exp(\beta_0) \exp(\beta_1 X_1) \exp(\beta_2 X_2) \cdots \exp(\beta_k x_k) \cdots \exp(\beta_K X_K)}$$

$$= \frac{\exp(\beta_0) \exp(\beta_1 X_1) \exp(\beta_2 X_2) \cdots \exp(\beta_k x_k) \exp(\beta_k) \cdots \exp(\beta_K X_K)}{\exp(\beta_0) \exp(\beta_1 X_1) \exp(\beta_2 X_2) \cdots \exp(\beta_k x_k) \cdots \exp(\beta_K X_K)}$$

$$= \exp(\beta_k).$$

That is, $O_{x_k+1} = \exp(\beta_k)O_{x_k}$, or, $\exp(\beta_k)$ is the factor (Long, 1997), or multiplicative, change in the odds for each unit increase in X_k, net of other regressors.

More on the Odds Ratio. Because the odds ratio is such a staple of interpretation in logistic regression, it is worth some elaboration. Consider the zero-order relationship between substance abuse and violence for our 4095 couples: Of the 3857 couples with no substance abuse, 485 reported violence. For this group, the probability of violence is $485/3857 = .126$, and their odds of violence is therefore $.126/(1 - .126) = .144$. Among the 238 couples with substance abuse problems, 70 reported violence. Their probability of violence is therefore $70/238 = .294$, implying an odds of violence of $.294/(1 - .294) = .416$. To quantify the "effect" of substance abuse on violence, we compute the ratio of these odds, or $.416/.144 = 2.889$. That is, substance abuse raises the odds of violence by a factor of 2.889. Or, the odds of violence is 2.889 times higher for those with substance abuse problems. Notice that it is incorrect to say that those with substance abuse problems are "2.889 times as likely" to be violent, since this suggests that their *probability* of violence is 2.889 times higher. In fact, their probability of violence is only $.294/.126 = 2.333$ times higher. The ratio of probabilities, called the *relative risk*, is equivalent to the odds ratio only if the probabilities are both very small. The reason for this is that for two people, a and b, their odds ratio can be written

$$\psi_{a,b} = \frac{\pi_a/(1 - \pi_a)}{\pi_b/(1 - \pi_b)} = \frac{\pi_a}{\pi_b}\frac{1 - \pi_b}{1 - \pi_a}.$$

The first term in the rightmost expression is the relative risk. If both probabilities are small, the second term in this expression will be close to 1, in which case the odds ratio is approximately equal to the relative risk.

Logistic regression effects can also be expressed in terms of *percent changes* in the odds. The percent change in the odds for a unit increase in X_k is

$$\% \, O_{x_k+1} = 100\left(\frac{O_{x_k+1} - O_{x_k}}{O_{x_k}}\right) = 100\left(\frac{O_{x_k+1}}{O_{x_k}} - 1\right) = 100[\exp(\beta_k) - 1].$$

For example, based on the logit estimates in Table 7.1, every year longer the couple had been together at time 1 lowers the odds of violence by $100[\exp(-.046) - 1] = -4.5$, or about 4.5%. Each unit increase in *male isolation* increases the odds of violence by $100[\exp(.02) - 1] = 2.02$, or 2.02%. Now if two people are c units apart on X_k, where c is any value, their odds ratio is $\psi_{x_k+c} = \exp(\beta_k c)$, and the percent change in the odds for a c-unit increase in X_k is $100[\exp(\beta_k c) - 1]$. With respect to *relationship duration*, again, being together 10 years longer lowers the odds of violence by $100[\exp(-.046 \times 10) - 1] = -36.872$, or about 37%. Or, a 5-unit increase in *male's isolation* elevates the odds of violence by $100[\exp(.02 \times 5) - 1] = 10.517$, or about 10.5%.

Confidence intervals for odds ratios can be found by exponentiating the endpoints of confidence intervals for the coefficients. As an example, a 95% confidence interval for the effect of *relationship duration* on the logit of violence, from Table 7.1, is

$-.046 \pm 1.96(.005) = (-.056, -.036)$. A 95% confidence interval for the ratio of the odds of violence for those who are a year apart on *relationship duration* is then $[\exp(-.056), \exp(-.036)] = (.946, .965)$. In other words, we can be 95% confident that each year longer the couple had been together at time 1 lowers the odds of violence by between 3.5 and 5.4%.

Odds ratios are also useful in estimating changes in the probability of event occurrence with changes in predictors once a baseline probability has been calculated. For example, let our baseline couple be married, nonminority, alcohol/drug-free, and average in *relationship duration, female's age at union, male isolation*, and *economic disadvantage*. This couple's odds of violence is $\exp(-2.151) = .1164$. Now, the probability is equal to $O_i/(1 + O_i)$, so the couple's probability of violence is $.1164/(1 + .1164) = .1043$. If this couple were to develop a substance abuse problem, we would estimate that their odds of violence would become $(.1164) [\exp(1.029)] = .3257$. This means that their probability of violence would be $.3257/(1 + .3257) = .2457$, or they would experience a $.1414$ increase in the probability of violence. In that logit coefficients are especially interpretable, logistic regression will be the focus of the rest of this chapter and the next chapter as well.

Standardized Coefficients. In linear regression, standardized coefficients are often employed to compare the relative impacts of different predictors in the same equation. The standardized coefficient is the product of the unstandardized coefficient times the ratio of the standard deviation of X_k to the standard deviation of Y. In logistic (or probit) regression, calculating a standardized coefficient is not as straightforward. Based on the latent-variable development of these models, we would need an estimate of the standard deviation of Y^*, in which case the standardized coefficient would be

$$b_k^s = b_k \frac{s_{x_k}}{s_{y^*}}.$$

But Y^* is unobserved, so its standard deviation is not readily estimated [but see Long (1997) for a suggested estimation procedure]. One solution (found in earlier versions of SAS) is to *partially standardize* the coefficients by multiplying them by $s_{x_k}/\sigma_\varepsilon$, where σ_ε is the standard deviation of the error term in the latent-variable equation. In logit, σ_ε is equal to about 1.814, the square root of $\pi^2/3$ (in probit, it is equal to 1). Applying this transformation to the coefficients for *relationship duration* $(SD = 12.823)$, *female's age at union* $(SD = 7.081)$, and *male's isolation* $(SD = 6.296)$ in the logit equation for violence, we get partially standardized coefficients of $-.325, -.105$, and $.069$, respectively. Consistent with the magnitudes of the unstandardized coefficients, the partially standardized coefficients point to the effect of *relationship duration* as being the largest of the three in magnitude.

Numerical Problems. Estimation via maximum likelihood is frequently plagued by numerical difficulties. Some of these are also common to estimation with least squares. For example, multicollinearity creates the same types of problems in logistic regression and other binary response models that it does in OLS: inflation in the magnitudes of estimates as well as in their standard errors, or in the extreme case, counterintuitive signs of coefficients (Schaefer, 1986). Collinearity diagnostics are not necessarily available in logit or probit software (e.g., none are currently provided in SAS's procedure LOGISTIC). However, in that collinearity is strictly a problem connected with the explanatory variables, it can also be addressed with linear regression software. In SAS, I use collinearity diagnostics in the OLS regression procedure (PROC REG) to evaluate linear dependencies in the predictors. The best single indicator of collinearity problems is the *VIF* for each coefficient (as discussed in Chapter 6). As mentioned previously, *VIF*'s greater than about 10 signify problems with collinearity.

Other problems are more unique to maximum likelihood estimation. The first pertains to zero cell counts. If the cross-tabulation of the response variable with a given categorical predictor results in one or more zero cells, it will not be possible to estimate effects associated with those cells in a logistic regression model. In an earlier article (DeMaris, 1995) I presented an example using the 1993 General Social Survey in which the dependent variable is *happiness*, coded 1 for those reporting being "not too happy," and 0 otherwise. Among categorical predictors, I employ *marital status*, represented by four dummy variables (*widowed, divorced, separated, never married*) with *married* as the reference group, and *race*, represented by two dummies (*black, other race*), with *white* as the reference group. Among other models, I try to estimate one with the interaction of *marital status* and *race*. The problem is that among those in the "other race" category who are separated, all respondents report being "not too happy," leaving a zero cell in the remaining category of the response. I was alerted that there was a problem by the unreasonably large coefficient for the "other race × separated" term in the model and by its associated standard error, which was about 20 times larger than any other. Running the three-way cross-tabulation of the response variable by both *marital status* and *race* revealed the zero cell. An easy solution, in this case, was to collapse the categories of race into "white" versus "nonwhite" and then to reestimate the interaction. If collapsing categories of a categorical predictor is not possible, it could be treated as continuous, provided that it is at least ordinal scaled (Hosmer and Lemeshow, 2000).

A problem that is much more rare occurs when one or more predictors perfectly discriminates between the categories of the response. (Actually, it's when some *linear combination* of the predictors, which might be just one predictor, discriminates the response perfectly.) Suppose, as a simple example, that all couples with incomes under $10,000 per year report violence and all couples with incomes over $10,000 per year report being nonviolent. In this case, *income* completely separates the outcome groups. Correspondingly, the problem is referred to as *complete separation*. When this occurs, the maximum likelihood estimates do not exist (Albert and Anderson, 1984; Santner and Duffy, 1986). Finite maximum likelihood estimates exist only when there is some overlap in the distributions of explanatory variables for groups defined by the response variable. If the overlap is only marginal—say, at a single or at a few tied values—a problem of *quasicomplete separation* develops. In either case, the analyst is again made aware that something is amiss by unreasonably large estimates, particularly of coefficient standard errors. SAS also provides a warning if the program can detect separation problems. Surprisingly, the suggested solution for this problem is to revert to OLS regression. One advantage of the LPM over logit or probit is that estimates of coefficients are available under complete or quasi-complete separation (Caudill, 1988).

An example of quasicomplete separation comes from a recent analysis of 1995 data from a national sample of American women (DeMaris and Kaukinen, 2003). Using logistic regression, we examined the impact of violent victimization on the tendency to engage in binge drinking among a sample of 7353 women from the NVAWS survey (described in Chapter 1). The coefficient for *needing dental care due to a physical assault* was -14.893 with a standard error of 1556.7, and SAS provided a warning that there was "possibly" a quasicomplete separation problem. Upon closer inspection, we saw that the dummy for *needing dental care* exhibited no variation among the binge drinkers—all had values of 0, whereas values of 0 and 1 were observed among the nonbinge drinkers. An easy solution to the problem was simply to eliminate this dummy from the set of regressors.

EMPIRICAL CONSISTENCY AND DISCRIMINATORY POWER IN LOGISTIC REGRESSION

In this final section of the chapter we take up the issues of empirical consistency and discriminatory power. Here, the focus is on the logistic regression model, as more work has been done in these areas on the logit model than on other specifications.

Empirical Consistency

Recall that empirical consistency refers to the property that Y behaves in accordance with model predictions. One way to assess this might be as follows. For each distinct covariate pattern in the data, compare the observed numbers of observations falling into the interest $(Y = 1)$ and reference $(Y = 0)$ categories on Y with the expected numbers from the hypothesized model. A chi-squared test could then inform us whether or not the fit of observed to expected frequencies is within sampling error. Unfortunately, with at least one continuous covariate in the model, the number of covariate patterns in the data is usually close to the sample size. In that case, there is only one observation in each covariate pattern, and the chi-squared statistic does not have the desired chi-squared distribution under the null hypothesis of a good model fit. This statistic would have the chi-squared distribution only when the expected frequencies are large (Hosmer and Lemeshow, 2000).

Hosmer and Lemeshow (2000) solve this problem by grouping the covariate patterns in such a way that the expected frequencies can become large as n increases, allowing the appropriate asymptotic principles to operate. In particular, the Hosmer–Lemeshow goodness-of-fit test for logistic regression groups observations by *deciles of risk*. That is, group 1 consists of the $n/10$ subjects with the lowest predicted probabilities of being in the interest category on Y; group 2 consists of the $n/10$ subjects with the next-lowest predicted probabilities of being in the interest category on Y, and so on. Once these 10 groups have been identified, the expected number of observations in the interest category on Y in each group is calculated as the sum of the $\hat{\pi}$ over all subjects in that group. Similarly, the expected number of cases in the reference category is the sum of $(1 - \hat{\pi})$ over all cases in the same group. The numbers of cases observed in the interest and reference categories are readily tallied from the data. The Hosmer–Lemeshow statistic is then the Pearson chi-squared statistic for the 10×2 table of observed and expected frequencies. Under the null hypothesis that the model is empirically consistent, this statistic has approximately a chi-squared distribution with 8 degrees of freedom (Hosmer and Lemeshow, 2000). A significant χ^2 indicates a model that is not empirically consistent. Hosmer and Lemeshow (2000) suggest that a conservative rule regarding the sample size needed for this test is that all expected frequencies should exceed 5.

Table 7.3 shows the results of employing this test with the logit model in Table 7.1. The table shows each decile of risk, along with the number of couples in each group (which should be generally around $4095/10 \approx 410$), the number of violent couples, the expected number of violent couples based on the model, the number of nonviolent couples, the expected number of nonviolent couples based on the model, and finally, the Hosmer–Lemeshow χ^2. As is evident, the χ^2 is just significant at $p = .046$. This suggests that the model is not quite empirically consistent. We will see in Chapter 8 that this is due to the omission of some important effects from the model.

Table 7.3   Observed and Expected Frequencies of Violent and Nonviolent Couples within Deciles of Risk According to the Logit Model of Table 7.1

Decile of Risk	N	Violent		Nonviolent	
		Obs.	Exp.	Obs.	Exp.
1	413	19	14.34	394	398.66
2	412	30	24.50	382	387.50
3	409	24	33.95	385	375.05
4	412	34	42.55	378	369.45
5	410	42	49.33	368	360.67
6	409	68	55.83	341	353.17
7	411	69	62.72	342	348.28
8	410	65	69.64	345	340.36
9	410	72	80.35	338	329.65
10	399	132	121.66	267	277.34
Total	4095	555	554.87	3540	3540.13
Hosmer–Lemeshow χ^2		15.743			
df		8			
p		.046			

Discriminatory Power

Here I consider two different approaches to assessing discriminatory power: the classification table and analogs of the OLS R^2.

Classification Tables. One means of assessing the ability of the model to discriminate among categories of the response is to examine whether it can accurately classify observations into each category of Y. Following the biostatistics literature, I refer to observations that fall into the interest category as *cases* and those falling into the reference category as *controls* (Hosmer and Lemeshow, 2000). A classification table for logistic regression is a cross-tabulation of case versus control status based on model predictions, against whether or not observations are actually cases or controls. Table 7.4 presents the classification table for couple violence based on the logit model in Table 7.1. The model-based classification procedure is as follows. We pick a criterion value for $\hat{\pi}$, and if an observation's model-generated $\hat{\pi}$ is greater than that criterion, it is classified as a case; otherwise, it is classified as a control. By default, the criterion is usually taken to be .5, and this is the value used to construct Table 7.4.

We see that of the 555 couples actually observed to be violent, only 6 were classified as violent by the model. The *sensitivity* of classification, or the proportion of actual cases that are classified as cases, is 6/555, or 1.08%. The *specificity* of classification, or the proportion of actual controls that are classified as controls, is 3537/3540, or 99.92%. The *false positive rate*, or the proportion of actual controls that are classified as cases, is 1 − specificity, or .08%. Notice that the sensitivity is higher than the false positive rate, and this is typically what we find. If the model affords no improvement in prediction of the response over what could be achieved by random guessing, these will be the same. In this particular example, specificity is very high but sensitivity is abysmally low. However, the *proportion correctly classified* is (6 + 3537)/4095 = .8652, or 86.52% of observations. The proportion

Table 7.4   Classification Table for Violence Based on the Logit Model in Table 7.1

Classified	Observed		Total
	Violent	Nonviolent	
Violent	6	3	9
Nonviolent	549	3537	4086
Total	555	3540	4095
Criterion	.50		
Sensitivity	1.08%		
Specificity	99.92%		
False positive rate	.08%		
Percent correctly classified	86.52%		
Percent correct by chance	76.57%		

classified correctly can be quite misleading, as it is here, because it depends both on the proportion of cases falling into the interest category on Y and on the classification criterion. Because the probability of violence in the sample is relatively low, very few cases' predicted values meet the .5 criterion. Hence, almost all of the cases are predicted to be controls. As most of the cases are, indeed, controls, we get correct predictions most of the time. But ideally, we want good prediction of cases as well as good prediction of controls.

We can get a sense of how good the percent correctly classified is by considering what that would be if we ignored the model. Of course, we could just predict that everyone is nonviolent and be correct 86.45% of the time. However, this misclassifies all of the violent couples. The preferred chance classification rule is one that maximizes prediction of both controls *and* cases. Following the reasoning I articulated in prior work (DeMaris, 1992), the chance classification rate is figured as follows. In that 13.55% of couples in the sample are violent, we predict, with probability equal to .1355, that a couple is violent. On the other hand, we predict, with probability equal to $1 - .1355 = .8645$, that a couple is nonviolent. What is the chance that we will make a correct prediction? Our prediction is correct if an actual case was predicted to be a case and an actual control was predicted to be a control. Now, since couples are actually violent with probability equal to .1355 and actually nonviolent with probability equal to .8645, the probability of a correct prediction is

$$P(\text{case predicted to be a case})P(\text{observation is a case})$$
$$+ P(\text{case predicted to be a control})P(\text{observation is a control}).$$

In this case, we have $.1355^2 + .8645^2 = .7657$, or 76.57% of couples will be correctly predicted, based only on the marginal (i.e., sample) probability of violence. In general, the chance correct prediction rate is $p^2 + (1 - p)^2$, where p is the sample proportion in the interest category on Y.

If a lower criterion value is employed, we can achieve greater sensitivity, although at the expense of specificity. For example, if the criterion is .1355, the marginal proportion of violent couples, sensitivity increases to 66.67%, but specificity drops to 58.19%, and the false positive rate is 41.81%. The percent correctly classified, moreover, falls to only 59.34%. We could continue varying the criterion value in this manner, each time examining properties of the classification. In fact, this is the strategy behind the *receiver operating characteristic* (ROC) *curve* (Hosmer and Lemeshow, 2000; Kramar et al., 2001). This is a plot of the sensitivity against the false positive rate resulting from the criterion being varied throughout the range 0 to 1. Such a plot is shown in Figure 7.6 for the logit model of *couple violence*. Ideally, the plot should form a bow-shaped curve over the 45° line in the center of the plot, the line representing no improvement in prediction afforded by the model. The key statistic in evaluating the quality of the ROC is the area under the curve (AUC). This area is interpreted as *the likelihood that a case will have a higher $\hat{\pi}$ than a control across the range of criterion values investigated* (Hosmer and Lemeshow, 2000). Hosmer and Lemeshow (2000) suggest the following guidelines regarding AUC:

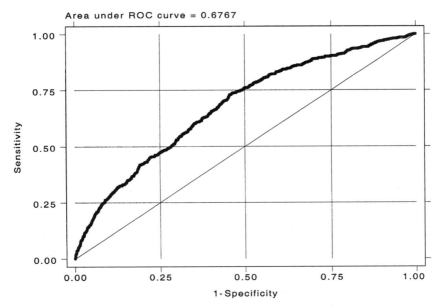

Figure 7.6   Receiver operating characteristic curve for the logit model in Table 7.1.

AUC = .5   The model has no discriminatory power.

.7 < AUC < .8   The model has acceptable discriminatory power.

.8 < AUC < .9   The model has excellent discriminatory power.

AUC > .9   The model has outstanding discriminatory power.

As the AUC in Figure 7.6 is .6767, the logit model in Table 7.1 does not quite have acceptable discriminatory power.

Analogs of R^2. In linear regression, R^2 is the most commonly used measure for assessing the discriminatory power of the model. R^2 possesses three properties that make it especially attractive for this purpose. First, it is standardized to fall in the range [0,1], equaling 0 when the model affords no predictive efficacy over the marginal mean, and equaling 1 when the model perfectly accounts for, or discriminates among, the responses. Second, it is nondecreasing in **x**, meaning that it cannot decrease as regressors are added to the model. Third, it can be interpreted as the proportion of variation in the response accounted for by the regression. Although many R^2 analogs have been suggested for logistic regression (see, e.g., Long, 1997; Mittlboeck and Schemper, 1996), they fail to satisfy one or more of these properties. As a consequence, none is in standard use. In this section I discuss a handful of such measures, beginning with the two that I think are best.

The first issue to be considered is: What is the theoretical criterion being estimated? This depends on the nature of the binary response. If the response is a proxy for a latent scale, Y^*, the quantity of interest could be considered to be the variation in Y^* accounted for by the regression. Recall from Chapter 3 that P^2 is the population proportion of variation in Y that is variation in the linear predictor. In that the logistic regression coefficients are estimates of the effects of the regressors on Y^*, an estimator of the P^2 for the regression of Y^*, suggested by McKelvey and Zavoina (1975), is

$$R_{MZ}^2 = \frac{V\left(\sum b_k X_k\right)}{V\left(\sum b_k X_k\right) + \pi^2/3}, \tag{7.20}$$

where the b_k are the sample logistic regression coefficients, and therefore $V(\sum b_k X_k)$ is an estimate of the variance of the linear predictor for the regression of Y^*. The term $\pi^2/3$ is the variance of ε for this regression, because the error is assumed to follow the standard logistic distribution. Therefore, the denominator of equation (7.20) is an estimate of $V(Y^*)$. Because the numerator and denominator are each consistent for their population counterparts, R_{MZ}^2 is a consistent estimator of the P^2 for Y^*. For the logit model in Table 7.1, $R_{MZ}^2 = .127$. R_{MZ}^2 is the measure that I recommend if one is interested in the variation accounted for in the latent scale underlying the binary response. An extensive simulation found R_{MZ}^2 to be least biased and closest to the actual parameter value across a range of conditions, compared to several other estimators of P^2 in logistic regression (DeMaris, 2002c). Although R_{MZ}^2 has an explained variance interpretation and is bounded by 0 and 1, it is not necessarily nondecreasing in **x**.

On the other hand, suppose that the response is a naturally dichotomous variable, with no underlying continuous referent. It turns out that a generalization of the variance-decomposition principle invoked in Chapters 2 and 3 to derive P^2 can be drawn upon to decompose the variation in a binary variable. From Greene (2003), a general expression for the decomposition of the variance of Y in a joint distribution of Y and **x** is

$$V(Y) = V_{\mathbf{x}}[E(Y|\mathbf{x})] + E_{\mathbf{x}}[V(Y|\mathbf{x})].$$

That is, the variance in Y equals the variance of the conditional mean of Y given **x** plus the mean of the conditional variance of Y given **x**. Dividing through by $V(Y)$ results in

$$1 = \frac{V_{\mathbf{x}}[E(Y|\mathbf{x})]}{V(Y)} + \frac{E_{\mathbf{x}}[V(Y|\mathbf{x})]}{V(Y)}. \tag{7.21}$$

Applying these principles to the linear regression model leads to the expression for P^2 (see Chapter 3). Now in linear regression, $V(Y|\mathbf{x}) = V(\varepsilon) = \sigma^2$, and the average of σ^2 over **x** is just σ^2. So the second term on the right-hand side of (7.21) is just $\sigma^2/V(Y)$ for the linear regression model. Recall that the variance of a binary response, however, is $\pi(1 - \pi)$, which is a function of the conditional mean. Therefore, the conditional variance of Y given **x**, $V(Y|\mathbf{x})$, is $\pi(1 - \pi)|\mathbf{x}$. Hence, a binary-response analog of P^2, which I denote by Δ, is

$$\Delta = 1 - \frac{E_{\mathbf{x}}[\pi(1 - \pi)|\mathbf{x}]}{\pi(1 - \pi)}, \tag{7.22}$$

where π in the denominator is the population marginal probability that $Y = 1$ (its sample counterpart is p).

I refer to Δ as the *explained risk* of an event and show that it is bounded by 0 and 1 (DeMaris, 2002c). Like P^2, Δ reflects the model's ability to account for the response. When the model affords no improvement in predicting the event of interest compared to using the marginal probability to predict Y, Δ will be 0. At the other extreme, if all conditional probabilities are either 0 or 1, the occurrence/nonoccurrence of an event is predicted with certainty. In this case, the conditional variance of Y_i is 0 for all i, and Δ equals 1. A consistent estimator of Δ is

$$\hat{\Delta} = 1 - \frac{\left(\sum_{i=1}^{n} \hat{\pi}_i \, (1 - \hat{\pi}_i) |\mathbf{x}_i| \right)/n}{p(1-p)},$$

where the numerator of the second expression to the right of the equals sign is just the average estimated conditional variance of Y over all n cases. Like R^2_{MZ}, $\hat{\Delta}$ has an explained variance interpretation and is bounded by 0 and 1 but is not necessarily nondecreasing in $\mathbf{x}$. My simulation results showed that $\hat{\Delta}$ was the best estimator of Δ among the several R^2 analogs investigated (DeMaris, 2002c). Both R^2_{MZ} and $\hat{\Delta}$ require casewise calculations and are not currently available in conventional software. (A program in SAS that calculates these as well as six other R^2 analogs is available from the author by request, however.) For the logit model in Table 7.1, $\hat{\Delta} = .051$.

Three other measures are worth mentioning because they are all nondecreasing in $\mathbf{x}$ as well as bounded by 0 and 1. All are based on the likelihood function, and as such, have much broader applicability than R^2_{MZ} and $\hat{\Delta}$. These can be employed with any model employing maximum likelihood estimation. They are also readily calculated from standard logistic regression output. The first is the *likelihood-ratio index* (Long, 1997):

$$R^2_{\mathrm{L}} = \frac{-2 \log L_0 - (-2 \log L_1)}{-2 \log L_0},$$

the numerator of which is just the model χ^2, described above. As mentioned above, $-2 \log L_0$ is analogous to *TSS* and $-2 \log L_1$ is analogous to *SSE*; hence R^2_{L} is a direct counterpart of the OLS R^2. However, it does not have an explained variance interpretation. Rather, it represents the proportionate reduction in minus twice the log-likelihood (a measure of total uncertainty in Y) when the likelihood function is evaluated at the MLEs for the parameters rather than the MLE for an intercept-only model. I have found that R^2_{L} is a comparatively good estimator of explained risk (DeMaris, 2002c). For the logit model in Table 7.1, R^2_{L} is

$$R^2_{\mathrm{L}} = \frac{3249.56 - 3054.132}{3249.56} = .060.$$

The second measure is the *generalized R^2* (Allison, 1995; Maddala, 1983). Its formula is

$$R^2_{\mathrm{G}} = 1 - \left(\frac{L_0}{L_1}\right)^{2/n}.$$

It can be shown that R_G^2 is identically equal to the OLS R^2 in a linear regression model with normally distributed errors. (A proof is available from the author on request.) Because the likelihood functions in this equation may be difficult to work with, the measure can also be computed as (Allison, 1995)

$$R_G^2 = 1 - \exp\left(\frac{-\chi^2}{n}\right),$$

where in this case, χ^2 is the model χ^2 discussed above. For the logit model in Table 7.1, R_G^2 is

$$R_G^2 = 1 - \exp\left(\frac{-195.394}{4095}\right) = .0466.$$

It turns out that R_G^2 has an upper bound of less than 1. In particular, the upper bound is $1 - (L_0)^{2/n}$, suggesting the scaled measure, R_{GSC}^2, proposed by Cragg and Uhler (1970) as well as others (Maddala, 1983; Nagelkerke, 1991):

$$R_{GSC}^2 = \frac{R_G^2}{1 - (L_0)^{2/n}},$$

or in terms of ease of computation, we have

$$R_{GSC}^2 = \frac{1 - \exp(-\chi^2/n)}{1 - \exp[(2/n)\log L_0]}.$$

The scaling ensures that R_{GSC}^2 lies between 0 and 1. Although both R_G^2 and R_{GSC}^2 are comparable to R^2 in the linear regression model, they do not have an explained variance interpretation in logistic regression. For the logit model in Table 7.1, R_{GSC}^2 is

$$R_{GSC}^2 = \frac{.0466}{1 - \exp[(2/4095)(-1624.763)]} = \frac{.0466}{.548} = .085.$$

All five R^2 analogs just discussed are shown in Table 7.1, at the bottom of the logit model column (R_L^2, R_G^2, and R_{GSC}^2 are also shown for probit). Which to employ depends on how couple violence should be conceptualized. Physical aggression should probably be considered a continuous variable that ranges from displacement aggression (e.g., kicking in doors) to relatively minor violence (e.g., pushing and shoving), to severe acts (e.g., beating someone up or attacking a person with a weapon). Due to the crudeness of measurement in this case, however, this range of acts is simply mapped into *violence*, a binary yes–no response. In that the continuous response of *physical aggression* is really the focus, however, R_{MZ}^2 is the preferred measure. It indicates that about 13% of the variance in *physical aggression* is accounted for by the logistic regression. Were we to be solely interested in the event of whether or not a couple is reported as being violent, however, $\hat{\Delta}$ suggests that only about 5% of this phenomenon is accounted for by the model.

EXERCISES

7.1 The following (x,y) pairs were obtained for five individuals: $(1.5, 1)$, $(2, 1)$, $(1.5, 0)$, $(1.75, 0)$, $(4, 0)$. Maximum likelihood estimates for a logistic regression of Y on X produced the following equation: $\ln \hat{O} = 1.7184 - 1.0627X$. Using this equation and a calculator, give the log of the likelihood function for this model, evaluated at the MLEs of the parameter estimates.

7.2 Give the model χ^2 for the model in Exercise 7.1.

7.3 Give R_L^2, R_G^2, and R_{GSC}^2 for the model in Exercise 7.1.

7.4 Using a calculator and the equation in Exercise 7.1, give R_{MZ}^2 and $\hat{\Delta}$ for the model in Exercise 7.1.

7.5 For the low-risk couple described in this chapter, give the probability of violence, according to the scobit model in Table 7.2.

7.6 For the low-risk couple described in this chapter, give the probability of violence according to the complementary log-log model in Table 7.2.

7.7 For the 4095 couples in the NSFH, define a high-risk couple as a cohabiting, minority couple in which at least one partner has a problem with alcohol or drugs, who is 1 standard deviation below the mean on *relationship duration* and *female's age at union*, and 1 standard deviation above the mean on *male's isolation* and *economic disadvantage*. Give the estimated probability of violence for such a couple based on the logit model in Table 7.1.

7.8 Give the estimated probability of violence for the high-risk couple in Exercise 7.7 based on the probit model in Table 7.1.

7.9 Give the partial derivative of the probability of violence, with respect to relationship duration, based on both logit and probit, for the high-risk couple in Exercise 7.7.

7.10 Recall that a "baseline" couple (having all covariates in Table 7.1 set to zero) is estimated to have odds of violence equal to .1164 and a probability of violence of .1043. Give the change in probability of violence expected for an additional year of being together, according to the logit model in Table 7.1.

7.11 Based on the probit model in Table 7.1, give the baseline probability of violence, as well as the change in the probability of violence for an additional year of being together, for a baseline couple.

7.12 Use partial derivatives to estimate the changes in probability calculated in Exercises 7.10 and 7.11. How close are the approximations in this case?

7.13 A simplified logit model for couple violence for the 4095 NSFH couples is ln $\hat{O}$ = −1.948 + 1.0563 *alcohol/drug problems* + .0307 *economic disadvantage*.

(a) Interpret all three of the parameter estimates in terms of the odds of violence, recalling that *economic disadvantage* is a centered variable.

(b) Show that equations (7.15) and (7.16) imply that the model is *interactive* in the regressors in their effects on $\pi_{violent}$. This can be shown by showing that a 1-unit increase in *economic disadvantage* has a different impact on $\pi_{violent}$ for those with alcohol/drug problems than for those without such problems. (*Hint*: Use those without alcohol/drug problems, who have mean *economic disadvantage*, as the baseline group.)

Use the following additional analyses of couple violence for the 4095 NSFH couples for Exercises 7.14 to 7.20:

Predictor	Logit 1	Logit 2	Probit 1	Probit 2
Intercept	.0011	−.4657*	−.1295	−.3810***
Average age	−.0449***	−.0332***	−.0231***	−.0168***
First union	−.2394*	−.2212*	−.1268*	−.1146*
Number of children	−.0403	−.1027*	−.0255	−.0569**
Female traditional	−.0773	−.1472	−.0439	−.0747
Male traditional	.2444*	.1898	.1391*	.1071
Both traditional	−.0528	−.0822	−.0251	−.0348
Conflict over money		.3220***		.1812***
Conflict over time		.1942***		.1082***
Conflict over sex		.1735***		.0951**
−2 ln L	3120.892	2973.667	3122.895	2976.173

Partial covariance matrix for logit 2:

	Money	Time	Sex
Money	.0027	−.0008	−.0005
Time		.0025	−.0010
Sex			.0027

7.14 Excluding the intercept, interpret all regressor coefficients in logit 1 with respect to the odds of couple violence.

7.15 Give model χ^2's for all four models and test whether the addition of the three conflict variables adds significantly to model fit for both the logit and probit analyses.

7.16 Give R_L^2, R_G^2, and R_{GSC}^2 for all four models.

7.17 Based on logit 1, give:

   (a) The impact on the odds of violence for a 10-year increase in average age.

   (b) The impact on the odds of violence for having three additional children.

   (c) The ratio of the odds of violence for couples in which only the female is traditional vs. couples in which only the male is traditional.

7.18 Find the predicted probability of violence for a couple whose average age is 25, who is in their first union, who has three children, and among whom only the male is traditional, based on both logit 1 and probit 1.

7.19 What is the difference in $\pi_{violent}$, compared to the couple in Exercise 7.18, for having been in a prior union, based on both logit 1 and probit 1?

7.20 Using the results for logit 2, test whether the effects of *conflict over money, time,* and *sex* on the log odds of violence are different from each other.

Use software along with the couples dataset to answer Exercises 7.21 to 7.26.

7.21 The variable FAGRESS represents whether or not the female partner has hit, shoved, or thrown things at the male partner in the past year, and is coded 1 ("yes") and 2 ("no"). Regress this variable on SEPARATE, CHILDREN, MAGUNION, CONFLICT (the average of MFIGHTS and FFIGHTS), and DURYRS, using both logistic and probit regression. Show the logit and probit estimates and model χ^2's, indicating significance levels with asterisks. Interpret the effect of *relationship duration* in the logit and probit models.

7.22 Find the predicted probability of *female aggression* at 1 standard deviation above the mean of *conflict*, for those who never lived apart because of disagreements, and who are average on all other predictors, for both the logit and probit models in Exercise 7.21.

7.23 What is the change in probability of *female aggression* for a standard deviation increase in the level of *conflict*, at the $\hat{\pi}$ calculated in Exercise 7.22, based on both logit and probit?

7.24 Based on the logit results in Exercise 7.21: What is the percent difference in the odds of *female aggression* for those who have lived apart for a time because of disagreements vs. those who have not? By what percent do couples' odds of *female aggression* change for a standard deviation increase in the *male's age at union formation*?

7.25 What is the discriminatory power of the model, based on both R_L^2 and R_{MZ}^2, for both the logit and probit models in Exercise 7.21?

7.26 Using the *couples dataset*, create a *marital status* variable with the following levels (dummy names):

- Couples currently cohabiting unmarried (COHABIT)
- Married couples, both in a first marriage (FIRSTMAR)
- Married couples, husband married before (HUSBRE)
- Married couples, wife married before (WIFERE)
- Married couples, both married before (BOTHRE)

(*Hint*: You can create MARHIST using SAS code, as in: IF MARCOHAB > 1 THEN MARHIST = 1; * cohabitors; IF MARCOHAB = 1 AND MTI-MARR > 1 AND FTIMARR = 1 THEN MARHIST = 2; * husband remarried; and so on.) Then, using logistic regression:

(a) Test whether *marital status* has a significant effect on the probability of *female aggression*, ignoring other covariates.

(b) Test all possible contrasts between pairs of *marital status* categories, again, ignoring other covariates. Show the odds ratios (of *female aggression*) for each of these pairs. (*Hint*: This is most easily done via repeated estimation of the model after changing the contrast category for the *marital-status* dummies.)

(c) Test whether the effect of *marital status* is still significant after controlling for the covariates in the model of Exercise 7.21.

Use the following data for Exercises 7.27 to 7.30. These data are from Spector and Mazzeo (1980), as reported in Aldrich and Nelson (1984). They are from a study that examined the effect of a teaching method (PSI) on performance in an intermediate macroeconomics class. The other regressors are entering GPA (GPA) and entering knowledge of the material (TUCE).

OBS	GPA	TUCE	PSI	GRADE	OBS	GPA	TUCE	PSI	GRADE
1	2.66	20	0	C	17	2.75	25	0	C
2	2.89	22	0	B	18	2.83	19	0	C
3	3.28	24	0	B	19	3.12	23	1	B
4	2.92	12	0	B	20	3.16	25	1	A
5	4.00	21	0	A	21	2.06	22	1	C
6	2.86	17	0	B	22	3.62	28	1	A
7	2.76	17	0	B	23	2.89	14	1	C
8	2.87	21	0	B	24	3.51	26	1	B
9	3.03	25	0	C	25	3.54	24	1	A
10	3.92	29	0	A	26	2.83	27	1	A
11	2.63	20	0	C	27	3.39	17	1	A
12	3.32	23	0	B	28	2.67	24	1	B
13	3.57	23	0	B	29	3.65	21	1	A
14	3.26	25	0	A	30	4.00	23	1	A
15	3.53	26	0	B	31	3.10	21	1	C
16	2.74	19	0	B	32	2.39	19	1	A

7.27 Estimate the probability of getting an A (π_A) as a linear function of GPA, TUCE, and PSI, using OLS. Interpret the coefficients. Give $\hat{\pi}_A$ for OBS 11.

7.28 Estimate π_A as a linear function of GPA, TUCE, and PSI, using WLS. This is done as follows: From the OLS regression in Exercise 7.27, save the fitted values (i.e., the $\hat{\pi}_A$). Recode any $\hat{\pi}_A \leq 0$ as .001. Recode any $\hat{\pi}_A \geq 0$ as .999. Create the estimated conditional variance of Y for each case as $\hat{\pi}_A(1 - \hat{\pi}_A)$. The weight for WLS is then the reciprocal of this estimated conditional variance. Give $\hat{\pi}_A$ for OBS 11, based on the WLS estimates.

7.29 Estimate π_A as a function of GPA, TUCE, and PSI, using logistic regression. Interpret all estimates in terms of odds. Give the $\hat{\pi}_A$ for OBS 11. Give R_L^2 for the model.

7.30 Estimate π_A as a function of GPA, TUCE, and PSI, using both probit and complementary log-log models. Give $\hat{\pi}_A$ for OBS 11, based on each set of estimates.

CHAPTER 8

Advanced Topics in Logistic Regression

CHAPTER OVERVIEW

In Chapter 7 we saw that the logistic regression model was particularly advantageous with respect to the interpretation of coefficients. In this chapter, then, we consider the logistic regression model in greater detail, continuing our investigation of couple violence in the NSFH. I begin by outlining the techniques used to investigate interaction effects. First, I sketch out how we compare models across groups: in this case, minority vs. nonminority couples. This involves a logistic regression analog of the Chow test that was covered in Chapters 3 and 4. I then consider modeling interaction effects involving specific regressors, using cross-product terms. From this I move to a discussion of modeling nonlinearity in the relationship between regressors and the logit, detailing a general means of assessing whether linearity is plausible. I then address the testing of coefficient changes across models, demonstrating an analog of the procedure discussed in Chapter 3. I conclude the discussion of binary logistic regression by reinvestigating the discriminatory power and empirical consistency of the final model of interest. The narrative then moves to an explication of logistic regression with multinomial responses. First, I develop the multinomial logistic regression model for the case in which the categories of the response are not ordered, and consider both limited- and full-information maximum likelihood approaches. Finally, I take advantage of the ordinal nature of the response and introduce the ordered logit model.

MODELING INTERACTION

An artifact of the logistic regression model, as well as the other probability models considered, is that the regressors are automatically interactive with respect to

Regression with Social Data: Modeling Continuous and Limited Response Variables,
By Alfred DeMaris
ISBN 0-471-22337-9   Copyright © 2004 John Wiley & Sons, Inc.

probabilities. (Showing this was the theme of Exercise 7.13.) However, to model interaction over and above what is incorporated into the nature of the logit link, or to model interaction in the odds or the log odds, we have two choices. Either we can compare the model over levels of an explanatory variable or we can utilize cross-product terms, similar to the approaches taken in linear regression.

Comparing Models across Groups

Until now I have been utilizing a combined sample of minority and nonminority couples to investigate models of couple violence. The dummy variable reflecting minority status in the model simply allows minorities to have a different baseline log odds of violence than that for nonminorities. Otherwise, the effects of the other regressors are assumed to be the same for minorities, as opposed to nonminorities. Although I have no reason to suspect otherwise, it would be fruitful to provide a statistical justification for combining the models for both groups. In Chapters 3 and 4 we saw that the Chow test is used for this purpose in linear regression. There is an analog of this test for logistic regression, recently outlined by Allison (1999).

Recall that the Chow test in linear regression assumes that the equation error variance is the same across groups. The following explication similarly assumes equal error variance. In this case, however, the error variance in question pertains to the latent-scale formulation of the logistic regression model. That is, if

$$Y_i^* = \sum \beta_k X_{ik} + \varepsilon_i$$

is the equation that underlies the binary response for the first group, and

$$Y_i^* = \sum \gamma_k X_{ik} + \upsilon_i$$

underlies the binary response for the second group, the assumption is that $V(\varepsilon_i) = V(\upsilon_i)$. [It is not necessary to assume that a latent variable underlies the response in order to motivate the homogeneity-of-variance assumption; see Allison (1999) for details.] Unfortunately, there is no simple means of testing this assumption [see Allison (1999) for suggested techniques, however]. In the present example, we simply assume that the error variances are equal. The Chow test analog for logistic regression involves estimating the model for the combined sample and then for each sample separately. For two groups, the test statistic is then $\chi^2 = -2\ln L_c - [-2\ln L_1 + (-2\ln L_2)]$, where $\ln L_c$ is the fitted log-likelihood for the combined sample, $\ln L_1$ the fitted log-likelihood for group 1, and $\ln L_2$ the fitted log-likelihood for group 2. Under the null hypothesis that $\gamma = \beta$, that is, that regressor effects are the same across groups, χ^2 has a chi-squared distribution with degrees of freedom equal to the difference in the number of parameters estimated in the combined versus the separate sample approaches. As with linear regression, the intercept can be constrained to be the same by omitting the dummy for group in the combined-sample model. Or, it can be allowed to differ across groups by including the group dummy in the combined-sample model.

Example. Table 8.1 presents the results of estimating logistic regression models for *couple violence* using the combined sample of 4095 couples, as well as using separate samples of 1216 minority and 2879 nonminority couples. The first model is just a repeat of the logit model in Table 7.1 and represents a model in which the intercept is unconstrained across groups. The second model constrains the intercept, in addition to the regressors, to be the same for minorities and nonminorities. The third and fourth columns represent the models for minorities and nonminorities, respectively. A glance at the coefficients in the last two columns suggests that there are some differences in regressor effects across groups. But differences in coefficients are to be expected, due purely to sampling variability. The test statistic for coefficient invariance across groups, allowing the intercept to be unconstrained, is $\chi^2 = 3054.132 - (1006.921 + 2043.172) = 4.039$. The degrees of freedom for the test are figured as follows. Including the intercept, we estimate eight parameters in the combined model, whereas we estimate seven parameters in each of the submodels. The test therefore has $(7 + 7) - 8 = 6$ degrees of freedom. The result is nonsignificant ($p = .671$). If we allow the intercept to be constrained, we have $\chi^2 = 3058.240 - (1006.921 + 2043.172) = 8.147$. With seven degrees of freedom, this test is also nonsignificant ($p = .32$). Based on either result, I conclude that there is insufficient evidence to suggest that regressor effects are any different for minorities versus nonminorities. As was the case in linear regression, an equivalent test can be fashioned using cross-product terms. Testing the difference between the model in the first column in Table 8.1 versus a model in which the cross-products of minority status with all 6 other regressors has been added gives us the unconstrained-intercept test above. Adding the dummy for minority status plus the six cross-product terms to the model in the second column of Table 8.1, and testing the difference in resulting models, produces the constrained-intercept test.

Table 8.1   Regression Results for the Logit Model in Table 7.1 Estimated for the Combined Sample as Well as Separately for Minority and Nonminority Couples

| Predictor | Combined Sample | | Minority Couples | Nonminority Couples |
	Intercept Unconstrained	Intercept Constrained		
Intercept	−2.151***	−2.081***	−1.892***	−2.154***
Relationship duration	−.046***	−.046***	−.039***	−.050***
Cohabiting	.810***	.836***	.706*	.960**
Minority couple	.221*			
Female's age at union	−.027***	−.026***	−.017	−.033***
Male's isolation	.020*	.019*	.027*	.017
Economic disadvantage	.023*	.030***	.017	.032*
Alcohol/drug problem	1.029***	1.011***	1.244***	.946***
−2 log L	3054.132	3058.240	1006.921	2043.172
Number of parameters	8	7	7	7
n	4095	4095	1216	2879

* $p < .05$. ** $p < .01$. *** $p < .001$.

Examining Variable-Specific Interaction Effects

Interactions between specific variables in their effects on the response are best investigated with cross-product terms. In that the model for the probabilities is already interactive in the regressors, what is achieved by adding a cross-product if the probability is our focus? The main advantage is that using a cross-product term allows interaction effects in the probabilities to be disordinal, whereas without it, the interaction is constrained to be ordinal. Recall from Chapter 3 that the descriptors *ordinal* and *disordinal* refer to degrees of interaction. When the impact of the focus variable differs only in magnitude across levels of the moderator, the interaction is ordinal. If the nature (or direction) of the impact changes over levels of the moderator, the interaction is disordinal. Without a cross-product term, the partial slope of X_k on the probability is, as noted earlier, $\beta_k[\pi(\mathbf{x}) (1 - \pi(\mathbf{x}))]$. [I'm using the notation $\pi(\mathbf{x})$ here to emphasize that π changes with $\mathbf{x}$.] Because $[\pi(\mathbf{x}) (1 - \pi(\mathbf{x}))] \geq 0$, the impact of X_k on the probability always has the same sign, regardless of the settings of the X's in the model (i.e., $\beta_k[\pi(\mathbf{x}) (1 - \pi(\mathbf{x}))]$ takes on whatever sign β_k takes). Hence, the effect of X_k can differ only in magnitude, but not direction, with different values of the X's. With a cross-product in the model of the form $\gamma X_k X_j$, however, the partial slope for X_k becomes $(\beta_k + \gamma X_j) [\pi(\mathbf{x}) (1 - \pi(\mathbf{x}))]$. In this case, because β_k and γ could be of opposite signs, the impact of X_k on the probability could change direction at different levels of X_j, producing a disordinal interaction.

The interaction model for the log odds, with two predictors, X and Z, is

$$\ln O = \beta_0 + \beta_1 X + \beta_2 Z + \gamma XZ.$$

Interpretation is similar to that for the linear regression model, except that in this case, the response is the log odds of event occurrence. With X as the focus variable, its effect can be seen by factoring out its common multipliers:

$$\ln O = \beta_0 + \beta_2 Z + (\beta_1 + \gamma Z)X.$$

The impact of X on the odds is therefore $\exp(\beta_1 + \gamma Z)$, or, following the framework I have previously articulated (DeMaris, 1991), it is $\exp(\beta_1) [\exp(\gamma)]^z$. This last expression suggests that each unit increase in Z magnifies the *impact* of X by $\exp(\gamma)$.

Table 8.2 presents logistic regression models for *couple violence* allowing *alcohol/drug problem* to interact with *economic disadvantage* in their effects on the log odds of violence. There is solid justification for expecting these factors to interact. The impact of having an alcohol or drug problem on intimate violence may be exacerbated in disadvantaged neighborhoods. Living amidst economic disadvantage is likely to enhance the stress associated with substance abuse. Additionally, the social isolation and absence of community monitoring that typically characterizes such neighborhoods may create a context in which "high" or inebriated partners feel freer to vent their frustrations physically on one another (Miles-Doan, 1998; Sampson et al., 1997). Similarly, living in a disadvantaged neighborhood may be more stressful for those with substance abuse problems.

Table 8.2   Logistic Regression Models for the Interaction between Economic Disadvantage and Alcohol/Drug Problem in Their Effects on Violence

Predictor	Model 1	Model 2	Model 3
Intercept	−2.148***	−1.180***	−2.065***
Relationship duration	−.046***	−.046***	−.046***
Cohabiting	.830***	.830***	.830***
Minority couple	.216*	.216*	.216*
Female's age at union	−.027***	−.027***	−.027***
Male's isolation	.021*	.021*	.021*
Economic disadvantage	.016	.085**	
Alcohol/drug problem	.968***		1.321***
Alcohol/drug free		−.968***	
Economic disadvantage − SD			.016
Economic disadvantage × alcohol/drug problem	.069*		
Economic disadvantage × alcohol/drug Free		−.069*	
Economic Disadvantage − SD × alcohol/drug problem			.069*

*$p < .05$. ** $p < .01$. *** $p < .001$.

Model 1 in Table 8.2 includes the cross-product of *alcohol/drug problem* with *economic disadvantage*, the latter being a centered regressor. The effect of *alcohol/drug problem* on the log odds of *violence* is .968 + .069 *economic disadvantage*. The effect on the odds of violence is exp(.968 + .069 *economic disadvantage*). Thus, having a substance abuse problem raises the odds of violence by exp(.968) = 2.633 for those in neighborhoods of average economic disadvantage levels. For those a unit higher in economic disadvantage, the effect of *alcohol/drug problem* on the odds is exp(.968 + .069) = exp(1.037) = 2.821. Or, the effect of having an alcohol or drug problem is magnified by a factor of exp(.069) = 1.071 for each 1-unit increase in *economic disadvantage* [see DeMaris (1991) for additional use of this multiplicative framework for interpreting interactions]. For those who are a standard deviation (5.128) higher in *economic disadvantage*, the effect of *alcohol/drug problem* is exp[.968 + .069(5.128)] = exp(1.321) = 3.747. Or, with *economic disadvantage* as the focus, the impact of *economic disadvantage* on the log odds is .016 + .069 *alcohol/drug problem*. For those without substance abuse problems, each unit increase in *economic disadvantage* magnifies the odds of violence by a factor of exp(.016) = 1.016, a nonsignificant effect. For those with substance abuse problems, the effect is exp(.016 + .069) = exp(.085) = 1.089.

Targeted Centering

The significant interaction effect means that the effect of *alcohol/drug problem* (*economic disadvantage*) *changes* over levels of *economic disadvantage* (*alcohol/drug problem*). But is the effect of, say, *economic disadvantage* significant for those *with* alcohol or drug problems? Or, is the effect of *alcohol/drug problem* significant for those at, say, 1 standard deviation above mean *economic disadvantage*? What is needed is a test for the significance of the focus variable at a particular level of the moderator. In Chapter 3 I presented a formula for finding the variance of the partial slope for the impact of X at a particular level, z, of Z. If the estimated partial slope is written $b_1 + gz$, the formula for $V(b_1 + gz)$, given by equation (3.10), is

$$V(b_1 + gz) = V(b_1) + 2z \operatorname{Cov}(b_1, g) + z^2 V(g).$$

Here I outline a technique, first introduced in Chapter 6, that I refer to as *targeted centering*, which obviates the need to compute such a variance by hand. We simply code the variables involved in creating the cross-product term so that the effect of interest is the main effect of one of the variables. To discern whether *economic disadvantage* is significant for couples with alcohol or drug problems, we simply recode the dummy for *alcohol/drug problem* so that *having* an alcohol or drug problem is the reference category. The new dummy is called "alcohol/drug free" in Table 8.2. Then the cross-product of *economic disadvantage* with *alcohol/drug free* is formed to capture the interaction effect. Model 2 in Table 8.2 shows the results. The main effect on the log odds of violence of *economic disadvantage* is now the effect for those *with* alcohol or drug problems. The value of .085 agrees with what was shown above for those who have substance abuse problems. Now we see that the effect of *economic disadvantage* is, indeed, significant for this group.

Model 3 shows the effect of applying targeted centering to *economic disadvantage*. In that the standard deviation of *economic disadvantage* is 5.128, "economic disadvantage $-SD$" is *economic disadvantage* -5.128. [In that the variable is already centered, *economic disadvantage* $-$ (mean $+ SD$) reduces to *economic disadvantage* $- SD$.] The cross-product of *alcohol/drug problem* with (*economic disadvantage* $-SD$) is formed and included in the model. Then the effect of *alcohol/drug problem* is: $1.321 + .069$ (*economic disadvantage* $-SD$). For couples whose *economic disadvantage* is 1 standard deviation above the mean, *economic disadvantage* $-SD$ equals zero. Hence, the effect of *alcohol/drug problem* at 1 standard deviation above mean *economic disadvantage* is just the main effect of *alcohol/drug problem*, or 1.321, which, as shown, is very significant. (This effect also agrees with the calculations above.) As a final note, targeted centering can be employed in any regression equation, linear or otherwise.

MODELING NONLINEARITY IN THE REGRESSORS

Although the logistic regression model is nonlinear in the regressors with respect to probabilities, it is assumed to be linear in the log odds. However, this may not be a reasonable assumption. For example, in the case of *couple violence*, it might be expected that the log odds of *couple violence* would initially drop markedly with increasing relationship duration. Couples are likely to learn to use more constructive conflict-resolution techniques in their relationships over time and should become less likely to be violent as they spend a longer time together. Violent couples are also more likely to dissolve their relationships, so over time, we expect only the better-adjusted couples to stay together. Nevertheless, beyond a certain relationship duration, increasing duration would not be expected to keep having a beneficial effect on inhibiting violence. Rather, the effect of increasing duration should lessen in magnitude for couples who have been together for a long time. Although such a trend might be fitted with a quadratic term, we begin without assuming any particular parametric form that a nonlinear trend might take.

Testing for Nonlinearity

We can explore whether there is nonlinearity in the relationship between the log odds and a continuous regressor using a technique suggested by Hosmer and Lemeshow (2000). They recommend partitioning a continuous regressor into quartiles or quintiles of its sample distribution. That is, for a continuous regressor, X, and using quintiles, we recode X so that the value 1 represents the 20% of the sample with the lowest scores on X, the value 2 represents the 20% of the sample with the next-lowest scores on X, and so on, until finally, the value 5 represents the 20% of the sample with the highest scores on X. We then dummy up the quintiles and examine the relationship between the log odds of event occurrence and the quintile dummies, while controlling for other model effects. Table 8.3 shows this technique applied to *relationship duration*.

Table 8.3   Models Investigating the Nonlinear Effect of Relationship Duration on the Log Odds of Violence

Predictor	Model 1	Model 2	Model 3	Model 4
Intercept	-1.529***	-1.027***	-2.310***	-2.837***
Cohabiting	.759**	.773**	.769**	.679**
Minority couple	.244*	.247*	.242*	.242*
Female's age at union	$-.028$***	$-.028$***	$-.029$***	$-.029$***
Male's isolation	.021*	.021**	.021**	.021**
Economic disadvantage	.023*	.023*	.023*	.023*
Alcohol/drug problem	1.036***	1.030***	1.040***	1.040***
Duration quintile 2	$-.059$			
Duration quintile 3	$-.493$***			
Duration quintile 4	-1.211***			
Duration quintile 5	-1.349***			
Duration quintile		$-.371$***		
Relationship duration			$-.054$***	
(Relationship duration)^2			.001**	
Relationship duration $-$ SD				$-.029$***
(Relationship Duration $-$ SD)^2				.001**
Model χ^2	207.054***	196.495***	204.127***	204.127***
df	10	7	8	8

* $p < .05$. ** $p < .01$. *** $p < .001$.

Model 1 presents the regression of the log odds of *couple violence* on the covariates in Table 8.1, with *relationship duration* quintiles 2 to 5 represented as dummies. The lowest quintile of *relationship duration* is the reference category. The dummy coefficients indicate that the reduction in the log odds of violence associated with quintiles 4 and 5 is substantially larger than the reductions associated with quintiles 2 and 3, suggesting a nonlinear trend. However, to observe this trend more readily, we can plot the relationship between the log odds of violence and *duration quintile*, controlling for the other covariates. This is accomplished simply by using the dummies and the intercept to estimate the log odds for each quintile. That is, for married, nonminority, alcohol- and drug-free couples at average *female age at union, male isolation,* and *economic disadvantage,* the estimated log odds of violence is -1.529 for those in the lowest duration quintile. For comparable couples in the second quintile, the log odds is $-1.529 - .059 = -1.588$. For comparable couples in the third quintile, the log odds is $-1.529 - .493 = -2.022$, and so on. The last two log odds are -2.74 and -2.878. These values are then plotted against *duration quintile*, whose values range from 1 through 5. The result is shown in Figure 8.1.

The plot suggests, as suspected, that the logit of violence declines with increasing duration until quintile 4, at which point it levels off. This type of trend can be captured nicely with a quadratic term. However, there is also the suggestion that the

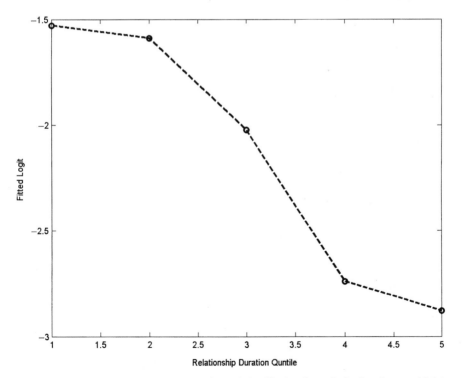

Figure 8.1 Plot of log odds of violence against relationship duration quintile, based on model 1 in Table 8.3.

decline begins gradually and then accelerates between quintiles 2 and 4 before it levels off. That is, the entire curve actually appears to have three bends, suggesting that perhaps a cubic polynomial in *relationship duration* would provide a good fit. Nevertheless, we first need to test whether the trend in *relationship duration* is significantly nonlinear. For this purpose, we can test model 1 in Table 8.3 against model 2, which includes *duration quintile*, coded 1 to 5, modeled as a linear effect. The test statistic is $\chi^2 = 207.054 - 196.495 = 10.559$. The test has 3 degrees of freedom and is significant at $p = .014$. I conclude that the relationship between the log odds of violence and *relationship duration* is not linear. To model the trend in Figure 8.1, I elected to include both *duration-squared* and *duration-cubed* in the model, in addition to *duration*, all using the original continuous coding. For this purpose, relationship duration is first centered, thus minimizing collinearity problems in the estimates due to higher-order terms. It turns out that the cubic term was not significant (results not shown) but the quadratic term was. So I chose the quadratic model as best capturing the nonlinearity in *relationship duration*. The results are shown as model 3 in Table 8.3.

Targeted Centering in Quadratic Models

The estimates in model 3 suggest that the partial slope for *relationship duration* is $-.054 + 2(.001)$ *relationship duration*. The implication is that *relationship duration* reduces the log odds of violence, but at a declining rate the longer couples have already been together. That is, the effect of an instantaneous increase in *relationship duration* depends on the value of *relationship duration*. The main effect of *relationship duration*, $-.054$, is the partial slope (or partial derivative) of duration at *mean* duration, which is 14.9 years. This is clearly significant. What about the effect at a standard deviation above the mean? The unrounded estimates for *duration* and *duration-squared* are actually $-.0535$ and $.000968$, respectively, and a standard deviation of *relationship duration* is 12.8226 years. The effect is, therefore, $-.0535 + 2(.000968) (12.8226) = -.0287$, or $-.029$. To assess whether this effect is significant, we would normally want to calculate its standard error using covariance algebra. Letting X represent *relationship duration*, the estimated partial slope can be denoted $b + 2gx$. Its variance is (as the reader can verify with covariance algebra)

$$V(b + 2gx) = V(b) + 4x\,\text{Cov}(b,g) + 4x^2\,V(g).$$

Once again, however, we can use targeted centering to avoid computation of the standard error by hand. In this case, we simply create a new "centered" variable, *relationship duration* $- 12.8226$, shown as "relationship duration $- SD$" in Table 8.3, along with its square, (*relationship duration* $- SD$)^2, and use these in the model to capture the effect of *relationship duration*. The results are shown in model 4 in Table 8.3. The partial slope of relationship duration is now $-.029 + 2(.001)$ (*relationship duration* $- SD$). When *relationship duration* is at 1 standard deviation above the mean, (*relationship duration* $- SD$) equals zero, and the effect is $-.029$, as calculated above. As is evident, this effect is also quite significant.

TESTING COEFFICIENT CHANGES IN LOGISTIC REGRESSION

In Chapter 3 we saw that we could test for the significance of changes in coefficient values across nested linear regression models using the procedure outlined by Clogg et al. (1995). In the same article, the authors detailed the procedure for testing coefficient changes in any form of the generalized linear model. Here, I discuss that procedure in the context of logistic regression. As an example, Table 8.4 presents two models for couple violence. Model 1 represents the results of refinements in this chapter, with respect to interaction and nonlinear effects, made to the base model of Chapter 7. It is essentially the logit model from Table 7.1 to which have been added the quadratic effect of *relationship duration*, plus the interaction between *economic disadvantage* and *alcohol/drug problem*. The model suggests that a number of factors affect the log odds of couple violence. Of interest, however, are the mechanisms responsible for these effects.

Why, for example, do substance abuse problems elevate the odds of violence? One possible explanation is that alcohol and drugs promote relationship dissension, particularly of a more hostile nature (Leonard and Senchak, 1996). Arguments may erupt over a partner's drinking or drug use. Or, the substance-abusing partner may, under the influence of alcohol or drugs, become more belligerent than he or she normally would. At any rate, I suggest that part of the effect of substance abuse on violence is due to its tendency to increase hostile argumentation, which, in turn, elevates the risk of violence. Similar arguments could be made for the other predictors. Each may affect the risk of violence because it elevates the level of hostile argumentation in the

Table 8.4 Comparing Coefficients for Focus Variables before vs. after Adding Conflict Management Mediators

Predictor	Model 1	Model 2	$\hat{\delta}$	z for $\hat{\delta}$
Intercept	−2.308***	−2.553***		
Cohabiting	.789**	.867***	−.077	−2.745
Minority couple	.238*	.206	.031	3.109
Female's age at union	−.030***	−.016	−.014	−14.004
Male's isolation	.021**	.013	.008	16.062
Economic disadvantage	.016	.014	.002	1.638
Alcohol/drug problem	.978***	.654***	.323	18.417
Relationship duration	−.054***	−.052***	−.001	−2.089
(Relationship duration)^2	.00098**	.0015***	−.0005	−21.873
Economic disadvantage × alcohol/drug problem	.070*	.061*	.009	1.719
Open disagreement		.064***		
Positive communication		−.441***		
Model χ^2	210.319***	453.310***		
df	9	11		
Hosmer–Lemeshow χ^2 (8)		12.470		
R^2_{MZ}		.239		
$\hat{\Delta}$		.131		

* $p < .05$. ** $p < .01$. *** $p < .001$.

relationship. Model 2 explores this possibility by adding two measures of relationship dissension. *Open disagreement* taps the frequency of open disagreements, and *positive communication* is a scale that measures the extent to which disagreements are conducted calmly. It is evident that the effect of *alcohol/drug problem* (at mean *economic disadvantage*), as well as the effects of several other factors (e.g., *female's age at union, male's isolation*) have been noticeably reduced after controlling for these measures of verbal conflict. However, are these coefficient changes significant?

Variance–Covariance Matrix of Coefficient Differences

In general, the test for the significance of coefficient changes in logistic regression is as follows. Once again, we define a baseline model with p parameters as

$$\ln O = \alpha + \sum \beta_p^* X_p,$$

whereas the full model with q added parameters is:

$$\ln O = \alpha + \sum \beta_p X_p + \sum \gamma_q Z_q.$$

We are interested in whether the β_k^*, the coefficients of the focus variables in the baseline model, are significantly different than the β_k, the coefficients of the same focus variables in the full model, after the other variables (the $Z_1, Z_2, \ldots, Z_q$) have been included. Therefore, we wish to test whether the coefficient differences, $\delta_k = \beta_k^* - \beta_k$, for $k = 1, 2, \ldots, p$, are different from zero. Under the assumption that the full model is the true model that generated the data, the statistic $d_k/\hat{\sigma}_{d_k}$ (where $d_k = b_k^* - b_k$ is the sample difference in the kth coefficient and $\hat{\sigma}_{d_k}$ is the estimated standard error of the difference) is distributed asymptotically as standard normal (i.e., it is a z test) under H_0: $\delta_k = 0$ (Clogg et al., 1995).

Unfortunately, the standard errors of the d_k are not a standard feature of logistic regression software. However, they can be recovered via a relatively straightforward matrix expression. If we let $V(\hat{\delta})$ represent the estimated variance–covariance matrix of the coefficient differences, the formula for this matrix is

$$V(\hat{\delta}) = V(\hat{\beta}) + V(\hat{\beta}^*)[V(\hat{\beta})]^{-1}V(\hat{\beta}^*) - 2V(\hat{\beta}^*),$$

where $V(\hat{\beta})$ is the sample variance–covariance matrix for the b_k's in the full model, $V(\hat{\beta}^*)$ is the sample variance–covariance matrix for the b_k^*'s in the reduced model, and $[V(\hat{\beta})]^{-1}$ is the inverse of the variance–covariance matrix for the b_k's in the full model (Clogg et al., 1995). [A copy of an SAS program that estimates the reduced and full models, computes $V(\hat{\delta})$, and produces z tests for coefficient changes across models is available on request from the author.]

The third column of Table 8.4 shows the coefficient changes for all variables in model 1 after *open disagreement* and *positive communication* have been added. Column 4 shows the z tests for the significance of the changes. We see that all changes are quite significant except those for the effects of *economic disadvantage* and the interaction of *economic disadvantage* with *alcohol/drug problem.* In particular, part of the effect of *alcohol/drug problem* (at mean *economic disadvantage*) appears to be mediated by *open disagreement* and *positive communication.* A final comment before closing this section: The reader will notice that even relatively small coefficient changes across models turn out to be significant. The standard errors of the coefficient changes tend to be quite small in this test, primarily because coefficients for the same regressors in baseline and full models are not independent (Clogg et al., 1995). As in other such scenarios involving dependent samples (e.g., the paired *t* test) standard errors of differences between dependent estimates tend to be smaller, inflating the values of test statistics.

Discriminatory Power and Empirical Consistency of Model 2

We saw in Chapter 7 that the base logit model in Table 7.1 was not quite empirically consistent, according to the Hosmer–Lemeshow statistic. Nor did it have impressive discriminatory power, according to either the explained-variance or ROC criteria. However, that model was missing some important effects, which, hopefully, have been incorporated into model 2 in Table 8.4. Reassessing empirical consistency and discriminatory power for this model produced the following results, shown at the bottom of the model 2 column. The Hosmer–Lemeshow statistic is 12.47, which, with 8 degrees of freedom, is no longer significant ($p = .13$). The model now appears to be empirically consistent or to have an adequate fit. About 24% of the variance in the underlying *physical aggression* scale is now accounted for by the model. Or, the model explains about 13% of the *risk* of violence. The ROC for model 2 is shown in Figure 8.2. As is evident, the area under

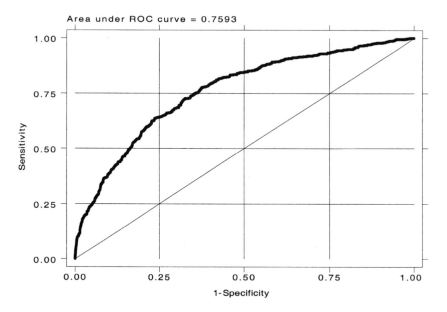

Figure 8.2   Receiver operating characteristic curve for model 2 in Table 8.4.

the curve is now .7593, which, according to the guidelines articulated in Chapter 7, represents acceptable discriminatory power.

MULTINOMIAL MODELS

Response variables may consist of more than two values but still not be appropriate for linear regression. Unordered categorical, or nominal, variables are those in which the different values cannot be rank ordered. Ordered categorical variables have values that represent rank order on some dimension, but there are not enough values to treat the variable as continuous (e.g., there are fewer than, say, five levels of the variable). Logistic regression models are easily adaptable to these situations and are addressed in this section of the chapter.

Unordered Categorical Variables

Until now I have been treating intimate violence as a unitary phenomenon. However, in that violence by males typically has graver consequences than violence by females (Johnson, 1995; Morse, 1995) it may be important to make finer distinctions. For this reason, I distinguish between two types of violence in couples. I refer to the first as "intense male violence," which reflects any one of the following scenarios: The male is the only violent partner, both are violent but he is violent more often, or both are violent but only the female is injured. All other manifestations of violence are referred to as "physical aggression." Both types of violence are contrasted with "nonviolence" (or, more accurately, "the absence of reported violence"), the third category of the response variable that I term *couple violence profile*. Of interest now is the degree to which the final model for violence, model 2 in Table 8.4, discriminates among "intense male violence," "physical aggression," and "nonviolence." I begin by treating *couple violence profile* as unordered categorical. That is, these three levels are treated as qualitatively different types of intimate violence (or the lack of it). However, it can be argued that they represent increasing degrees of violence severity, with "intense male violence" being more severe than "physical aggression," which is obviously more severe than "nonviolence." In a later section, these three categories are therefore treated as ordered.

Of the 4095 couples in the current example, 3540, or 86.4%, are "nonviolent"; 406, or 9.9%, have experienced "physical aggression"; and 149, or 3.6%, are characterized by "intense male violence." There are three possible nonredundant odds that can be formed to contrast these three categories. Each of these is conditional on being in one of two categories of *couple violence profile* (Theil, 1970). For example, there are 3946 couples who experienced either "nonviolence" or "physical aggression." Given location in one of these two categories, the odds of "physical aggression" is 406/3540 = .115. This odds is also the ratio of the probability of "physical aggression" to the probability of "nonviolence," or .099/.864 = .115. Similarly, given that a couple is characterized by either "nonviolence" or "intense male violence," the odds of "intense male violence" is 149/3540 (= .036/.864) = .042. Only two of the odds are independent: once they are recovered, the third is just the ratio of the first two. Thus, given some type of violence, the odds that it is "intense male violence" is $.036/.099 = .364$. In general, for an M-category variable, there are $M(M-1)/2$ nonredundant odds that can be contrasted, but only $(M-1)$ independent odds.

Modeling $(M-1)$ Log Odds

As before, we typically wish to model the log odds as functions of one or more explanatory variables. However, this time we require $(M-1)$ equations, one for each independent log odds. Each equation is equivalent to a binary logistic regression model, in which the response is a conditional log odds—the log odds of being in one vs. another category of the response variable, given location in one of these two categories. Each odds is the ratio of the probabilities of being in the respective categories. Equations for all of the other $M(M-1)/2 - (M-1)$ dependent log odds are functions of the parameters for the independent log odds, and therefore do not need to be estimated from the data. Typically, we choose one category, say the Mth, of the response variable as the baseline, and contrast all other categories with it (i.e., the probability of being in the baseline category forms the denominator of each odds). With $\pi_1, \pi_2, \ldots, \pi_M$ representing the probabilities of being in category 1, category 2, ..., category M, of the response variable, respectively, the multinomial logistic regression model with K predictors is

$$\log \frac{\pi_1}{\pi_M} = \beta_0^1 + \beta_1^1 X_1 + \cdots + \beta_K^1 X_K$$

$$\log \frac{\pi_2}{\pi_M} = \beta_0^2 + \beta_1^2 X_1 + \cdots + \beta_K^2 X_K \qquad (8.1)$$

$$\vdots$$

$$\log \frac{\pi_{M-1}}{\pi_M} = \beta_0^{M-1} + \beta_1^{M-1} X_1 + \cdots + \beta_K^{M-1} X_K,$$

where the superscripts on the betas indicate that effects of the regressors can change, depending on which log odds is being modeled.

Estimation. As before, parameters are estimated via maximum likelihood. In that the model consists of a series of binary logistic regression equations, one method of estimating the model, particularly in the absence of multinomial logistic regression software, is via *limited information maximum likelihood* (LIML) *estimation*. In the current example, this would be accomplished by selecting all cases characterized by either "intense male violence" or "nonviolence" and ignoring those exhibiting "physical aggression." One would then estimate a binary logistic regression for "intense male violence" versus "nonviolence," using only the cases selected. Next, one would select only the cases characterized by either "physical aggression" or "nonviolence." For this group, one would estimate a binary logistic regression for "physical aggression" versus "nonviolence." The two resulting sets of estimates would then constitute the multinomial logistic regression estimates. This approach, originally proposed by Begg and Gray (1984), produces estimates that are consistent and asymptotically normal. However, they are not as efficient as those produced by maximizing the joint likelihood function for *all* the parameters (across the $M - 1$ equations), given the data [see Hosmer and Lemeshow (2000) for an expression for this likelihood]. The latter approach is what is commonly employed for estimating the multinomial logistic regression model. Maximization of the joint likelihood function using all the data is referred to as *full information maximum likelihood* (FIML) *estimation*.

In SAS, one can use the procedures LOGISTIC or CATMOD for estimating the FIML model. As SAS's CATMOD (employed for this chapter) automatically chooses the highest value of the response variable as the baseline, one controls the choice of baseline by coding the variable accordingly. In the current example, I wanted "nonviolence" to be the baseline category, so the variable *couple violence profile* was coded 0 for "intense male violence," 1 for "physical aggression," and 2 for "nonviolence." Table 8.5 presents the results of estimating the multinomial model of *couple violence profile* using the regressors in model 2 of Table 8.4. Shown are the equations for the two independent log odds—contrasting "intense male violence" with "nonviolence" and contrasting "physical aggression" with "nonviolence"—based on either FIML or LIML. The estimates in the equation for the third logs odds, which contrasts "intense male violence" with "physical aggression," are simply the differences in the

Table 8.5 Multinomial Logistic Regression Models for Violence Profile as a Function of Couple Characteristics

Predictor	FIML Estimates		LIML Estimates	
	IM vs. NV	PA vs. NV	IM vs. NV	PA vs. NV
Intercept	−4.024***	−2.810***	−4.046***	−2.794***
Cohabiting[ab]	1.368***	.591	1.430***	.645*
Minority couple	.255	.191	.251	.149
Female's age at union	−.029*	−.011	−.029*	−.010
Male's isolation	.028	.008	.030	.008
Economic disadvantage	.037*	.005	.035*	.006
Alcohol/drug problem[b]	.867**	.579**	.913***	.585**
Relationship duration[b]	−.066***	−.049***	−.068***	−.048***
(Relationship duration)^2[b]	.0005	.0016***	.0005	.0016***
Economic disadvantage × alcohol/drug problem	.048	.067	.055	.072*
Open disagreement[b]	.058**	.065***	.058**	.069***
Positive communication[b]	−.536***	−.408***	−.534***	−.406***
Model χ^2	475.738***		224.211***	296.482***
df	22		11	11
R_L^2	.122		.180	.113

[a] Significant discriminator of intense male violence (IM) versus physical aggression (PA).
[b] Significant global effect on both log odds.
* $p < .05$. ** $p < .01$. *** $p < .001$.

first two sets of estimates. To see why, let π_0 be the probability of "intense male violence," π_1 be the probability of "physical aggression," and π_2 be the probability of "nonviolence." Then, with K predictors of the form $X_1, X_2, \ldots, X_K$ in the model, the equation for the odds of "intense male violence" vs. "physical aggression" is

$$\frac{\pi_0}{\pi_1} = \frac{\pi_0/\pi_2}{\pi_1/\pi_2} = \frac{\exp(\beta_0^1) \exp(\beta_1^1 X_1) \cdots \exp(\beta_K^1 X_K)}{\exp(\beta_0^2) \exp(\beta_1^2 X_1) \cdots \exp(\beta_K^2 X_K)}$$

$$= \exp(\beta_0^1 - \beta_0^2) \exp[(\beta_1^1 - \beta_1^2)X_1] \cdots \exp[(\beta_K^1 - \beta_K^2)X_K].$$

Of course, this third equation can easily be generated via SAS simply by changing the coding of the response. In this case, I would switch the coding so that 1 represents "nonviolence" and 2 represents "physical aggression." Then the first set of estimates shown in the output would be for the log odds of "intense male violence" versus "physical aggression." However, it is usually less confusing to show only the equations for the independent log odds, after selecting the most appropriate group as the baseline category.

Interpretation of Coefficients. Model coefficients are interpreted just as they are in the binary case, except that now, more than two outcome categories are being compared. For example, according to the FIML estimates, cohabitors have higher odds of "intense male violence" (versus "nonviolence") than married by a factor of $\exp(1.368) = 3.927$. They are not at higher risk than married, however, for "physical aggression." Their odds of "physical aggression," compared to the odds for married, is inflated by $100[\exp(.591) - 1] = 80.6\%$, but this effect is not significant. Given violence, the odds that it is "intense male violence" versus "physical aggression" is $\exp(1.368 - .591) = \exp(.777) = 2.175$ times higher for cohabitors than for married. Is this effect significant? One could pose the question in one of two ways. First, we could ask whether the effect of cohabiting on the log odds of "intense male violence" is different from its effect on "physical aggression." This would involve a test of the difference between the coefficients for cohabiting in each equation. Or we can estimate the log odds of "intense male violence" vs. "physical aggression" (the third, noninde-pendent equation) and ask whether the cohabiting effect is significant in *that* equation. In fact, it is (results not shown). The latter perspective is equivalent to asking whether cohabiting discriminates "intense male violence" from "physical aggression." In fact, it is the only significant discriminator in the model, as indicated in the table footnote.

Several other effects in the FIML model are significant. Having an alcohol or drug problem and more frequent disagreements elevates the odds of both "intense male violence" and "physical aggression." More positive communication and a longer relationship duration reduce the odds of both types of violence. The female's age at union, living in an economically disadvantaged neighborhood, and the nonlinear effect of relationship duration, however, affect only the odds of one type of violence. Nevertheless, none of these variables' effects are significantly different for "intense male violence" as opposed to "physical aggression."

The LIML estimates are generally quite close to the FIML ones. This should usually be the case, as the LIML estimates are nearly as efficient as their FIML counterparts (Begg and Gray, 1984). One might wonder why we would bother with LIML if the FIML software is available, as it generally is. As Hosmer and Lemeshow (2000) point out, the LIML approach has some specific advantages. First, it allows the model for each log odds to be different if we so choose, that is, to contain different regressors or different functions of regressors, an approach not possible with FIML. Second, it allows one to take advantage of features that may be offered in binary logistic regression software but not multinomial logistic regression software. Examples are weighting by case weights or diagnosing influential observations. Third, means of assessing empirical consistency, such as the Hosmer–Lemeshow χ^2, are not yet well developed for the multinomial model (Hosmer and Lemeshow 2000). However, using the LIML approach, empirical consistency can be assessed for each equation separately, as discussed in Chapter 7. In fact, for the model in Table 8.5, the Hosmer–Lemeshow χ^2 is 11.823 for the equation for "intense male violence" and 6.652 for the equation for "physical aggression." Both values are nonsignificant, suggesting an acceptable model fit.

Inferences. In multinomial logistic regression, there are several statistical tests of interest. First, as in binary logistic regression, there is a test statistic for whether the model as a whole exhibits any predictive efficacy. The null hypothesis is that all $K(M-1)$ of the regression coefficients (i.e., the betas) in equation group (8.1) equal zero. Once again, the test statistic is the model chi-squared, equal to $-2\log(L_0/L_1)$, where L_0 is the likelihood function evaluated for a model with only the MLEs for the intercepts and L_1 is the likelihood function evaluated at the MLEs for the hypothesized model. This test is not automatically output in CATMOD. However, as the program always prints out $-2\log L$ for the current model, it can be readily computed by first estimating a model with no predictors and then recovering $-2\log L_0$ from the printout (it is the value of "$-2\log$ likelihood" for the last iteration on the printout). This test can then be computed as $-2\log L_0 - (-2\log L_1)$. For the model in Table 8.5, $-2\log L_0$ was 3895.2458, while $-2\log L_1$ was 3419.5082. The test statistic was therefore $3895.2458 - 3419.5082 = 475.7376$, with $11(2) = 22$ degrees of freedom, a highly significant result. (The LIML equations each have their own model χ^2, as shown in the table.)

Second, the test statistic, using FIML, for the global effect on the response variable of a given predictor, say X_k, is not a single-degree-of-freedom test statistic as in the binary case. For multinomial models, there are $(M-1)$ β_k's representing the global effect of X_k, one for each of the log odds in equation group (8.1). Therefore, the test statistic is for the null hypothesis that all $M-1$ of these β_k's equal zero. There are two ways to construct the test statistic. One is to run the model with and without X_k and note the value of $-2\log L$ in each case. Then, if the null hypothesis is true, the difference in $-2\log L$ for the models with and without X_k is asymptotically distributed as chi-squared with $M-1$ degrees of freedom. This test requires running several different models, however, excluding one of the predictors on each run. Instead, most software packages, including SAS, provide an asymptotically equivalent Wald chi-squared test statistic [see Long (1997) for its formula] that performs the same function. Predictors having significant global effects on violence types, according to this test, are flagged with a superscript b in Table 8.5.

A third test statistic is for the effect of a predictor on a particular log odds. This is simply the ratio of a given coefficient to its asymptotic standard error, which, as in the binary case, is a z-test statistic. Fourth, it may be desirable to test effects of predictors on the nonindependent log odds—the odds of "intense male violence" versus "physical aggression," in the current example. As noted above, it is a simple matter to obtain these tests, simply by rerunning the program and changing the coding of the response variable. Fifth, tests of nested models with FIML are accomplished the same as in the binary case. That is, if model B is nested inside model A (because, for example, the predictors in B are a subset of those in A), $-2\log(L_B/L_A) = -2\log L_B - (-2\log L_A)$ is a chi-squared test for the significance of the difference in fit of the two models.

Finally, there is a test of collapsibility of outcome categories. Two categories of the outcome variable are *collapsible* with respect to the predictors if the predictor set is unable to discriminate between them. In the current example, it was seen that only one predictor—cohabiting—discriminates significantly between "intense male violence" and "physical aggression." A global chi-squared test for the collapsibility of these two categories of violence can be conducted as follows. First, I select only the couples experiencing one of the other of these types of violence, a total of 555 couples. Then I estimate a binary logistic regression model for the odds of "intense male violence" versus "physical aggression." The test of collapsibility is the usual likelihood-ratio chi-squared test that all of the betas in this binary model are zero. Under the null hypothesis that the predictors do not discriminate between these types of violence, this statistic is asymptotically distributed as chi-squared (Long, 1997). The test turns out to have the value 23.51, which, with 11 degrees of freedom, is just significant at $p < .05$. Apparently, the model covariates do generally discriminate between "intense male violence" and "physical aggression," with cohabiting being the primary discriminator, as noted above.

Estimating Probabilities. The probabilities of being in each category of the response are readily estimated, based on the sample log odds. That is, if U is the estimated log odds of "intense male violence" for a given couple and V is the estimated log odds of "physical aggression" for that couple, the estimated probabilities of each response for that couple are

$$P(\text{intense male violence}) = \frac{e^U}{1 + e^U + e^V},$$

$$P(\text{physical aggression}) = \frac{e^V}{1 + e^U + e^V}, \tag{8.2}$$

$$P(\text{nonviolence}) = \frac{1}{1 + e^U + e^V}.$$

Table 8.6 Predicted Probabilities for Intense Male Violence (IM), Physical Aggression (PA), and Nonviolence (NV), as a Function of Alcohol/Drug Problems

Settings of Other Covariates	Alcohol/Drug Problems	P(IM)	P(PA)	PI(NV)
Low risk	Yes	.00322	.01943	.97735
	No	.00174	.01540	.98286
Average risk	Yes	.03701	.09342	.86957
	No	.01659	.05585	.92756
High risk	Yes	.50758	.33663	.15579
	No	.36520	.29371	.34109

Table 8.6 presents the probabilities for each category of *couple violence profile*, based on the FIML estimates in Table 8.5. The focus in Table 8.6 is on the effect, in particular, of *alcohol/drug problem*. This is evaluated at three settings of the other covariates: low risk, average risk, and high risk. Low risk represents a covariate profile that predicts low probabilities of violence. For this profile, I set *cohabiting* and *minority couple* each to 0; *male's isolation, economic disadvantage*, and *open disagreement* to 1 standard deviation below the mean; and *female's age at union, relationship duration*, and *positive communication* to 1 standard deviation above the mean. For average risk, all covariates (apart from *alcohol/drug problem*) are set to zero. High-risk couples have *cohabiting* and *minority couple* each set to 1; *male's isolation, economic disadvantage*, and *open disagreement* set to 1 standard deviation above the mean; and *female's age at union, relationship duration*, and *positive communication* set to 1 standard deviation below the mean. (Remember that the model also contains an interaction between *alcohol/drug problem* and *economic disadvantage* as well as a quadratic effect of *relationship duration*.) Let's calculate the probabilities for the first row of Table 8.6 to see how equations (8.2) work. Recalling that 0 represents "intense male violence" and 1 represents "physical aggression," and employing several decimal places for the coefficients, we have the following estimated log odds (numbers in parentheses are the standard deviations of the continuous regressors). The log odds of "intense male violence" for low-risk couples with alcohol or drug problems is

$$\ln O_0 = -4.024 - .0293(7.0807) + .0275(-6.2958) + .0372(-5.1283)$$
$$+ .8672 - .0662(12.8226) + .000475(12.8226^2) + .0479(1)(-5.1283)$$
$$+ .0575(-4.0237) - .5359(1.3788) = -5.7148,$$

and the log odds of "physical aggression" for low-risk couples with alcohol or drug problems is

$$\ln O_1 = -2.8099 - .0114(7.087) + .00815(-6.2958) + .00532(-5.1283)$$
$$+ .579 - .0489(12.8226) + .00162(12.8226^2) + .0665(1)(-5.1283)$$
$$+ .0654(-4.0237) - .4084(1.3788) = -3.9182.$$

The estimated probabilities for each category of *couple violence profile* are, therefore,

$$P(\text{intense male violence}) = \frac{\exp(-5.7148)}{1 + \exp(-5.7148) + \exp(-3.9182)} = .0032,$$

$$P(\text{physical aggression}) = \frac{\exp(-3.9182)}{1 + \exp(-5.7148) + \exp(-3.9182)} = .0194,$$

$$P(\text{nonviolence}) = \frac{1}{1 + \exp(-5.7148) + \exp(-3.9182)} = .9774.$$

At this point it is instructive to compare the odds ratios generated by the FIML model in Table 8.5 with the probabilities in Table 8.6, for a couple of reasons. For one, odds ratios based on the FIML model must agree with odds ratios generated from ratios of probabilities in Table 8.6. This helps us understand the meaning of the odds ratios. Second, odds ratios convey a different impression than probabilities, particularly in multinomial models. It is therefore important to understand the differences between the "story" told by odds ratios and the "story" told by the probabilities. For example, the probabilities of each type of violence are increased markedly for high-risk as opposed to low- or average-risk couples. For high-risk couples with alcohol or drug problems, the probabilities associated with either type of violence are especially high. In fact, the chances of a high-risk couple with alcohol or drug problems experiencing "intense male violence" are better than 50–50. One might be tempted to infer that alcohol or drug problems are especially predictive of violence for high-risk couples. As high-risk couples are those who are, among other things, 1 standard deviation above the mean on *economic disadvantage*, we might conclude that alcohol or drug problems have the strongest effect on violence at high (i.e., a standard deviation above mean) *economic disadvantage*.

The odds ratios, however, present a different picture. Consider the effect of *alcohol/ drug problem* on "intense male violence" versus "physical aggression." That is, given violence, what is the impact of *alcohol/drug problem* on the odds that it is of the "intense-male" type? Employing four decimal places for increased accuracy, the coefficient for the main effect of *alcohol/drug problem* (from Table 8.5) is .8672 − .5790 = .2882. The coefficient for *economic disadvantage* × *alcohol/drug problem* is .0479 − .0665 = − .0186. Therefore, the effect of *alcohol/drug problem* on the log odds of "intense male violence" (versus "physical aggression") is .2882 − .0186 *economic disadvantage*. Or, the effect on the odds is

$$\Psi_{\text{alc/drug}} = \exp(.2882)[\exp(-.0186)]^{\text{ecndisad}}$$

$$= 1.33402(.98157)^{\text{ecndisad}}.$$

This suggests that each unit increase in *economic disadvantage reduces* the effect of *alcohol/drug problem* by a factor of .98157, approximately a 2% reduction. In other words, the effect of *alcohol/drug problem* is actually *diminishing* with greater *economic disadvantage*. In particular, at 1 standard deviation below mean *economic disadvantage*, the effect of *alcohol/drug problem* is

$$\psi_{alc/drug} = 1.33402(.98157)^{-5.1283} = 1.46755.$$

At mean *economic disadvantage* the effect is

$$\psi_{alc/drug} = 1.33402.$$

At 1 standard deviation above mean *economic disadvantage*, the effect is

$$\psi_{alc/drug} = 1.33402(.98157)^{5.1283} = 1.21264.$$

These figures agree with the probabilities in Table 8.6. Taking ratios of probabilities, the effect of *alcohol/drug problem* for low-risk couples (who are at -1 *SD* on *economic disadvantage*) is

$$\psi_{alc/drug} = \frac{.00322/.01943}{.00174/.01540} = 1.46674.$$

The effect of *alcohol/drug problem* for average-risk couples (who are at mean *economic disadvantage*) is

$$\psi_{alc/drug} = \frac{.03701/.09342}{.01659/.05585} = 1.33369,$$

and the effect of *alcohol/drug problem* for high-risk couples (who are at $+1$ *SD* on *economic disadvantage*) is

$$\psi_{alc/drug} = \frac{.50758/.33663}{.36520/.29371} = 1.21266.$$

These values agree with those based on the FIML model, within rounding error. The point of this exercise is that odds ratios reflect a comparison of *odds* rather than probabilities. In that capacity, they can indicate a declining effect even in the presence of increasing probabilities, because it is the *ratio* of probabilities (i.e., the odds) that is being measured. Thus, even though the probability of "intense male violence" is increasing dramatically for those with alcohol or drug problems across degrees of couple risk, given violence, the odds that it is of the "intense-male" type is declining. Because the presentation of the equations for the nonindependent log odds can lead to some confusion of this type, I prefer to focus on the equations for the independent log odds. If an appropriate baseline category is chosen, these are usually sufficient to describe the results.

Ordered Categorical Variables

When the values of a categorical variable are ordered, it is usually wise to take advantage of that information in model specification. For example, the trichotomous categorization of violence used for the analyses in Tables 8.5 and 8.6 represents different degrees of violence severity, as mentioned previously. In this section, therefore, I treat *couple violence profile* as an ordinal variable. The ordered logit model is a variant of logistic regression specifically designed for ordinal-level dependent variables. Although there is more than one way to form logits for ordinal variables [see, e.g., Agresti (1984, 1989) for other formulations], I focus on *cumulative logits*. These are especially appropriate if the dimension represented by the ordinal measure could theoretically be regarded as continuous (Agresti, 1989). As I have already argued, this is the case for *couple violence profile*. Cumulative logits are defined as follows. Suppose that the response variable consists of J ordered categories coded $1, 2, \ldots, J$. The jth cumulative odds is the ratio of the probability of being in category j or lower on Y to the probability of being in category $j + 1$ or higher. That is, if $O_{\leq j}$ represents the jth cumulative odds, and π_j is the probability of being in category j on Y, then

$$O_{\leq j} = \frac{\pi_1 + \pi_2 + \cdots + \pi_j}{\pi_{j+1} + \pi_{j+2} + \cdots + \pi_J}.$$

Cumulative odds are therefore constructed by utilizing $J - 1$ bifurcations of Y. In each one, the probability of being lower on Y (the sum of probabilities that $Y \leq j$) is contrasted with the probability of being higher on Y (the sum of probabilities that $Y > j$). This strategy for forming odds makes sense only if the values of Y are ordered. With regard to violence, the first cumulative odds, $O_{\leq 0}$, is the ratio of the probability that *couple violence profile* is 0 ("intense male violence") to the probability that *couple violence profile* is 1 ("physical aggression") or 2 ("nonviolence"). Using the marginal probabilities given above of each type of violence, the marginal sample value is $.036/(.099 + .864) = .037$. The second cumulative odds, $O_{\leq 1}$, is the ratio of the probability that *couple violence profile* is 0 or 1 to the probability that it is 2, with marginal value $(.036 + .099)/.864 = .156$. In other words, each odds is the odds of more severe vs. less severe violence, with "more severe" and "less severe" being defined using different values of j—the *cutpoint* (Agresti, 1989)—in either case. The jth cumulative logit is just the log of this odds. For a J-category variable, there are a total of $J - 1$ such logits that can be constructed. These logits are ordered, because the probabilities in the numerator of the odds keep accumulating as we go from the first through the $(J - 1)$th logit. That is, if U_j is the jth cumulative logit, then it is the case that $U_1 \leq U_2 \leq \cdots \leq U_{J-1}$.

One model for the cumulative logits, based on a set of K explanatory variables, is

$$\log O_{\leq j} = \beta_0^j + \beta_1^j X_1 + \beta_2^j X_2 + \cdots + \beta_K^j X_K, \tag{8.3}$$

where the superscripts on the coefficients of the regressors indicate that the effects of the regressors can change, depending on the cutpoint, j. This model has heuristic

Table 8.7   Ordered Logit Models for Violence Profile as a Function of Couple Characteristics

Predictor	Intense Male Violence vs. Other Response	Violence vs. Nonviolence	More vs. Less Violence
Intercept	−4.008***	−2.553***	
Intercept$_1$			−4.106***
Intercept$_2$			−2.528***
Cohabiting	1.202***	.867***	.909***
Minority couple	.207	.206	.182
Female's age at union	−.029*	−.016	−.017*
Male's isolation	.026	.013	.013
Economic disadvantage	.036*	.014	.018
Alcohol/drug problem	.724**	.654***	.639***
Relationship duration	−.058***	−.052***	−.052***
(Relationship duration)^2	.0001	.0015***	.0014***
Economic disadvantage × alcohol/drug problem	.012	.061*	.048
Open disagreement	.041	.064***	.062***
Positive communication	−.443***	−.441***	−.435***
Model χ^2	177.792***	453.110***	460.817***
Model df	11	11	11
Score test			12.457
Score df			11
R_L^2	.139	.140	.118

* $p < .05$. ** $p < .01$. *** $p < .001$.

value as a starting point. However, strictly speaking, it is not legitimate for an ordinal response. The reason has to do with the fact that at any given setting of the covariate vector, $\mathbf{x}$, it must be the case that $P(Y \leq j \mid \mathbf{x}) \leq P(Y \leq j + 1 \mid \mathbf{x})$ for all j. If covariates' effects can vary over cutpoints, however, it is possible that $P(Y \leq j \mid \mathbf{x}) > P(Y \leq j + 1 \mid \mathbf{x})$ for some j, which is logically untenable if Y is ordered. (I wish to thank Alan Agresti for bringing this model flaw to my attention.) At any rate, model (8.3) is easily estimated using binary logistic regression software, as it is just a binary logistic regression based on bifurcating Y at the jth cutpoint. Table 8.7 presents the results of estimating this model for *couple violence profile*. Estimates in the second column, for the log odds of "violence" versus "nonviolence," are just the estimates from model 2 in Table 8.4, repeated here for completeness. Estimates in the first column are for the log odds of "intense male violence" versus any other response.

Invariance to the Cutpoint.   For the most part, results suggest that regressors have the same effects on the log odds of more severe versus less severe violence, regardless of the cutpoint used to make this distinction. For example, a unit increase in *open disagreement* elevates the odds of "intense male violence" versus any other response by a factor of $\exp(.041) = 1.042$, whereas it raises the odds of "any violence" versus "no violence" by a factor of $\exp(.064) = 1.066$. Or each unit increase in *positive communication* lowers the odds of "intense male violence" by $\exp(-.443) = .642$, whereas it lowers the odds of "any violence" by $\exp(-.441) = .643$, a virtually indistinguishable difference in effects. If the effects of predictors are invariant to the cutpoint, a more parsimonious specification of equation (8.3) is possible. This is

$$\log O_{\leq j} = \beta_0^j + \beta_1 X_1 + \beta_2 X_2 + \cdots + \beta_K X_K. \tag{8.4}$$

This is the *ordered logit* or *proportional odds model* (Agresti, 2002). In this model, the effects of predictors are the same, regardless of the cutpoint for the odds. That is, each unit increase in a given predictor, say X_k, multiplies the odds by a proportionality constant of $\exp(\beta_k)$, regardless of the cutpoint chosen. The results of estimating this model (using procedure LOGISTIC in SAS) are shown in the last column of Table 8.7. Notice that the intercept is allowed to depend on the cutpoint, so there are two intercepts in the equation. (In fact, there are two different equations, but the coefficients are being constrained to be the same in each.) In that predictors are assumed to be invariant to the cutpoint, there is only one set of regression coefficients. Effects are interpreted just as in binary logistic regression, except that the response is the log odds of "more severe" versus "less severe" violence, rather than, as in Table 8.4, violence per se. Thus, *cohabiting* is seen to raise the odds of "more severe" violence by $\exp(.909) = 2.482$, or about 148%, whereas each unit increase in *positive communication* lowers the odds of "more severe" violence by about 35%.

Test of Invariance. In the first two columns of Table 8.7, where effects are allowed to depend on the cutpoint, some predictors appear to have different effects on the odds of "intense male violence" compared to violence per se. For example, the effect of *cohabiting* on the odds of "intense male violence" is $\exp(1.202) = 3.327$, whereas its effect on the odds of "any violence" is only $\exp(.867) = 2.380$. Moreover, other regressors, for example, *female's age at union* or *economic disadvantage*, have significant effects on only one of the odds. Are these variations significantly different or just the result of sampling error? This can be tested using the *score test for the proportional odds assumption* (provided automatically in SAS). The test statistic tests the null hypothesis that regressor effects are the same across all $J - 1$ possible cutpoints. That is, H_0 is that for each of the K regressors in the model, $\beta_k^j = \beta_k$ for $j = 1, 2, \ldots, J - 1$. Under the null hypothesis, the score statistic is asymptotically distributed as chi-squared with degrees of freedom equal to $K(J - 2)$. This is the difference in the number of parameters required to estimate the model in equation (8.3) versus equation (8.4): $K(J - 1) - K = K(J - 1 - 1) = K(J - 2)$. As shown in Table 8.7 for the current example, its value is 12.457, which, with 11 degrees of freedom, is not significant. This suggests that there is insufficient evidence to reject the proportional odds assumption. Apparently, the more parsimonious proportional-odds model appears reasonable for the data.

Estimating Probabilities with the Proportional Odds Model. In the event that estimates of $P(Y = j)$ for $j = 1, 2, \ldots, J$, based on the proportional odds model, are of interest, they are relatively straightforward to calculate. We draw on the probability rule that for an integer-valued random variable, Y, $P(Y=j) = P(Y \leq j) - P(Y \leq j-1)$. Now, note that, according to the proportional odds model,

$$\log O_{\leq j} = \log \frac{P(Y \leq j)}{P(Y > j)} = \beta_0^j + \mathbf{x}^{i\prime}\boldsymbol{\beta}, \tag{8.5}$$

where, in this case, $\mathbf{x}^{i\prime}\boldsymbol{\beta} = \beta_1 X_1 + \beta_2 X_2 + \cdots + \beta_K X_K$. Thus, equation (8.5) implies that

$$\frac{\exp(\log O_{\leq j})}{1 + \exp(\log O_{\leq j})} = \frac{P(Y \leq j)/P(Y > j)}{1 + P(Y \leq j)/P(Y > j)} = \frac{P(Y \leq j)/P(Y > j)}{[P(Y > j) + P(Y \leq j)]/P(Y > j)}$$

$$= \frac{P(Y \leq j)/P(Y > j)}{1/P(Y > j)} = P(Y \leq j).$$

Therefore, for $j = 2, 3, \ldots, J-1$, the formula for $P(Y=j)$ is

$$P(Y=j) = \frac{\exp(\log O_{\leq j})}{1 + \exp(\log O_{\leq j})} - \frac{\exp(\log O_{\leq j-1})}{1 + \exp(\log O_{\leq j-1})}. \tag{8.6}$$

However, for $j=1$, the lowest ordered value of Y, the formula for $P(Y=1)$ is

$$P(Y=1) = \frac{\exp(\log O_{\leq 1})}{1 + \exp(\log O_{\leq 1})} \tag{8.7}$$

since $P(Y \leq 0) = 0$. And for $j=J$, the highest-ordered value of Y, the formula for $P(Y=J)$, is

$$P(Y=J) = 1 - P(Y \leq J-1) = 1 - \frac{\exp(\log O_{\leq J-1})}{1 + \exp(\log O_{\leq J-1})}. \tag{8.8}$$

As an example, suppose that Y is a four-category ordered response, with values $j = 1, 2, 3, 4$, and the model is

$$\log O_{\leq 1} = -3.2 + .45X,$$
$$\log O_{\leq 2} = -2.8 + .45X,$$
$$\log O_{\leq 3} = -1.25 + .45X.$$

For $X = 1.75$, the four probabilities are

$$P(Y=1) = \frac{\exp[-3.2 + .45(1.75)]}{1 + \exp[-3.2 + .45(1.75)]} = .082,$$

$$P(Y = 2) = \frac{\exp[-2.8 + .45(1.75)]}{1 + \exp[-2.8 + .45(1.75)]} - \frac{\exp[-3.2 + .45(1.75)]}{1 + \exp[-3.2 + .45(1.75)]}$$

$$= .118 - .082 = .036,$$

$$P(Y = 3) = \frac{\exp[-1.25 + .45(1.75)]}{1 + \exp[-1.25 + .45(1.75)]} - \frac{\exp[-2.8 + .45(1.75)]}{1 + \exp[-2.8 + .45(1.75)]}$$

$$= .386 - .118 = .268,$$

$$P(Y = 4) = 1 - .386 = .614.$$

It is easily seen that the four probabilities sum to 1, and that $P(Y \leq j) \leq P(Y \leq j + 1)$ for $j = 1, 2, 3$.

Alternatives to the Proportional Odds Model. If the score test for the proportional odds assumption is significant, so that the proportional odds assumption is untenable, what should one do? In this case the researcher has several options. First, he or she can use the proportional odds model anyway, especially if noninvariant effects are only peripheral to the study. As an example, if I were interested primarily in how *open disagreement* and *positive communication* affect more severe violence, net of other factors, it is clear that the proportional odds model in the last column of Table 8.7 summarizes these variables' effects on the cumulative logits in an elegant fashion. Particularly when there are several categories of the response variable, equation (8.4) is a substantially more parsimonious description of the data than any of the other multinomial approaches. Nevertheless, it is frequently desirable to use a different modeling strategy when invariance is rejected. So one alternative is to choose the most informative bifurcation of the response variable and proceed with binary logistic regression. For example, either the first or second column in Table 8.7 could be a legitimate model to estimate. Or, if it is especially important to preserve the distinctions among different response categories, the multinomial analysis presented in Table 8.5 is a workable strategy. Even if the response variable is ordinal, use of the multinomial approach is always legitimate. At worst, we are only wasting some information by not taking advantage of the ordering information in the values of the response. Of course, if the response variable has at least five levels, its sample distribution is not too skewed, and the sample is large, the researcher may just want to treat it as continuous and employ OLS.

Empirical Consistency and Discriminatory Power in Multinomial Models. As indicated above, measures of empirical consistency (or goodness of fit) are not as well developed for multinomial models as they are for, say, binary logistic regression. Therefore, one means of assessing fit for the multinomial logistic regression of unordered responses has already been suggested above. This is to use the Hosmer–Lemeshow χ^2 statistic along with the LIML estimation technique to evaluate the empirical consistency of each separate equation. For ordered outcomes, Lipsitz et al. (1996) have recently developed a goodness-of-fit test that is an extension of the Hosmer–Lemeshow statistic to an ordinal response. The test is readily conducted using existing software for the proportional odds model (e.g., SAS's PROC LOGISTIC).

The reader is referred to the authors' article for further details. Discriminatory power is easier to handle. Any of the measures such as R_L^2, R_G^2, or R_{GSC}^2, which are based on the log-likelihood, can be used with multinomial models. In Tables 8.5 and 8.7, I have relied on R_L^2 to indicate the predictive efficacy of the models. According to this statistic, discriminatory power for the multinomial models of *couple violence profile* is modest, at best.

EXERCISES

8.1 Let P_0 = the model-estimated probability, for a given case, of "intense male violence," P_1 = the probability of "physical aggression," and P_2 = the probability of "nonviolence." Using equation group (8.2), show that these three probabilities sum to 1 for any given case.

8.2 Show how equation group (8.2) generates each probability defined in Exercise 8.1. That is, show that equation group (8.2) is the correct formula for generating each probability. [*Hint:* Substitute $\ln(P_0/P_2)$ for U and $\ln(P_1/P_2)$ for V in equation group (8.2) to recover the probabilities.]

8.3 Based on model 1 in Table 8.2, give the odds ratio for those with vs. without alcohol or drug problems at 1 standard deviation below, $\frac{1}{2}$ standard deviation above, and $1\frac{1}{2}$ standard deviations above mean *economic disadvantage* ($SD = 5.1283$). In particular, in each case, by what factor is the impact of *alcohol/drug problem* reduced/inflated as a function of *economic disadvantage*?

8.4 Based on model 1 in Table 8.3, give the estimated probability of violence at each duration quintile for cohabiting, minority couples, with an alcohol or drug problem, who are 1 standard deviation above the mean on *male's isolation* ($SD = 6.2958$) and *economic disadvantage* ($SD = 5.1283$) and 1 standard deviation below the mean on *female's age at union* ($SD = 7.0807$).

8.5 Based on model 3 in Table 8.3, give the effect (i.e., partial derivative) of *relationship duration* on the log odds of violence at .5, 1.5, and 2 standard deviations ($SD = 12.8226$) above mean *relationship duration*.

8.6 Based on the FIML estimates in Table 8.5, give the equation for the log odds of "intense male violence" versus "physical aggression."

8.7 Using the FIML model in Table 8.5, verify the probabilities in the last row of Table 8.6, within rounding error.

8.8 Give the effect of *alcohol/drug problem* on the odds of "intense male violence" versus "nonviolence," at mean *economic disadvantage*, 1 standard deviation below mean *economic disadvantage*, and 1 standard deviation above mean *economic disadvantage*, based on the FIML estimates in Table 8.5. Use .8672 for the coefficient of *alcohol/drug problem*, .0479 for the coefficient for the interaction term, and 5.1283 as the *SD* of *economic disadvantage*.

8.9 Estimate the same effects as in Exercise 8.8, but this time using the probabilities in Table 8.6, showing that these effects agree with those in Exercise 8.8, within rounding error.

8.10 Based on the proportional odds model in Table 8.7, give the odds of "intense male violence" versus any other response, and the odds of "violence" versus "nonviolence," for cohabiting, minority couples, with alcohol or drug problems, at mean levels of the continuous regressors.

8.11 Using the principle, for a discrete variable, Y, that $P(Y = j) = P(Y \leq j) - P(Y \leq j - 1)$, along with the formulas in equations (8.6) to (8.8), give the estimated probabilities of "intense male violence," "physical aggression," and "nonviolence," for the couples of Exercise 8.10, based also on the proportional odds model in Table 8.7.

8.12 Using the *students dataset*, regress MISSG (a dummy for having missing data on EXAM2 that reflects dropping the class before the second exam) on the EXAM1 score and its square, *student classification* (CLASSIF), and STAT-MOOD, using logistic regression. Missing imputation: Substitute 76.9774766 for missing data on EXAM1. Then:

(a) Interpret all coefficients, including the intercept and the main effect of EXAM1.

(b) Center EXAM1 and form EXAM1^2 from the centered term. Rerun the regression and interpret the effect of EXAM1 in this model.

(c) Use targeted centering to estimate and test the effect (i.e., partial derivative) of EXAM1 on the log odds of being missing on EXAM2 at 1 standard deviation below the mean of EXAM1.

(d) Compute the standard error for the test in part (c) using the formula for $V(b + 2gx)$ given in the chapter, and compare it to the standard error in part (c). For $V(b)$, $V(g)$, and $Cov(b,g)$, use the variance–covariance matrix of parameter estimates based on the *centered* (*not* target-centered) version of the model for EXAM1.

8.13 Using the *students dataset*, estimate the proportional odds model for FIRS-TEX, the ordinal version of EXAM1, as a function of COLGPA, SCORE, STATMOOD, and RATIO. Missing imputation: Follow the instructions for Exercise 6.20. Then:

(a) Test whether the proportional odds hypothesis is supported.

(b) Interpret the effects of all regressors (excluding the intercept) on the odds of a higher, as opposed to a lower, grade.

8.14 Based on the estimates from Exercise 8.13:
   (a) Give and interpret the estimated cumulative logit for each possible bifurcation of FIRSTEX for someone with a college GPA of 3.2, a math diagnostic score of 39, a STATMOOD of 3, and a RATIO of 25.
   (b) Give the estimated probabilities of an A, B, C, D, and F, for the student
   in part (a).

*Use the following results for Exercises 8.15 to 8.19. The variable PORNLAW in the
GSS98 dataset was recoded as PORN18 as follows: 1 = "no prohibition of any
kind," 2 = "prohibit its distribution to all ages," and 3 = "prohibit its distribution to
persons under 18." A multinomial logistic regression of PORN18 on FREQBARS
(frequency of going to bars, coded 1 to 6), AGE (in years), MALE (a dummy for
being male), RELOSITY (religiosity score; the sum of standardized items with the
high score being most religious), and CONSERV (coded 1 to 7, with the high score
being most politically conservative) produced the following results for n = 879 cases
using PROC CATMOD in SAS. The first parameter estimate listed for a regressor is
for the log odds of PORN18 = 1 versus PORN18 = 3; the second estimate is for the
log odds of PORN18 = 2 vs. PORN18 = 3:*
Intercept-only model: $-2 \ln L = 1446.8468.$
Wald $\chi^2_{(2)}$ for global effects of each regressor:

Intercept	39.15
FREQBARS	9.03
AGE	37.41
MALE	9.23
RELOSITY	46.16
CONSERV	8.04

Hypothesized model: $-2 \ln L = 1264.5865.$
Effects in hypothesized model:

Effect	Parameter	Estimate	Std. Error
Intercept	1	−4.2337	.8784
	2	−1.8312	.3925
FREQBARS	3	.0266	.1014
	4	−.1507	.0514
AGE	5	.0222	.0105
	6	.0286	.00473
MALE	7	−.3028	.3715
	8	−.4796	.1589
RELOSITY	9	−.2397	.0944
	10	.2915	.0495
CONSERV	11	.1376	.1335
	12	.1661	.0600

8.15 Give the model χ^2 for the hypothesized model and assess its discriminatory power using an appropriate measure. Is the model significant?

8.16 Construct a table of regression estimates along the lines of the first two columns of Table 8.5, starring the significant coefficients, and noting which regressors have significant global effects on both log odds.

8.17 Interpret the effects of the regressors (excluding the intercept) on each odds. Does it appear, from the nature of the regressor effects, that PORN18 might be treated as ordinal and estimated with the proportional odds model? Why or why not?

8.18 Give the estimated equation for the log odds of "no prohibition" versus "prohibition for all," based on the results for PORN18. Interpret all regressor (excluding the intercept) effects in terms of odds.

8.19 Give the predicted probabilities associated with each response to PORN18 for a 21-year-old man of average *religiosity*, with FREQBARS = 3 and CONSERV = 4.

Use the following results for Exercises 8.20 to 8.23. The variable DIVLAW (should divorce be easier or more difficult to obtain?) in the GSS98 dataset is coded 1 = "make divorce easier to obtain," 2 = "make divorce more difficult to obtain," 3 = "keep it as is." A multinomial logistic regression of DIVLAW on AGE (in years), EDUCAT (education in years of schooling), MALE (a dummy for being male), RELOSITY (religiosity score; the sum of standardized items with the high score being most religious), BLACK (a dummy for being black, with white as the contrast group), and OTHRACE (a dummy for being other than black or white, with white as the contrast group) produced the following results, for n = 879 cases, using PROC CATMOD in SAS. The first parameter estimate listed for a regressor is for the log odds of DIVLAW = 1 vs. DIVLAW = 3; the second parameter estimate is for the log odds of DIVLAW = 2 versus DIVLAW = 3:
Intercept-only model: $-2\ln L = 1734.7917$.
Wald $\chi^2_{(2)}$ for global effects of each regressor:

Intercept	30.31
AGE	17.13
EDUCAT	16.17
MALE	2.88
RELOSITY	52.07
BLACK	55.61
OTHRACE	16.78

Hypothesized model: $-2 \ln L = 1575.1477$.
Effects in hypothesized model:

Effect	Parameter	Estimate	Std. Error
Intercept	1	3.2549	.7097
	2	3.1548	.5994
AGE	3	−.0283	.00688
	4	−.0112	.00552
EDUCAT	5	−.1535	.0404
	6	−.1124	.0334
MALE	7	−.3396	.2253
	8	−.0574	.1879
RELOSITY	9	.00313	.0609
	10	.3241	.0536
BLACK	11	1.3872	.3197
	12	−.3424	.3164
OTHRACE	13	1.1421	.4322
	14	−.0789	.4301

8.20 Construct a table of regression estimates along the lines of the first two columns of Table 8.5, starring the significant coefficients, and noting which regressors have significant global effects on both log odds. At the bottom of the table, include the model χ^2, its *df*, and a measure of discriminatory power.

8.21 Interpret the effects of the regressors (excluding the intercept) on each odds. Does it appear from the nature of the regressor effects that DIVLAW might be treated as ordinal and estimated with the proportional odds model? Why or why not?

8.22 Give the estimated equation for the log odds of "making divorce easier" versus "making it more difficult." Interpret the effects in terms of odds. Also, what is the odds ratio for blacks versus those of other races?

8.23 Give the predicted probabilities associated with each response to DIVLAW for a 45-year-old black female with 12 years of *education* and a *religiosity score* of 5.

8.24 Use the *kids dataset* to estimate a binary logistic regression equation for MULTIPLE (whether or not the focal child has had more than one sex partner in the last month) as a function of FCAGE2, MALE, NONINTAC, MSEXATT, FSEXATT, MSTYLE1, MSTYLE2, FSTYLE1, FSTYLE2, PERMISIV, and the interaction of PERMISIV with MALE. Center PERMISIV to guard against collinearity problems.

(a) Interpret the interaction effect by examining how the impact of PERMISIV differs for male and female focal children, and by examining whether

PERMISIV has a significant effect on having multiple partners, separately within each gender.

(b) Perform Allison's (1999) Chow test analog to discern whether the model in part (a), minus the interaction term, differs on the whole between male and female children.

(c) Estimate the discriminatory power of the interaction model investigated in part (a).

8.25 Use the *students dataset* for the following exercise. Missing imputation: Do not impute any missing values for this problem; use listwise deletion.

(a) ORDEXAM1 is a trichotomous version of EXAM1 with three values: A (>89), B (>79), and C or worse (≤ 79). Estimate a multinomial logistic regression model for this variable, with C as the baseline category, as a function of SCORE, MALE, STATMOOD, COLGPA, RMAJOR (dummied up, with "sociology majors" as the reference group), and PREVMATH. Interpret the coefficient estimates in terms of odds.

(b) Give the model χ^2 for this model.

(c) What variables significantly discriminate an A grade from a B grade?

(d) Test the collapsibility of the A and B categories.

(e) Estimate the ordered logit model for this response, interpret the coefficients and their significance levels, and assess whether the hypothesis of ordinality (proportional odds) is supported.

CHAPTER 9

Truncated and Censored Regression Models

CHAPTER OVERVIEW

In this chapter we once again focus on continuous (or, at least, approximately inter-val-level) response variables, but take up the subject of regression models for trun-cated or censored data. Truncated and censored responses arise when the data observed on a continuous response are not fully representative of the range of values of the response in the target population. I begin the chapter by defining truncation, censoring, and a special variant of truncation, known as incidental truncation, that gives rise to sample-selection bias. I then introduce the truncated regression model and show how it is estimated via maximum likelihood. The discussion then moves to the more commonly used censored regression, or tobit model. For this model I discuss estimation via maximum likelihood, various ways of interpreting model coefficients, and an analog of the R^2 used in OLS. An alternative to the tobit model is also presented that relaxes one of its major constraints. Finally, I take up sample-selected regression, presenting both two-step and maximum-likelihood estimation techniques for this model. Substantive applications are illustrated throughout the chapter to enhance understanding of each technique.

TRUNCATION AND CENSORING DEFINED

Truncation

In contrast to approaches in other chapters, I begin this one with a more abstract set of examples. First, consider a normally distributed variable, X, with a mean of 5 and

Regression with Social Data: Modeling Continuous and Limited Response Variables,
By Alfred DeMaris
ISBN 0-471-22337-9   Copyright © 2004 John Wiley & Sons, Inc.

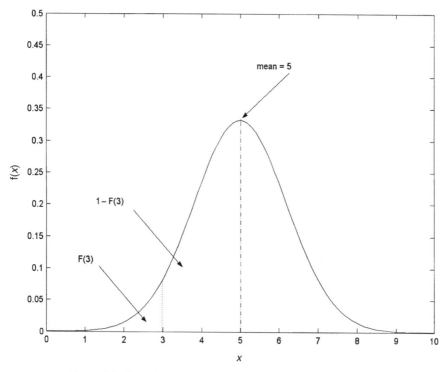

Figure 9.1   Normal density with mean = 5 and standard deviation = 1.2.

a standard deviation of 1.2. Figure 9.1 shows the density of X. Notice the dotted line
at the point $x = 3$. Recall that f(x), the density of x, is the point on the curve corresponding to x. For example, f(3) is the point on the curve corresponding to 3. F(x),
on the other hand, is the area under the curve to the left of x. Hence, F(3) is the area
to the left of 3 in Figure 9.1, while $1 - F(3)$ is the area to the right of 3. If the population of X values were the target population, a random sample taken from this distribution would be representative of that target population.

However, suppose that we were to draw a sample *under the condition that X be
greater than 3*. It is then not possible to observe values of X less than or equal to 3 in
the sample, and we say that the density (or distribution) from which we are sampling
is *truncated* at 3. That is, the sample is a random sample in the usual sense, but the
"population" from which it is being drawn is now truncated at 3. The population, or
density, that applies in this case is depicted in Figure 9.2. The sample is no longer representative of the target population shown in Figure 9.1; rather, it is representative of
the population in Figure 9.2. Moreover, the density in Figure 9.2 is not simply the part
of Figure 9.1 that is to the right of 3. Rather, in order for the resulting function,
denoted $f(x \mid x > 3)$, to remain a density, it must be rescaled so that the total area under
the curve is, again, 1. This is done by dividing the original density by $1 - F(3)$, the
area to the right of 3. That is, in general, if X is normal with density function f(x),

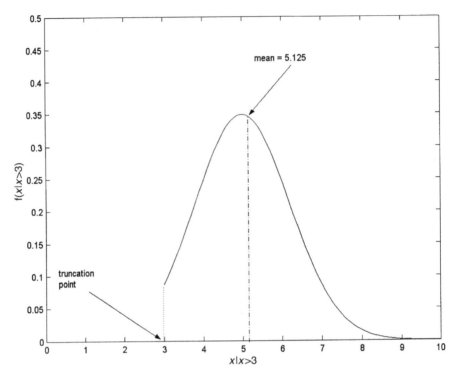

Figure 9.2   Truncated normal density with mean = 5.125 and standard deviation = 1.083.

mean μ, standard deviation σ, and truncation from below at the point c, then $f(x|x>c)$ is

$$f(x|x>c) = \frac{f(x)}{1 - F(c)}. \tag{9.1}$$

In the case of truncation from above, we obtain $f(x|x<c)$ by dividing $f(x)$ by $F(c)$ (see Greene, 2003; Wooldridge, 2000).

What does truncation do to the mean and variance of the new (i.e., truncated) random variable? From Greene (2003, p. 759) we have the following moments of the truncated normal distribution. *If X is normal with mean μ and standard deviation σ, and X is truncated from below at c, then*

$$E(X|\text{truncation}) = \mu + \sigma\lambda(\alpha), \tag{9.2}$$

$$V(X|\text{truncation}) = \sigma^2[1 - \delta(\alpha)], \tag{9.3}$$

where

$$\alpha = \frac{c - \mu}{\sigma}, \tag{9.4}$$

$$\lambda(\alpha) = \frac{\phi(\alpha)}{1 - \Phi(\alpha)}, \tag{9.5}$$

$$\delta(\alpha) = \lambda(\alpha)[\lambda(\alpha) - \alpha], \tag{9.6}$$

and where $\phi(\alpha)$ and $\Phi(\alpha)$ are the standard normal density and distribution functions, respectively. Moreover, note that $0 < \delta(\alpha) < 1$ for all values of α. Notice that α is just the z-score for c, that is, it measures the number of standard deviations that c is away from μ. The term $\lambda(\alpha)$ is called the *inverse Mills ratio* (IMR), or *hazard function*, for the standard normal distribution (Greene, 2003). As it plays an integral part in all of the models in this chapter, it is worth examining in closer detail.

Figure 9.3 depicts the major components of the IMR. Shown is the standard normal distribution with mean 0 and standard deviation 1. Letting $\alpha = .5$ in this case, we see that $\phi(.5)$ is the density associated with .5 and is equal to

$$\phi(.5) = \frac{1}{\sqrt{2(3.14159)}} \exp\left[-\frac{1}{2}(.5^2)\right] = .352$$

and $\Phi(.5)$ is the shaded area to the left of .5 under the curve, which is equal to .691 (using a table of areas under the standard normal curve). One minus $\Phi(.5)$, therefore,

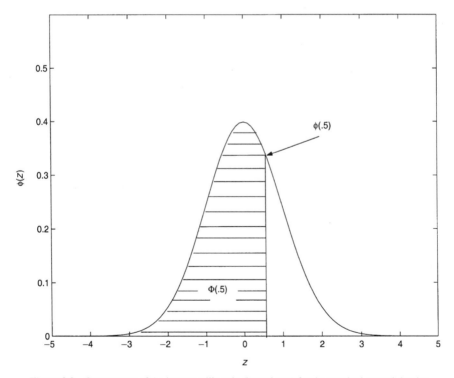

Figure 9.3 Components of the inverse mills ratio (hazard rate) for the standard normal density.

is the unshaded area that lies to the right of .5 and is equal to .309. The IMR for .5, denoted $\lambda(.5)$, is the ratio $\phi(.5)/[1 - \Phi(.5)]$, or $.352/.309 = 1.139$. In general, as z becomes increasingly negative, $\phi(z)$ approaches zero while $1 - \Phi(z)$ approaches 1, so $\lambda(z)$ approaches zero. As z becomes increasingly positive, on the other hand, both $\phi(z)$ and $1 - \Phi(z)$ approach zero. It turns out, however, that $1 - \Phi(z)$ approaches zero more rapidly; hence $\lambda(z)$ actually increases without bound. In other words, $\lim_{z \to \infty} \lambda(z) = \infty$. So the theoretical range of $\lambda(z)$ is $(0,\infty)$; however, given that virtually all of a standard normal variable lies within 4 standard deviations of its mean, the practical range of the IMR is approximately $(0, 4.2)$. (I postpone further interpretation of the IMR until we get to sample-selection models.)

Returning to the distribution in Figure 9.2, α is $(3 - 5)/1.2 = -1.667$, $\phi(-1.667)$ is .0994, and $1 - \Phi(-1.667)$ is .9522. Hence, $\lambda(\alpha) = .0994/.9522 = .1044$, and the mean of the distribution is therefore $5 + 1.2(.1044) = 5.125$. Now $\delta(\alpha) = .1044$ $[.1044 - (-1.667)] = .185$. So the variance of the distribution is $(1.2^2)(1 - .185) = 1.083$. As is evident from the formulas for the truncated mean and variance, and as shown in this example, truncation from below results in an increase in the mean but a reduction in the variance. [Truncation from above results in a reduction in both the mean and the variance; see Greene (2003) for details.]

Censoring

Suppose that we sample from the untruncated distribution of X shown in Figure 9.1 but that for all X below the value of 3, we simply record X as being equal to 3. In this case the sample is drawn from the full target population, but *it is not fully representative of that population with respect to the values of X*, since a portion of the values have been "censored" at 3. That is, for those observations, all we know is that X is less than 3. We say that the sample is *censored* at 3. Censoring, unlike truncation, does not represent a limitation in the population from which the data were drawn. Rather, it represents a limitation in the measurement of the variable of interest. If the data were not censored, they would be representative of the target population with respect to X (Greene, 2003).

The density that applies to censored data is actually a mixture of discrete and continuous densities and for the current example is shown in Figure 9.4. For values of X above 3, the density is just the $f(x)$ shown in Figure 9.1. However, for $X = 3$, the density is the probability that X is less than 3. (Recall from Chapter 1 that for a continuous X, $P[X < 3]$ and $P[X \le 3]$ are the same.) That is, we assign the area $F(3)$ to be the density corresponding to the value 3. This is depicted as a solid bar above the value of 3 in Figure 9.4. Again, I draw on a theorem presented in Greene (2003, p. 763) to calculate the moments of the censored normal variable. *If X is normal with mean μ and standard deviation σ, and X is censored from below at the threshold c, then*

$$E(X|\text{censoring}) = \Phi(\alpha)c + [1 - \Phi(\alpha)] [\mu + \sigma\lambda(\alpha)], \qquad (9.7)$$

$$V(X|\text{censoring}) = \sigma^2[1 - \Phi(\alpha)] [(1 - \delta(\alpha)) + [\alpha - \lambda(\alpha)]^2\Phi(\alpha)], \qquad (9.8)$$

where α, $\lambda(\alpha)$, and $\delta(\alpha)$ are defined as in expressions (9.4) to (9.6). For the current example, then, $E(X|\text{censoring at 3}) = (.0478)(3) + .9522[5 + 1.2(.1044)] = 5.024$; and

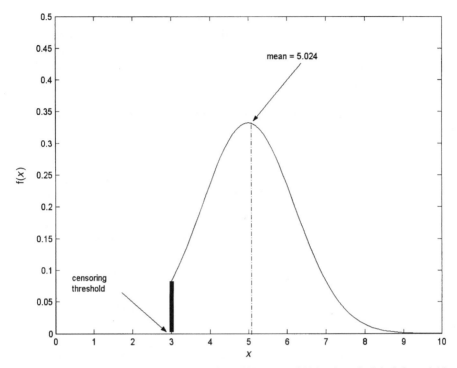

Figure 9.4   Density of the censored normal variable with mean = 5.024 and standard deviation = 1.15.

$V(X | \text{censoring at } 3) = 1.2^2(.9522)[(1 - .185) + (-1.667 - .1044)^2(.0478)] = 1.323,$
in which case $\sigma = 1.15$.

Both truncation and censoring involve limits on the response variable that are based on the values of the response variable itself. Another form of truncation, known as *incidental truncation*, involves limits on the response variable that are imposed by the values of another variable. This other variable can be denoted Z^*. When Z^* is greater than some threshold, c, values of X are observed. Otherwise, they are missing. This is the conceptual underpinning for statistical solutions to the problem of sample-selection bias, that is, the bias in OLS estimates of a population regression in which there is self-selection into the sample (Heckman, 1979). I shall postpone discussion of incidentally truncated data until I have introduced the simulation data below.

Simulation

For the purposes of introducing each of the models in this chapter, I created some simple data that follow the assumptions required for each model. With this approach, I intend to use ideal conditions to illustrate the advantages of truncated, censored, and sample-selected regression models over linear regression estimated via OLS, so that the reader can clearly see how each model is supposed to work. (Subsequently, of course, we will "dirty things up" with real data and consider how these models actually work in practice.) Hence I constructed a data set consisting of 1000 observations with the following variables: a variable ε that is normally distributed with mean 0 and variance 4, a variable u that is normally distributed with mean 0 and variance 1, a variable X that is normally distributed with mean 3 and variance 1.75 but is assumed to be fixed over repeated sampling, a variable W equal to $1 + .75X + u$, and a variable Y equal to $-2 + 1.5X + \varepsilon$. Moreover, ε and u were created so that their joint distribution is bivariate normal, and their correlation, ρ, is .707. Otherwise, ε and u are both uncorrelated with X.

What we essentially have, then, is a random sample of $n = 1000$ observations on two variables of primary interest: X and Y. Moreover, Y follows a linear regression on X with an intercept of -2 and a slope of 1.5. The conditional error, ε, is normally distributed with a mean of zero and a variance of 4. The unconditional mean of Y is $E(Y) = E(-2 + 1.5X + \varepsilon) = -2 + 1.5E(X) + E(\varepsilon) = -2 + 1.5(3) + 0 = 2.5$. The unconditional variance of Y can be recovered by noting that if $Y = -2 + 1.5X + \varepsilon$, then $V(Y) = V(-2 + 1.5X + \varepsilon) = V(-2 + 1.5X) + V(\varepsilon) = (1.5)^2 V(X) + V(\varepsilon) = 2.25$ $(1.75) + 4 = 7.9375$. Moreover, the contribution due to the structural part of the model, the linear predictor, is $2.25(1.75) = 3.9375$, which represents 49.6% of the total variance of Y. Hence, P^2, the population R^2 for the regression, is .496. The regression model for Y is one of the equations we will be trying to estimate in what follows.

To create truncated and censored data, I proceeded as follows. To truncate or censor 40% of the observations on Y, I created a variable $Y^* = Y - c$, where c is the value (1.859) representing the 40th percentile of the sample distribution on Y. This simply sets the truncation point and censoring thresholds at zero rather than at c. Then I created a truncated version of Y, Y_t, by setting Y_t to missing when $Y^* \leq 0$, and setting $Y_t = Y^*$ otherwise. To create a censored version of Y, Y_c, I set Y_c to 0 whenever $Y^* \leq 0$, and I set $Y_c = Y^*$ otherwise. Notice that this changes the underlying regression model slightly. By subtracting 1.859 from both sides of the equation for Y, we see that $Y^* = -3.859 + 1.5X + \varepsilon$. This is the equation we are trying to estimate in the truncated and censored regression examples. In the sample, we have 601 observations on Y_t, with the other 399 observations missing on Y_t. We have 1000 observations on Y_c, but for 399 of them the value of Y_c is zero. For the other 601 observations, Y_c takes on positive values.

Incidentally truncated data were created in a somewhat different manner. First, I chose a new value of c (3.634) that constituted the 60th percentile for the sample distribution of W. I then constructed a variable Z^* equal to $W - 3.634$. In this case, Z^* is less than or equal to zero for 60% of the cases. I then created a dummy variable, Z, equal to 0 whenever $Z^* \leq 0$, and equal to 1 otherwise. I then set Y to missing if $Z = 0$; otherwise, Y is left as is. In the current example, then, only 400 cases have valid scores on Y. The variable Y here is said to be incidentally truncated. I refer to Z^* as the *selection propensity*. When Z^* is above zero, the case is selected into the current sample (of responses) and we observe Y. Otherwise, we do not observe Y for that case, although we do observe X. This is the model employed to understand and correct for self-selection bias (more on this below). Recall that Y has a mean of 2.5 and a variance of 7.9375. What does incidental truncation do to the population mean and variance of the truncated response? As before, I draw on a result presented in

Greene (2003, p. 781) for the moments of the incidentally truncated bivariate normal distribution: *If y and z have a bivariate normal distribution with means* μ_y *and* μ_z, *standard deviations* σ_y *and* σ_z, *and correlation* ρ, *then*

$$E(y|z>a) = \mu_y + \rho\sigma_y\lambda(\alpha_z), \tag{9.9}$$

$$V(y|z>a) = \sigma_y^2[1 - \rho^2\delta(\alpha_z)], \tag{9.10}$$

where

$$\alpha_z = \frac{a - \mu_z}{\sigma_z}, \tag{9.11}$$

$$\lambda(\alpha_z) = \frac{\phi(\alpha_z)}{1 - \Phi(\alpha_z)}, \tag{9.12}$$

$$\delta(\alpha_z) = \lambda(\alpha_z)[\lambda(\alpha_z) - \alpha_z]. \tag{9.13}$$

In the current example, recall that ε and u are distributed as bivariate normal with a correlation of .707. Also, since $Z^* = W - 3.634 = 1 + .75X + u - 3.634$, the equation for Z^* is $-2.634 + .75X + u$. And, of course, the equation for Y is $-2 + 1.5X + \varepsilon$. By theorem, a linear function of a normal random variable is also normally distributed (Hoel et al., 1971), so Z^* and Y, both being linear functions of normally distributed errors, are also normally distributed, with a correlation (ρ) of .852 (found using covariance algebra). This also implies that Z^* and Y are bivariate normally distributed (Hoel et al., 1971). We can therefore apply the theorem above to find the mean and variance of Y, where $Z^* = $ "z" above and $0 = $ "a." Using principles of expectation, along with covariance algebra and some simple arithmetic, we also have $\alpha_z = .273$, $\mu_y = 2.5$, $\sigma_y = 2.817$, $\lambda(\alpha_z) = .98$, and $\delta(\alpha_z) = .692$. Thus,

$$E(Y|Z^*>0) = 2.5 + .852(2.817)(.98) = 4.852,$$

$$V(Y|Z^*>0) = 2.817^2[1 - .852^2(.692)] = 3.949.$$

Once again, we see that truncation has increased the mean, but reduced the variance, of Y. At this point, we have the tools needed to understand the regression models presented in this chapter.

TRUNCATED REGRESSION MODEL

The truncated regression model is based on the following idea. First, we assume that there is an "underlying" regression model for Y in the target population that follows the classic linear regression assumptions. That is, we assume that

$$Y_i = \mathbf{x}^{i\prime}\boldsymbol{\beta} + \varepsilon_i, \tag{9.14}$$

where as in previous chapters, $\mathbf{x}^{i\prime}\boldsymbol{\beta}$ represents $\beta_0 + \beta_1 X_{i1} + \beta_2 X_{i2} + \cdots + \beta_K X_{iK}$, and ε_i is normally distributed with mean zero and variance σ^2. However, the sample is restricted so that we only select cases in which Y is greater than c, where c is some constant. A substantive example would be a study of factors affecting college GPA using a random sample of sociology majors at a particular university. In that one must have a minimum GPA of 3.0 to have a declared major in sociology at this institution, the sample is necessarily restricted to those with GPAs of 3.0 or higher and is therefore not representative of college students in general.

Now, suppose that one were to use the truncated response to estimate the regression model for Y using OLS. What are we actually estimating? First, understand that using a sample drawn from a truncated population implies that what is being estimated is the regression function in the *truncated population*, not the regression function in the general population. Then, note that Y_i is greater than c only if $\mathbf{x}^{i\prime}\boldsymbol{\beta} + \varepsilon_i > c$, or only if $\varepsilon_i > c - \mathbf{x}^{i\prime}\boldsymbol{\beta}$. Therefore, the conditional mean of Y in the truncated population is

$$E(Y_i \mid Y_i > c, \mathbf{x}^i) = E(\mathbf{x}^{i\prime}\boldsymbol{\beta} + \varepsilon_i \mid \varepsilon_i > c - \mathbf{x}^{i\prime}\boldsymbol{\beta})$$

$$= \mathbf{x}^{i\prime}\boldsymbol{\beta} + E(\varepsilon_i \mid \varepsilon_i > c - \mathbf{x}^{i\prime}\boldsymbol{\beta}).$$

To derive $E(\varepsilon_i \mid \varepsilon_i > c - \mathbf{x}^{i\prime}\boldsymbol{\beta})$, we apply equation (9.2), with c equal, in this case, to $c - \mathbf{x}^{i\prime}\boldsymbol{\beta}$, and μ equal to the mean of ε_i, which is zero. By equation (9.2) we have that

$$E(\varepsilon_i \mid \varepsilon_i > c - \mathbf{x}^{i\prime}\boldsymbol{\beta}) = 0 + \sigma\lambda(\alpha_i) = \sigma\lambda(\alpha_i), \tag{9.15}$$

where

$$\alpha_i = \frac{c - \mathbf{x}^{i\prime}\boldsymbol{\beta} - 0}{\sigma} = \frac{c - \mathbf{x}^{i\prime}\boldsymbol{\beta}}{\sigma}$$

and

$$\lambda(\alpha_i) = \frac{\phi(\alpha_i)}{1 - \Phi(\alpha_i)}.$$

Therefore, the conditional mean of Y is

$$E(Y_i \mid Y_i > c, \mathbf{x}^i) = \mathbf{x}^{i\prime}\boldsymbol{\beta} + \sigma\lambda(\alpha_i). \tag{9.16}$$

To answer the previous question, then, equation (9.16) is what we are inadvertently trying to estimate with OLS. But using OLS to estimate this conditional mean will result in biased and inconsistent estimators of $\boldsymbol{\beta}$, due to the term $\lambda(\alpha_i)$, the IMR, in the true model for the conditional mean. In that we have no sample measure of the IMR, using OLS essentially results in omitted-variable bias.

Estimation

In both the censored and sample-selected regression models discussed below, we have $\mathbf{x}^i$ for all observations, so it is possible to estimate $\lambda(\alpha_i)$ consistently using OLS and then to include it in the model as an extra term (see below). However, truncation results in the exclusion of $\mathbf{x}^i$ as well as Y_i for cases in the sample. Therefore, this two-step approach is not possible here. Instead, we resort to estimation via maximum likelihood. Based on the expression in equation (9.1), the likelihood function for the truncated regression model (Breen, 1996) is

$$L(\beta,\sigma \,|\, y_i, \mathbf{x}^i) = \prod_{i=1}^{n} \frac{(1/\sigma)\, \phi[(y_i - \mathbf{x}^{i\prime}\beta)/\sigma]}{1 - \Phi(\alpha_i)} .$$

Maximizing this function with respect to β and σ gives us the MLEs. Unlike OLS estimates, the resulting MLEs are consistent, asymptotically efficient, and asymptotically normally distributed.

Simulated Data Example

Recall the simulation of truncated data discussed above. The underlying regression model is $Y^* = -3.859 + 1.5X + \varepsilon$. Table 9.1 presents the results of estimating this model with OLS using both the full ($n = 1000$) and truncated ($n = 601$) samples. With the full sample, OLS estimates should demonstrate the usual properties of being unbiased and efficient. As is evident from the "OLS: full sample" column, these parameter estimates are quite close to their true values. In particular, the slope is estimated as 1.459 compared to its true value of 1.5, and σ is estimated as 1.996 whereas its true value is 2. The truncated data, on the other hand, result in OLS estimates that are quite far from their true values. In fact, all estimates are too small in magnitude, a general result of using OLS with a truncated response. For example, the slope is estimated as .847 and σ is estimated as 1.551. The next column, headed "MLE: truncated sample," gives the maximum likelihood estimates for the truncated regression model. (LIMDEP was used to estimate truncated regression models for this chapter.) Notice again that these are quite close to the true values: the slope is estimated as 1.438 and σ is estimated as 2.034.

Table 9.1   Regression with Simulated Data, Showing Effects of Truncation and Censoring on Parameter Estimates

Regressor	Parameters	True Values	OLS: Full Sample[a]	OLS: Truncated Sample[b]	MLE: Truncated Sample[b]	OLS: Censored Sample[a]	Tobit Model[a]
Intercept	β_0	−3.859	−3.686	−.556	−3.631	−1.154	−3.717
X	β_1	1.500	1.459	.847	1.438	.881	1.465
	σ	2.000	1.996	1.551	2.034	1.464	2.018
	P^2	.496	.482	.270		.386	.478

[a] $n = 1000$.
[b] $n = 601$.

Table 9.2   Unstandardized OLS and ML Estimates for the Truncated Regression Model of Exam 1 Scores

Regressor	OLS Estimates	ML Estimates
Intercept	23.728*	4.348
College GPA	7.484***	9.437***
Math diagnostic score	.889***	1.164***
Attitude toward statistics	.175	.224
$\hat{\sigma}$	7.833	8.645
R^2	.262	

Note: $n = 149$.

$*\, p < .05.\ **\, p < .01.\ ***\, p < .001.$

Application: Scores on the First Exam

To show an application of truncated regression, I employ an artificial data scenario for heuristic purposes. Using the *students dataset*, I once again (as in Chapter 3) examine the regression of *first exam score*, but this time I only sample students with scores of 70 or better. There are 149 such cases out of the 214 students who took that exam. Normally, there would be no reason to limit the response variable in this manner unless, say, we only had information on explanatory variables for students with grades of at least 70, or a similar constraint. At any rate, Table 9.2 shows the results of regressing *first exam score* on *college GPA, math diagnostic score*, and *attitude toward statistics*, using both OLS and MLE on the truncated sample. Once again, with the exception of the intercept, all parameter estimates from OLS are smaller in magnitude than those from the truncated regression model. Interpretation of the coefficients in either model pertains to the entire population of students taking introductory statistics at BGSU, not just to those with first exam scores above 70. Thus, the coefficient for *math diagnostic score* in the truncated regression model suggests that each unit higher a student scores on the math diagnostic is associated with being 1.164 points higher on the first exam on average, net of the other regressors. The other coefficients are interpreted in a similar fashion.

CENSORED REGRESSION MODEL

To motivate the censored regression model we once again begin with the presumption of an underlying regression model that pertains to the target population. This time we denote the underlying response as Y_i^*. The model for Y_i^* in the target population is

$$Y_i^* = \mathbf{x}^{i\prime}\boldsymbol{\beta} + \varepsilon_i, \tag{9.17}$$

where, as before, ε_i is assumed to be normally distributed with mean 0 and variance σ^2. However, in our sample, the response is censored at a lower threshold of c, where, without loss of generality, c is typically taken to be zero. [See Long (1997) for a discussion of the case in which censoring involves an upper threshold, and for a more general development in which the threshold can be any value. Regardless of the value of c, though, the response can always be rescaled so that the censoring threshold is zero, as was done above in creating the simulated data.] Thus, the observed response, Y_i, is defined as

$$Y_i = \begin{cases} 0 & \text{if } Y_i^* \leq 0, \\ Y_i^* & \text{if } Y_i^* > 0. \end{cases} \qquad \begin{matrix} (9.18) \\ (9.19) \end{matrix}$$

In other words, what we observe are only zero or positive responses, whereas the model of interest, in equation (9.17), pertains to an outcome that can, in theory at least, take on a wider range of values. The task, as with truncated data, is to estimate the parameters of equation (9.17) using incomplete information on the response. A naive approach might be either to attempt to estimate equation (9.17) with all of the observations using OLS, or to estimate equation (9.17) using OLS after limiting the sample to those with positive scores on the response. As we will see below, either approach will result in biased and inconsistent estimates of the conditional mean function.

Social Science Applications

The social science literature is replete with examples of response variables that are treated as censored. An interesting use of the tobit model is Fair's (1978) analysis of the impact of various factors on *leisure time spent in extramarital affairs* among a sample of married women. In one of his analyses, the dependent variable was a measure of the *rate of extramarital sex per length of marriage*, and constructed as [(*number of different extramarital partners × frequency of sexual relations with each*)/*number of years married*]. Those reporting no affairs had values of zero and were treated as censored cases. In this instance, we might define Y* as the *desired rate of extramarital sex*, which is observed in the form of an actual rate only when some threshold value is crossed. Another example is Walton and Ragin's (1990) study of *protest severity* in response to austerity programs imposed in debtor nations to stabilize the economy. In their sample of 60 debtor nations, 26 had positive values on the dependent variable of *protest severity*, an index based largely on journalistic accounts of protest activities in each country. Countries with no recorded austerity protests were assigned a value of 0 on the response. These cases were those whose protests were "... not severe enough to be recorded in international media" (Walton and Ragin, 1990, p. 884).

Mean Functions

Given the model in equations (9.17) to (9.19), let's derive the functions for the conditional mean of Y, given $Y_i^* > 0$, and for the conditional mean of Y among all observations.

Regression Function for $E(Y_i|Y_i^* > 0, \mathbf{x}^i)$. The regression function for the conditional mean of the positive responses is as follows. The model implies that Y is positive only if $Y_i^* = \mathbf{x}^{i\prime}\boldsymbol{\beta} + \varepsilon_i > 0$, which implies that $\varepsilon_i > -\mathbf{x}^{i\prime}\boldsymbol{\beta}$. Thus:

$$E(Y_i|Y_i^* > 0, \mathbf{x}^i) = E(\mathbf{x}^{i\prime}\boldsymbol{\beta} + \varepsilon_i|\varepsilon_i > -\mathbf{x}^{i\prime}\boldsymbol{\beta})$$
$$= \mathbf{x}^{i\prime}\boldsymbol{\beta} + E(\varepsilon_i|\varepsilon_i > -\mathbf{x}^{i\prime}\boldsymbol{\beta}) = \mathbf{x}^{i\prime}\boldsymbol{\beta} + \sigma\lambda(\alpha_i). \tag{9.20}$$

This result follows from applying equation (9.15), where $c = 0$. Notice that this is just the truncated regression model again, since we are conditioning on Y^* being greater than zero. In this case, however, α_i simplifies to

$$\alpha_i = \frac{0 - \mathbf{x}^{i\prime}\boldsymbol{\beta}}{\sigma} = \frac{-\mathbf{x}^{i\prime}\boldsymbol{\beta}}{\sigma}, \tag{9.21}$$

and therefore

$$\lambda(\alpha_i) = \frac{\phi(-\mathbf{x}^{i\prime}\boldsymbol{\beta}/\sigma)}{1 - \Phi(-\mathbf{x}^{i\prime}\boldsymbol{\beta}/\sigma)} = \frac{\phi(\mathbf{x}^{i\prime}\boldsymbol{\beta}/\sigma)}{\Phi(\mathbf{x}^{i\prime}\boldsymbol{\beta}/\sigma)}, \tag{9.22}$$

with the last expression resulting from the symmetry of the normal distribution.

Regression Function for $E(Y_i|\mathbf{x}^i)$. The regression function for Y among all observations can be obtained by realizing that equation (9.17) implies that Y^* is normally distributed with mean $\mathbf{x}^{i\prime}\boldsymbol{\beta}$ and variance σ^2. We can then apply equation (9.7) with "X" equal to Y^* and "c" $= 0$ to arrive at (Greene, 2003)

$$E(Y_i|\mathbf{x}^i) = \Phi\left(\frac{\mathbf{x}^{i\prime}\boldsymbol{\beta}}{\sigma}\right)[\mathbf{x}^{i\prime}\boldsymbol{\beta} + \sigma\lambda(\alpha_i)]. \tag{9.23}$$

This is the *censored regression*, or *tobit* model. Here, again, the symmetry of the normal distribution allows us to write

$$1 - \Phi\left(\frac{-\mathbf{x}^{i\prime}\boldsymbol{\beta}}{\sigma}\right)$$

as

$$\Phi\left(\frac{\mathbf{x}^{i\prime}\boldsymbol{\beta}}{\sigma}\right).$$

It should now be clear that attempts to estimate regression models for Y using OLS with either all of the data or only the cases in which Y is positive will result in biased and inconsistent estimators due to the additional terms shown in equations (9.20) and (9.23).

Estimation

Unlike the case in truncated regression, censored regression models involve measures of the explanatory variables for all cases. With this information it would actually be possible to estimate $\lambda(\alpha_i)$ for each case based on a probit analysis of the probability that

Y^* is uncensored. This estimate could then be substituted for $\lambda(\alpha_i)$ in equation (9.20) so that the model could be estimated with OLS [see Breen (1996) for further details]. This two-step technique is not necessary in this case, however, as the log-likelihood function for the tobit model is quite well-behaved. That function is (Greene, 2003)

$$\ln L(\beta,\sigma \mid y_i, x^i) = \sum_{y_i > 0} -\frac{1}{2}\left[\ln(2\pi) + \ln\sigma^2 + \left(\frac{y_i - x^{i\prime}\beta}{\sigma}\right)^2\right] + \sum_{y_i = 0}\ln\left[1 - \Phi\left(\frac{x^{i\prime}\beta}{\sigma}\right)\right].$$

In this expression, the first sum represents the contribution of the uncensored observations to the log-likelihood, whereas the second sum represents the contribution of the censored observations. Maximizing this function with respect to the parameters (β and σ) of the model results in estimators with the usual asymptotic properties of MLEs (Greene, 2003). It should be mentioned that tobit estimates based on the normal distribution may be inconsistent if the equation disturbance is nonnormal. Estimation under different distributional assumptions for the error term is possible in programs such as LIMDEP or PROC LIFEREG in SAS.

Interpretation of Parameters

The effects of individual regressors can be interpreted with respect to three different mean functions in the tobit model (McDonald and Moffitt, 1980). The first is for the mean of Y^*, the underlying response; the second, shown in expression (9.20), is for the mean of the positive observations; and the third, shown in equation (9.23), is for the mean of *all* of the observed responses, the Y_i.

Based on equation (9.17), the effect of the jth regressor on the mean of Y^* is

$$\frac{\partial}{\partial X_j}E(Y^* \mid x^i) = \beta_j;$$

that is, the regression coefficients have the usual interpretation as the change in the mean of Y^* for a unit increment in X_j, net of the other regressors.

For the second function, the marginal, or partial, effect of X_j is

$$\frac{\partial}{\partial X_j}E(Y_i \mid Y_i^* > 0, x^i) = \beta_j\{1 - z_i\lambda(z_i) - [\lambda(z_i)]^2\}, \tag{9.24}$$

where

$$z_i = \frac{x^{i\prime}\beta}{\sigma},$$

and $\lambda(z_i) = \lambda(\alpha_i)$ is defined in equation (9.22).

The marginal effects for the third function are of the form

$$\frac{\partial}{\partial X_j}E(Y_i \mid x^i) = \Phi(z_i)\beta_j. \tag{9.25}$$

The marginal effects of independent variables shown in equations (9.24) and (9.25) are no longer constant, but depend on the values of all variables in the models, including X_j. The reason for this is that they are functions of z_i, which changes across covariate patterns. Most often, however, these effects would be evaluated at the means of the regressors. That is, for z_i we substitute

$$\bar{z} = \frac{\bar{x}'\hat{\beta}}{\hat{\sigma}}, \tag{9.26}$$

with $\bar{x}$ being the vector of regressor means.

McDonald and Moffitt's Decomposition. McDonald and Moffitt (1980) suggested a useful decomposition for the effect of the jth regressor on $E(Y_i|\mathbf{x}^i)$. They showed that equation (9.25) could be expressed as

$$\frac{\partial}{\partial X_j} E(Y_i|\mathbf{x}^i) = \Phi(z_i) \frac{\partial}{\partial X_j} E(Y_i|Y_i > 0, \mathbf{x}^i) + E(Y_i|Y_i > 0, \mathbf{x}^i) \frac{\partial}{\partial X_j} \Phi(z_i). \tag{9.27}$$

Note that $\Phi(z_i)$ is the probability of being uncensored for the ith case. Thus, equation (9.27) shows that the effect of X_j on the mean of the observed Y_i is a weighted sum of its effect on the mean of the positive observations, weighted by the probability of being uncensored, and its effect on the probability of being uncensored, weighted by the mean of the positive observations. It is therefore possible to partition the effect of X_j into its constituent parts. McDonald and Moffitt further simplified this partition by showing that the proportion of X_j's effect due to its effect on the mean of the positive observations is simply $1 - z_i\lambda(z_i) - [\lambda(z_i)]^2$. Once this is estimated, the proportion of X_j's effect due to its influence on being uncensored is just 1 minus this value. Again, this partition is usually accomplished at the mean of the regressors by substituting $\bar{z}$, defined in equation (9.26), for z_i in this expression. Examples are given in the applications below.

Analog of R^2

A useful measure of discriminatory power for the tobit model has been proposed by Laitila (1993). His pseudo-R^2, denoted R_p^2, is essentially identical to R_{MZ}^2, presented in Chapter 7 in connection with logistic regression. As in that situation, the measure captures the variance in the underlying continuous variable, Y^*, accounted for by the structural part of the model. Denoting the tobit coefficient estimates by b_k, we have

$$R_p^2 = \frac{V\left(\sum b_k X_k\right)}{V\left(\sum b_k X_k\right) + \hat{\sigma}^2},$$

where the numerator is just the variance of the linear predictor, and the denominator is an estimate of the variance of Y^* and consists of the variance of the linear predictor plus the estimated variance of the conditional errors. Laitila (1993) comments that this measure is bounded by 0 and 1 and has an explained-variance interpretation with respect to Y^*. However, it may decrease with the addition of regressors to the model, particularly when they are irrelevant to the response. The calculation of R_p^2 is demonstrated in the applications below.

Alternative Specification

The tobit model imposes a constraint that may or may not always be reasonable. Consider the probability that a case is uncensored. In the tobit model, that probability is

$$P(Y_i > 0 | \mathbf{x}^i) = \Phi\left(\frac{\mathbf{x}^{i\prime}\boldsymbol{\beta}}{\sigma}\right),$$

whereas the mean of the positive values, from expression (9.20), is

$$E(Y_i | Y_i^* > 0, \mathbf{x}^i) = \mathbf{x}^{i\prime}\boldsymbol{\beta} + \sigma\lambda(\alpha_i). \tag{9.28}$$

We see that because $\boldsymbol{\beta}$ is the same in both functions, the coefficients governing whether a case is uncensored are constrained to be proportional to those governing the mean of Y, given that Y is uncensored. That is, $\boldsymbol{\beta}$ is proportional to $\boldsymbol{\beta}/\sigma$, with $1/\sigma$ being the proportionality constant. This precludes some interesting possibilities. For example, suppose that we are investigating the relationship between the age of an offender and the length of sentence given, for a sample of criminal offenders. In that one has to get caught in order to be sentenced, it may be that older offenders are more experienced at crime and therefore less likely to get caught. However, given that they are caught, they may get longer sentences because they are perceived to be more incorrigible than younger offenders. In other words, age may have opposite effects on the likelihood of being sentenced vs. the length of sentence given. The tobit model, on the other hand, forces age to have the same kind of effect on each component of the process.

To free this constraint, Cragg (1971) proposed an alternative model, in which the probability of being uncensored and the mean of the uncensored observations are allowed to be governed by different parameters. In this model, the probability of being uncensored is assumed to follow a probit model with parameter vector $\boldsymbol{\delta}$:

$$P(Y_i > 0 | \mathbf{x}^i) = \Phi(\mathbf{x}^{i\prime}\boldsymbol{\delta}),$$

where since σ and $\boldsymbol{\delta}$ are not separately identifiable in probit, σ is assumed to equal 1. The mean function for the positive responses is then characterized by the truncated regression model in equation (9.28), employing a different parameter vector, $\boldsymbol{\beta}$. If the constraint is imposed that $\boldsymbol{\delta} = \boldsymbol{\beta}/\sigma$, the Cragg specification reduces to the tobit model (Greene, 2003; Smith and Brame, 2003). To estimate the model we simply use a probit model to estimate whether a case is uncensored, using all of the observations. We then estimate a truncated regression model for the conditional mean of the response, using only the uncensored observations. A test statistic for whether the tobit model is valid, compared to the Cragg model, that is, a test statistic for the null hypothesis that $\boldsymbol{\delta} = \boldsymbol{\beta}/\sigma$, is constructed as

$$\chi^2_{(\Delta df)} = -2\ln L_{\text{tobit}} - [-2\ln L_{\text{probit}} + (-2\ln L_{\text{truncated}})],$$

where Δdf is the difference in the number of parameters estimated between the Cragg and tobit models. If this result is significant, the Cragg model is to be preferred. An example is given below.

Simulated Data Example

As was the case with the truncated regression reported above for the simulated data, the underlying model is $Y* = -3.859 + 1.5X + \varepsilon$. The last two columns of Table 9.1 show the results of estimating the underlying model using OLS on all of the observations, compared to using the tobit model. (The tobit model was estimated using PROC LIFEREG in SAS.) As is evident, the OLS estimates are all too low, whereas the tobit estimates, like those for the truncated regession model, are close to the true parameter values. We can apply the McDonald–Moffitt decomposition to partition the effect of X into its effect on the probability of being uncensored and its effect on the conditional mean of the positive responses. For these data, $\bar{x} = 3.024$, so $\bar{z} = [-3.717 + 1.465(3.024)]/2.018 = .353$. Now $\phi(.353) = .375$, whereas $\Phi(.353) = .638$. Hence, $\lambda(.353) = .375/.638 = .588$. The proportion of X's effect that is due to its effect on the conditional mean of the positive Y's is, therefore, $1 - .353(.588) - .588^2 = .447$. So about 45% of X's effect is due to its impact on the conditional mean, while 55% is due to its impact on the probability of being uncensored. Finally, the OLS estimate of P^2 is also an underestimate of the parameter. Laitila's pseudo-R^2, on the other hand, at .478, is fairly close to the true value of .496.

In this particular case, because the data were truly generated by the tobit model, the Cragg specification should be no improvement over tobit. Or, to put it differently, the tobit model should fit no worse than the Cragg model. Minus twice the log-likelihood for the tobit model is 3018.49, while $-2\ln L$ for the probit model of whether a case is uncensored is 974.923, and $-2\ln L$ for the truncated regression model (shown in the column "MLE: truncated sample" in Table 9.1) is 2043.234. The test statistic is therefore a chi-squared variate equal to $3018.49 - (974.923 + 2043.234) = .333$. The degrees of freedom for the test are figured as follows. We estimate two parameters for the probit model (intercept and slope) and three parameters for the truncated regression model (intercept, slope, and σ), whereas we estimate just three parameters in tobit (intercept, slope, and σ). The difference in parameters estimated is $5 - 3 = 2$. As a χ^2 equal to .333 with 2 df is quite insignificant, the hypothesis that $\delta = \beta/\sigma$ would not, in this case, be rejected—a correct decision.

Applications of the Tobit Model

Two response variables that can exhibit considerable censoring are *depressive symptomatology* and the *severity of physical assaults*. In both cases, paper-and-pencil measures may simply not be sensitive enough to capture scores at the lower end of the construct. Therefore, both variables will typically exhibit a large number of zero scores.

Depressive Symptomatology. Table 9.3 presents the results of regressing *depressive symptomatology*—tapped by the Center for Epidemiological Studies Depression Scale

Table 9.3 Unstandardized OLS and ML Estimates for Censored Regression Models of Depressive Symptomatology

Regressor	OLS Estimates	Tobit (ML) Estimates	
		Model 1	Model 2
Intercept	9.842***	7.667***	3.608
Female	2.685	3.371*	−.532
Open disagreement[a]	6.202**	6.753**	
Female × open disagreement	6.157*	6.494*	
Unstable relationship	.807	1.668	1.668
Relationship duration	.083	.077	.077
(Open disagreement + 1 SD)			6.753**
Female × (open disagreement + 1 SD)			6.494*
$\hat{\sigma}$	14.542	16.305	16.305
R^2	.148		
R_p^2		.145	.145

Note: $n = 416$.

[a] Centered variable.

* $p < .05$. ** $p < .01$. *** $p < .001$.

(CESD) (Mirowsky and Ross, 1984)—on *gender* and characteristics of the relationship for the couples in the *couples dataset*. Of the 416 couples, 64, or 15.4%, had scores of zero on the scale and are considered censored cases. The CESD was measured for the main respondent, who could have been either the man or the woman. The purpose of the current analysis was to examine whether relationship conflict has the same effect on depression for women as it does for men. I expect that *open disagreement* will have a stronger effect on depression for women, because they are typically more sensitive than men are to the tenor of the relationship. Tobit estimates for a model with *gender, open disagreement*, and their interaction—plus controls (a dummy for being in an *unstable relationship* and a continuous measure of *relationship duration*)—are shown in the column labeled "model 1." OLS estimates are also shown, for comparison purposes. As is evident, *gender, open disagreement*, and their interaction are all significant. The tobit effects are somewhat larger than those estimated by OLS. As expected, *open disagreement* has a stronger effect on *depressive symptomatology* for women than for men. According to Tobit model 1, for men the effect is simply the main effect of *open disagreement*, 6.753. For women, the effect is 6.753 + 6.494 = 13.247, a substantially larger value.

As *open disagreement* is centered, the main effect of "female" suggests that at average levels of disagreement, women have mean *depressive symptomatology* that is about 3.4 points higher than men's. But is there a gender difference in depression when there is little open disagreement? This question is easily answered by employing targeted centering. One standard deviation of *open disagreement* is .601. Let's calculate the gender difference in *depressive symptomatology* at 1 standard deviation below mean disagreement, or at a value of −.601 for centered *open disagreement*. Subtracting −.601 from *open disagreement* is equivalent to adding .601 to *open disagreement*, hence the new variable is shown as "open disagreement + 1 *SD*" in Table 9.3. We then create the cross-product of *female* × (*open disagreement* + 1 *SD*) and include these variables in the model in place of *open disagreement* and *female* × *open disagreement*. The results are shown in model 2. The main effect of "female" is now the effect of being female among couples who are 1 *SD* below the mean of *open disagreement*. The effect is −.532 and is no longer significant. Apparently, when there is little conflict in the relationship, there is no gender difference in *depressive symptomatology*.

Additional Issues. Laitila's R_p^2 is easily calculated, since the variance of the linear predictor (not shown) is 45.024, and the estimated error variance is $16.305^2 = 265.853$. R_p^2 is thus $45.024/(45.024 + 265.853) = .145$. Using McDonald and Moffitt's decomposition of effects (calculations are left as an exercise for the reader), we find that 46% of regressor effects are due to their impact on the probability of exhibiting any depressive symptomatology, while 54% are due to their influence on mean *depressive symptomatology* for those above the threshold. In other words, slightly more than half of the effects of the explanatory variables is due to their influence on the conditional mean. The test of tobit against the Cragg specification (not shown) in this case is highly significant ($p < .0001$), suggesting that the tobit constraint is not justified for these data. Inspection of the Cragg coefficients (not shown) suggests that they are all substantially larger than those in Table 9.3; however, only the effect of *open disagreement* even approaches significance ($p < .07$) in the Cragg model.

Physical Assault Severity. Table 9.4 presents the results of regressions of *physical assault severity*, measured using the conflict tactics scale (CTS) (Straus, 1979), for

Table 9.4 Unstandardized OLS and ML Estimates for Censored Regression Models of Physical Assault Severity

Regressor	OLS Estimates	Tobit (ML) Estimates
Intercept	91.552***	60.581*
Child abuse index	11.405***	15.013***
Age	−.143	−.329
Education	−9.290**	−12.715**
Nonwhite	−16.642*	−12.670
Previously married[a]	6.355	12.341
Never married[a]	−16.465	−22.328
Income	−3.046*	−2.158
$\hat{\sigma}$	138.559	173.329
R^2	.048	
R_p^2		.049

Note: $n = 1779$.
[a] Currently married is the reference category.
* $p < .05$. ** $p < .01$. *** $p < .001$.

1779 women in the *victims dataset* (NVAWS) who had been the victims of either physical assault, sexual assault, stalking, or threats. Predictors include the *child abuse index* (interval variable coded 1 to 9, with higher scores indicating more severe abuse experienced as a child), *age* (in years), *education* (interval-level measure coded from 1 = "no schooling" to 7 = "postgraduate"), *nonwhite* (dummy for being nonwhite, as opposed to white), *previously married*, *never married* (currently married is the reference category for both dummies), and *annual income* (interval variable coded from 1 = "less than $5,000" to 10 = "over $100,000"). In this example, 30.7% of the cases are censored with values of zero for *physical assault severity*. Both OLS and tobit estimates are shown.

Some effects (e.g., those for *child abuse* and for *education*) are estimated to be larger in magnitude in tobit, while others (e.g., for being *nonwhite* and for *income*) are larger using OLS. The tobit results suggest that *child abuse* enhances *physical assault severity*, whereas *education* reduces it. Aside from these effects, none others are significant. McDonald and Moffitt's decomposition (calculations are again left as an exercise) suggests that 63.5% of regressor effects are due to their influence on reporting any violence, while the other 36.5% are due to their impact on the average level of violence among those reporting it. In this particular case, the Cragg specification could not be estimated, because the truncated part of the model would not converge, even with the tobit estimates as start values. According to Wooldridge (2000), however, one way to evaluate the appropriateness of the tobit specification informally is to compare the $\hat{\delta}_j$ from the probit analysis to $\hat{\beta}_j / \hat{\sigma}$ from tobit. If these are not too different from each other, the tobit model might be justified. Examining the significant effects in this example, we compare effects for *child abuse* (.087 in tobit; .045 in probit) and for *education* ($-.073$ in tobit; $-.04$ in probit). As these are almost twice as large in tobit than in probit, we conclude, again that the tobit specification may not be justified.

SAMPLE-SELECTION MODELS

The sample-selection model consists of two equations. The substantive equation describes the response of interest:

$$Y_i^* = \mathbf{x}^{i\prime}\boldsymbol{\beta} + \varepsilon_i, \tag{9.29}$$

where ε_i is assumed to be normally distributed with mean 0 and variance σ_ε^2. However, whether Y_i^* is actually observed or not depends on a second variable, Z_i^*, the selection propensity, whose equation is

$$Z_i^* = \mathbf{w}^{i\prime}\boldsymbol{\gamma} + u_i, \tag{9.30}$$

where u_i is normally distributed with mean 0 and variance σ_u^2, and $\mathbf{w}$ includes $\mathbf{x}$ as a proper subset; that is, all elements of $\mathbf{x}$ are in $\mathbf{w}$, but $\mathbf{w}$ may contain elements that are not in $\mathbf{x}$ (Wooldridge, 2000). Now Y_i^* is observed only if Z_i^* is greater than zero. That is, the observed variable, Y_i, is such that

$$Y_i \text{ is missing if } Z_i^* \leq 0,$$
$$Y_i = Y_i \text{ if } Z_i^* > 0.$$

Further, the joint distribution of ε and u is bivariate normal, with correlation ρ. The expected value of Y_i is, therefore (Greene, 2003),

$$E(Y_i | Z_i^* > 0) = E(Y_i | u_i > -\mathbf{w}^{i\prime}\boldsymbol{\gamma})$$
$$= \mathbf{x}^{i\prime}\boldsymbol{\beta} + E(\varepsilon_i | u_i > -\mathbf{w}^{i\prime}\boldsymbol{\gamma}) = \mathbf{x}^{i\prime}\boldsymbol{\beta} + \rho\sigma_\varepsilon \lambda(\alpha_u),$$

or

$$E(Y_i | Z_i^* > 0) = \mathbf{x}^{i\prime}\boldsymbol{\beta} + \theta\lambda(\alpha_u), \tag{9.31}$$

where

$$\theta = \rho\sigma_\varepsilon,$$

$$\alpha_u = \frac{-\mathbf{w}^{i\prime}\boldsymbol{\gamma}}{\sigma_u},$$

and

$$\lambda(\alpha_u) = \frac{\phi(\mathbf{w}^{i\prime}\boldsymbol{\gamma}/\sigma_u)}{\Phi(\mathbf{w}^{i\prime}\boldsymbol{\gamma}/\sigma_u)}.$$

Equation (9.31) follows from (9.9) by letting "y" equal ε, "z" equal u, and "a" equal $-\mathbf{w}^{i\prime}\boldsymbol{\gamma}$, and recognizing that μ_ε is zero here. As in previous models, attempts to estimate the mean function for the observed Y using OLS and the variables in $\mathbf{x}$ result in omitted-variable bias, due to omission of the term $\lambda(\alpha_u)$.

Conceptual Framework

The sample-selection model is employed to adjust for nonrandom selection into the current sample. Self-selection into the sample is a pervasive condition in the social sciences (Berk, 1983; Breen, 1996). Individuals selectively choose to respond to surveys, to respond to particular questions in surveys, to be followed up in subsequent waves of a survey, and so on. Such selectivity is a problem to the extent that unobserved elements that determine selection into the sample also affect the substantive outcome, Y_i (more on this below). As an example, DeMaris et al. (2003) tested an integrated model of domestic violence using a sample of couples remaining intact over both waves of the NSFH. In that the authors only included couples with non-missing responses on the questions about intimate violence, their final sample was affected by three different sources of selectivity: panel attrition, couple attrition (through divorce or separation), and item nonresponse. A total of 45% of couples were excluded due to these sources of selectivity. In all likelihood, unobserved character-istics of couples which influenced their refusal to answer items on violence, their failure to remain together as a couple, or their inability to be resurveyed also affect their proclivity for violence. The authors therefore employed a sample-selection model to adjust for this problem.

As another example of selection effects: providing information on particular items may be dependent on individuals' choosing to engage in an activity, a choice that can be considered the product of a selection propensity. For example, studies of earnings necessarily include valid data from only those electing to participate in the labor force. However, the response variable can be considered to be the *hourly wage offer*, which is relevant to all those who are of working age. For those who work, the *hourly wage offer* is the actual wage earned. For those who are not working, the *hourly wage offer* is unobserved (Wooldridge, 2000). Hence, estimates of regressor effects using only working people are likely to be biased. As Wooldridge (2000, p. 558) explains: "Because working may be systematically correlated with unob-servables that affect the wage offer, using only working people . . . might produce biased estimators of the parameters in the wage function." Researchers should be judicious, however, in how they define the target population. Otherwise, virtually every analysis would seem to need correction for sample selectivity. Excluding, for exam-ple, single people from a regression analysis of the propensity to dissolve an exist-ing marital union (e.g., Booth et al., 1984) doesn't result in selection bias provided that it is the population of *currently married couples* that is of interest. As Stolzenberg and Relles (1990, p. 408) observe: ". . . the severity of censoring, and probably the severity of censoring bias, is affected by substantive decisions about the population to which one wishes to draw inferences."

Estimation

Although equation (9.30) posits a continuous selection propensity, in fact, all that is observed is whether or not a case in selected into the current sample. Therefore the model that is actually estimated employs a selection equation based on a binary indi-cator, Z_i:

$$Z_i = \begin{cases} 0 & \text{if } Z_i^* \leq 0, \\ 1 & \text{if } Z_i^* > 0 \end{cases}$$

and the condition for observing the outcome of interest is

$$Y_i \text{ is missing if } Z_i = 0,$$

$$Y_i = Y_i^* \text{ if } Z_i = 1.$$

The selection equation then becomes (based on the reasoning articulated in Chapter 7)

$$P(Z_i = 1 \mid \mathbf{w}^i) = \Phi(\mathbf{w}^{i\prime}\boldsymbol{\gamma}), \tag{9.32}$$

where, due to identifiability requirements, σ_u^2 is now assumed to equal 1. The model can be estimated with maximum likelihood [see Breen (1996) for an expression for the log-likelihood]. An alternative procedure that has also been employed extensively is the Heckman two-step, or *Heckit* (Greene 2003, p. 784) *estimator*. It is constructed as follows. First, estimate (9.32) as a probit model, employing the full sample of cases. Then using $\mathbf{w}^i$ and $\hat{\gamma}$ from that analysis, estimate $\lambda(\alpha_u)$ as

$$\hat{\lambda}(\alpha_u) = \frac{\phi(\mathbf{w}^{i\prime}\hat{\gamma})}{\Phi(\mathbf{w}^{i\prime}\hat{\gamma})}.$$

Finally, estimate equation (9.31) with OLS using only those with nonmissing scores on Y_i and adding $\hat{\lambda}(\alpha_u)$ as a regressor in the equation. This results in a consistent estimate of $\boldsymbol{\beta}$. However, the estimates of σ_ε^2 and coefficient standard errors produced by OLS software are not correct because the error term is heteroscedastic. Greene (2003) provides expressions for $\hat{\sigma}_\varepsilon^2$ and for the correct asymptotic variance–covariance matrix of parameter estimates and has incorporated them into the Heckit procedure in LIMDEP. This procedure also generates an estimate of ρ equal to $\hat{\theta}/\hat{\sigma}_\varepsilon$, but as this is not a sample correlation, the estimate can fall outside the range $[-1, 1]$. When it does, LIMDEP reports it as 1 or -1 (see the example below). It should be noted here that sample-selection models are not limited to probit and linear regression for the selection and substantive equations, respectively. Programs such as LIMDEP offer a variety of alternative specifications for both equations [see, e.g., Greene (1998) for details].

Nuances

Some elements of the sample-selection model are characterized by nuances that need further explanation. First, notice that selection bias comes about because of a nonzero correlation, ρ, between ε and u, since the impact of the omitted term in equation (9.31) is $\theta = \rho\sigma_\varepsilon$, and σ_ε would never be zero. What does this correlation mean? As in other situations involving correlated errors (e.g., seemingly unrelated regression equations or factor-analysis models) ρ represents *a residual correlation between two outcomes that remains after all observable effects have been accounted for*. In this particular case, ρ is the correlation between Z^* and Y^* that is not accounted for by their mutual dependence on a set of observable explanatory variables. As an example, in the simulated data above, $\text{Corr}(Z^*,Y^*) = .852$, while $\rho = .707$. Thus, only $1 - (.707/.852)$, or 17% of the Z^*–Y^* correlation, is due to the explanatory variable, X, while the remainder is due to the correlation between the two disturbance terms. If the correlation between Z^* and Y^* *were* accounted for entirely by the explanatory variables in both equations, ρ would be zero and selection bias would not be a problem. This is tantamount to the situation of *exogenous* selection (Wooldridge, 2000), or selection based on the independent variables, which causes no problems of bias (Wooldridge, 2000). On the other hand, ρ is nonzero to the extent that latent characteristics of observations affect both individuals' propensity to respond and their score on the substantive outcome [see also Berk (1983)] and this is what induces selection bias.

Second, $\mathbf{w}$ and $\mathbf{x}$ can be the same set of regressors, but it is better if we have at least one factor in $\mathbf{w}$ that is not in $\mathbf{x}$. Otherwise, the coefficients in equation (9.31) are identified only because $\hat{\lambda}(\alpha_u)$ is a nonlinear translation of the model's regressors. Even so, having $\mathbf{w}$ and $\mathbf{x}$ be the same regressors causes substantial collinearity problems, since $\mathbf{x}$ and $\hat{\lambda}(\alpha_u)$ will then tend to be highly correlated. Ideally, we want to choose the unique element of $\mathbf{w}$ to be a factor that is unrelated to the outcome.

Third, how is $\lambda(\alpha_u)$ to be interpreted? This term can be considered the *hazard of exclusion* (Berk, 1983). The higher its value, the more a given case possesses characteristics associated with *exclusion* from the sample. Why? Once again, regard Figure 9.3. Remember that as $z = \mathbf{w}^{i\prime}\gamma$ becomes increasingly positive, equation (9.32) tells us that $\Phi(\mathbf{w}^{i\prime}\gamma)$, the probability of being selected into the sample, also increases, and $\phi(\mathbf{w}^{i\prime}\gamma)$ decreases to zero, which means that $\lambda(\alpha_u)$ shrinks toward zero. As $\mathbf{w}^{i\prime}\gamma$ becomes increasingly negative, $\Phi(\mathbf{w}^{i\prime}\gamma)$, the probability of being selected, shrinks toward zero, while $\phi(\mathbf{w}^{i\prime}\gamma)$ also decreases to zero but at a slower rate, which implies that $\lambda(\alpha_u)$ becomes large. Therefore, larger values of $\lambda(\alpha_u)$ reflect a lower probability of inclusion into the sample.

Fourth, θ, the "effect" of the hazard of exclusion, can be misleading, since it is opposite to intuition. For one thing, it should probably be thought of only as an association parameter, since the hazard of exclusion does not actually "cause" Y^*. Additionally, the sign of this effect is the same as the sign of ρ, since σ_ε is always positive. If ρ is positive, for instance, whatever unobserved factors raise the probability of selection also elevate the outcome. A positive "effect" of the hazard of exclusion in this case indicates that the tendency to be *included*—not excluded— is associated with a higher mean outcome. This is a subtle point that can easily cause confusion. As an example, Berk's (1983) exposition of sample selection effects considered potential bias in the regression model for *satisfaction with jury duty* brought about by using only the sample that responded to a mail survey. The effect of $\hat{\lambda}(\alpha_u)$ in the model for *satisfaction with jury duty* was seen to be negative, implying a negative value for ρ. The temptation is to conclude that exclusion was associated with less satisfaction, or that the dissatisfied were less likely to respond. Yet a negative ρ means that, net of observed covariates, the tendency to *respond* was associated with less satisfaction; in other words, the dissatisfied were *more* likely to respond.

Fifth, an examination of the nature of the bias associated with omitting $\lambda(\alpha_u)$ reveals the conditions under which sample selectivity does, and does not, create problems. For simplicity of exposition, let's assume that there is only one regressor, although the principles generalize to any number of regressors. The substantive equation is

$$Y^* = \beta_0 + \beta_1 X + \varepsilon,$$

where ε is normal with zero mean and variance σ_ε^2. The selection equation is

$$Z^* = \gamma X + u,$$

where u is normal with zero mean and variance 1. We define $Z = 1$ if $Z^* > 0$, and $Z = 0$ otherwise. Furthermore, $Y = Y^*$ is observed only when $Z = 1$. The mean of Y is then

$$E(Y \mid X, Z = 1) = \beta_0 + \beta_1 X + \rho \sigma_\varepsilon \lambda,$$

where $\lambda = \phi(\gamma X)/\Phi(\gamma X)$. The equation for Y is then

$$Y = \beta_0 + \beta_1 X + \rho \sigma_\varepsilon \lambda + \upsilon,$$

where $Cov(X, \upsilon) = 0$. If we estimate Y using OLS and omit λ, what is b_1 consistent for? We have

$$b_1 = \frac{cov(X, Y)}{s_x^2}.$$

so

$$\text{plim } b_1 = \frac{Cov(X, Y)}{\sigma_x^2} = \frac{Cov(X, \beta_0 + \beta_1 X + \rho \sigma_\varepsilon \lambda + \upsilon)}{\sigma_x^2}$$

$$= \frac{\beta_1 \sigma_x^2 + \rho \sigma_\varepsilon Cov(X, \lambda)}{\sigma_x^2} = \beta_1 + \frac{\rho \sigma_\varepsilon Cov(X, \lambda)}{\sigma_x^2}. \qquad (9.33)$$

The rightmost expression in equation (9.33) suggests that bias is a function of ρ and the association between X and λ. Most important, both must be nonzero for bias to be a problem. This means that for any particular focus variable, X_k, selection bias is present whenever ρ is nonzero *and* X_k is significantly correlated with λ—demonstrated by X_k having a significant effect in equation (9.32). If one or the other of these conditions fails to hold, selection bias is not a problem for one's analysis, at least in regard to the X in question.

Simulation

Recall the incidentally truncated simulation data discussed above. As noted there, the underlying substantive model is $Y^* = -2 + 1.5X + \varepsilon$, and 60% of the cases were truncated incidentally. In this particular simple example, $\mathbf{w} = \mathbf{x}$, since they are both the same regressor, X. Also, as $\rho = .707$ and $\sigma_\varepsilon = 2$, $\theta = \rho\sigma_\varepsilon = .707(2) = 1.414$. Table 9.5 shows the true values of the parameters, as well as the estimated values based on OLS, the Heckit procedure, and ML estimation of the sample-selection model for the 399 cases with no missing Y values. Only the substantive estimates, not the selection equation estimates, are shown for the sample-selection models. Also, notice the absence of lambda for the ML results. The direct incorporation of ρ and $\mathbf{w}^{i\prime}\gamma$ into the likelihood function obviates the need to include lambda as a separate regressor. In that there was a positive correlation between ε and u, and X had a positive effect on both selection and outcome, what do we expect the nature of the bias of the OLS slope to be? This is

Table 9.5 Regression with Simulated Data, Showing Effects of Incidental Truncation on Parameter Estimates

Regressor	Parameters	True Values	OLS	Heckit	MLE
Intercept	β_0	−2.000	1.623	−3.032	−1.488
X	β_1	1.500	.825	1.644	1.370
Lambda	θ	1.414		2.093	
	σ	2.000	1.769	2.333	2.044
	P^2	.496	.203	.213	
	ρ	.707		.897	.694

Note: n = 399.

a little tricky. We have to remember that X's effect on the hazard of exclusion is *opposite* its effect on the probability of inclusion, which in this case implies a *negative* Cov(X,λ). Hence, by expression (9.33), the OLS b_1 should be an *under*estimate of β_1. This is, in fact, what Table 9.5 shows, as the OLS estimate is .825, whereas the Heckit estimate of 1.644 and the MLE of 1.37 are both much closer to the true value of 1.5. Although the Heckit and ML estimates of the slope are about equally distant from the true parameter value, the ML estimates for β_0, σ, and ρ are considerably better than the Heckit estimates. When possible, ML is to be preferred over the Heckman two-step procedure. However, whereas the Heckit model can always be estimated, the ML technique may at times fail to converge (see the applications below).

Applications of the Sample-Selection Model

I illustrate two applications in this section. In the first, I examine selection effects in a regression of *academic self-esteem at time 2* (T2) on measures of *prior (T1) academic self-esteem, academic achievement*, and controls for a sample of 423 students in introductory sociology at Bowling Green State University (Bradley, 2000). The data were collected during the course of a semester, with the outcome variable measured several weeks after the explanatory variables. A total of 649 students were initially enrolled in the study, but only 423 provided nonmissing data on the response. Hence, 35% of the data were incidentally truncated. For this example, I expect that ρ will be positive: unmeasured correlates of inclusion in the sample (e.g., diligence in attending class and in responding to the survey) should be associated with higher *academic self-esteem.*

The second example, based on the NVAWS, is a regression of *posttraumatic stress disorder symptoms* (PTSD) for a sample of 331 women who had been victimized by their current intimate partner. This is a subset of a larger sample of 1829 women who have ever been victimized by an intimate partner; however, only those currently partnered were asked about PTSD. Because the response is relevant to all women victimized by intimate partners, and because fully 82% of the sample is incidentally truncated, selection effects pose a serious threat to inference. Predictors employed in this analysis are largely measures of the severity of victimization perpetrated by any adult, including the current partner. Again, I expect a positive ρ, since unmeasured factors associated with being in a currently abusive relationship (e.g., growing up in a violent or unstable family) should also be predictive of higher levels of PTSD.

Academic Self-Esteem. Table 9.6 presents the regression models for *T2 academic self-esteem* (interval variable with higher scores reflecting greater academic self-esteem), showing estimates from the Heckman two-step procedure as well as ML, both estimated using LIMDEP, with uncorrected OLS (or, simply, "OLS") for comparison. Predictors include *high school GPA, college GPA, high school grades* (coded in half-point increments from 0 = "mostly F's" to 4 = "mostly A's"), *male* (dummy for being male), *T1 academic self-esteem* (interval variable with higher scores reflecting greater academic self-esteem), and *T1 test anxiety* (interval variable with higher scores reflecting greater test anxiety). As a unique predictor for the selection models, I chose an indicator of whether the student *pays his or her own tuition* as a variable affecting selection, but not *self-esteem*. The reasoning is that students who pay their own tuition are much more likely to attend every class. Since missing data on the response is due primarily to absenteeism, paying one's own tuition should be a predictor of inclusion in the valid sample. There is no reason, on the other hand, why paying one's own tuition would per se boost *self-esteem*.

The only factor significantly related to selection is *college GPA*, with higher GPAs presaging a greater probability of inclusion. Lambda, the hazard of exclusion in the Heckit model, has a positive slope, as expected. Similarly, the ML estimate of

Table 9.6 Unstandardized OLS and ML Estimates of Sample-Selection Models of T2 Academic Self-Esteem

Regressor	Uncorrected OLS[a]	Heckman Two-Step[b] Selection	Heckman Two-Step[b] Response	ML[b] Selection	ML[b] Response
Intercept	10.518***	−.991	−8.002	−1.003	9.473
Pays own tuition		−.080		−.105	
High school GPA	.851	.140	1.830	.141	.907
High school grades	−.306	.061	.235	.060	−.275
College GPA	.873**	.237**	2.644	.240**	.972
Male	.678	.006	.694	.006	.679
T1 academic self-esteem	.591***	−.0003	.585***	.0001	.590***
T1 test anxiety	−.091***	.002	−.074	.002	−.090***
Lambda			14.086		
$\hat{\sigma}$	3.343		10.799		3.366
R^2	.573		.576		
$\hat{\rho}$			1.000		.237

[a] $n = 423$.

[b] $n = 649$.

* $p < .05$. ** $p < .01$. *** $p < .001$.

ρ is .237. However, neither $\hat{\theta}$ nor $\hat{\rho}$ (reported as 1.000 in Heckit) is significant, suggesting that there may not be much of a selection problem in these data. This is also evident in the relative lack of differences between OLS and ML estimates of the parameters (the Heckit estimates are notoriously poor if selectivity is minimal; more about this below). Nonetheless, for instructive purposes, consider the change in the estimate for *college GPA* from the OLS to the ML estimates. Panel (a) of Figure 9.5

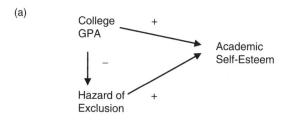

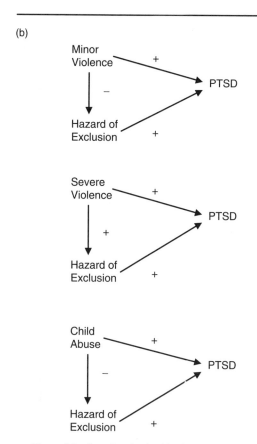

Figure 9.5 Sample-selection bias in the analyses of academic self-esteem and PTSD.

illustrates the effect of having omitted the correction for selectivity in the model. As *college GPA* has a positive effect on *academic self-esteem* and a negative effect on the hazard of exclusion (since it has a positive effect on the probability of *inclusion*), and as the hazard of exclusion has a positive "effect" on *academic self-esteem*, the impact of *college GPA* is underestimated without correcting for selectivity. The effect of *college GPA* therefore goes from .873 in OLS to 2.644 in Heckit and .972 in ML, although neither of the latter estimates is significant. At any rate, as selectivity does not seem to be a problem in this analysis; the OLS estimates would be preferred over the others.

PTSD. Table 9.7 shows OLS and Heckit estimates for the regression of PTSD. Predictors include dummies for having experienced *minor physical assault*, *severe physical assault*, having had to take *time off due to the physical assault*, and having experienced *rape with penetration*, as well as the interval variables *annual income* and the *child abuse index*, as described above. In this particular case, the ML procedure did not converge, a pitfall of this technique, as noted above. The unique predictor of selection chosen here was an indicator of being *married to the partner*. As marriage implies more barriers to the termination of relationships, I would expect those who are married to the current partner to be more likely to be found in currently abusive partnerships. On the other hand, there is no particular reason why marital status per se should affect PTSD. The selection model results reveal that being *married to the partner* is, indeed, predictive of inclusion in the current sample. So are having experienced *minor physical*

Table 9.7   Unstandardized OLS Estimates of Sample-Selection Models of PTSD

Regressor	Uncorrected OLS[a]	Heckman Two-Step[b] Selection	Heckman Two-Step[b] Response
Intercept	31.727***	−1.582***	9.329
Married to partner		.491***	
Minor physical assault	−1.464	.727***	6.793
Severe physical assault	5.657***	−.350***	1.937
Took time off due to assault	6.590***	−.002	5.929**
Child abuse index	1.092**	.060***	1.631***
Rape with penetration	3.453*	−.123	1.312
Income	−.963**	−.027	−1.206***
Lambda			12.741**
$\hat{\sigma}$	13.994		17.631
R^2	.209		.228
$\hat{\rho}$			.723

[a] $n = 331$.

[b] $n = 1829$.

* $p < .05$. ** $p < .01$. *** $p < .001$.

assault and a greater severity of *child abuse*. Interestingly, having experienced *severe physical assault* lowers the probability of inclusion. This is reasonable, since those experiencing severe assault, which was most likely at the hands of an intimate partner, might be especially likely to avoid ending up in another (i.e., a current) abusive liaison.

In this example we find that lambda is also quite significant, and, as expected, $\hat{\rho}$ is positive with a value of .723. These findings suggest that selectivity bias could be a problem in this analysis. In particular, the coefficients for *minor assault, severe assault*, and *child abuse* should all be affected. Panel (b) in Figure 9.5 illustrates the nature of the biases. We should find the effects of *minor physical assault* and *child abuse* to be suppressed by omission of lambda, while the effects of *severe violence* should be confounded with the hazard. A comparison of OLS coefficients with those of the Heckit procedure show that these expectations are borne out. The coefficient for *minor physical assault* increases from -1.464 in OLS to 6.793 in Heckit, although neither coefficient is significant. The coefficient for *child abuse* increases from 1.092 in OLS to 1.631 in Heckit, with both estimates being significant. The coefficient for *severe physical assault*, on the other hand, is reduced to insignificance after controlling for the hazard of exclusion, going from 5.657 in OLS to 1.937 in Heckit. Apparently, the impact of *severe physical assault* is overestimated in OLS when failing to correct for the fact that severely victimized women are more likely to be excluded from the sample, and the hazard of exclusion is associated with a greater mean PTSD.

Caveats Regarding Heckman's Two-Step Procedure

Stolzenberg and Relles's (1990, 1997) simulations compare the Heckman procedure to uncorrected OLS under a range of conditions. They find that, *on average*, the Heckman procedure performs no better than uncorrected OLS. However, Heckit appears to reduce bias consistently when two conditions are simultaneously met: (1) ρ is very high and (2) **x** and **w** are very highly correlated (Stolzenberg and Relles, 1990). They note further that substantive equations with high R^2's can tolerate considerable sample selectivity without showing much bias; and if the selection model itself exhibits poor discriminatory power, bias is also likely to be minimal. In sum, they suggest that if bias is very severe and the sample is large, Heckit probably improves the estimates. But if bias is only moderate, or if samples have "only a few hundred cases," there is substantial risk that the Heckman procedure will make the estimates worse (Stolzenberg and Relles, 1997, p. 503). One way to approach suspected selection bias is to estimate the sample-selection model with ML if possible, or with the Heckit procedure if not. If one's focus variables are significant predictors of inclusion in the sample, *and* if $\hat{\theta}$ (Heckit) or $\hat{\rho}$ (ML) are significant, sample-selection corrections are appropriate. If these conditions do not obtain, correction for selectivity may not be warranted. Moreover, when using the Heckit approach, the test of $\hat{\theta}$ should employ the asymptotically correct standard error found in programs such as LIMDEP or STATA rather than relying on the t test in standard OLS software.

EXERCISES

9.1 Suppose that Y is normally distributed with mean 12 and standard deviation 4. If the distribution is truncated at $y = 8$, what are the mean and standard deviation of the truncated distribution?

9.2. Suppose that Y is normally distributed with mean 12 and standard deviation 4. If the distribution is censored at $y = 8$, what are the mean and standard deviation of the censored variable?

9.3 Suppose that y and z have a bivariate normal distribution with $\mu_y = 12$, $\mu_z = 8$, $\sigma_y = 4$, $\sigma_z = 2$, and $\rho = .5$. Suppose further that y is observed only when $z > 5$; that is, the distribution of y is incidentally truncated at $z = 5$. What are the mean and standard deviation for the incidentally truncated distribution of y?

9.4 Note that the inverse Mills ratio is expressed as $\lambda(z) = \phi(z)/[1 - \Phi(z)]$. However, for $-z$ we have the IMR as

$$\lambda(-z) = \frac{\phi(-z)}{1 - \Phi(-z)} = \frac{\phi(z)}{\Phi(z)} = \lambda(\alpha),$$

where $\alpha = -z$, which is the form it takes in the censored (from below) and sample-selected regression models discussed in this chapter. [Note that $\phi(z)/\Phi(z)$ in certain contexts is also denoted simply as "λ".] Illustrate the practical range of the IMR (in this latter form) by evaluating it at z equal to -4, 0, and 4.

9.5 Using equation (9.16) and the ML estimates in Table 9.2 for the truncated regression model of *exam 1 score*, give the predicted mean *exam 1 score* for a student in the population of students scoring at least 70 on *exam 1*, with a *college GPA* of 3.2, a *math diagnostic score* of 43, and an *attitude* score of 10.

9.6 Using equation (9.16) and the ML estimates in Table 9.2 for the truncated regression model of *exam 1 score*, give the predicted mean *exam 1 score* for a student, in the population of students scoring at least 70 on exam 1, with a *college GPA* of 2.5, a *math diagnostic score* of 37, and an *attitude* score of -2.

9.7 Based on equation (9.23) and the estimates in model 1 of Table 9.3, give the predicted observed CESD score for a woman 1 standard deviation below the mean on *open disagreement*, in an *unstable relationship*, with a *relationship duration* of 12 years.

9.8 Based on equation (9.23) and the tobit estimates in Table 9.4, give the predicted observed *physical assault severity* score for a 35-year-old woman with *child abuse index* = 4, who has an *education* score of 5, is married and white, and whose *income* score is 2.

9.9 Means for the regressors used in Table 9.3 are *open disagreement* = 0 (centered variable), *female* = .553, *unstable relationship* = .361, and *relationship* *duration* = 16.155. Verify the McDonald–Moffitt decomposition of regressor effects for this application, as discussed in the text.

9.10 Means for the regressors used in Table 9.4 are *child abuse index* = 1.776, *age* = 39.868, *education* = 4.786, *nonwhite* = .201, *previously married* = .334, *never married* = .153, *income* = 4.031. Verify the McDonald–Moffitt decomposition of regressor effects for this application, as discussed in the text.

Exercises 9.11 to 9.13 involve the following (x,y) pairs: (1.5, 0), (2.2, 0), (1.8, 0), (3.5, 1.25), (.82, 3.42), (3.85, 8.65). Also, tobit estimates for the regression of y on x yield $\hat{\beta}_0 = -4.85827$, $\hat{\beta}_1 = 2.24803$, $\hat{\sigma} = 4.53832$. Denote this information as Data One.

9.11 Using Data One, calculate Laitila's R_p^2.

9.12 Using Data One, calculate $[\text{corr}(y, \hat{y})]^2$, where $\hat{y}$ is the predicted value for the observed y's, based on equation (9.23). Notice that this R^2 analog is for the observed y's, whereas R_p^2 is for Y^*.

9.13 Give the McDonald–Moffitt decomposition for the impact of x on y in Data One.

Exercises 9.14 to 9.16 involve the following (x,y) pairs: (8, 0), (6.2, 0), (7, .4), (5.5, .55), (5.8, .62), (2.3, .89). Also, tobit estimates for the regression of y on x yield $\hat{\beta}_0 = 1.38552$, $\hat{\beta}_1 = -.17864$, $\hat{\sigma} = .26422$. Denote this information as Data Two.

9.14 Using Data Two, calculate Laitila's R_p^2.

9.15 Using Data Two, calculate $[\text{corr}(y, \hat{y})]^2$, where $\hat{y}$ is the predicted value for the observed y's, based on equation (9.23).

9.16 Give the McDonald–Moffitt decomposition for the impact of x on y in Data Two.

9.17 In the analyses reported in Table 9.7, PTSD was treated as an uncensored, continuous response. Ignoring, for the moment, the problem of incidental truncation, suppose that we treat PTSD as censored at its minimum value of 21, noting that 29.3% of women score at this value. A tobit analysis for $n = 330$ women produces the following results:

	$\hat{\beta}$	Mean of X
Intercept	29.708	
Minor physical assault	−3.390	.924
Severe physical assault	6.939**	.455
Took time off due to assault	8.472***	.242
Child abuse index	1.244**	2.139
Rape with penetration	4.393*	.333
Income	−1.275**	3.933
$\hat{\sigma}$	17.531	

(a) Give the predicted observed PTSD score for a woman who has experienced *minor* and *severe physical assault, rape with penetration*, and *taking time off because of the physical assault*, and who has average *income* and *child abuse* scores.

(b) To what extent is the effect of *child abuse* due to its effect on elevating average PTSD for those above the censoring threshold, as opposed to its effect on raising the risk of reporting any PTSD symptoms at all?

9.18 Suppose that $y^* = \beta_0 + \beta_1 x_1 + \beta_2 x_2 + \beta_3 x_3 + \beta_4 x_4 + \varepsilon$ and $z^* = \gamma_0 + \gamma_1 x_1 + \gamma_2 x_2 + \gamma_3 x_3 + \gamma_4 x_4 + \upsilon$, where $\mathrm{Cov}(x_j,\varepsilon) = \mathrm{Cov}(x_j,\upsilon) = \mathrm{Cov}(\varepsilon,\upsilon) = 0$, all variables are standardized, all x's are positively correlated, and all parameters except β_3, β_4, γ_3, and γ_4 are positive, with the latter being negative. But you estimate $y^* = \beta_0 + \beta_1 x_1 + \beta_2 x_2 + e$ and $z^* = \gamma_0 + \gamma_1 x_1 + \gamma_2 x_2 + u$. What is ρ_{eu}? [*Hint*: Compute $\mathrm{corr}(e,u) = \mathrm{cov}(e,u)$ using covariance algebra, and evaluate the sign of the resulting expression.]

9.19 Suppose that $y^* = \beta_0 + \beta_1 x_1 + \beta_2 x_2 + \beta_3 x_3 + \varepsilon$ and $z^* = \gamma_0 + \gamma_1 x_1 + \gamma_2 x_2 + \gamma_3 x_3 + \upsilon$, where $\mathrm{Cov}(x_j,\varepsilon) = \mathrm{Cov}(x_j,\upsilon) = \mathrm{Cov}(\varepsilon,\upsilon) = 0$, all variables are standardized, all x's are positively correlated, and all parameter values are positive. But you estimate $y^* = \beta_0 + \beta_1 x_1 + \beta_2 x_2 + e$ and $z^* = \gamma_0 + \gamma_1 x_1 + \gamma_2 x_2 + u$. What is ρ_{eu}? (*Hint*: Follow procedure for Exercise 9.18.)

9.20 Refer to equation (9.33) and the accompanying discussion of sample-selection bias. Suppose that the models for y^*, z^* are $y^* = 2 + 3.2x + \varepsilon$ if $z^* > 0$; y^* is missing otherwise; $z^* = .8x + \upsilon$. Also, $\sigma^2_{y^*} = 17.64$, the squared correlation of y^* with x is .726, $\sigma_x = 1.25$, $\rho = .66$, and $\mathrm{E}(\lambda) = 3 - 1.5x$. What is the bias in b_1 as an estimate of β_1 if no correction is made for incidental truncation, where bias(b_1) is defined in Exercise 9.21? Will b_1 over- or underestimate β_1? [*Hint*: In general, in the simple linear regression of y on x, $\beta_1 = \mathrm{Cov}(x,y)/\sigma^2_x$.]

9.21 Refer to equation (9.33) and the accompanying discussion of sample-selection bias. Suppose that the models for y^* and z^* are: $y^* = 2 + 3.2x + \varepsilon$ if $z^* > 0$; y^* is missing otherwise; $z^* = .8x + \upsilon$. Also, $\sigma_\varepsilon = 2.2$, $\sigma_x = 1.25$, $\rho = .66$, and $\mathrm{Cov}(x,\lambda) = .75$. What is the bias in b_1 as an estimate of β_1 if no correction is made for incidental truncation [where bias$(b_1) = \mathrm{plim}\, b_1 - \beta_1$]? Will b_1 over- or underestimate β_1? (*Hint*: See the hint for Exercise 9.20.)

9.22 Use the female subsample of the *couples dataset* to estimate a tobit model for CESD as a function of FEDUC, IHTOT2, VIOLENT, UNSTABLE, DISAGMT, and PRESCHDN. Then:

(a) Interpret the significant effects.

(b) Give R^2_p for the underlying *depressive symptomatology* score.

(c) Give the McDonald–Moffitt decomposition of regressor effects.

(d) Give the predicted value of observed CESD at the mean of the regressors.

9.23 In the *students dataset*, 15.7% of the 235 students are missing EXAM2 scores because they dropped out of the course before the second exam. If the target population is all students enrolling in introductory statistics at BGSU, a regression of EXAM2 using only those with valid scores could be biased by self-selection. Estimate a sample-selection model for EXAM2 regressed on EXAM1, COLGPA, SCORE, and STATMOOD using either the ML or Heckit procedure, along with uncorrected OLS. Use CLASSIF as the unique regressor for the selection equation. Missing imputation: Substitute the parenthetical values for missing data on each variable indicated: EXAM1 (76.9774766), COLGPA (3.0827835), and SCORE (40.9358974). Evaluate the extent and nature of any selection bias found.

9.24 Use the *GSS98 dataset* to estimate the regression of INCOME on EDUCAT, RESPAGE, and MALE using OLS. In that 34.7% of the 2832 respondents are missing on INCOME, correct for selection bias with either the ML or Heckit procedure, employing CONSERV as the unique regressor for the selection equation. Missing imputation: Substitute the parenthetical values for missing data on each variable indicated: EDUCAT (13.251) and CONSERV (4.098). Evaluate the extent and nature of any selection bias found.

9.25 Use the *introductory sociology dataset* to estimate the regression of *T2 test anxiety* (TESTANX2) on HSGPA, HSGRADE, COLLGPA, MALE, ACADSE1, and TESTANX1. Because 36.5% of the 649 students are missing on TESTANX2, correct for selection bias with either the ML or Heckit procedure, employing PAYOWN as the unique regressor for the selection equation. Evaluate the extent and nature of any selection bias found.

CHAPTER 10

Regression Models for an Event Count

CHAPTER OVERVIEW

In this chapter we return to the study of discrete response variables, in particular, count responses, variables that represent the number of occurrences of some event. The number of events that occur to each case in a given observation period, or in a given observation space, is assumed to be governed by an underlying rate of event occurrence. In this chapter we focus on the outcome of this process, the count of events over the period or the space. In Chapter 11 we focus on the waiting time from the beginning of risk for an event *until* event occurrence, or the survival time in the "nonevent" state. These chapters are therefore tied together by their common reliance on an unobserved continuous rate of event occurrence. Examples of count variables are *the number of articles published during graduate school* by male and female biochemists (Long, 1990), *the number of domestic violence incidents* reported for a given offender in a 6- to 22-month follow-up of domestic violence offenders (Sherman et al., 1992), or *the number of police contacts* for a sample of juveniles aged 8 to 26 (Land et al., 1996).

I begin by defining count data and presenting probability distributions that are commonly associated with count responses. I then consider why OLS is not optimal for these types of data, and present instead the Poisson regression model. The likelihood function for the model is illustrated, along with schemes for interpreting the coefficients, and analogs of R^2. Truncated, censored, and sample-selected variants of the model are also discussed. Due to the restrictive property of Poisson regression that the conditional mean and variance of the dependent variable be identical, the Poisson model is rarely adequate for count data, which are frequently

Regression with Social Data: Modeling Continuous and Limited Response Variables,
By Alfred DeMaris
ISBN 0-471-22337-9   Copyright © 2004 John Wiley & Sons, Inc.

overdispersed; that is, the variance exceeds the mean. For this reason, I then consider a few variations on the Poisson model. One is the negative binomial regression model, which is appropriate when data are generally overdispersed. I also consider a version of this model designed for truncated count distributions. Finally, for cases in which overdispersion is due to an excessive number of zero counts, I consider zero-inflated Poisson and negative binomial models, as well as the hurdle model.

DENSITIES FOR COUNT RESPONSES

An *event count* is *the number of occurrences of an event within a fixed domain of observation* (King, 1988). The domain can either be temporal or spatial. Hence, the number of partners one has had sex with in the past month is an event count, where the event is having a sex partner, and the domain is one month. From above, the number of domestic violence incidents in a 6- to 22-month period is an event count in which the event is the incident and the domain varies between 6 and 22 months. The number of taverns per city in a given state is, similarly, an event count, with the event being the occurrence of a tavern and the domain being city size, most likely expressed as the number of inhabitants. An example of an event count from the *students dataset* is the *number of previous college-level math courses* taken by each student. In this case, the domain is assumed to be the student's college career. This variable, for 230 students, ranges from 0 to 7, has a sample mean of 1.274, and a sample variance of 1.274. The sample distribution on the variable is shown in Figure 10.1.

We notice right away that the distribution is right-skewed and that there is a sizable proportion of students, about 24%, who have had no previous math classes. These features are fairly typical of count data. Often, our first impulse with this type of response might be to use linear regression with OLS estimation. For reasons to be detailed below, this would be a poor choice of estimator. Another strategy might be to collapse the variable into a dichotomy, contrasting taking one or fewer classes with taking more than one class. One might then use logistic regression to analyze this binary version of the variable. But this would be quite wasteful of information, since we would be treating 0 or 1 class as the same answer, and we would also be lumping two through seven classes into the same category. Fortunately, neither strategy is necessary, as count variables are well represented by certain specific discrete densities, which can then be used as the basis of a likelihood function for a regression model.

Poisson Density

For a random variable, Y, the Poisson density is

$$f(y \mid \mu) = \frac{e^{-\mu}\mu^{y}}{y!}$$

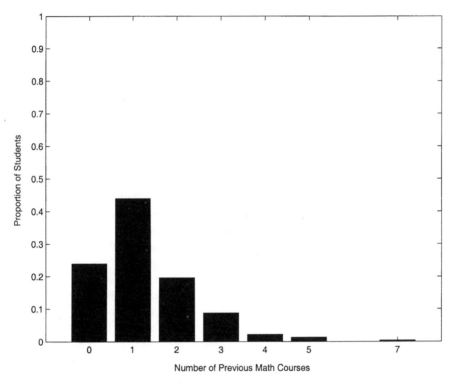

Figure 10.1   Distribution on number of previous math courses for 230 students in introductory statistics (mean = 1.2739; variance = 1.2740).

for $y = 0, 1, 2, \ldots$. The parameter μ is both the mean and the variance of Y. Figure 10.2 shows a Poisson density with mean and variance equal to $\mu = 1.274$. Notice that it resembles the sample distribution of number of previous math courses in Figure 10.1. The probability of a given value for Y is readily computed using this density. For example, the probability that $Y = 5$, according to this density, is

$$f(5 \mid \mu = 1.274) = \frac{e^{-1.274}1.274^5}{5!} = .0078.$$

The sample proportion of students having five previous math classes, on the other hand, is .022. The Poisson density is constrained so that its variance equals its mean, a property called *equidispersion* (Long, 1997). The next density relaxes that restriction.

Negative Binomial Density

A second density that is very important for count data is the negative binomial density. This density can be expressed in a variety of different ways (see, e.g., Hoel et al., 1971). Moreover, Cameron and Trivedi (1998) note that there are 13 distinct

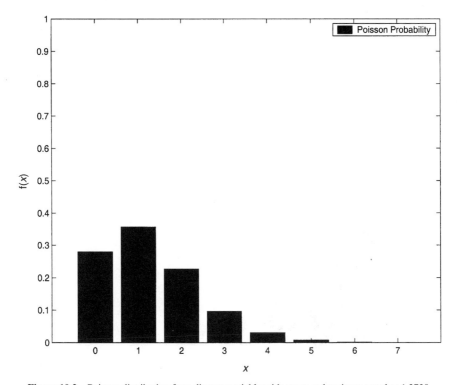

Figure 10.2   Poisson distribution for a discrete variable with mean and variance equal to 1.2739.

stochastic mechanisms for generating it. One such mechanism is as follows. Define
r to be an integer and p to be a number between 0 and 1. Then the negative binomial
density arises if we let y be the number of failures encountered before achieving the
rth success in a series of trials, where p is the probability of a success:

$$f(y \mid r, p) = \frac{(r+y-1)!}{y!\,(r-1)!} p^r (1-p)^y \tag{10.1}$$

for $y = 0, 1, 2, \ldots$ (Hoel et al., 1971). The negative binomial density is said to be
overdispersed, meaning that the variance exceeds the mean. This is evident, since for
the negative binomial density,

$$E(y) = r\frac{1-p}{p} \quad \text{and} \quad V(y) = r\frac{1-p}{p^2}.$$

In that, for p between 0 and 1, $p^2 < p$, the variance is larger than the mean.
Figure 10.3 shows a negative binomial distribution with parameters $r = 2$ and
$p = .61089$. As the reader can verify, the distribution again has a mean of 1.274, but
this time its variance is 2.0853. For this distribution, the probability that $Y = 5$ is

$$f(5 \mid r = 2, p = .61089) = \frac{(2+5-1)!}{5!\,(2-1)!} .61089^2 (1 - .61089)^5 = .02.$$

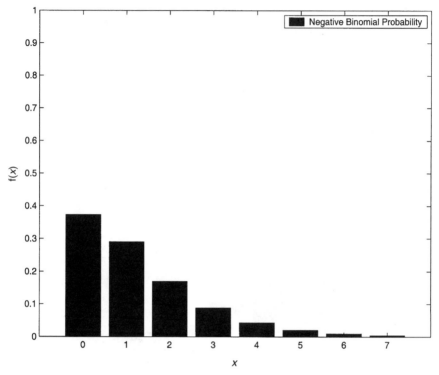

Figure 10.3 Negative binomial distribution for a discrete variable with mean equal to 1.2739 and variance equal to 2.0853.

MODELING COUNT RESPONSES WITH POISSON REGRESSION

Problems with OLS

With Y as an event count, the linear regression model is $Y = \sum \beta_k X_k + \varepsilon$. There are several problems with this formulation, especially when using OLS estimation. First, the usual assumption that the errors are normally distributed fails, since Y is typically nonnormal, as is evident in Figure 10.1. With large samples, however, the parameter estimates are approximately normally distributed, by the CLT, so inferences regarding the regression parameters are still possible. Nevertheless, there are more serious difficulties with using a linear function. The mean structure—the nature of the function relating the conditional mean to the linear predictor—is probably misspecified with a linear model, given that a negative mean is not possible. Recall that as mentioned in Chapter 7, the right-hand side of the linear regression equation is not constrained to be positive. If the mean structure is misspecified, OLS estimators are inconsistent (Cameron and Trivedi, 1998). The OLS estimator also assumes a homoscedastic error structure. This is problematic if the data are generated by a count distribution such as the Poisson or negative binomial. In such distributions, the variance is a function of the mean, and as such, varies as the mean varies. If the errors are really heteroscedastic, the standard error estimates produced by OLS are biased.

Poisson Regression Model

The *Poisson regression model* (PRM) assumes that the occurrence of events within a given domain is governed by an unobserved, continuous process. Although the process is unobserved, its end product—an event count over the domain of interest—is observed. The process is encapsulated in μ_i, the rate of event occurrence for the ith case. The model makes two assumptions about this process (King, 1989). The *independence assumption* is that the probability of a subsequent event is independent of the occurrence of a previous event. (*Contagion* is the situation in which the probability of a subsequent event is enhanced by a previous event, whereas *negative contagion* obtains when this probability is reduced by a previous event.) The *homogeneity assumption* is that the rate of event occurrence is constant over the domain of interest, which is usually time. (In Chapter 11 we relax this particular assumption about event processes.)

The Poisson model assumes that the number of events for the ith case follows a Poisson distribution with parameter μ_i, and that μ_i is an exponential function of the covariates. Hence, the density for Y_i is

$$f(y_i \mid \mathbf{x}^i, \boldsymbol{\beta}) = \frac{e^{-\mu_i}\mu_i^{y_i}}{y_i!}$$

for $y_i = 0, 1, 2, \ldots$ and the specification for μ_i is $\mu_i = \exp(\sum \beta_k X_{ik})$. The rate of event occurrence, μ_i, is conditional on the covariates for each case. Hence, μ_i is both the conditional mean and the conditional variance of Y_i. The exponential specification ensures that the mean is always positive. Also, recall from Chapter 1 that in the generalized linear model, if the response has a Poisson distribution, the link function is the log of the mean, which implies an exponential function for the mean itself [i.e., $\ln \mu_i = \sum \beta_k X_{ik}$ implies that $\mu_i = \exp(\sum \beta_k X_{ik})$].

If the domain for the event process is not the same size for all cases, this needs to be controlled, since a greater number of events would be expected the larger the domain size. The *domain size* is the length of time that cases are at risk for events, or the size of the geographical area over which events are counted, and so on. With varying domain sizes, the solution is to model the expected count *per domain size* (Cameron and Trivedi, 1998; King, 1988). If E_i represents the *exposure*, or domain size, for the ith case, the model becomes

$$\frac{\mu_i}{E_i} = \exp\left(\sum_{k=1}^{K} \beta_k X_{ik}\right),$$

which implies that

$$\mu_i = E_i \exp\left(\sum_{k=1}^{K} \beta_k X_{ik}\right) = \exp(\ln E_i) \exp\left(\sum_{k=1}^{K} \beta_k X_{ik}\right)$$

$$= \exp\left[\left(\sum_{k=1}^{K} \beta_k X_{ik}\right) + \ln E_i\right]. \tag{10.2}$$

This suggests that the log of E_i should be entered into the model with its coefficient constrained to 1. In some software (e.g., SAS) E_i is referred to as an *offset* when it is to be entered into the model in this way. However, as others have noted (King, 1988), there is no harm in allowing the coefficient for ln E_i to be estimated. In fact, in the author's experience, the model is less likely to have convergence problems if this parameter is estimated rather than constrained.

Estimation. The parameters of the model—the β's—can be estimated via maximum likelihood. The likelihood function follows from the joint density of the observed y_i. Letting **y** represent the vector, or set, of observed event counts, **x** represent the vector of explanatory variables (including, possibly, ln E_i), and β represent the vector of parameters, we have

$$L(\beta \mid y, x) = \prod_{i=1}^{n} \frac{e^{-\mu_i} \mu_i^{y_i}}{y_i!},$$

which implies that the log-likelihood is

$$\ln[L(\beta \mid y, x)] = \sum_{i=1}^{n} (y_i \ln \mu_i - \mu_i - \ln y_i!) = \sum_{i=1}^{n} \left[y_i \left(\sum_{k=0}^{K} \beta_k X_{ik} \right) - \exp\left(\sum_{k=0}^{K} \beta_k X_{ik} \right) - \ln y_i! \right]$$

The MLEs of the betas are the b_k that maximize this function. As with all MLEs, the resulting estimates are consistent as well as asymptotically unbiased, efficient, and normally distributed.

Inferences in the Poisson Regression Model. As in logistic regression, the test for the significance of the model as a whole is performed with the model chi-squared statistic. The null hypothesis is $H_0: \beta_1 = \beta_2 = \cdots = \beta_K = 0$. The alternative hypothesis is that at least one of these betas is not zero. The test statistic is $\chi^2 = -2 \ln L_0 - (-2 \ln L_1)$, where L_0 is the likelihood function for a model with only an intercept and L_1 is the likelihood function for the hypothesized model, evaluated at the MLEs of the parameters. Under the null hypothesis, this statistic has a chi-squared distribution with K degrees of freedom. Comparison of nested models is achieved with the nested chi-squared test. If model 2 is nested within model 1, the test for whether the nesting constraints are valid is $\Delta\chi^2 = -2 \ln L_2 - (-2 \ln L_1)$, with L_2 the likelihood for the nested model, evaluated at its MLEs. Under the null hypothesis that the constraints are valid, this statistic has the chi-squared distribution with degrees of freedom equal to the number of constraints imposed to produce the nested model. Finally, individual regression coefficients can be tested with z-tests of the form $z = b_k / \hat{\sigma}_{b_k}$.

Interpretation of Regression Coefficients. As in logistic regression, the betas no longer can be interpreted as the change in the mean for a unit increase in the predictors, due to the nonlinear functional form of the model. This is easily seen by attempting to simplify the expression for the change in the mean when one of the regressors, say X_j, increases by 1 unit, holding all other $(K-1)$ regressors constant.

Letting $\mathbf{x}_{-j}$ be the collection of all other regressors:

$$E(Y \mid x_j + 1, \mathbf{x}_{-j}) - E(Y \mid x_j, \mathbf{x}_{-j}) = \exp[\beta_0 + \beta_1 X_1 + \cdots + \beta_j(x_j + 1) + \cdots + \beta_K X_K]$$
$$- \exp(\beta_0 + \beta_1 X_1 + \cdots + \beta_j x_j + \cdots + \beta_K X_K).$$

It should be clear that this expression does not simplify further. The partial derivative of μ_i with respect to X_j, on the other hand, is

$$\frac{\partial \mu_i}{\partial X_j} = \frac{\partial}{\partial X_j}[\exp(\beta_0 + \beta_1 X_{i1} + \cdots + \beta_j X_{ij} + \cdots + \beta_K X_{iK})]$$
$$= \beta_j \exp(\beta_0 + \beta_1 X_{i1} + \cdots + \beta_j X_{ij} + \cdots + \beta_K X_{iK}) = \beta_j \mu_i. \quad (10.3)$$

This shows that the effect of the jth regressor is not constant, but rather, depends on the levels of all covariates, since $\beta_j \mu_i$ changes with μ_i. However, if we divide expression (10.3) by μ_i, we can isolate β_j. As the partial derivative is the change in the response for an infinitesimal increase in the predictor at a given predictor value, β_j can be interpreted as the proportional change in μ_i (as a proportion of μ_i) with an infinitesimal increase in X_j at x_j.

A more appealing interpretation can be found by exponentiating β_j. Called the *factor change* by Long (1997, p. 225), $\exp(\beta_j)$ is the multiplicative change in the expected count for each unit increase in X_j, net of the other regressors. To see this, consider the ratio of expected counts for those who are 1 unit apart on X_j:

$$\frac{E(Y \mid x_j + 1, \mathbf{x}_{-j})}{E(Y \mid x_j, \mathbf{x}_{-j})} = \frac{\exp(\beta_0 + \beta_1 X_1 + \cdots + \beta_j(x_j + 1) + \cdots + \beta_K X_K)}{\exp(\beta_0 + \beta_1 X_1 + \cdots + \beta_j x_j + \cdots + \beta_K X_K)}$$

$$= \frac{\exp(\beta_0) \exp(\beta_1 X_1) \cdots \exp(\beta_j x_j) \exp(\beta_j) \cdots \exp(\beta_K X_K)}{\exp(\beta_0) \exp(\beta_1 X_1) \cdots \exp(\beta_j x_j) \cdots \exp(\beta_K X_K)}$$

$$= \exp(\beta_j).$$

In other words, $E(Y \mid x_j + 1, \mathbf{x}_{-j}) = \exp(\beta_j) E(Y \mid x_j, \mathbf{x}_{-j})$, which shows that a 1-unit increase in x_j multiplies the expected count by $\exp(\beta_j)$, controlling for the other regressors. The *proportionate* change in the expected count for a unit increase in X_j is

$$\frac{E(Y \mid x_j + 1, \mathbf{x}_{-j}) - E(Y \mid x_j, \mathbf{x}_{-j})}{E(Y \mid x_j, \mathbf{x}_{-j})} = \frac{E(Y \mid x_j + 1, \mathbf{x}_{-j})}{E(Y \mid x_j, \mathbf{x}_{-j})} - 1 = \exp(\beta_j) - 1, \quad (10.4)$$

and therefore the *percent* change in the expected count for a 1-unit increase in X_j is $100[\exp(\beta_j) - 1]$. If X_j is a dummy variable, $\exp(\beta_j)$ represents the ratio of expected counts for those in the interest, versus the reference, categories, and expression (10.4) represents the proportion by which being in the interest category compared to the reference category increases or decreases the expected count.

A number of statisticians give a *unit proportional impact* interpretation to β_j. For example, Cameron and Trivedi (1998, p. 81) say: "The coefficient β_j equals the proportionate change in the conditional mean if the jth regressor changes by one unit." As expression (10.4) shows, this is not technically correct. However, it is approximately correct whenever a unit change in X_j equals a very small change in that variable, which is what statisticians have in mind when they imbue β_j with this interepretation. In the latter case, β_j should be a value close to zero, and $\exp(\beta_j)$ is then approximately equal to $1 + \beta_j$. For example, if $\beta_j = .05$, then $\exp(.05) = 1.051$. In this scenario, $\exp(\beta_j) - 1 \approx 1 + \beta_j - 1 = \beta_j$, and the unit proportional impact interpretation for β_j is then appropriate. Otherwise, expression (10.4) provides the correct proportionate change in the expected count for unit increases in X_j.

Example. Table 10.1 presents both OLS and PRM estimates for the regression of *number of previous math courses* on several characteristics of students: *age over 21* (a dummy for whether students are over 21), *male* (a dummy for being male), *social sciences major* (a dummy for majoring in a social science other than sociology, with sociology majors as the contrast group), *other major* (a dummy for majoring in other than a social science field, with sociology majors as the contrast group), *classification* (student classification), *high school GPA*, and *college GPA*. Substantively, the models provide similar conclusions about the nature of regressor effects. Both models are significant as a whole. The F statistic for the model estimated with OLS is 7.67; with 7 and 222 degrees of freedom, this is significant at the .0001 level. Similarly, the model χ^2 for the PRM of 43.862, with 7 degrees of freedom, is also

Table 10.1 Unstandardized OLS and PRM Estimates for the Regression of Number of Previous Math Courses

Regressor	OLS		PRM		
	b	t	b	z	$\exp(b)$
Intercept	−1.551	−2.311	−1.951	−3.287	.142
Age over 21	.265	1.595	.187	1.330	1.206
Male	.436	2.877	.313	2.548	1.368
Social sciences major[a]	−.052	−.255	−.028	−.154	.972
Other major[a]	−.024	−.144	.037	.258	1.038
Classification	.402	4.143	.336	3.723	1.399
High school GPA	.547	3.486	.395	2.914	1.484
College GPA	−.190	−1.323	−.174	−1.375	.840
F/model χ^2	7.670		43.862		
R^2	.195				
r^2			.207		
R_L^2			.067		
R_D^2			.179		

Note: $n = 230$.

[a] Sociology major is the reference category.

significant at the .0001 level. Three predictors are significant in both models, and in similar directions. Males, students with higher classifications (e.g., seniors, as opposed to sophomores) and those with higher high school GPAs all have a higher expected number of previous math courses, compared to others. The coefficients cannot be directly compared, however. For example, the OLS results suggest that men have a mean number of *previous math courses* that is higher by .436, compared to females. The PRM model's coefficient suggests that men's expected count of *previous math courses* is 36.8% higher than women's. The coefficient for *high school GPA* in OLS implies that each unit increase in *high school GPA* adds .547 to the expected count. The PRM coefficient, on the other hand, suggests that the expected count increases by 48.4% for each unit increase in *high school GPA*.

The linear regression model gives negative predicted mean math-course counts—which are clearly untenable values—at certain covariate patterns. For example, a student who is a sophomore female social science major, under 21 years of age, with a *high school GPA* of 2.5 and a *college GPA* of 3.5 has a predicted mean math-course count of $-1.551 - .052 + .402(2) + .547(2.5) - .19(3.5) = -.097$. In the PRM, on the other hand, this student's expected count is $\exp[-1.951 - .028 + .336(2) + .395(2.5) - .174(3.5)] = .395$. The PRM also allows us to generate a predicted probability for any given count of *previous math courses*, based on the parameter estimates. Using the formula for the Poisson probability, the predicted probability of a particular value of Y for the ith case is

$$\hat{f}(y_i \mid \mathbf{x}^i, \hat{\boldsymbol{\beta}}) = \frac{e^{-\hat{\mu}_i}\hat{\mu}_i^y}{y!}.$$

Notice that this depends on the case's covariates, since μ_i varies with the covariates. Hence, for any given value of Y, there are potentially n different predicted probabilities for that value. As an example, let's calculate the probability of having had three previous math courses for a senior male social science major, over 21 years of age, with a *high school GPA* of 2.8 and a *college GPA* of 3.1. First, the expected math-course count for this student is $\exp[-1.951 + .187 + .313 - .028 + .336(4) + .395(2.8) - .174(3.1)] = 1.54$. Then, the predicted probability of having had three math courses for this student is

$$\hat{f}(3) = \frac{e^{-1.54}(1.54)^3}{3!} = .13.$$

Empirical Consistency and Discriminatory Power

Empirical Consistency. There are few formal tests of empirical consistency for count models [but see Cameron and Trivedi (1998) and Greene (2003) for some suggested approaches]. However, one means of informally assessing whether the data behave according to model predictions is to compare the observed sample proportions of cases having each value of Y with the mean predicted probability of each value of Y, where the mean is taken over all n cases (Cameron and Trivedi, 1998; Long, 1997). The mean predicted probability that $Y = y$, based on a PRM, is denoted

$\hat{p}_y^p$, and is calculated as

$$\hat{p}_y^p = \frac{1}{n}\sum_{i=1}^{n}\frac{e^{-\hat{\mu}_i}\hat{\mu}_i^y}{y!}.$$

As is evident, $\hat{p}_y^p$ is simply the average predicted probability of y across all n cases. Figure 10.4 shows a plot of $\hat{p}_y^p$, based on the PRM in Table 10.1, against the observed proportions of students having each number of previous math courses. The PRM appears to fit the sample proportions quite closely for counts of 2 or more. However, the fit appears to be poor for counts of zero or one previous math course. Apparently, the PRM overpredicts zero counts and underpredicts counts of 1. Below we consider whether the PRM is really an appropriate model for this response.

Discriminatory Power. As is the case in logistic regression, there is no single counterpart in count models to the R^2 in linear regression for measuring discriminatory power. Hence, I will discuss three different R^2 analogs that have been proposed. First, we should recall from Chapter 7 that the R^2 in linear regression exhibits two properties that are highly desirable in any measure of discriminatory power: it falls

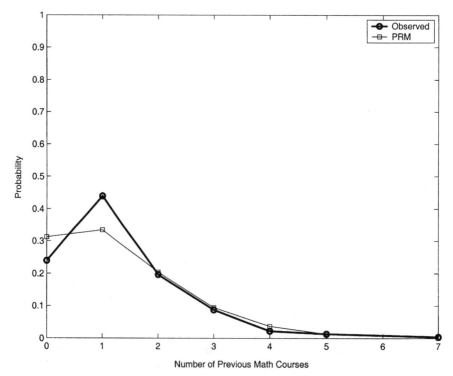

Figure 10.4 Observed versus mean predicted probabilities for number of previous math courses, with predictions based on the PRM.

within the range of 0 to 1, and it is nondecreasing as regressors are added to a model. The first measure I consider is the *likelihood-ratio index*, also considered for logistic regression models. The formula in count models is essentially the same:

$$R_{\mathrm{L}}^2 = 1 - \frac{\ln \mathrm{L}_1}{\ln \mathrm{L}_0},$$

where L_1 is the likelihood for the estimated model, evaluated at its MLEs, and L_0 is the likelihood for the intercept-only model. This measure is nondecreasing as predictors are added, since the likelihood never decreases as parameters are added to the model. (If the likelihood *could* decrease with addition of predictors, $\ln \mathrm{L}_1$, which is a negative value, could become *larger* in magnitude, implying a smaller R_{L}^2.) In logistic regression, this measure is also bounded by 0 and 1. However, in count models, this measure cannot attain its upper bound of 1 (Cameron and Trivedi, 1998), and so may underestimate the discriminatory power of any particular model.

A second analog employed by some statisticians (Land et al., 1996) is the correlation between Y and its predicted value according to the model. Recall that this gives us the R^2 for linear regression. Hence, this measure is $r^2 = [\mathrm{corr}(y, \hat{\mu})]^2$. Although this measure is bounded by 0 and 1, it is not necessarily nondecreasing with the addition of parameters. The advantage to these first two measures, on the other hand, is that they are readily calculated from output produced by count-model software.

The third measure, proposed by Cameron and Windmeijer (1997), is the deviance R^2. It is defined as follows. First, we define the Kullback–Leibler (KL) divergence, a measure of the discrepancy between two likelihoods. Let $\mathbf{y}$ be the vector of observed counts and $\hat{\boldsymbol{\mu}}$ be the vector of predicted counts based on a given model. Further, let $\ell(\hat{\boldsymbol{\mu}}_0, \mathbf{y})$ be the log-likelihood for the intercept-only model, $\ell(\hat{\boldsymbol{\mu}}, \mathbf{y})$ the log-likelihood for the hypothesized model, and $\ell(\mathbf{y}, \mathbf{y})$ the maximum log-likelihood achievable. This last would be the log-likelihood for a saturated model, one with as many parameters as observations. Then the KL divergence between saturated and intercept-only models, $K(\mathbf{y}, \hat{\boldsymbol{\mu}}_0)$, equals $2[\ell(\mathbf{y}, \mathbf{y}) - \ell(\hat{\boldsymbol{\mu}}_0, \mathbf{y})]$. This represents an estimate of the information on $\mathbf{y}$, in sample data, that is "potentially recoverable by inclusion of regressors" (Cameron and Windmeijer, 1997, p. 333) and corresponds to the *TSS* in linear regression. The information on $\mathbf{y}$ that remains after regressors are included in the model is the KL divergence between saturated and fitted models, $K(\mathbf{y}, \hat{\boldsymbol{\mu}})$, which is equal to $2[\ell(\mathbf{y}, \mathbf{y}) - \ell(\hat{\boldsymbol{\mu}}, \mathbf{y})]$. This is analogous to the *SSE* in linear regression. Finally, the deviance R^2 is

$$R_{\mathrm{D}}^2 = 1 - \frac{K(\mathbf{y}, \hat{\boldsymbol{\mu}})}{K(\mathbf{y}, \hat{\boldsymbol{\mu}}_0)}.$$

The reader should recognize that the right-hand side of R_{D}^2 is analogous to $1 - SSE/TSS$, the R^2 in linear regression. In this application, however, R_{D}^2 does not have an "explained variance" interpretation. Rather, it is "the fraction of the maximum potential likelihood gain (starting with a constant-only model) achieved by the fitted model" (Cameron and Windmeijer, 1997, p. 338). R_{D}^2 possesses both of the other properties of a desirable R^2 analog: It is bounded by 0 and 1 and it is nondecreasing with the addition of parameters. Moreover, in logit models, where $\ell(\mathbf{y}, \mathbf{y}) = 0$, this measure reduces to

$$R_D^2 = 1 - \frac{K(\mathbf{y}, \hat{\boldsymbol{\mu}})}{K(\mathbf{y}, \hat{\boldsymbol{\mu}}_0)} = 1 - \frac{-2\ell(\hat{\boldsymbol{\mu}}, \mathbf{y})}{-2\ell(\hat{\boldsymbol{\mu}}_0, \mathbf{y})} = 1 - \frac{\ell(\hat{\boldsymbol{\mu}}, \mathbf{y})}{\ell(\hat{\boldsymbol{\mu}}_0, \mathbf{y})} = R_L^2.$$

The only drawback to R_D^2 is that it can be quite tedious to compute, and is not automatically provided in software for count models. An exception is LIMDEP, which provides R_D^2 for PRM models. For other models, however, the LIMDEP user has to program the calculations for R_D^2.

Measures of discriminatory power for both OLS and PRM models are shown in the bottom of Table 10.1. The OLS analysis suggests that 19.5% of the variation in *number of previous math* courses is accounted for by the model. The three R^2 analogs for the PRM have widely differing values, with r^2, at .207, being closest to the OLS R^2. R_L^2's value of .067 suggests that the model is weak, but this measure is likely to underestimate discriminatory power. The superior measure is R_D^2, which, with a value of .179, is similar to the OLS R^2 in suggesting a moderately efficacious model.

Testing the Equidispersion Hypothesis. The PRM is the most basic of the count-data models. Typically, it is not the appropriate model for the data because the equidispersion hypothesis fails. And most often, this is because the data are overdispersed. Use of the PRM in the presence of overdispersion results in inefficient estimators and downwardly biased estimates of standard errors (Cameron and Trivedi, 1998). Cameron and Trivedi (1998) suggest that a quick diagnostic for this condition is simply to examine the sample unconditional mean and variance of Y. If $s_y^2 < \bar{y}$, the data are probably underdispersed, whereas if $s_y^2 > 2\bar{y}$, the data are probably overdispersed. In the latter case, the factor of 2 is suggested, since the inclusion of covariates will tend to reduce the conditional variance of Y in comparison with the unconditional variance. It is the conditional variance of Y in relation to the conditional mean that matters. For *number of previous math courses*, recall that s_y^2 and $\bar{y}$ are both 1.274. Given that the conditional variance should be reduced even further by the regressors, it is likely that the data are actually underdispersed.

One way to test for equidispersion is to compare the PRM to a known alternative model that allows for under- or overdispersion, which includes the PRM as a special case. Then a nested chi-squared test is a test for equidispersion. We consider this type of test below. The drawback, however, is that such a test requires an assumption of a particular alternative parametric form for the density of Y. A test of the equidispersion hypothesis that does not require this assumption has been proposed by Cameron and Trivedi (1990). Their approach only requires that we specify the nature of the relationship between the mean and variance of Y. The null hypothesis for the test is that the mean and variance are equal. That is, H_0 is $V(Y) = E(Y)$. The alternative hypothesis, H_1, is that the variance is a function of the mean. Two possible functions that have been considered are $V(Y) = E(Y) + \alpha E(Y)$ and $V(Y) = E(Y) + \alpha[E(Y)^2]$. In either case, if α is negative, the data are underdispersed, and if α is positive, the data are overdispersed. Letting μ stand for $E(Y)$, the null and alternative hypotheses are reexpressed as

$$H_0: E[(Y - \mu)^2] = \mu,$$

$$H_1: E[(Y - \mu)^2] = \mu + \alpha\mu \quad \text{or} \quad E[(Y - \mu)^2] = \mu + \alpha\mu^2.$$

This implies the formulation

$$H_0: E[(Y - \mu)^2 - Y] = 0,$$

$$H_1: E[(Y - \mu)^2 - Y] = \alpha\mu \quad \text{or} \quad E[(Y - \mu)^2 - Y] = \alpha\mu^2.$$

This formulation suggests that a test for whether $\alpha = 0$ in the linear regression of $(Y - \mu)^2 - Y$ on $\alpha\mu$ or $\alpha\mu^2$ is a test for equidispersion (Cameron and Trivedi, 1990). With $\hat{\mu}_i$ equal to the fitted values from the PRM, Cameron and Trivedi's test involves performing a linear regression of $[(y_i - \hat{\mu}_i)^2 - y_i]/\sqrt{2}\hat{\mu}_i$ on $\alpha[g(\hat{\mu}_i)/\sqrt{2}\hat{\mu}_i] + e$, using OLS, where $g(\hat{\mu}_i)$ is either $\hat{\mu}_i$ or $\hat{\mu}_i^2$. (Note that this is a *no-intercept* model.) The t test for $\hat{\alpha}$ from the regression is the test of H_0. For the PRM in Table 10.1, the coefficients for both $\hat{\mu}_i$ and $\hat{\mu}_i^2$ were negative ($-.245$, and $-.143$, respectively) and significant, suggesting that the data are underdispersed. King (1989) discusses generalized event count models that can handle underdispersed data. As such models are not always readily available in commercial software (but see LIMDEP's gamma model for an exception), I will not discuss them further. Below I discuss models for handling overdispersion, the more common situation.

Tobit versus Count-Data Models. At times there may be some confusion about whether the data call for a tobit model or a count-data model, particularly when the minimum value of Y is zero. The author has seen the tobit model used on a count response with the rationale that the count is a proxy for an underlying continuous variable that is modeled more appropriately using linear regression. Here I briefly articulate the differences between these modeling approaches. First, a latent continuous variable can be said to underlie the observed response in both cases: Y^* in the tobit model and μ in the PRM (King, 1989). Nevertheless, there are clear differences between these models. In tobit, the latent response is determined by a linear regression, and negative values of Y^* are reasonable. In the PRM, the latent response is an exponential function of the regressors, and negative values of μ are not possible. In tobit, $Y^* = Y$ once the censoring threshold has been crossed. In the PRM, μ is never synonymous with Y. In tobit, zeros represent censored values of Y^*—the zero simply means that Y^* is below the threshold. In the PRM, zeros are legitimate counts and do not represent censoring. Perhaps most important, in tobit, the response is either continuous or a proxy for a continuous variable. In the PRM, the response is a *count*. In short, whenever the response is a count, a count-data model such as the PRM is the appropriate model.

Truncated PRM

It is commonly the case that count variables are sampled from truncated distributions. A *zero-truncated sample* occurs when cases enter the sample conditional on having experienced at least one event (Long, 1997). For example, in the *kids dataset*, a question asks about the number of lifetime sex partners that each offspring has had. Because the data are limited to households of offspring who have initiated sexual intercourse, the sample is selective of sexually active offspring. Hence, counts of zero are not observed. The mean and variance of *number of lifetime sex partners* for the 357 offspring in the sample are 6.698 and 36.15, respectively, with a range of 1 to 20 partners. (Notice that the data appear to be overdispersed; this will be addressed when we discuss the negative binomial regression model below.)

To understand the rationale for the truncated model, consider first the following probability rule: If event B is a subset of event A, the probability of (A *and* B) is just the probability of B itself. For instance, in a random draw from a deck of cards, the probability that the card is a king *and* the king of spades is the same as the probability that the card is the king of spades, since the event "king of spades" is a subset of the event "king." Therefore, by the rules of conditional probability, the probability of drawing the king of spades *given* that the card is a king equals P(king of spades *and* king)/P(king) = P(king of spades)/P(king) = (1/52)/(4/52) = 1/4, which is quite intuitive. In a similar vein, for Y defined as a count variable *limited to positive values*, the event that Y equals any particular value 1, 2, 3, . . . , is a subset of the event that Y is greater than zero. Thus, $P(Y = y \mid Y > 0) = P(Y = y$ and $Y > 0)/P(Y > 0) = P(Y = y)/P(Y > 0)$.

In the zero-truncated PRM, the density of Y is therefore adjusted by the probability that Y is a positive count. That is, since $P(Y = 0) = e^{-\mu}\mu^0/0! = \exp(-\mu)$, the probability that Y is a positive count is $1 - \exp(-\mu)$. Therefore, the density of Y for the truncated PRM is

$$f(y_i \mid y_i > 0, \mathbf{x}^i, \boldsymbol{\beta}) = \frac{e^{-\mu_i}\mu_i^{y_i}}{y_i!(1 - e^{-\mu_i})}. \tag{10.5}$$

As before, the likelihood function is formed by making the substitution $\mu_i = \exp(\sum \beta_k X_{ik})$, and taking the product of the densities over the n cases in the sample (this is left as an exercise for the reader). Grogger and Carson (1991) give the conditional mean and variance of Y for the truncated PRM: The conditional mean is $\mu_i/(1 - e^{-\mu_i})$. In that the denominator is less than 1, the truncated mean is larger than the untruncated mean. The conditional variance is

$$\frac{\mu_i}{1 - e^{-\mu_i}}\left(1 - \frac{\mu_i e^{-\mu_i}}{e^{\mu_i} - 1}\right).$$

Notice that because the term in parentheses is always less than 1, the conditional variance in the truncated PRM is smaller than the conditional mean; hence equidispersion no longer holds for the truncated PRM.

Example: Number of Lifetime Sex Partners. The column labeled "PRM" in Table 10.2 presents estimates of a truncated PRM for *the number of lifetime sex partners* of 357 focal children in the NSFH, as a function of characteristics of both the child and his or her parents. Predictors include father's and mother's *education* (in years of schooling attained), a *parental monitoring* index (interval variable, with higher scores indicating greater supervision of the child's activities in wave 1), father's and mother's *sexual permissiveness* (interval variables, with higher scores indicating greater sexual permissiveness), the child's *age at first intercourse* (in years), *male* (a dummy for male children), *child's sexual permissiveness* (interval variable, with higher scores indicating greater permissiveness), and the cross-product of *male* with *child's sexual permissiveness*.

The model is significant overall, with a model chi-squared of 767.711 ($p < .0001$). The likelihood-ratio index, with a value of .262, suggests that the model has moderate discriminatory power. In this instance, the model has a control for differential exposure. The *exposure* factor is the log of the number of years of sexual activity, defined as the difference between the child's age at the time of the wave 2 survey and the child's age at initiation of sexual activity. This factor is quite significant and positive, as would be expected. Other significant factors are the father's and mother's *education*, the *age of the child at first intercourse, being male*, and the *child's sexual permissiveness*. Interestingly, father's and mother's *educations* have opposite effects on the average number of partners, with father's *education* diminishing the expected count, and mother's *education* raising it. More intuitive are the results for *age at first intercourse*. Each additional year the child waits before initiating sex reduces the mean number of partners by $100[\exp(-.056) - 1]$, or about

Table 10.2 Unstandardized Truncated PRM and NBRM Estimates for the Regression of Number of Lifetime Sex Partners for Focal Children in the NSFH

Regressor	PRM	NBRM
Intercept	.458	.966
Father's education	−.022*	−.021
Mother's education	.045***	.055**
Parental monitoring	.125	.013
Father's sexual permissiveness	.002	−.013
Mother's sexual permissiveness	.012	.014
Age at first intercourse	−.056***	−.101***
Male	.517***	.542***
Child's sexual permissiveness	.060***	.071***
Male × child's sexual permissiveness	−.028	−.019
Exposure	.536***	.484***
Overdispersion parameter		.391***
Model χ^2	767.711***	1109.866***
R_L^2	.262	.379
Equidispersion χ^2		342.156***

Note: n = 357.

* $p < .05.$ ** $p < .01.$ *** $p < .001.$

$5\frac{1}{2}\%$. Males' expected number of partners is $\exp(.517) = 1.677$ times greater than females'. As expected, the more sexually permissive the child, the greater his or her estimated number of sex partners. This last factor may well be endogenous to the number of partners. That is, those who have had more partners probably become more permissive so that their attitudes are consistent with their behavior. As noted above, the data are probably overdispersed. We address this problem shortly. Estimated probabilities for the truncated model are calculated by substituting $\hat{\mu}_i$, based on the sample regression function, into equation (10.5).

Censoring and Sample Selection

Censored and sample-selected versions of the PRM are similar in principle to the models discussed in Chapter 9. Censoring from above is a common occurrence with count data. This applies whenever values of Y above a certain number are all collapsed into one category. For example, the NSFH asks respondents in intimate relationships how many times they have "hit or thrown things at" a partner in the past year. Responses are recorded as 0, 1, 2, 3, and 4 or more times. All frequencies above 4 have been recorded as 4, and we therefore have a count variable censored from above at the value 4. The PRM is readily adjusted for censoring by making the necessary alterations to the likelihood function. Cameron and Trivedi (1998) provide the details.

Sample selection bias is addressed by assuming that a latent propensity to respond determines whether or not a count response is observed for the ith case in the sample. However, all we observe is whether or not a count is recorded for the ith case. Denote the observed count by Y_1 and the binary indicator of whether a count is observed by Y_2. The selection model is the probit model for Y_2, while the substantive model is the count model for Y_1. Selection effects are handled by assuming that the model for the observed count includes a disturbance term having a bivariate normal distribution with the disturbance in the probit selection model. This specification allows the formation of a likelihood function based on the joint density of Y_1 and Y_2. This full-information technique allows maximum-likelihood estimation of the parameters of the count model in the presence of selection effects. A two-step estimator analogous to the Heckman approach discussed in Chapter 9 is also possible. Cameron and Trivedi (1998) discuss selection models at some length. LIMDEP allows estimation of count models in the presence of both censoring and sample selection.

COUNT-DATA MODELS THAT ALLOW FOR OVERDISPERSION

The PRM is typically an inadequate model for count data because such data are usually overdispersed. In this section we consider some models that are designed to fit overdispersed data. The first is the *negative binomial regression model* (NBRM). In the PRM we assumed that the unobserved rate of event occurrence, μ_i, for each case was determined exactly by the regressors. We then assumed that the density of y_i was

Poisson with parameter μ_i, and this assumption formed the basis for the likelihood function. In the NBRM, we relax the assumption that the regressors perfectly determine the expected event count and allow a disturbance term into the relationship. We further specify a continuous density for the disturbance term, so that the density of y_i becomes a mixture of two densities, one discrete and one continuous. The discrete density is then integrated over the continuous density to produce the NBRM. As we will see below, this strategy allows the conditional mean and variance of Y to differ.

A second approach to addressing overdispersion is to recognize that this problem frequently arises because the data observed contain a substantially higher proportion of zero counts than would be predicted by the PRM. We therefore consider two types of models that allow for excess zero counts. The *zero-inflated* PRM and NBRM models make a distinction between zeros that arise probabilistically, in the context of the PRM or NBRM stochastic process, versus zeros that arise because certain cases are precluded from having positive counts. In contrast, the *hurdle* model treats all cases as being at risk for having positive counts, but allows for the process generating subsequent counts, given at least one count, to be different from the process that generates positive counts, in general.

Negative Binomial Regression Model

The NBRM arises as a natural consequence of allowing a random disturbance term in the relationship between the rate of event occurrence (i.e., the conditional mean of Y_i) and the regressors. That is, we model $E(Y_i) = \theta_i$ as

$$\theta_i = \exp[(\textstyle\sum \beta_k X_{ik}) + \varepsilon_i]$$
$$= \exp(\textstyle\sum \beta_k X_{ik}) \exp(\varepsilon_i)$$
$$= \mu_i \upsilon_i,$$

where $\upsilon_i = \exp(\varepsilon_i)$. Observe now that the conditional mean of Y_i, θ_i, is determined both by the model covariates, which determine μ_i, and by a multiplicative disturbance term, υ_i. The model regressors, in the form $\sum \beta_k X_{ik}$, constitute *observed heterogeneity*, meaning measured characteristics that induce variation in μ_i across cases. The disturbance term, on the other hand, is a measure of *unobserved heterogeneity*, which as we will see leads to overdispersion in Y. For the NBRM to be identified, we must assume that $E(\upsilon_i) = 1$, in which case $E(\theta_i) = E(\mu_i \upsilon_i) = \mu_i E(\upsilon_i) = \mu_i$ (Cameron and Trivedi, 1998; Long, 1997). The density of Y, conditional on the regressors and υ_i, is now

$$f(y_i \mid \mathbf{x}^i, \upsilon_i, \boldsymbol{\beta}) = \frac{e^{-\theta_i} \theta_i^{y_i}}{y_i!} = \frac{e^{-\mu_i \upsilon_i}(\mu_i \upsilon_i)^{y_i}}{y_i!}. \tag{10.6}$$

However, since υ_i is unobserved, this density cannot be used to construct the likelihood function. (Remember that the only unknowns in the likelihood function must be the model parameters, not unobservable variables.) The solution is to assume that υ_i has a particular parametric density and then to "integrate it out" of density (10.6) in order to arrive at the marginal density for Y; that is, the density that is no longer conditional on υ_i. The usual assumption is that υ_i has a gamma density with parameter α^{-1}. This is a continuous right-skewed density that resembles a chi-squared variable. In fact, the chi-squared density is a special case of the gamma density (Hoel et al., 1971).

Constructing the NBRM Density. "Integrating out" υ_i means that we take the average value of density (10.6) over the distribution of υ_i [see Greene (2003) for the details of the integration]. The advantage of assuming a gamma density for υ_i here is that this integration then has a closed-form solution. The resulting marginal density of Y_i, given the regressors and α, is the negative binomial density (Cameron and Trivedi, 1998):

$$f(y_i \mid \mathbf{x}^i, \boldsymbol{\beta}, \alpha) = \frac{\Gamma(\alpha^{-1} + y_i)}{\Gamma(\alpha^{-1})\Gamma(y_i + 1)} \left(\frac{\alpha^{-1}}{\alpha^{-1} + \mu_i} \right)^{\alpha^{-1}} \left(\frac{\mu_i}{\mu_i + \alpha^{-1}} \right)^{y_i}, \qquad (10.7)$$

where as before, μ_i equals $\exp(\sum \beta_k X_{ik})$. The gamma function, $\Gamma(\cdot)$, in this expression is defined by an integral with no closed-form solution (Hoel et al., 1971). However, it turns out that $\Gamma(a) = (a - 1)!$ if a is an integer. The term α^{-1} is not typically an integer. But if it were, given that y_i is an integer, density (10.7) would have the same form as the negative binomial density in expression (10.1), where $r = \alpha^{-1}$ and $p = \alpha^{-1}/(\alpha^{-1} + \mu_i)$. The product of density (10.7) over all n sample cases is the likelihood function, which is then maximized with respect to α and $\boldsymbol{\beta}$ to find the MLEs. Because the conditional mean of Y_i in density (10.7) is still $\mu_i = \exp(\sum \beta_k X_{ik})$, the betas still have the same interpretations as given to those in the PRM. (This holds true for all count models discussed in this chapter.) Estimated probabilities for each count are calculated by substituting $\hat{\mu}_i = \exp(\sum \hat{\beta}_k X_{ik})$ and $\hat{\alpha}$ into expression (10.7). Greene (1998) presents a convenient recursion formula that can be programmed into LIMDEP for the calculation of these probabilities.

Testing for Overdispersion. The conditional variance of Y_i in the NBRM is $\mu_i + \alpha\mu_i^2$. The parameter α, called the *overdispersion parameter*, is always greater than or equal to zero. This means that the conditional variance is normally greater than the conditional mean. If α equals zero, the conditional variance is equal to the conditional mean and the NBRM reduces to the PRM. That is, the PRM is nested inside the NBRM, and therefore a test for overdispersion is a test for whether $\alpha = 0$. This can be performed using either a nested chi-squared test or a Wald test of the form $z = \hat{\alpha}/\sigma_{\hat{\alpha}}$. The two tests are asymptotically equivalent (Cameron and Trivedi, 1998). However, in that α cannot be less than zero, the distribution of these test statistics is nonstandard. Thus, when performing the chi-squared test at a given level of significance, say δ, we use the critical value of 2δ for the test statistic as the criterion. For the Wald test, we simply use the critical value corresponding to δ rather than $\delta/2$. For example, performing the chi-squared or Wald test at the .05 level for H_0: $\alpha = 0$ involves using the critical χ^2 value corresponding to the .1 level, or the critical z value corresponding to the .05 level, and so on (Cameron and Trivedi, 1998).

Example: Number of Days of Depression. Figure 10.5 shows the number of days during the past week on which respondents "could not shake off the blues, even with help from your family or friends," for the 416 main respondents in the *couples dataset*. This item is one of 12 similar items that constitute the short form of the Center for Epidemiological Studies Depression (CESD) scale. Although these items are typically used in a scale, here I focus simply on a count of the number of days on which respondents experienced this particular symptom of depression. The variable has a mean of .839 and a variance of 2.926. These values suggest that the data are overdispersed. We notice also that there is an exceptionally high proportion of zero counts: 71% of respondents report no days in which they experienced this symptom. Both phenomena, overdispersion and excess zeros, suggest that the PRM will not be an appropriate model for these data. One hypothesis of interest with respect to this variable is that women will tend to be psychologically more adversely affected by relationship problems than men, or conversely, more psychologically protected by a good relationship. This is reasonable given that (1) women are socialized to be more sensitive to relationship issues than men are, and (2) women tend to respond to stress more with depressive symptomatology, whereas men tend to respond more with abuse of drugs and alcohol.

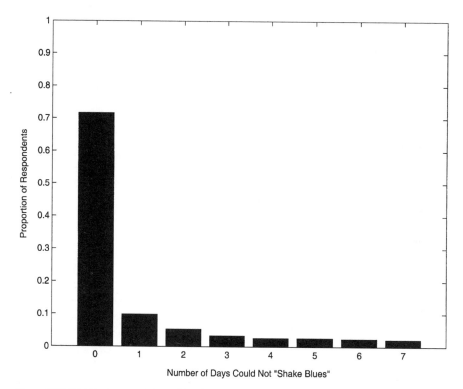

Figure 10.5 Distribution on number of days in the past week respondents could not "shake off the blues" for 416 couples in the NSFH.

The PRM estimates for $Y = $ *number of days could not "shake blues"* (*depression days*, for short), based on several couple characteristics, are shown in the first column of Table 10.3. Predictors not previously described are a dummy variable reflecting the male being the main respondent of the face-to-face interview (*male main respondent*), *the male's age at inception of the union*, the total *household income* for the couple, *the number of children* under 18 in the household, a *relationship happiness* score (interval variable based on both partners' reports, ranging from 1 = "very unhappy" to 7 = "very happy"), and the cross-products of *male* with both *open disagreement* and *relationship happiness*. Of particular interest are the effects of *open disagreement* and *relationship happiness*, as well as their interactions with *being male*. The model suggests that *open disagreement* has a significant positive effect on *depression days*: Each 1-unit increase in *disagreements* increases mean *depression days* by a factor of $\exp(.934) = 2.545$ for women. For men, each 1-unit increase in *disagreement* increases the expected count of *depression days* by $\exp(.934 - .418) = 1.675$. As expected, the impact of *disagreement* is significantly weaker for men than for women. On the other hand, the effect of *happiness* in reducing the mean number of *depression days* is stronger for men than for women, contrary to hypothesis. For women, the factor change for unit increases in *happiness* is $\exp(-.095) = .909$, and not significant. For men, it is significantly greater in magnitude than for women, at $\exp(-.095 - .195) = .748$. The results also suggest that a greater *number of children* reduces *depression days*. (The significant effect for *being male* is not interpretable, since it represents the gender difference at zero *disagreements*, a value outside the observed range of this predictor.)

Table 10.3 Unstandardized PRM and NBRM Estimates for the Regression of Number of Days in the Past Week Respondents Could Not "Shake Off the Blues"

Regressor	PRM	NBRM
Intercept	−1.271*	−1.280
Male main respondent	1.697*	2.195
Male's age at union inception	.006	.007
Union duration	.003	.001
Household income	−.005	−.004
Number of children	−.128*	−.109
Open disagreements	.934***	.951**
Relationship happiness	−.095	−.111
Male × open disagreements	−.418**	−.410
Male × relationship happiness	−.195*	−.283
Overdispersion parameter		3.438***
Model χ^2	172.203***	443.810***
R_{L}^2	.126	.326
Equidispersion χ^2		271.607***

Note: $n = 416$.

$* p < .05.\ ** p < .01.\ *** p < .001.$

The second column of the table shows the NBRM estimates. The first item of interest is the test for whether the data are overdispersed, which is the test for H_0: $\alpha = 0$. The value of α is estimated as 3.438. The Wald test statistic, with a z-value of 5.319 ($p < .0001$), strongly rejects H_0. The nested chi-squared test statistic, shown as "equidisperion χ^2" in the bottom of the table, is a one-degree-of-freedom chi-squared equal to the difference in model chi-squareds between the PRM and the NBRM. Its value is $443.810 - 172.203 = 271.607$, as shown in the table. With a p-value less than .0001, this test statistic also results in a sound rejection of the equidispersion hypothesis. We notice now that although the coefficients of the NBRM are comparable in value with those in the PRM, all are nonsigificant except for the effect of *open disagreement*. The latter suggests that unit increases in *open disagreement* increase the expected count of *depression days* by a factor of $\exp(.951) = 2.588$, or 159%. Recall that standard errors for the PRM are downwardly biased in the presence of overdispersion. Hence, using the appropriate model has resulted in larger standard errors, diminishing the size of test statistics for the individual coefficients. Although fewer coefficients are significant in the NBRM, compared to the PRM, R_L^2 has increased substantially. This is due to the addition of the overdispersion parameter to the likelihood function rather than to an enhanced ability of the regressors to account for the response.

Truncated NBRM. A truncated version of the NBRM can also be estimated for data limited to positive counts. Again, the density function for the response is adjusted by the probability of a positive count. The probability of a zero count in the NBRM is

$$f(0 \,|\, \mathbf{x}^i, \boldsymbol{\beta}, \alpha) = \frac{\Gamma(\alpha^{-1} + 0)}{\Gamma(\alpha^{-1})\Gamma(0 + 1)} \left(\frac{\alpha^{-1}}{\alpha^{-1} + \mu_i} \right)^{\alpha^{-1}} \left(\frac{\mu_i}{\mu_i + \alpha^{-1}} \right)^0 = \left(\frac{\alpha^{-1}}{\alpha^{-1} + \mu_i} \right)^{\alpha^{-1}} = (1 - \alpha\mu_i)^{-\alpha^{-1}},$$

and therefore the probability of a positive count is

$$1 - (1 - \alpha\mu_i)^{-\alpha^{-1}}.$$

The density of the truncated NBRM is

$$f(y_i \,|\, y_i > 0, \mathbf{x}^i, \boldsymbol{\beta}, \alpha) = \frac{1}{1 - (1 - \alpha\mu_i)^{-\alpha^{-1}}}$$

$$\frac{\Gamma(\alpha^{-1} + y_i)}{\Gamma(\alpha^{-1})\Gamma(y_i + 1)} \left(\frac{\alpha^{-1}}{\alpha^{-1} + \mu_i} \right)^{\alpha^{-1}} \left(\frac{\mu_i}{\mu_i + \alpha^{-1}} \right)^{y_i}. \tag{10.8}$$

As always, the product of this density over all n cases is the likelihood function, which is maximized to find the MLEs. Estimated probabilities for each count are calculated by substituting $\hat{\mu}_i = \exp(\sum \hat{\beta}_k X_{ik})$ and $\hat{\alpha}$ into expression (10.8). In untruncated models, employing the PRM with overdispersed data does not interfere with obtaining consistent estimates of the parameters. However, as Grogger and Carson (1991) point out, this property does not carry over to truncated data. Ignoring overdispersion in a truncated response by inappropriately applying the

PRM results in inconsistent parameter estimates, as well as biased estimates of standard errors.

Lifetime Number of Sex Partners, Revisited. The second column of Table 10.2 presents the truncated NBRM for *the lifetime number of sex partners* for the NSFH focal children. Both the Wald and the nested chi-squared tests for the overdispersion parameter suggest that the hypothesis of equidispersion should be rejected. In this case, there is little substantive change in the conclusions except that the coefficient for *father's education* is no longer significant. Again, with the addition of the overdispersion parameter, R_L^2 has increased somewhat.

Zero-Inflated Models

Zero-inflated models, defined by Greene (1994), Lambert (1992), and Long (1997), among others, account for excess zeros by distinguishing between two different types of zero counts. Borrowing terminology employed for the analysis of contingency tables, I refer to these as *structural* versus *sampling* zeros (Agresti, 2002). Structural zeros come from a population that is not at risk for the events of interest, often because they are logically precluded from experiencing such events. Sampling zeros come from a different population that *is* at risk for experiencing events, but people with zero counts simply have not experienced any events within the domain of observation, due to the stochastic nature of the event process. As an example, suppose that we were to ask a sample of adolescents: "How many different sex partners have you had in the past month?" Zero counts would arise for two reasons. One population of adolescents has not yet initiated sexual activity and so are logically precluded from having any sex partners. Their zeros are therefore structural zeros. The other population has initiated sexual activity, but certain adolescents have just not engaged in sexual activity with anyone in the past month. Their zeros are sampling zeros; they are subject to an event process that eventuates in some probability of having a zero count. Similarly, our example of the number of days in the previous week on which respondents could not "shake the blues" can be seen as arising from a zero-inflated event process. Some people are simply not prone to melancholy or depression because they do not respond in that manner to stress. Other people are indeed prone to depression but have not experienced any depression days in the past week. In short, whenever some of the zero counts in a sample can come from a subpopulation of those who for some reason are precluded from experiencing the events of interest, a zero-inflated model may be appropriate.

ZIP Model. The *zero-inflated Poisson* (ZIP) *model* applies the Poisson model to the population of those who are at risk for the events in question, and a separate binary response model to model the probability of being in the structural-zero group. The probability of a zero count is then a weighted average of the probability that $Y = 0$ in each group, where the weights are the probabilities of belonging to each group. Let P_0 represent the structural-zero population and P_+ represent the population at risk for at least one event. Also, let ψ_i represent the probability that the *i*th case is in the structural-zero group, and $(1 - \psi_i)$ represent the probability of being in the at-risk group. The probability of a zero count, according to the ZIP, is

$$f(0 \mid \mathbf{x}^i, \boldsymbol{\beta}) = \psi_i P(Y_i = 0 \mid P_0) + (1 - \psi_i) P(Y_i = 0 \mid P_+)$$

$$= \psi_i (1) + (1 - \psi_i) \exp(-\mu_i)$$

$$= \psi_i + (1 - \psi_i) \exp(-\mu_i). \tag{10.9}$$

Notice that the probability of a zero for those in P_0 is 1. This is referred to as a *degenerate probability distribution* (Cameron and Trivedi, 1998). For Y greater than zero, the probability of any given count is equal to the probability of being in the at-risk group times the probability of having that count, given that one is at risk for events:

$$f(y_i \mid \mathbf{x}^i, \boldsymbol{\beta}) = (1 - \psi_i) \frac{e^{-\mu_i} \mu_i^{y_i}}{y_i!} \qquad \text{for } y_i = 1, 2, 3, \ldots, \tag{10.10}$$

where $\mu_i = \exp(\sum \beta_k X_{ik})$. Together, expressions (10.9) and (10.10) constitute a density since the probabilities for $Y = 0, 1, 2, \ldots$ sum to 1 (the proof is left as an exercise).

The probability of being in P_0, ψ_i, is governed by a separate binary response model. Most often, a logit model is used, in which the covariates are the same as those in the Poisson model (although they need not be), but the parameters are, of course, different. Assuming the regressors are the same in both models, the model for ψ_i is

$$\ln \frac{\psi_i}{1 - \psi_i} = \sum \gamma_k X_{ik}. \tag{10.11}$$

The ZIP model formulated in this way requires twice as many parameters as the PRM. Lambert (1992) suggested that if the same covariates affect both ψ_i and μ_i, it would be natural to reduce the number of parameters by formulating ψ_i as a function of μ_i. However, I agree with Long (1997, pp. 243–244) that ". . . it is difficult to imagine a social science application in which one would expect the parameters in the binary process to be a simple multiple of the parameters in the Poisson process." Therefore, I do not cover this more simplistic model here. The interested reader is referred to Lambert (1992) for that coverage.

Estimation. Estimation of the ZIP model proceeds by maximizing the ZIP likelihood function with respect to the parameters [Cameron and Trivedi (1998) present the log-likelihood function for the model]. Estimated probabilities for counts under the ZIP model are calculated by employing the MLEs to construct $\hat{\psi}_i$ and $\hat{\mu}_i$ and then inserting these terms into expressions (10.9) and (10.10) to recover the probabilities. The appropriate formulas are

$$\hat{\psi}_i = \frac{\exp\left(\sum \hat{\gamma}_k X_{ik}\right)}{1 + \exp\left(\sum \hat{\gamma}_k X_{ik}\right)}, \tag{10.12}$$

$$\hat{\mu}_i = \exp\left(\sum \hat{\beta}_k X_{ik}\right). \tag{10.13}$$

The conditional mean and variance of Y in the ZIP model are (Long, 1997)

$$E(Y_i|\mathbf{x}_i) = (1 - \psi_i)\mu_i,$$
$$V(Y_i|\mathbf{x}_i) = (1 - \psi_i)(\mu_i + \psi_i\mu_i^2).$$

In that $\mu_i + \psi_i\mu_i^2$ is greater than μ_i unless $\psi_i = 0$, the conditional variance exceeds the conditional mean, and overdispersion is therefore accommodated.

Comparing PRM and ZIP Models. There is no nested chi-squared test for comparing the ZIP and PRM because the PRM is not nested inside the ZIP. In order for the PRM to be nested within the ZIP, ψ_i would have to equal zero, in which case the ZIP would reduce to the PRM. However, there is no simple set of parameter constraints that can achieve this result. The natural constraint would be to set $\gamma = 0$ in equation (10.11). However, under this condition ψ_i is $\exp(0)/[1 + \exp(0)] = \frac{1}{2}$, which is not the desired result. In light of this, we can employ a test proposed by Vuong (1989) that compares nonnested models. Let $\hat{f}_1(y_i|\mathbf{x}^i)$ be the predicted probability that $Y_i = y_i$ for the ith case under model 1, with $\mathbf{x}_i$ that case's vector of regressor values, and let $\hat{f}_2(y_i|\mathbf{x}^i)$ be the predicted probability under model 2. Furthermore, let

$$m_i = \ln\frac{\hat{f}_1(y_i|\mathbf{x}^i)}{\hat{f}_2(y_i|\mathbf{x}^i)},$$

and let the mean and standard deviation of m_i over all n cases be $\overline{m}$ and s_m, respectively. Then the Vuong test statistic for testing model 1 against model 2 is

$$V = \frac{\overline{m}}{s_m/\sqrt{n}},$$

which is asymptotically distributed as standard normal. If model 1 is the ZIP and model 2 is the PRM, the ZIP is favored if V is greater than 1.96, whereas the PRM is favored if V is less than -1.96.

ZINB Model. The *zero-inflated negative binomial* (ZINB) *model* is developed in a comparable manner to the ZIP. For the NBRM, the probability of a zero count is $(1 - \alpha\mu_i)^{-\alpha^{-1}}$. Therefore, the ZINB model for the zero and positive counts is

$$f(0|\mathbf{x}^i,\boldsymbol{\beta}) = \psi_i + (1-\psi_i)(1 - \alpha\mu_i)^{-\alpha^{-1}},$$

$$f(y_i|\mathbf{x}^i,\boldsymbol{\beta},\alpha) = (1-\psi_i)\frac{\Gamma(\alpha^{-1}+y_i)}{\Gamma(\alpha^{-1})\Gamma(y_i+1)}\left(\frac{\alpha^{-1}}{\alpha^{-1}+\mu_i}\right)^{\alpha^{-1}}\left(\frac{\mu_i}{\mu_i+\alpha^{-1}}\right)^{y_i} \quad \text{for } y_i = 1, 2, 3, \ldots,$$

where ψ_i, as in the ZIP, is determined by a logit model with a separate parameter set. These densities are used to construct the likelihood function to obtain the MLEs. As usual, plugging $\hat{\psi}_i$, $\hat{\mu}_i$, and $\hat{\alpha}$ into these expressions provides predicted probabilities for each count, where $\hat{\psi}_i$ and $\hat{\mu}_i$ are estimated according to equations (10.12) and

(10.13), based on the MLEs for the ZINB model. The conditional mean and variance for the ZINB model are (Long, 1997)

$$E(Y_i \,|\, \mathbf{x}^i) = (1 - \psi_i)\mu_i,$$

$$V(Y_i \,|\, \mathbf{x}^i) = (1 - \psi_i)\mu_i[1 + \mu_i(\psi_i + \alpha)].$$

Once again, it is evident that the conditional variance exceeds the conditional mean. The ZIP is nested inside the ZINB, so a nested chi-squared test can help choose between these models. However, the NBRM is not nested inside the ZINB model, so again, the Vuong statistic can be used to compare them, as in the case for the ZIP and PRM.

Depression Days, Continued. Table 10.4 presents both ZIP and ZINB models for the number of days respondents could not "shake the blues." Two columns are shown for each model. The logit column displays parameter estimates for the logistic regression of whether a case is a structural zero, while the PRM and NBRM columns provide the estimates for the relevant count models for those at risk for days of depression. The reader should notice that the logit and count-model coefficients are generally of opposite signs. This is sensible, since attributes that reduce the likelihood of being in the structural-zero group tend to enhance the expected event count, and vice versa. In the ZIP, the only factor that predicts being in the structural-zero group is *open disagreement*, with more disagreements reducing the likelihood of being in this group. The PRM for those with positive counts suggests, again, that disagreements have a

Table 10.4 Unstandardized ZIP and ZINB Estimates for the Regression of Number of Days in the Past Week Respondents Could Not "Shake Off the Blues"

	ZIP		ZINB	
Regressor	Logit	PRM	Logit	NBRM
Intercept	.551	−.103	.216	−.461
Male main respondent	−.202	1.773*	−.083	1.900
Male's age at union inception	−.001	.007	−.002	.009
Union duration	−.001	.001	−.001	.002
Household income	−.002	−.006*	−.002	−.007
Number of children	.151	−.023	.167	−.012
Open disagreements	−.901**	.484***	−.877*	.556**
Relationship happiness	.260	.005	.272	.014
Male × open disagreements	.072	−.459*	−.050	−.550
Male × relationship happiness	.112	−.127	.142	−.111
Overdispersion parameter				.278
Model χ^2		471.786***		480.358***
R_L^2		.347		.353
Equidispersion χ^2				8.572**
Vuong statistic		16.144***		5.828***

Note: $n = 416$.

$* \, p < .05. \; ** \, p < .01. \; *** \, p < .001.$

stronger positive effect on the expected event count for women than for men, as hypothesized. Additionally, those with a higher *household income* have significantly fewer expected depression days, net of other effects. The Vuong statistic, with a value of 16.144, strongly suggests that the ZIP is an improvement over the ordinary PRM. R_L^2 suggests that the ZIP has moderate discriminatory power.

The ZINB model, on the other hand, suggests that *open disagreement* constitutes the only factor that affects either the likelihood of being in the structural-zero group, or the expected event count, given that one is at risk for *depression days*. The Vuong statistic for the ZINB versus the ordinary NBRM, at 5.828, indicates that the ZINB is to be preferred. Whether the ZINB is to be preferred to the ZIP is not quite as straightforward. The equidispersion χ^2 is significant, but the Wald test for the dispersion parameter is not. As these tests are asymptotically equivalent, they should agree; however, at times there will be such disparities. The safest conclusion is that although there is some evidence that *disagreement* has a stronger effect on *depression days* for women than men, the only robust effect is that open *disagreement* raises both the risk for *depression days* and the expected count of *depression days*, net of other factors.

Figure 10.6 illustrates another means of comparing all four models investigated for *depression days*, as suggested by Long (1997). It shows a plot of observed minus predicted probabilities for the PRM, NBRM, ZIP, and ZINB models of *depression*

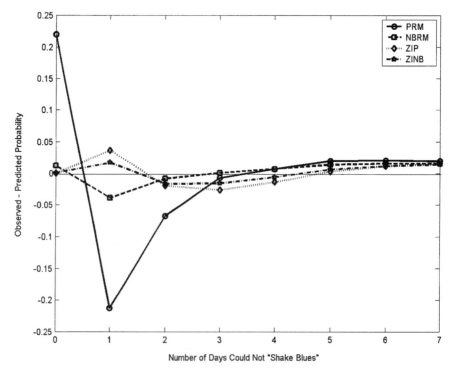

Figure 10.6 Observed–predicted probabilities for number of days could not "shake blues," based on PRM, NBRM, ZIP, and ZINB models.

days. That is, let p_y be the sample proportion exhibiting a count of y depression days and $\hat{p}_y^M$ be the mean predicted probability of a count of y according to a count model, where M is, variously, PRM, NBRM, ZIP, or ZINB. Then each line in Figure 10.6 is a plot of $p_y - \hat{p}_y^M$ over the values of $Y = $ *number of depression days.* The ideal model is the one whose plot is closest to the zero line—the horizontal line in the middle of the graph. The PRM is clearly a poor-fitting model, particularly for 0, 1, or 2 *depression days.* The NBRM is somewhat better but still has a noticeable overprediction for a count of one day. Both the ZIP and the ZINB models appear to have similarly good fits to the data, with ZINB being slightly superior to ZIP.

Hurdle Models

The hurdle model was introduced by Mullahy (1986) in the context of a discussion of modified count models. In contrast to the approach taken by the zero-inflated model, the hurdle model assumes that *all* cases are at risk for events and that the probability of a zero count is governed by a stochastic process. However, the hurdle model allows model regressors to have different effects on the likelihood of experiencing at least one event, as opposed to the likelihood of a subsequent event given that at least one event has occurred. The idea is that experiencing at least one event is a "hurdle" that requires a different parameterization than is the case for experiencing subsequent events. As an example, the NVAW survey asked women respondents to report the *number of physical assaults* they had ever experienced from any type of offender. For the 1343 minority women in the sample, the range was 0 to 124 assaults, with a mean of 2.081, a standard deviation of 10.035, and 71.7% reporting no assaults. For this variable, it is not reasonable to suppose that there is a population that is not at risk for assault. As this type of event is not under individual control, everyone is presumed at risk for an assault. However, the process generating subsequent assaults may well be different than the process affecting the likelihood of at least one assault. We will explore this using the Poisson hurdle model.

Parameterization of the Poisson Hurdle Model. In the *Poisson hurdle model,* allowance is made for the rate of event occurrence to be different for zero, as opposed to positive, counts. The model for a zero count is, therefore,

$$f(0 \mid \mathbf{x}^i, \boldsymbol{\gamma}) = \frac{e^{-\mu_{1i}} \mu_{1i}^0}{0!} = e^{-\mu_{1i}},$$

where $\mu_{1i} = \exp(\sum \gamma_k X_{ik})$. The model for positive counts is then a truncated PRM based on a different parameter set, adjusted for the probability of a nonzero count (Cameron and Trivedi, 1998). That is, for $Y > 0$,

$$f(y_i \mid \mathbf{x}^i, \boldsymbol{\beta}, \boldsymbol{\gamma}) = [1 - f(0 \mid \mathbf{x}_i, \boldsymbol{\gamma})][f(y_i \mid y_i > 0, \mathbf{x}_i, \boldsymbol{\beta})]$$

$$= (1 - e^{-\mu_{1i}}) \frac{e^{-\mu_{2i}} \mu_{2i}^{y_i}}{y_i!(1 - e^{-\mu_{2i}})},$$

where $\mu_{2i} = \exp(\sum \beta_k X_{ik})$. Now in the event that $\gamma = \beta$, we have that $\mu_{1i} = \mu_{2i}$, and the hurdle model reduces to the ordinary PRM, since the specification is now

$$f(0 \mid \mathbf{x}^i, \boldsymbol{\beta}) = \frac{e^{-\mu_{1i}} \mu_{1i}^0}{0!} = e^{-\mu_{1i}},$$

$$f(y_i \mid \mathbf{x}^i, \boldsymbol{\beta}) = (1 - e^{-\mu_{1i}}) \frac{e^{-\mu_{1i}} \mu_{1i}^{y_i}}{y_i!(1 - e^{-\mu_{1i}})} = \frac{e^{-\mu_{1i}} \mu_{1i}^{y_i}}{y_i!} \qquad \text{for } y_i = 1, 2, 3, \ldots.$$

This means that the PRM is nested inside the Poisson hurdle model (King, 1989) and a nested chi-squared test can be used to compare these models. The test is also a test for the constraint that $\gamma = \beta$, that is, that the covariates have the same effects on the probability of at least one count as they do on the probability of each additional count. The test has $K + 1$ degrees of freedom.

Estimation. The likelihood function for the hurdle model is worth considering in some detail, as it suggests how the model can be estimated, even in the absence of specialized software. Let 0 represent the zero counts in the sample and $y > 0$ represent the positive counts. Then the likelihood function is of the form

$$L(\boldsymbol{\beta}, \boldsymbol{\gamma} \mid \mathbf{y}, \mathbf{x}) = \prod_{y \in 0} P(y = 0) \prod_{y \in y > 0} P(y > 0) P(y \mid y > 0)$$

$$= \prod_{y \in 0} e^{-\mu_1} \prod_{y \in y > 0} (1 - e^{-\mu_1}) \frac{\mu_2^y}{(e^{\mu_2} - 1) y!},$$

which, in terms of the parameters is

$$L(\boldsymbol{\beta}, \boldsymbol{\gamma} \mid \mathbf{y}, \mathbf{x}) = \prod_{y \in 0} \exp\left[-\exp\left(\sum \gamma_k X_{ik}\right)\right] \prod_{y \in y > 0} \left\{ 1 - \exp\left[-\exp\left(\sum \gamma_k X_{ik}\right)\right] \right\} \qquad (10.14)$$

$$\times \frac{\prod_{y \in y > 0} \exp\left(y \sum \beta_k X_k\right)}{\left(\left\{\exp\left[\exp\left(\sum \beta_k X_k\right)\right]\right\} - 1\right) y!}. \qquad (10.15)$$

Of importance is that this partitions into two separate likelihoods that can be maximized independently. The reader should recognize expression (10.14) as the likelihood for a complementary log-log model of the probability that Y is greater than zero (see Chapter 7), whereas expression (10.15) is the likelihood for the truncated PRM. Therefore, the Poisson hurdle model can be estimated in two steps: a complementary log-log model for the probability of a positive count, followed by a truncated PRM for cases with positive counts.

Example: Number of Lifetime Physical Assaults. Table 10.5 presents estimates of the regular PRM and of the Poisson hurdle model for the *number of lifetime physical assaults* for the 1343 minority women in the *minority women dataset*. Explanatory variables are *education, income*, the *child abuse index* (as described previously), plus dummy variables for employment status (*other employment,*

Table 10.5 Unstandardized PRM vs. Poisson Hurdle Estimates for the Regression of Number of Lifetime Physical Assaults Experienced by Minority Women in the NVAW Survey

| | | Hurdle Model | |
| | PRM | P(Pos. Count) | Truncated PRM |
Regressor	b	b	b
Intercept	−.953***	−1.306	−4.285
Education	.006	.037	−.411
Income	−.018	.027	−.236
Abused as a child	.282***	.179***	1.493***
Other employment[a]	−.154**	−.117	−.243
Unemployment[a]	−.169*	.202	−1.259
Formerly married[b]	.647***	.469***	3.134
Never married[b]	−.257***	.125	−1.607
Raped	.299***	.408***	.631
Exposure	.283***	−.149	3.109
Model χ^2	2423.127***	122.890***	882.762***
R_L^2	.161	.077	.110
χ^2 for hurdle			4039.749***

Note: $n = 1343$.

[a] Employed either full- or part-time is the reference category.

[b] Currently married is the reference category.

* $p < .05$. ** $p < .01$. *** $p < .001$.

unemployment, with full- or part-time employment as the reference group), dummy variables for marital status (*formerly married, never married*, with currently married as the reference group), and a dummy for having been *raped* (either with or without penetration). The variable *exposure* is the log of the woman's age in years, the rationale being that the older a woman is, the longer she is at risk for assault. Exposure should be positively related to the number of assaults. Moreover, I would expect prior experience of victimization (e.g., having been abused as a child or having been raped) to raise the expected number of assaults. Research suggests that prior victimization may increase women's tolerance of abusive situations, including those that eventuate in physical aggression (Kalmuss, 1984).

The PRM suggests that several factors in addition to *exposure* affect the mean number of assaults: *child abuse, employment status, marital status*, and the experience of having been raped. However, significance tests for the PRM will be too liberal if the data are overdispersed, as these data appear to be, judging from the mean and standard deviation reported above. The next two columns report the Poisson hurdle results. Of interest is whether the hurdle model is an improvement over the PRM. The test is $-2 \ln L_1 - (-2 \ln L_2)$, where model 1 is the PRM and model 2 is the Poisson hurdle model. The log-likelihood for the hurdle model is simply the sum of the log-likelihoods for the complementary log-log part and the truncated PRM part of the model. The test statistic is therefore $-2[-6312.701 - (-3554.224 - 738.603)] = 4039.749$. This statistic has the chi-squared distribution with 10 degrees of freedom under the null hypothesis that $\gamma = \beta$ (i.e., that the hurdle offers no improvement over the PRM). It is highly significant ($p < .0001$), suggesting that the hurdle model is to be preferred. According to this model, the probability of being assaulted is elevated by having been abused as a child, having been raped, or having been married before. The last effect is probably due to the fact that an overwhelming proportion of physical assaults on women are perpetrated by intimate partners. Given that a woman has been assaulted, however, only having been abused as a child increases the expected number of assaults. This finding is consistent with others' work (see, e.g., Kalmuss, 1984).

EXERCISES

10.1 Recall that for a discrete variable, Y, the sum of $f(y)$ over the values of Y equals 1.0, where $f(y)$ is the density of Y. Show that the truncated Poisson density has this property.

10.2 Show that the ZIP density has the property mentioned in Exercise 10.1.

10.3 Suppose that Y has a Poisson density with mean equal to 3.2. Find (**a**) $P(Y = 0)$; (**b**) $P(Y = 3)$; (**c**) $P(Y = 10)$.

10.4 Suppose that Y has a Poisson density with mean equal to 1.5. Find (**a**) $P(Y = 0)$; (**b**) $P(Y = 3)$; (**c**) $P(Y = 10)$.

10.5 Suppose that Y has a negative binomial density with parameters $r = 4$, $p = .1$. Find (**a**) $P(Y = 0)$; (**b**) $P(Y = 3)$; (**c**) $P(Y = 10)$.

10.6 Suppose that Y has a negative binomial density with parameters $r = 1$, $p = .45$. Find (**a**) $P(Y = 0)$; (**b**) $P(Y = 3)$; (**c**) $P(Y = 10)$.

10.7 Suppose that Y has a negative binomial density and let $r = 1$. (**a**) Find p such that $E(Y) = 3.2$. Then find (**b**) $P(Y = 0)$; (**c**) $P(Y = 3)$; (**d**) $P(Y = 10)$.

10.8 For a PRM with covariates $X_1, X_2, \ldots, X_K$, show that the multiplicative change in $E(Y)$ for a c-unit increase in X_j is $\exp(\beta_j c)$.

10.9 Give the log of the likelihood function for the truncated PRM.

10.10 Prove that $1 - \mu e^{-\mu}/(e^{\mu} - 1)$ is always less than 1 for $\mu > 0$. (*Hint:* Start by assuming that the assertion is true, and manipulate the inequality until a more obviously correct assertion appears.)

10.11 Greene (1998) gives the following recursion formula for the probabilities of various Y-values under the PRM: $p_0 = \exp(-\mu)$; $p_j = (\mu/j)(p_{j-1})$ for $j = 1$,

2, ... Show that this recursion is equivalent to the formula $e^{-\mu}\mu^y/y!$ for $y = 0, 1, 2, 3, 4$.

10.12 Use Greene's recursion from Exercise 10.11 to find $\hat{f}(y \mid \mathbf{x}, \hat{\beta})$ for $y = 0, 1, 2, 3, 4$ if $\hat{\mu} = 1.75$.

10.13 Greene (1998) gives the following recursion formula for probabilities under the NBRM:

$$p_0 = \left(\frac{\alpha^{-1}}{\alpha^{-1}+\mu}\right)^{\alpha^{-1}}, \, p_j = \frac{\alpha^{-1}+j-1}{j}\frac{\mu}{\alpha^{-1}+\mu}p_{j-1}.$$

If $\hat{\alpha} = 3.5806$ and $\hat{\mu} = 1.75$, give the estimated probabilities for $y = 0, 1, 2, 3, 4$, based on the NBRM, using this recursion.

10.14 Sherman et al. (1992) randomly assigned 1200 domestic violence offenders to *arrest* versus *no arrest* treatments and then followed them for 6 to 18 months posttreatment. The response of interest was the *number of subsequent police reports* to a local women's shelter for the same offender. Employing the NBRM, their estimated model was

$$\ln \hat{\mu} = \hat{\beta}_0 + \mathbf{g}'\mathbf{x} + .198 \, arrest + .261 \, employed + .026 \, married$$
$$- .434 \, arrest * employed - .3 \, arrest * married,$$

where *arrest, employed,* and *married* are all dummy variables for these respective statuses, and $\mathbf{g}'\mathbf{x}$ represents the other terms in the model. Ln L for this model was -1218.77, while ln L for the model without the two cross-product terms was -1222.29.

(a) Test whether the interactions of *married* and *employed* with *arrest* are significant as a block.

(b) Interpret the interaction effects with respect to the conditional mean of *Y*.

10.15 King (1988) employed the PRM to examine several predictors of the *number of members of the U.S. House of Representatives who switched political parties* in a given year for the years 1802–1876. The *exposure* variable was the log of the *number of members of the House* in each year. His estimated model was $\ln \hat{\mu} = \hat{\beta}_0 + \mathbf{g}'\mathbf{x} + 3.49 \, exposure$.

(a) Let N = the *number of members of the House of Representatives* in a given year, and let the estimated equation, in general, be $\ln \hat{\mu} = \hat{\beta}_0 + \mathbf{g}'\mathbf{x} + d \, exposure$, and derive an expression for the proportionate change in the rate of event occurrence for each additional member in the House, net of the other covariates.

(b) Let $N = 250$, and estimate the proportionate change in the expected number of switchers for each additional member of the House of Representatives, using King's equation.

10.16 Show that $\sum_j P(Y=j) = 1$ for $j = 1, 2, \ldots$, in the truncated negative binomial density.

10.17 For the 416 couples in the *couples dataset*, a PRM for the *number of children under 18* in the household produces the following results:

$$\hat{\mu} = \exp[1.9939 - .5976 \ cohabiting - .0428 \ male's \ age \ at \ union$$
$$- .0389 \ union \ duration - .0054 \ household \ income],$$

where *cohabiting* is a dummy for cohabiting as opposed to being married, *male's age at union* and *union duration* are measured in years, and *household income* is in thousands of dollars. All effects are significant at $p < .05$.

(a) Interpret the *cohabiting* effect.
(b) Give the predicted probability of zero, one, and two children for a couple married for 12 years with a household income of $150,000, in which the male was 18 at inception of the marriage.
(c) $\text{Ln} \, L = -620.6443$ for the intercept-only model versus -555.5196 for the hypothesized model. Test the null hypothesis that all regression coefficients equal zero, and give R_L^2.

10.18 Cameron and Trivedi's (1990) regression-based test suggests that the response in exercise 10.17 is overdispersed. An NBRM for the same data produces the following estimates:

$$\hat{\mu} = \exp[2.1295 - .6295 \ cohabiting - .0457 \ male's \ age \ at \ union$$
$$- .0427 \ union \ duration - .0056 \ household \ income],$$
$$\hat{\alpha} = .1812, \ \hat{\sigma}_{\hat{\alpha}} = .0987, \ \text{and} \ \ln L = -552.5279.$$

(a) Do both Wald and likelihood-ratio chi-squared tests for equidispersion versus overdispersion. What do you conclude?
(b) Repeat Exercise 10.17(b) based on the NBRM model.

10.19 A ZIP model for the data in Exercises 10.17 and 10.18 produces the following results:

$$\ln \frac{\hat{\psi}}{1-\hat{\psi}} = -12.4046 + 2.1386 \ cohabiting + .2064 \ male's \ age \ at \ union$$
$$+ .2716 \ union \ duration + .0054 \ household \ income;$$
$$\hat{\mu} = \exp[.7758 - .3345 \ cohabiting - .0085 \ male's \ age \ at \ union$$
$$- .0225 \ union \ duration - .0085 \ household \ income],$$

and $V = 10.0016$ for ZIP/PRM.

(a) Interpret the model with respect both to the probability of being in the structural-zero group and to the expected count given that one is in the at-risk group.
(b) Which model is to be preferred: the ZIP or the PRM?
(c) If $\ln L = -487.8682$ for the ZIP model, give R_L^2 for the ZIP. [*Hint*: Use additional information from Exercise 10.17(c).]

10.20 Repeat Exercise 10.17(b) based on the ZIP results.

For Exercises 10.21 to 10.25, use the 1485 cases in the inmates' dataset to model the number of class III tickets (NUM3TIX) received by inmates as a function of LOG-TIME (the exposure factor), ETHNIC, EDUCCL, FIRSTARR, CENAGE, CENAGESQ, SMALL, MEDIUM. [Also, you need software (e.g., LIMDEP, STATA) for count models.]

10.21 Estimate the PRM for the *number of class III tickets*.
   (**a**) Show the regression estimates and their significance levels.
   (**b**) Perform Cameron and Trivedi's (1990) regression-based test for equidispersion and give your conclusion.
   (**c**) Estimate the discriminatory power of the model using r^2, R_L^2, and/or R_D^2.

10.22 Estimate the NBRM for these data.
   (**a**) Show the regression estimates and their significance levels.
   (**b**) Test the null hypothesis of equidispersion versus overdispersion via both Wald and likelihood-ratio tests.
   (**c**) Estimate the discriminatory power.

10.23 Estimate the ZIP model for these data.
   (**a**) Show estimates for $\ln[\psi/(1-\psi)]$ and for μ.
   (**b**) Which is to be preferred, ZIP or PRM?
   (**c**) Estimate the discriminatory power of the ZIP model.

10.24 Estimate the Poisson hurdle model for these data.
   (**a**) Show coefficient estimates both for the probability of a positive count and for the truncated PRM.
   (**b**) Test the hurdle model against the PRM.
   (**c**) Conceptually, which zero-altered model makes more sense here, the ZIP or the hurdle?

10.25 Construct a graph after the fashion of Figure 10.6 to compare the fit of PRM, NBRM, and ZIP to the observed sample proportions for the first 11 values of NUM3TIX (i.e., values 0 to 10).

CHAPTER 11

Introduction to Survival Analysis

CHAPTER OVERVIEW

In Chapter 10 the response of interest was the number of events occurring in some fixed period of time. This chapter is concerned, instead, with the *waiting*, or *survival*, time in the nonevent state until some event occurs. For example, rather than focusing on the number of subsequent domestic violence offenses committed in a given follow-up period by a sample of those arrested for domestic assault (Sherman et al., 1992), we might instead wish to model the time from release from arrest to the first incidence of recidivism. Or, equivalently, we may be interested in modeling the *risk* of recidivating at any given time, conditional on having survived up to that moment without recidivating. The corpus of tools for analyzing such data is typically collected under the rubric of *survival analysis*, and the risk of an event occurring at any given time is referred to as the *hazard* of the event. These terms stem from the biomedical literature, where many of the techniques were first developed and where the event of interest was often death. In the social sciences, another name for this body of techniques is *event history analysis*. These techniques have been used in the social science arena to study the occurrence of such events as the formation of the first marital or cohabiting union (Lamb et al., 2003), transitions out of cohabiting unions (DeMaris, 2001; Sanchez et al., 1998), the role of premarital cohabitation in marital disruption (Bennett et al., 1988), and gender differences in the promotion process in academic settings (Long et al., 1993), to cite a few examples.

Because event history data present some unique challenges for the data analyst, I begin this chapter by defining terms and acquainting the reader with the nature of survival data. I then illustrate the major concepts of survival analysis, such as survival and hazard functions, using the life table technique, and employ as an application the event of disruption from a marital or cohabiting union. I then consider regression models for event history data, beginning with a parametric model but moving quickly

Regression with Social Data: Modeling Continuous and Limited Response Variables,
By Alfred DeMaris
ISBN 0-471-22337-9   Copyright © 2004 John Wiley & Sons, Inc.

to the much more commonly employed Cox semiparametric regression model. Again, the event of union disruption is used to illustrate the major ideas. The Cox model is explored in some detail, along with strategies for dealing with time-varying covariates, nonproportional effects of covariates, and left truncation. Plots of survival curves and hazard functions along with examples from published research are used throughout to help the reader assimilate the techniques as they are used in practice.

NATURE OF SURVIVAL DATA

Key Concepts in Survival Analysis

Survival, or event history, data consist of observations on when events occur to a sample of people over time. An *event* is a qualitative change in state. In the study of existing marriages, for example, the event of divorce is a change from the married to the unmarried state. In the study of academic promotions, the event of promotion is a change in state from, say, associate professor to full professor, and so on. In the ideal scenario we would follow a collection of people from the time when they first become susceptible to a change in state until they all eventually experience that change. Some events lend themselves readily to this protocol. A sample of patients diagnosed with a terminal illness, for example, may be followed from diagnosis with the illness until death, an event that is sure to occur to all patients if the study continues for sufficient time. Hence survival times can usually be observed for everyone in the sample.

However, many events in the social sciences, such as first marriage, divorce, or pregnancy, are not inevitable; a number of people fail to experience the event regardless of study length. In these cases the survival time is said to be *right-censored*: All that is known is that survival time is greater than the last recorded time for a particular person. For example, if a sample of marriages is followed for 20 years and a given marriage is still intact at the end of the study, that marriage's survival time is censored at 20 years. This means that the couple's survival time is not exactly observed but is known only to be greater than 20 years. A critical assumption is that censoring is *noninformative*; that is, the process leading to being censored is independent of the hazard of event occurrence. If this assumption holds, censored cases are representative of all other persons surviving up until the same time, controlling for relevant covariates (Collett, 1994). In studies of marital dissolution, for example, couples' survival times are typically considered censored upon one partner becoming widowed. As the probability of dying is not normally presumed to be affected by marital instability, this type of censoring is typically considered noninformative.

I have been referring to "following" people or cases over time as though survival data are exclusively *prospective*. This is not at all the case. Much of the data on event histories in the social sciences are collected *retrospectively*. For example, the NSFH interview schedule contains long sections exploring respondents' marital and cohabitation histories, job histories, fertility histories, and so on. As long as significant events can be associated with specific dates in a case's biography, event histories can be constructed after the fact.

A word is in order about the use of the term *time*. In event history analysis, time typically refers to the duration in a given state rather than calendar time. Nevertheless, sometimes "time" refers to calendar time. In fact, a critical concept in survival analysis is the calendar time at which a case is first exposed to the risk of an event. I refer to this time as the *inception of risk*. Once the inception of risk is given, survival time is then calculated as the calendar time at which the event of interest occurs minus the inception of risk. If cases are lost to follow-up or the study ends before an event occurs, the time at which the case was last observed minus the inception of risk gives the censored survival time for that case.

In this chapter I moreover assume that time is a continuous variable, which gives rise to *continuous-time survival models*. In practice, time is never measured so precisely as to be truly continuous, but it can be treated as such if measured finely enough—say in days, weeks, or months. Use of a continuous-time model is predicated, however, on the notion that an event can occur at any given time or after any given amount of time has elapsed. However, some events can only occur at particular calendar times. For example, promotion in academic rank typically occurs only at the beginning of the school year. In this type of process, survival time is a discrete variable and the appropriate analytic technique is a *discrete-time survival model*. Even if the true event-generating process is a continuous one, however, a discrete-time analysis may still be appropriate if events are only known to occur in some interval of time. Discrete-time analyses are considered in Chapter 12.

Nature of Event Histories

To further tease out critical concepts in survival analysis, I direct the reader to Figure 11.1, which provides a schematic of various event histories in a given study. For simplicity, it is assumed in the figure that the event of interest is a single, nonrepeatable event. Examples would be the first instance of sexual intercourse or the first marriage. Although either of these events may be repeated several times, the first instance of either is a one-time affair. The horizontal axis represents survival time. Each line stands for a person's event history and represents the time period from inception of risk until either censoring or termination in the event of interest. These periods are also referred to as *episodes* or *spells*. Each line moves from left to right in the figure, with vertical bars on the left indicating known inceptions of risk and vertical bars on the right indicating right-censoring times. An arrowhead at the end of the line indicates that the person experienced the event of interest at a given time. Solid lines represent durations observed in a given study, while dotted lines reflect survival time that is outside the study. Individuals a, b, c, and e experience inception of risk at t_0, which is also the beginning of observation, or *start time* for the study. I consider these people first, as they are the most common kinds of observations. Case a represents an observation that is right-censored by the ending date of the study, denoted t_1. Case b, in contrast, is right-censored by virtue of either being lost to follow-up, or by experiencing a different type of event that removes him or her from the risk for the event of interest (e.g., the death of a spouse removes couples from the risk for divorce). Case c experiences the event of interest during the observation period and is referred to as being *uncensored*.

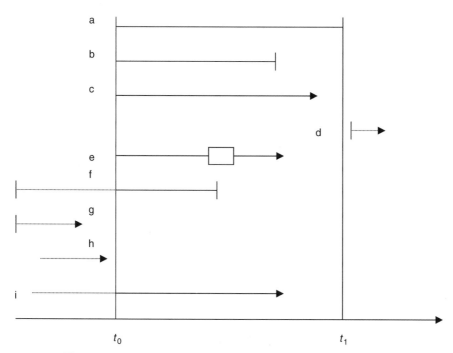

Figure 11.1 Schematic of event history data for single nonrepeatable events.

Case e experiences a different type of event—symbolized by a rectangle—that affects his or her risk for the event of interest. Influential events of this type are typically modeled as time-varying covariates, a topic taken up below. In sum, cases a, b, c, and e are all easily accommodated in standard survival analyses.

Other cases pose particular problems that may or may not be tractable. Case f has inception of risk at some time prior to its coming under observation at t_0. The duration over which f is at risk prior to t_0 is indicated by the dotted line. For case f, inception of risk is known. Case i is similar except that inception of risk is unknown. Both f and i are said to have *left-truncated* survival times. In survival analysis people are said to have left-truncated survival times whenever they have been at risk for the event of interest for some period prior to the start time of the study. As in previous chapters, truncation refers to the situation in which respondents are observed only when their responses are above or below some threshold value. In this instance, observation of survival times for left-truncated cases is predicated on their survival in the nonevent state up to the start of the study. Hence, if t_b is the duration of survival before the start of the study and T denotes survival time, in general, left-truncated observations are observed only when $T > t_b$. Left truncation is fairly common in panel studies in which respondents in a given state at time 1 are tracked in subsequent surveys to see if they experience the event of interest. As an example, the author followed married as well as unmarried cohabiting couples in wave 1 of the NSFH to explore whether intimate violence reported at time 1 predicted separation or divorce by wave 2 (DeMaris, 2000). As virtually all of these unions were begun prior to the date of the wave 1 survey, the data were characterized by considerable left truncation. Below I discuss the difficulties associated with left truncation, along with simple remedies for the problem when the inception time is known, as with case f. When this time is unknown, the problem is considerably less tractable.

The remaining cases pose more serious problems for survival models. Case d experiences the event of interest but inception of risk and time of event occurrence both take place after the study is completed. This type of case is referred to as being *fully right-censored* (Yamaguchi, 1991) and is not amenable to analysis using survival techniques. Similarly, cases g and h experience the event of interest before the start of the study. As an example, suppose that we were to follow a group of 12-year-olds to observe the length of time before they smoke their first cigarette. However, some children in the sample have already begun smoking but cannot remember the date on which that occurred. Assuming that inception of risk begins at birth, all we know of such children is that survival time is *less than* 12 years. These types of cases are accordingly known as *left-censored* survival times (Collett, 1994; Hosmer and Lemeshow 1999) and are also not very amenable to survival modeling.

A final concept of central importance that can be gleaned from Figure 11.1 is the *risk set*. This is the set of people who are at risk for event occurrence at any given time t. For example, the risk set at time t_0 in the figure consists of cases a, b, c, e, f, and i. Immediately after case f has been censored, the risk set consists of cases a, b, c, e, and i. Immediately before case c experiences the event, however, the risk set only consists of cases a and c, since cases b, e, f, and i have either experienced the event (cases e and i) or have been censored (cases b and f) at an earlier time.

Critical Functions of Time: Density, Survival, Hazard

In survival models, three functions of time are particularly important: the density function, the survival function, and the hazard function. The three are also closely interrelated, so that given any two, the third is readily calculated. First, we note that survival time, denoted by T, is a random variable ranging from zero to infinity, which like any other variable, has population distribution and density functions. Certain densities not featured elsewhere in this book are especially important in survival models. Examples are the exponential, Weibull, Gompertz, and log-logistic densities. (The exponential density was introduced in the Chapter 1 appendix.) As the exponential function is very easy to work with, I use it to illustrate the three functions discussed in this section. If survival time has an exponential distribution, its density function is

$$f(t) = \lambda \exp(-\lambda t),$$

where λ is a positive constant. For example, if survival time in days after contracting some disease has an exponential distribution with $\lambda = .35$, the density function at a time of 15 days is

$$f(15) = .35 \exp[-(.35)(15)] = .0018.$$

Recall that as the density of a continuous variable is not a probability, this value has no intuitive interpretation.

In general, the *distribution* function for survival time is

$$F(t) = P(T \leq t) = \int_0^t f(u)\, du.$$

As is evident, $F(t)$ is just the probability that T is less than or equal to t, and the integration symbol simply denotes that this function is found by *integrating*, or summing up, the area under the density curve between 0 and t. For the exponential distribution, we have

$$F(t) = 1 - \exp(-\lambda t).$$

In the example just given, the distribution function evaluated at 15 days is

$$F(15) = 1 - \exp[-(.35)(15)] = .995.$$

This means that the probability of dying within 15 days' time is .995.

The survival function, denoted $S(t)$, is simply $P(T > t)$, that is, the probability of surviving *beyond* time t. For the exponential density,

$$S(t) = 1 - P(T \leq t) = 1 - [1 - \exp(-\lambda t)] = \exp(-\lambda t).$$

To continue with the example, the probability of surviving beyond 15 days is

$$S(15) = \exp[-(.35)(15)] = .0052.$$

The interpretation is that there is only a .52% chance of surviving beyond 15 days.

Of central importance in survival analysis is the *hazard function*. In fact, the log of the hazard function is most often the response variable in survival regression, giving rise to models known as *hazard models*. Ideally, what we would like to model is the probability of experiencing the event of interest exactly at time t. (In discrete-time models, we *can* employ this as the response variable.) However, the probability that a continuous variable—in this case, time—takes on any specific value in its range is zero. So, instead, we define the hazard as follows. First, let Δt represent some small increment in time. Then consider the probability that the event occurs in some small time interval, say between t and $t + \Delta t$, conditional on its not having occurred yet. This probability is denoted

$$P(t \leq T < t + \Delta t \mid T > t).$$

If we further divide this probability by Δt, we would have the conditional probability of event occurrence in some small time interval *per unit of time*, or the *rate* of event occurrence in a small time interval. The hazard at time t, denoted $h(t)$, is then the limit of this rate as the time interval shrinks to zero. Formally, we have

$$h(t) = \lim_{\Delta t \to 0} \frac{P(t \leq T < t + \Delta t \mid T \geq t)}{\Delta t}. \tag{11.1}$$

The hazard in continuous time can be interpreted as the *instantaneous rate of event occurrence*, or as Blossfeld et al. (1989, p. 31) describe, it is "... the instantaneous probability that episodes in the interval $[t, t + \Delta t]$ are terminating provided that the event has not occurred before the beginning of this interval."

Relationships between the hazard, survival, and density functions follow from the foregoing definitions. Now, the density function is the first derivative of the distribution function. Thus, the density function can be written as

$$f(t) = F'(t) = \lim_{\Delta t \to 0} \frac{F(t + \Delta t) - F(t)}{\Delta t} = \lim_{\Delta t \to 0} \frac{P(t \leq T < t + \Delta t)}{\Delta t},$$

and by the rules for conditional probabilities,

$$P(t \leq T < t + \Delta t \mid T \geq t) = \frac{P(t \leq T < t + \Delta t \cap T \geq t)}{P(T \geq t)} = \frac{P(t \leq T < t + \Delta t)}{P(T \geq t)}.$$

Therefore, $h(t)$ can be written as

$$h(t) = \lim_{\Delta t \to 0} \frac{P(t \leq T < t + \Delta t \mid T \geq t)}{\Delta t} = \lim_{\Delta t \to 0} \frac{P(t \leq T < t + \Delta t) / P(T \geq t)}{\Delta t}$$

$$= \lim_{\Delta t \to 0} \frac{P(t \leq T < t + \Delta t) / \Delta t}{P(T \geq t)} = \frac{f(t)}{S(t)}. \tag{11.2}$$

In other words, the hazard function is the ratio of the density function to the survival function. Notice the similarity to the hazard function of the normal distribution— lambda—as defined in Chapter 9. Equation (11.2) implies further that $f(t) = h(t)S(t)$.

Employing equation (11.2), we find that the hazard function for the exponential distribution is

$$h(t) = \frac{\lambda \exp(-\lambda t)}{\exp(-\lambda t)} = \lambda.$$

As this is a constant rather than a function of t, the exponential distribution is said to have a *constant hazard*. In other words, the hazard of event occurrence remains constant over time. As this may be unrealistic for many applications, other distributions, such as the Weibull, allow for the hazard to change over time. However, the focus in this chapter is on a semiparametric modeling approach for the log hazard, or $\ln[h(t)]$, that does not depend on the choice of distribution for survival time. I therefore omit detailed discussion of more complex distributions.

Example: Dissolution of Intimate Unions

To illustrate the concepts just discussed, and in particular, to give a sense of the survival and hazard functions, I consider data drawn from both waves of the NSFH on the dissolution of married and unmarried cohabiting relationships. The example is based on a larger study (DeMaris, 2000) that examined the impact of intimate violence on union disruption. Whereas that study included all married and cohabiting couples who had been together for up to 20 years in wave 1, the present example is limited to the 1230 couples who had been together for at most three years in wave 1. Five to seven years later, in the wave 2 survey, the same couples were queried as to whether they were still together. If they had split up, they were asked to give the dates of separation or divorce. For consistency between cohabitors and marrieds, the date of separation was employed to index the event of *union disruption*. The few couples experiencing the death of either partner were considered censored as of the date of death. Because most couples had been in their unions for some time (anywhere from 0 to 3 years) before being initially surveyed, the data are left-truncated. Nevertheless, I will begin by ignoring this problem and treat couples as though they were all observed from inception of risk. Later I adjust the analyses for left truncation and compare results.

Nonparametric Estimation of S(t) and h(t). A first step in survival analysis typically involves examining survival and hazard functions for the sample as a whole, ignoring potential differences among people induced by explanatory variables. Table 11.1 presents nonparametric estimators of S(t) and h(t) for the 1230 couples

Table 11.1 Life-Table Estimates of Survival and Hazard Functions for the Event of Union Disruption

Time Interval	Number Failed	Number Censored	Risk Set	Conditional P(Failure)	Survival Function	Hazard Function
[0, 6)	52	0	1230.0	.0423	1.0000	.0072
[6, 12)	35	1	1177.5	.0297	.9577	.0050
[12, 18)	45	0	1142.0	.0394	.9293	.0067
[18, 24)	27	3	1095.5	.0246	.8926	.0042
[24, 30)	28	2	1066.0	.0263	.8706	.0044
[30, 36)	30	2	1036.0	.0290	.8478	.0049
[36, 42)	30	0	1005.0	.0299	.8232	.0051
[42, 48)	16	3	973.5	.0164	.7986	.0028
[48, 54)	17	2	955.0	.0178	.7855	.0030
[54, 60)	18	32	921.0	.0195	.7715	.0033
[60, 66)	25	161	806.5	.0310	.7565	.0052
[66, 72)	19	348	527.0	.0361	.7330	.0061
[72, ∞)	6	328	170.0	.0353	.7066	
Total	348	882				

in the sample. The method employed here is the *life-table* approach (Blossfeld et al., 1989). This technique involves partitioning survival time into a series of $q + 1$ nonoverlapping time intervals of the form $[0,a_1)$, $[a_1,a_2)$, . . ., $[a_q,\infty)$, where a_k indicates the kth value of time and then using the censored and uncensored cases in each interval to estimate the two functions. (Recall that "$[a_1,a_2)$" means that the interval includes a_1 but not a_2.) In the current case, since the follow-up interval between survey waves ranged between about five and seven years, survival time in months from the wave 1 interview until either disruption or censoring ranged from 0 to 86. This time period was then partitioned into 6-month intervals, except for the last period, which encompassed months 72 or later. The last interval, shown as "$[72,\infty)$" in the table, was made larger because there were so few disruptions after month 72.

We first estimate the conditional probability of failure for a given interval. Letting d_k represent the number of disruptions, or "failures" in the kth interval, w_k represent the number of cases censored in the kth interval, and R_k represent the number of individuals who had no event until the beginning of the kth interval, the conditional probability of failure for the kth interval is (Blossfeld et al., 1989):

$$\hat{\lambda}_k = \frac{d_k}{R_k - w_k/2}, \tag{11.3}$$

where the denominator is an estimator of the risk set for the kth interval. If there are no censored cases, $\hat{\lambda}_k$ is simply d_k/R_k. However, this simpler computation would tend to underestimate the actual hazard rate in the presence of censoring (Blossfeld et al., 1989). Hence, one-half of the number of censored cases is subtracted from R_k in (11.3). The reasoning here is that if censored observations are assumed to be uniformly distributed over the interval, the average size of the risk set in the interval is $R_k - w_k/2$ (Hosmer and Lemeshow, 1999). As an example, R_1 is 1230, and there are no censored cases, so the first conditional probability of failure is simply $52/1230 = .04227$ or $.0423$. In the second interval, we start with $1230 - 52 = 1178$ couples who have had no event so far. There is one censored case in this interval, so the risk set is $1178 - \frac{1}{2} = 1177.5$, and $\hat{\lambda}_2 = 35/1177.5 = .0297$. Now, the risk set for the next interval must remove the remaining censored cases plus the 35 uncensored cases, and there are no new censored cases, so it is calculated as $1177.5 - \frac{1}{2} - 35 = 1142$. The remaining computations proceed in a similar fashion.

The conditional probability of failure is used to construct estimates of both the survival and hazard functions. First, let $\hat{P}_k = 1 - \hat{\lambda}_k$ be the conditional probability of survival through the kth interval, given survival through the preceding interval. For example, $1 - .0423 = .9577$ is the conditional probability of surviving through the first interval, and $1 - .0297 = .9703$ is the conditional probability of surviving through the second interval. The survival function estimate, $\hat{S}_k$, is constructed as the product of these conditional probabilities. That is, the survival function estimate for the kth interval is

$$\hat{S}_k = (\hat{p}_k)(\hat{p}_{k-1})(\hat{p}_{k-2}) \cdots (\hat{p}_1).$$

In SAS's PROC LIFETEST, which was used for Table 11.1, the survival function is the proportion of the sample surviving to the *beginning* of the interval, hence $\hat{p}_1$ is always 1. For the second interval, $\hat{S}_2 = (1)(1 - .0423) = .9577$, which is the proportion surviving up to the second interval. The proportion surviving until the third interval is then $(1)(.9577)(1 - .0297) = .9293$, and so on.

Finally, an estimator of the average hazard rate at the midpoint of the kth interval is (Blossfeld et al., 1989)

$$\hat{h}_{m_k} = \frac{2\hat{\lambda}_k}{L_k(2 - \hat{\lambda}_k)},$$

where $L_k = a_k - a_{k-1}$ is the length of the kth interval. Thus,

$$\hat{h}_{m_1} = \frac{2(.0423)}{6(2 - .0423)} = .0072.$$

The other estimates in the column "hazard function" are computed in similar fashion. (No hazard estimate is shown for the 13th interval, as L_{13} cannot be computed.) Plots of the survival and hazard functions from Table 11.1 against the lower limits of the time intervals are shown in Figures 11.2 and 11.3. The survival

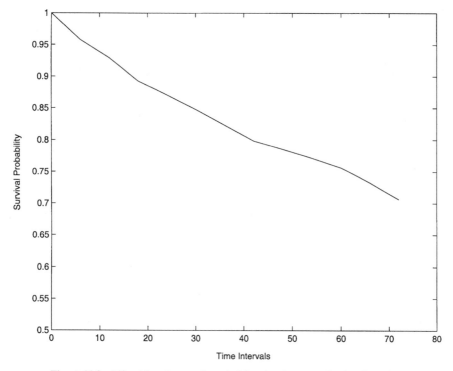

Figure 11.2 Life-table estimator of survival function for event of union disruption.

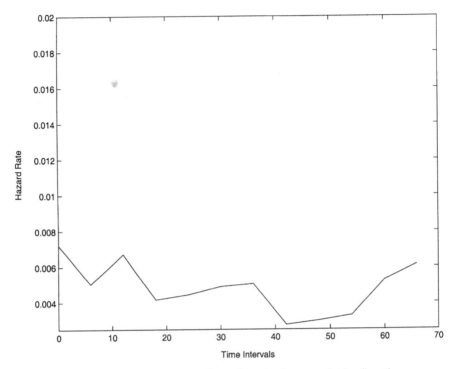

Figure 11.3   Life-table estimator of hazard function for event of union disruption.

function exhibits a relatively smooth decline from a high of 1 at the start of the process, to a minimum of .7066 for the interval with a lower limit of 72 months. Apparently, slightly over 70% of relationships survived for at least 72 months intact. The hazard function, on the other hand, displays a more erratic trend; there appears to be no clear upward or downward pattern. The greatest hazard of disruption is apparently in the first six-month interval, although the last interval's hazard is almost as high. Nonetheless, all of the hazards are quite low, with the highest being only .0072.

Examining Parametric Forms for the Hazard Function.   As Blossfeld et al. (1989) explain, nonparametric estimates of the survival function can be used to suggest a potential parametric model for the log of the hazard function. In particular, a plot of various functions of $\hat{S}(t)$ against time can be used to reveal which density function is most appropriate for survival time. As we will see below, if a parametric distribution for time can be identified, we can take advantage of maximum likelihood estimation in regression models for $\ln[h(t)]$, which confers certain benefits (e.g., asymptotic efficiency) over alternative estimation techniques. To understand the idea behind these transformations, we again examine the exponential density. Recall that its survival function is $S(t) = \exp(-\lambda t)$. Taking logs of both sides of this equation and multiplying both sides by -1, we have $-\ln S(t) = \lambda t$. That is, $-\ln S(t)$ is a linear function of time with an intercept of zero. Therefore, if time has an exponential distribution, a plot of $-\ln \hat{S}(t)$ against time should form a straight line through the origin.

Similarly, a Weibull distribution is supported if a plot of $\ln[-\ln \hat{S}(t)]$ against the log of time produces a straight line. On the other hand, if the function $\ln(\ln[\hat{S}(t)/\hat{S}(t+1)])$ has a linear relationship with time, a Gompertz distribution is supported. In this case, t and $t+1$ represent adjacent times or time intervals. Finally, if the function $\ln[(1 - \hat{S}(t))/\hat{S}(t)]$ has a linear relationship with the log of time, a log-logistic distribution for survival time is indicated. Rather than relying on visual inspection alone, one can regress the relevant transformation of $\hat{S}(t)$ on time and include quadratic or cubic terms in time to test for nonlinear effects. When this was done with the life-table estimates of $S(t)$ from Table 11.1, the exponential, Weibull, and log-logistic distributions were all supported. However, as the exponential distribution is the most parsimonious of the three, and as the plot in Figure 11.3 is consistent with a constant hazard, I will choose the exponential distribution for survival time, at least to start with.

REGRESSION MODELS IN SURVIVAL ANALYSIS

The hazard function depicted in Figure 11.3 assumes that all people have the same hazard at any given time. This is, of course unrealistic, since individual characteristics will typically raise or lower the hazards for certain people compared to others. One way to introduce individual *heterogeneity*, or variability, into the hazard function is to model $\ln[h(t)]$ as a function of a set of covariates representing people's attributes, giving rise to regression models for the log of the hazard function. In this section of the chapter I begin with a parametric model: in particular, the exponential model. As programs for the estimation of parametric models (e.g., PROC LIFEREG in SAS) usually model the log of survival time, rather than the log of the hazard, I begin with log survival-time models.

Accelerated Failure-Time Model

Letting T_i denote the survival time for the ith person, parametric models take the form

$$\ln T_i = \mathbf{x}^{i\prime}\boldsymbol{\beta} + \sigma\varepsilon_i, \tag{11.4}$$

where σ is a parameter to be estimated and ε_i is a random disturbance term whose density determines the parametric form of the model. For example, the exponential model is specified by constraining σ to equal 1 and assuming that ε_i has a Gumbel distribution (Allison, 1995), since this specification implies an exponential distribution for survival time. As noted, the exponential model assumes that the hazard of the event of interest is constant over time. A Weibull model would be specified by retaining the Gumbel distribution for ε_i but allowing σ to depart from the value of 1. If $\sigma > 1$, the hazard is decreasing with time. If σ is between .5 and 1, the hazard is increasing at a decreasing rate. If σ is between 0 and .5, the hazard is increasing at an increasing rate. And if $\sigma = .5$, the hazard is increasing at a constant rate (Allison,

1995). Models of the form in equation (11.4) are referred to as *accelerated failure-time* (AFT) *models*. The reason for this is that if we exponentiate both sides of equation (11.4), we get

$$T_i = \exp(\mathbf{x}^{i\prime}\boldsymbol{\beta} + \sigma\varepsilon_i) = e^{\beta_0} e^{\beta_1 X_1} \cdots e^{\beta_K X_K} e^{\sigma\varepsilon_i},$$

which makes it clear that increases in covariate values either accelerate or decelerate the survival (or, conversely, the failure) time.

Maximum Likelihood Estimation. Assuming that the disturbance term in equation (11.4) has a Gumbell distribution allows us to estimate the parameters of this equation via maximum likelihood. To form a general expression for the likelihood function for survival data, we reason as follows. First, let c_i be the censoring indicator for the ith person, coded 1 if the person is uncensored, and 0 if censored. (This seems like backward coding for a dummy variable but is conventional coding in survival analysis.) Then, for the ith uncensored case, let $f(t_i, \boldsymbol{\beta}, \mathbf{x}^i)$ be that case's density at time $T = t_i$ given parameter vector $\boldsymbol{\beta}$ and covariate vector $\mathbf{x}$. Similarly, for the ith censored case, let $S(t_i, \boldsymbol{\beta}, \mathbf{x}^i)$ be that case's survivor function, or probability of survival through time $T = t_i$, given parameter vector $\boldsymbol{\beta}$ and covariate vector $\mathbf{x}$. A general expression for the likelihood function for n sample cases is then (Hosmer and Lemeshow, 1999)

$$L(\boldsymbol{\beta} \mid \mathbf{t}, \mathbf{x}) = \prod_{i=1}^{n} \{[f(t_i, \boldsymbol{\beta}, \mathbf{x}^i)]^{c_i} [S(t_i, \boldsymbol{\beta}, \mathbf{x}^i)]^{1-c_i}\}. \tag{11.5}$$

Notice that all of the sample information regarding event and censoring times is being exploited in expression (11.5). Taking logs of both sides of this function, a general expression for the log of the likelihood function is

$$\ln L(\boldsymbol{\beta} \mid \mathbf{t}, \mathbf{x}) = \sum_{i=1}^{n} \{c_i \ln[f(t_i, \boldsymbol{\beta}, \mathbf{x}^i)] + (1 - c_i) \ln[S(t_i, \boldsymbol{\beta}, \mathbf{x}^i)]\} \tag{11.6}$$

To construct the likelihood for the exponential model, we must first specify its density and survival functions. First, to simplify notation somewhat, we rewrite the exponential model letting $Y_i = \ln T_i$ and imposing the constraint that $\sigma = 1$:

$$Y_i = \mathbf{x}^{i\prime}\boldsymbol{\beta} + \varepsilon_i. \tag{11.7}$$

Then note that in general, if X has a Gumbel distribution, its density function is

$$f(x) = \exp(x - e^x), \tag{11.8}$$

whereas its survival function is

$$S(x) = \exp(-e^x). \tag{11.9}$$

Recall that ε_i in equation (11.7) has a Gumbel distribution, and notice that this equation also implies that ε_i can be written as $y_i - \mathbf{x}^{i\prime}\boldsymbol{\beta}$. Then, substituting $y_i - \mathbf{x}^{i\prime}\boldsymbol{\beta}$ for x

in equations (11.7) and (11.8), and inserting both into equation (11.6), the log-likelihood function for the exponential model is:

$$\ln L(\boldsymbol{\beta}\,|\,\mathbf{t},\mathbf{x}) = \sum_{i=1}^{n} (c_i \ln\{\exp[y_i - \mathbf{x}^{i\prime}\boldsymbol{\beta} - \exp(y_i - \mathbf{x}^{i\prime}\boldsymbol{\beta})]\}$$
$$+ (1 - c_i)\ln\{\exp[-\exp(y_i - \mathbf{x}^{i\prime}\boldsymbol{\beta})]\}),$$

which after some algebraic manipulation reduces to

$$\ln L(\boldsymbol{\beta}\,|\,\mathbf{t},\mathbf{x}) = \sum_{i=1}^{n} [c_i(y_i - \mathbf{x}^{i\prime}\boldsymbol{\beta}) - \exp(y_i - \mathbf{x}^{i\prime}\boldsymbol{\beta})] \tag{11.10}$$

As always, maximizing function (11.10) with respect to $\boldsymbol{\beta}$ provides MLEs for the elements in $\boldsymbol{\beta}$. As the focus of this chapter is on the Cox model, discussed below, I postpone discussion of inferences until that section. However, the usual test statistics based on MLEs are applicable, including likelihood-ratio tests for the model as a whole, chi-squared difference tests for nested models, and Wald chi-squared tests for individual coefficients. These are similar in form to those for other likelihood-based models discussed in Chapters 7 to 10.

Application to Union Disruption. Model 1 in Table 11.2 presents the results of estimating the exponential model for the *log of survival time* in 1230 married and unmarried cohabiting unions in the NSFH. Several covariates that have been

Table 11.2 Parametric and Semiparametric Regression Models for the Hazard of Union Disruption

Predictor	Model 1[a]	Model 2[b]	Model 3[c]	Model 4[d]	Model 5[e]
Intercept	1.917***				
Relationship duration (months)	.029***	−.028***	−.021***	−.028***	−.022***
Female's age at union	.054***	−.053***	−.053***	−.074***	−.055***
Both in a first union	.792***	−.779***	−.779***	−.479***	−.621***
Alcohol or drug problem	−.376*	.360*	.373*	.363*	.368*
Open disagreement	−.030*	.029*	.030*	.031*	.031*
Conflict resolution style	.233***	−.228***	−.229***	−.184***	−.190***
Continuously cohabiting				1.193***	1.540***
Cohabiting to married				−.159	.321
Union birth				−1.457***	−.557***

Note: $n = 1230$.
[a] Exponential model.
[b] Cox model.
[c] Cox model adjusted for left truncation.
[d] Cox model adjusted for left truncation, with *cohabiting to married* and *union birth* treated as time invariant.
[e] Cox model adjusted for left truncation, with *cohabiting to married* and *union birth* as time-varying covariates.
* $p < .05$. ** $p < .01$. *** $p < .001$.

shown in past research to affect relationship dissolution are included. All covariates employed in models 1 to 3 were measured in the first wave of the survey and are fairly self-explanatory. *Relationship duration* in months taps the number of months from inception of the union (whether marital or cohabiting) until the date of the wave 1 interview. *Female's age at union* is a proxy for the couple's age at inception of the union, since partners' ages are highly correlated. *Both in a first union* is a dummy flagging couples in which neither partner has married or formed a cohabiting union before. *Alcohol or drug problem* is a dummy that identifies couples in which either partner has an alcohol or drug problem. *Open disagreement* is the frequency of open disagreements in the relationship over the past year, a continuous variable ranging from 6 to 31.2. *Conflict resolution style* is also a continuous variable, ranging from 2 to 10, tapping the extent to which disagreements have been calm and partners have avoided arguing heatedly or shouting.

Interpreting the Exponential Model. As is evident, all variables are significant predictors of survival time, with longer survival times evidenced by couples who had been together longer before wave 1, who were older when entering the union, who were in a first union, and who had calmer conflict resolution styles. Predictably, couples in which either partner had an alcohol or drug problem or those with more frequent disagreements had shorter survival times. It is straightforward to generate predicted survival times using the model. For example, consider a couple together for two years at wave 1, in which the woman was 21 at the start of the union, who were both in a first union, who had no substance abuse problems, whose *open disagreement* score was 10, and whose *conflict resolution* score was 5. Their predicted log-survival time in months is

$$\ln \hat{T} = 1.917 + .029(24) + .054(21) + .792 - .030(10) + .233(5) = 5.404,$$

which implies a survival time of 222.3 months, about $18\frac{1}{2}$ years.

Of perhaps more importance are the interpretations of the effects. Each coefficient represents the estimated additive change in *log-survival time* for a 1-unit increase in the relevant predictor, net of other covariates. Or, exponentiating the coefficient gives us the acceleration or deceleration in survival time for a 1-unit increase in the predictor. Hence, each year older the woman is before entering a union magnifies survival time by a factor of $\exp(.054) = 1.055$, or 5.5%. Similarly, those with substance abuse problems are estimated to have survival times that are lower by a factor of .687 compared to others. Or, their survival times are estimated to be reduced by about 31.3%. A convenient feature of the exponential model is that if the sign of the coefficient is reversed, we get the effect on the log of the hazard. This makes perfect intuitive sense, since a greater hazard of event occurrence should shorten survival time, and vice versa. Thus, the effect of relationship duration on the log hazard is $-.029$, while the effect of alcohol or drug problems is .376. Exponentiating these coefficients provides us with an estimate of the *hazard ratio*—the ratio of hazards—for those who are 1 unit apart on the relevant predictor. As an example, the hazard ratio for those with alcohol or drug problems versus others is exp(.376) = 1.456. Or, the hazard of disruption for couples with substance abuse problems is elevated by about 46%.

Cox Regression Model

One problem with parametric models is that their accuracy depends very much on whether the correct distribution is selected, and selection of the correct distribution, unless guided by strong theory, can be a challenging task. Recall that an exploration of the parametric form of the survival-time distribution for the union disruption problem using the techniques enumerated above failed to identify a single distribution as best. The exponential distribution was chosen for simplicity. But during model estimation a Lagrange multiplier chi-squared test for the constraint on σ (not shown) resulted in rejection of the null hypothesis that $\sigma = 1$. As the value of $\hat{\sigma}$ is 1.216, there is evidence that the hazard of union disruption is declining over time. (However, a declining hazard can also be an artifact of unmeasured heterogeneity, as explained in Chapter 12.) This suggests that the Weibull model may be more appropriate, although there is no guarantee that it is *the* correct distribution, either.

In a highly influential 1972 paper, the statistician D. R. Cox proposed a model and associated estimation technique that obviates the need to identify the appropriate distribution for survival time. Known widely as the *proportional hazards model*, this tool has come to dominate survival analysis in many fields. Although this technique has some drawbacks, its major advantage is that the researcher can be completely indifferent to the form of the survival-time distribution and still obtain good estimates of covariate effects. The estimation technique is called *partial likelihood estimation* and represents a rather remarkable insight on Cox's part. For this reason, it is worth considering in some detail.

Model and Its Interpretation. First, the Cox model takes the form

$$h_i(t) = h_0(t) \exp(\mathbf{x}^{i\prime}\boldsymbol{\beta}), \tag{11.11}$$

where $h_i(t)$ is the hazard of event occurrence at time t for the ith case. Note here that the covariate vector, $\mathbf{x}$, and parameter vector, $\boldsymbol{\beta}$, include only the explanatory variables; there is no constant term in either. The term $h_0(t)$ is a *baseline hazard function* that is left unspecified as to form. It is interpreted as the hazard function for a case whose covariate values are all zero. "Left unspecified" essentially means just that: We do not give it any particular form in the model. It could potentially represent any of the parametric survival-time functions discussed so far, as well as any number of others. As we will see, this function can be safely ignored in the estimation of the parameters. Covariates have the effect of raising or lowering the hazard from the baseline by some fixed amount. Moreover, $\exp(\beta_k)$ represents the ratio of hazards for people who are a unit apart on X_k, controlling for other effects. For example, suppose that there are two covariates in the model, X_1 and X_2. Further, consider the ratio of the hazards for individuals i and j who are 1 unit apart on X_1, controlling for X_2. We have

$$\frac{h_i(t)}{h_j(t)} = \frac{h_0(t)\exp[\beta_1(x_1 + 1)]\exp(\beta_2 X_2)}{h_0(t)\exp(\beta_1 x_1)\exp(\beta_2 X_2)} = \exp(\beta_1).$$

This means that the hazard for individual i is $\exp(\beta_1)$ times the hazard for individual j *at any given time*, or their hazards are proportional—with proportionality constant $\exp(\beta_1)$—over time. Formally, a proportional hazards model is one in which "the hazard for any individual is a fixed proportion of the hazard for any other individual" over time (Allison, 1995, p. 114). Despite this property giving the model its name, it is easy to modify the model to handle nonproportionality, as I show below. The model in log-hazard form is

$$\ln[h_i(t)] = \lambda_0(t) + \mathbf{x}^{i\prime}\boldsymbol{\beta},$$

where $\lambda_0(t) = \ln[h_0(t)]$.

Partial Likelihood Estimation. According to Blossfeld et al. (1989), the likelihood function for equation (11.11) can be factored into the product of separate terms as follows. First, suppose that k of the n people in the sample are uncensored. We then order the k event times such that $t(1) < t(2) < \cdots < t(k)$, where the notation "$t(i)$" represents the ith ordered event time. Additionally, let $R(t(i))$ denote the risk set at the ith ordered event time. As before, the risk set consists of all cases with survival or censored times greater than or equal to $t(i)$. Then the likelihood function for the Cox model is

$$L(\boldsymbol{\beta}, h_0(t) | \mathbf{t}, \mathbf{x}) = \prod_{i=1}^{k} \frac{\exp(\mathbf{x}^{i\prime}\boldsymbol{\beta})}{\sum_{o \in R(t(i))} \exp(\mathbf{x}^{o\prime}\boldsymbol{\beta})} \sum_{o \in R(t(i))} h_0(t(i))\exp(\mathbf{x}^{o\prime}\boldsymbol{\beta}) \prod_{i=1}^{k} S_0(t_i)^{\exp(\mathbf{x}^{i\prime}\boldsymbol{\beta})},$$

where $S_0(t_i)$ is the *baseline survival function* for the ith case. Cox termed the first factor the *partial likelihood* and treated it like an ordinary likelihood, while discarding the remaining terms. Thus, the partial likelihood (PL) function is

$$PL(\boldsymbol{\beta} | \mathbf{t}, \mathbf{x}) = \prod_{i=1}^{k} \frac{\exp(\mathbf{x}^{i\prime}\boldsymbol{\beta})}{\sum_{o \in R(t(i))} \exp(\mathbf{x}^{o\prime}\boldsymbol{\beta})}. \tag{11.12}$$

Cox proposed maximizing function (11.12) with respect to $\boldsymbol{\beta}$ to arrive at the parameter estimates. Because this approach discards information about $\boldsymbol{\beta}$ contained in the other terms in the likelihood function, the PL estimator is not fully efficient. However, the estimator still possesses the other desirable characteristics of ML estimation, such as consistency and asymptotic normality (Klein and Moeschberger, 1997).

Understanding the Partial Likelihood. Let's take a closer look at function (11.12) in order to understand what the terms inside the product operator represent. Each term is based on an application of conditional probability rules to the probability of an event at time $t(i)$. In particular, we ask: What is the "probability" that a person with covariate vector $\mathbf{x}^i$ "fails" at time $t(i)$ given that there is one failure at time $t(i)$? In that time is continuous, we are dealing with hazards instead of probabilities, but the rules are the same. We have

P(case $\mathbf{x}^i$ fails at $t(i)$ | one failure at $t(i)$)

$$
= \frac{\text{P(case } \mathbf{x}^i \text{ fails at } t(i) \cap \text{ one failure at } t(i))}{\text{P(one failure at } t(i))} = \frac{\text{P(case } \mathbf{x}^i \text{ fails at } t(i))}{\text{P(one failure at } t(i))}
$$

$$
= \frac{h_0(t)\exp(\mathbf{x}^{i\prime}\boldsymbol{\beta})}{\sum\limits_{o \in R(t(i))} h_0(t)\exp(\mathbf{x}^{o\prime}\boldsymbol{\beta})} = \frac{\exp(\mathbf{x}^{i\prime}\boldsymbol{\beta})}{\sum\limits_{o \in R(t(i))} \exp(\mathbf{x}^{o\prime}\boldsymbol{\beta})} \qquad (11.13)
$$

It should be clear that the numerator of expression (11.13) is the hazard for the case that fails at $t(i)$ and the denominator is the sum of the hazards for all those at risk at $t(i)$. As is evident here, the baseline hazard function cancels out of the numerator and denominator of expression (11.13) and can therefore be ignored. Finally, to see how function (11.12) is applied in practice, we take a very simplified example. Suppose that the sample consists of three observations on marital unions having survival times of 42, 56, and 86 months. The first two times are disruption times, while the last is a censored survival time. Further, assume there is only one covariate, X_i, in the model. The PL function for these three observations is (where subscripts represent the case numbers)

$$
PL(\boldsymbol{\beta}|\mathbf{t},\mathbf{x}) = \frac{\exp(\beta X_1)}{\exp(\beta X_1) + \exp(\beta X_2) + \exp(\beta X_3)} \cdot \frac{\exp(\beta X_2)}{\exp(\beta X_2) + \exp(\beta X_3)}.
$$

As is evident, censored observations (e.g., the third observation here) contribute only to the denominators in the partial likelihood, not to the numerators. In general, there are as many separate terms in this product as there are uncensored cases.

Advantages and disadvantages of PL. Several comments are in order about the PL technique. First, the PL function only employs information about the ordering of event times, not the times themselves. It is therefore not necessary that these times be measured precisely, as long as we are able to order them from smallest to largest. On the other hand, tied event times pose a problem, since the function assumes that the event times are distinct. Hosmer and Lemeshow (1999) discuss a handful of techniques that have been developed to handle ties, and these are implemented in mainstream software, such as SAS. An exact partial likelihood in the presence of ties is given in Kalbfleisch and Prentice (1980), whereas approximations to the exact likelihood have been developed by Breslow (1974) and Efron (1977). As the exact function requires longer computing time than either approximation, the latter have been viewed as more advantageous in application. However, given the rapid evolution of ever-speedier processors, the issue of computer time may ultimately diminish in importance. Of the two approximations, Efron's is recommended as being closer to the exact method (Hosmer and Lemeshow, 1999). Nevertheless, if there are too many tied survival times in one's data—say more than 5% of the observations are tied at any given time—the discrete-time techniques discussed in Chapter 12 should be considered (Yamaguchi, 1991).

Inferences for the Cox Model. Inferences for the Cox model are performed with tests that parallel the usual procedures for models estimated with maximum likelihood. First there is a likelihood-ratio test for the significance of the model as a whole. [Actually, Hosmer and Lemeshow (1999, p. 98) refer to the test as a "partial likelihood-ratio test," but I will use "likelihood-ratio test" for short.] The null hypothesis for the test is H_0: $\beta_1 = \beta_2 = \cdots = \beta_K = 0$ versus H_1: at least one β_k is not zero. The test statistic is the likelihood-ratio chi-squared ($LR\chi^2$):

$$LR\chi^2 = -2\ln\frac{L_0}{L_1} = -2\ln L_0 - (-2\ln L_1),$$

where L_0 is the partial likelihood for a model with no covariates and L_1 is the partial likelihood for the hypothesized model, evaluated at the partial-likelihood estimates (PLEs) of the parameters. Under the conditions that the sample size is large—in particular, that the number of uncensored cases is large (Hosmer and Lemeshow, 1999)—and the null hypothesis is true, this statistic is approximately distributed as chi-squared with K degrees of freedom. Also, there are $LR\chi^2$ tests for nested models. If model B is nested inside model A, a test for the constraints on A leading to B is $\Delta\chi^2 = LR\chi^2(A) - LR\chi^2(B)$. If the constraints are valid, $\Delta\chi^2$ has a chi-squared distribution with degrees of freedom equal to the number of constraints imposed (e.g., the number of parameters set to zero). As PL estimates are asymptotically normally distributed (Cox, 1975), tests for individual coefficients, referred to as Wald tests, take the form

$$\text{Wald } \chi^2 = \left(\frac{b_k}{\hat{\sigma}_{b_k}}\right)^2,$$

which are one-degree-of-freedom chi-squared tests under the hypothesis that $\beta_k = 0$. Finally, confidence intervals for the β_k, as well as the hazard ratios, $\exp(\beta_k)$, can be constructed using the standard errors of the coefficients, denoted $\hat{\sigma}_{b_k}$, and the critical values of the standard normal distribution. For example, a 95% confidence interval for β_k takes the form $b_k \pm 1.96\hat{\sigma}_{b_k}$. Exponentiating the lower and upper limits of this interval provides a 95% confidence interval for $\exp(\beta_k)$.

Cox Models for Union Disruption. Model 2 in Table 11.2 shows the results of estimating the Cox model for the log hazard of *union disruption*. As mentioned, there is no intercept in the model. The LRχ^2 statistic for the model as a whole is 166.826, which with six degrees of freedom is highly significant ($p < .0001$). The coefficients are quite close in magnitude to those of the exponential model, but opposite in sign, since we are modeling the log hazard rather than log survival time. As with the exponential model, exponentiating a coefficient provides an estimate of the hazard ratio for a unit difference on that predictor, net of other model covariates. Thus, being in a first union is associated with a hazard ratio of $\exp(-.779) = .459$, meaning that first unions have about a 54% lower hazard of union disruption at any given time compared to others. In a similar vein, exponentiating the coefficient for *alcohol or drug problem* suggests that relationships characterized by substance abuse have a 43% higher risk of union disruption at any given time. The standard error for *alcohol or drug problem* is .164. So a 95% confidence interval for the coefficient is $.36 \pm 1.96(.164) = (.039, .681)$. Or a 95% confidence interval for the ratio of hazards for those with, versus without, substance abuse problems is $[\exp(.039), \exp(.681)] = (1.040, 1.976)$. Note that the coefficients for model 2 cannot be employed to arrive at an estimator of a given couple's hazard of disruption. The reason is that an estimator of the baseline hazard function is missing in Cox models. However, we can estimate the impact on the hazard associated with given covariates. As indicated below, we can also use the estimates to construct estimated survival functions.

Adjusting for Left Truncation

As noted above, the sample of unions is characterized by various degrees of left truncation. Left truncation, also referred to as *delayed entry* into the risk set (Hosmer and Lemeshow, 1999) or *interrupted spells* (Hamerle, 1991), represents a form of sample selectivity. In that left-truncated cases have survived long enough to come under observation, they tend to overrepresent low-risk cases among any given cohort. This phenomenon can lead to a loss of estimator efficiency or even to biased estimates if uncorrected (Hamerle, 1991). Programs for AFT models typically do not allow adjustments for left truncation. But the Cox model is easily accommodated to left-truncated data, provided that inception of risk is known for each case, as is true of the current data. Essentially, the partial likelihood function is made conditional on having survived until the start time of the study [see Guo (1993) and Hamerle (1991) for technical details]. In SAS's Cox regression program, PHREG, left-truncated cases can be specified via the ENTRYTIME option on the model statement, as well as in other ways (see Allison, 1995). Model 3 in Table 11.2 shows the estimates for union disruption after adjusting for left truncation. There is little change in the coefficients, probably because of the prior restriction that at the beginning of observation (wave 1), couples had been together for no longer than three years. Left truncation is therefore not as extensive a problem here as it was in the full study [see DeMaris (2000) for details].

Another Nonparametric Estimator of S(t). The Cox model can also be used to obtain a nonparametric estimator of the survival function in the presence of left truncation. This provides an alternative estimator to the standard life-table approach

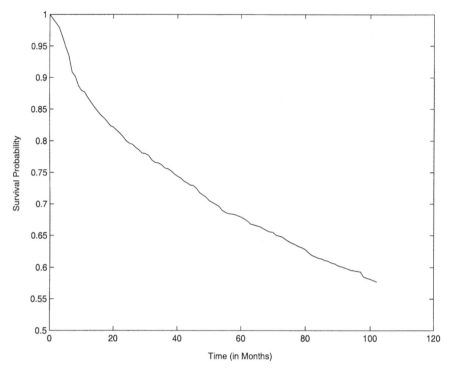

Figure 11.4　Kaplan–Meier estimator of survival function for event of union disruption, adjusted for left truncation.

offered by, say, SAS's procedure LIFETEST, which does not allow such adjustments. The procedure is to estimate the Cox model without any covariates, using the ENTRY-TIME option. The result is an alternative nonparametric estimator known as the *Kaplan–Meier estimator*. This estimator is particularly desirable when there are few tied survival times, because it is based on the exact event and censoring times in the data (Blossfeld et al., 1989). Figure 11.4 shows the nonparametric survival curve—ignoring the covariates—for the union disruption data using this approach.

Estimating Survival Functions in Cox Regression

One drawback to the Cox model is that there is no simple means for recovering the estimated survival function, since information about the survival times themselves is discarded in the estimation process. However, techniques have been developed for estimating the baseline survival function, $S_0(t)$, based on the partial likelihood parameter estimates. One method, according to Klein and Moeschberger (1997), is based on the *cumulative baseline hazard rate* at time t, denoted $H_0(t)$, which is a sort of cumulative sum of the hazards up to time t for all cases at risk. Once **b**, the vector of parameter estimates, is obtained by maximizing the PL likelihood, an estimate of $H_0(t)$ can be constructed. The baseline survivor function is then recovered via the formula $\hat{S}_0(t) = \exp[-\hat{H}_0(t)]$. Estimated survival functions at given values of the covariates can then be calculated as

$$\hat{S}(t \mid \mathbf{x} = \mathbf{x}^0) = \hat{S}_0(t)^{\exp(\mathbf{x}^{0\prime}\mathbf{b})}, \tag{11.14}$$

where $\mathbf{x}^0$ is a given setting of the covariates.

Figure 11.5 presents separate estimated survival functions for those in first unions and those with prior union histories, based on model 3 in Table 11.2. These functions are produced using the BASELINE statement in SAS's PHREG program. This feature estimates baseline survival functions and then applies equation (11.14) to produce separate survival curves for those in first unions versus others. The other model covariates are set to their mean values. The figure conveys quite forcefully that those in first unions have greater survival rates at any given time. For example, 83.5% of first unions are estimated to be intact after five years, compared with 68.5% of those with a prior union history; and 76.1% of first unions are predicted to survive for at least 102 months ($8\frac{1}{2}$ years), as opposed to 55.4% of those with a prior union history. The model, of course, creates these effects since the risk of union disruption is fixed at being 54% lower for

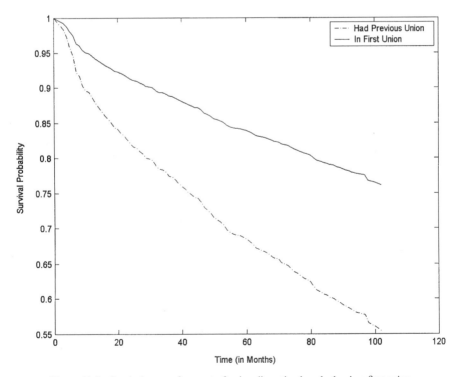

Figure 11.5   Survival curves for event of union disruption by whether in a first union.

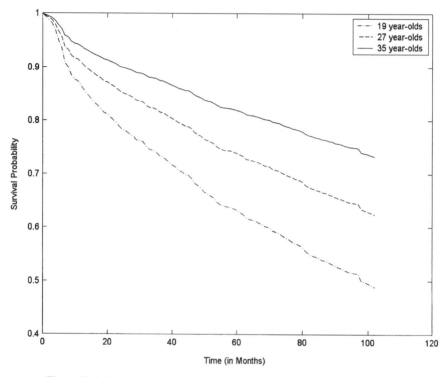

Figure 11.6   Survival curves for event of union disruption by female's age at union.

those in first unions. The reader may notice the appearance of an increasing gap between the survival probabilities of the two groups over time. However, this is just an artifact of the nonlinear transformation from log-hazards to survival rates rather than a differential effect of first-union status over time. In model 3, the effects of all covariates on the log hazard, including first-union status, are posited to be constant over time.

Figure 11.6 shows the effect of age at union formation on survival rates, with other covariates (including first-union status) set to their means. Separate survival curves are shown for three values for *female's age at union*: 19, 27, and 35. Again, the curves reveal the advantage in being older when forming a union: the survival probabilities for the three groups at five years are .632, .739, and .819, respectively.

Time-Varying Covariates

An advantage of the Cox model, and of survival analysis in general, is the ability to incorporate into the model predictor variables whose values vary over time. Called *time-varying covariates*, such explanatory variables are often a natural outcome of following cases over time. Many personal characteristics change over time, such as employment status, annual salary, marital status, number of children borne, whether pregnant, educational level attained, and so on. To effectively model the influence of these variables on the risk of event occurrence, we need to keep track of these changes. On the other hand, causal inferences for such variables must be tendered with caution, as these covariates are often subject to influence by the hazard process itself. Yamaguchi (1991) presents a lucid discussion of the varieties of time-varying covariates and their role in causal modeling. At any rate, the Cox model that incorporates time-varying covariates is

$$h_i(t) = h_0(t) \exp(x_t^{i'}\beta),$$

where the subscript t on the covariate vector indicates that one or more of the covariates may change in value over time. To simplify the notation in later models, however, I omit this subscript, with the understanding that the covariate vector may always include one or more time-varying covariates.

Time-varying covariates are readily specified in Cox regression software such as STATA or SAS. This is done in PHREG, for example, via programming statements after the initial specification of the model. As Allison (1995) admonishes, however, there is no way to check whether these covariate values are being implemented correctly. One advantage of the discrete-time approach discussed in Chapter 12 is that the data can be scrutinized to ensure that time-varying covariates are coded correctly.

In the union disruption data there are two time-varying covariates of interest. One pertains to the eventual marital status of couples who were cohabiting at time 1. Those cohabiting unmarried in wave 1 who subsequently marry should experience greater marital stability than those who remain unmarried. Marriage implies a greater commitment to the permanence of the relationship and is more legally binding on the partners. Therefore, the act of marrying, in and of itself, should reduce the risk of a breakup. Of course, it is also likely that those with higher-quality relationships to begin with are both more inclined to marry and less likely to break up. In this case the transition to marriage may be affected by the same process that affects the hazard of disruption and may not represent a distinct causal influence. In any case, it is desirable to distinguish cohabitors who remained unmarried throughout the follow-up period from those who married. Continuous cohabitation can be modeled via a time-invariant dummy. For cohabitors who married, we can construct a time-varying dummy variable that takes the value 0 until the month in which marriage occurs, at which point it changes to 1.

The other time-varying element of importance is the advent of a birth to the union. Several couples experienced the birth of one or more children to the union during the follow-up period. Having a child, or an additional child, should also mitigate against union disruption. Couples are motivated to preserve their unions when children are present due to the desirability of two-parent households for children's development. However, as in the case of the transition from cohabitation to marriage, fertility decisions may also be influenced by the hazard of disruption. Couples with troubled relationships may postpone childbearing in the anticipation that the union might end. Again, any causal role played by the event of a birth during the follow-up period should be tentatively entertained. Nevertheless, experiencing a birth can be coded as a time-varying covariate that takes the value 0 until the date of the first birth to the union, at which point it changes to 1. In creating this variable, however, we need to ensure that the variable is coded 1 only if the birth occurs prior to a disruption.

Models 4 and 5 in Table 11.2 add the two dummies for cohabiting status plus the dummy for a union birth. In model 4 these dummies are treated as time invariant. In other words, two dummies are created to represent the two types of cohabitation, and a dummy is created to represent the advent of a birth to the union prior to censoring or disruption. These dummies are treated as fixed over the follow-up period. In model 5, on the other hand, *cohabiting to married* and *union birth* are coded as time-varying covariates, as described above. The differential treatment of these variables across models is especially important for the cohabiting-status dummies. The method for coding the time-varying version of *cohabiting to married* necessarily lumps together cohabitors who later marry with married couples, until these cohabitors make the transition to marriage, at which point they form a distinct group. Thus, the risk of disruption for cohabitors who remain unmarried may appear lower than it actually is, due to the mixing of marrieds and cohabitors in the contrast category over time. Entering cohabiting status as time-invariant in model 4 therefore provides a check on the robustness of the cohabitation effects.

The results suggest that the substantive conclusions would be unchanged regardless of specification. According to either model, *continuous cohabitation* elevates the risk of disruption, compared to being married. Cohabitors who transition to marriage, on the other hand, experience no significant inflation in the risk for disruption compared to marrieds. Having a birth lowers the hazard of disruption compared to having no birth. In particular, model 5 suggests that *continuous cohabitation* elevates the risk for disruption by a factor of $\exp(1.54) = 4.665$, or 367%, compared to being married. Model 5 also suggests that there is about a 43% reduction in the hazard of disruption upon the advent of a *union birth*.

Handling Nonproportional Effects

As indicated, the model in equation (11.11) assumes proportionality in the hazards. When is this assumption violated? In truth, the model is already nonproportional as soon as we include time-varying covariates (Allison, 1995; Collett, 1994). The reason is that proportionality of hazards requires the effect of each covariate to be constant over time and also requires the number of units apart of any two people to be constant over time. The latter is violated with time-varying covariates, since these will change at different rates for different individuals, thereby changing the disparities in their covariate values. Regardless, the primary violation of the proportional hazards assumption occurs when effects of covariates depend on time. Nonproportional effects are readily tested and modeled via cross-product terms in the Cox model. We simply add the cross-product of the covariate in question, which can either be time-invariant or time-varying, with survival time. Hence, if the model for the log-hazard is nonproportional in X_j, it takes the form

$$\ln[h_i(t)] = \lambda_0(t) + \beta_1 X_1 + \cdots + \beta_j X_j + \gamma X_j T + \cdots + \beta_K X_K.$$

The effect of X_j on the log-hazard is then $\beta_j + \gamma T$, which clearly varies over time. At $T = 0$, or the inception of risk, the effect is β_j. At $T = t$, however, it is $\beta_j + \gamma t$, and so on. A general test for nonproportionality of covariate effects would be conducted by adding the block of interactions of all model covariates with time to the proportional hazards model, and testing for a significant improvement in fit via the $\Delta \chi^2$ statistic.

Table 11.3 presents the results of estimating nonproportional effects for two types of covariates in the union-disruption model. The first is *alcohol or drug problem*. A reasonable hypothesis is that the deleterious impact on relationship stability of substance abuse problems recorded at a given point in the relationship dampens over time. This can be due to the fact that couples become inured to such addictions on the part of their partner. Or, this could be the result of attempts on the part of the substance abuser to seek treatment for the problem. The second factor investigated is *cohabiting status*. As relationships endure over longer periods, the formalization of unions via marriage may become less important to their stability, implying a declining effect over time for the cohabitation dummies.

Model 1 examines the substance abuse effect by adding an interaction term of the form *alcohol or drug problem * time*. As expected, the effect is significant and negative. The impact of *alcohol or drug problem* on the log-hazard of *union disruption* is seen to be $.972 - .014$ *time*. The value of .972 suggests that the hazard ratio for those with vs. without substance abuse problems is $\exp(.972) = 2.643$ at union inception. After five years, however, the effect on the log-hazard drops to $.972 - .014(60) = .132$,

Table 11.3   Cox Models for Union Disruption Illustrating the Handling of Nonproportional Effects of Covariates

Predictor	Model 1	Model 2	Model 3[a]	Model 4[b]
Relationship duration (months)	−.022***	−.024***	−.029***	−.025***
Female's age at union	−.055***	−.055***	−.062***	−.056***
Both in a first union	−.618***	−.611***	−.728***	−.639***
Alcohol or drug problem	.972**	.965**	.348*	.377*
Continuously cohabiting	1.540***	2.230***		
Cohabiting to married[c]	.319	.296		
Union birth[c]	−.568***	−.603***	−.719***	−.606***
Open disagreement	.031*	.031*	.030*	.031*
Conflict resolution style	−.188***	−.184***	−.236***	−.185***
Alcohol or drug problem × time	−.014*	−.014 [d]		
Continuously cohabiting × time		−.017**		
Cohabiting to married × time		−.0003		
Likelihood ratio χ^2	319.722***	329.853***	178.307***	127.310***
Degrees of freedom	10	12	7	7

Note: $n = 1230$.

[a] Unstratified model.

[b] Model is stratified by marital status.

[c] Time-varying covariates.

[d] $p = .051$.

$* p < .05. ** p < .01. *** p < .001.$

implying a hazard ratio of only 1.141. Model 2 adds the interaction of *continuously cohabiting * time* and *cohabiting to married * time* to model 1. Testing the additional terms as a block results in a $\Delta\chi^2 = 329.853 - 319.722 = 10.131$, which, with 2 *df*, is significant at $p = .0063$. As predicted, both interaction terms are negative, although only the first is significant. Hence, the effect on the log-hazard of disruption for *continuously cohabiting*, compared to being *married*, is 2.23 at inception of the union, suggesting a hazard ratio of $\exp(2.23) = 9.3$. In other words, at the start of a union, those in strictly cohabiting unions have about nine times the risk of disrupting compared to marrieds. After five years, however, this risk ratio drops to $\exp[2.23 - .017(60)] = 3.353$.

Stratified Models

In the event that one suspects nonproportionality in the effect of a nominal covariate and that covariate's effect per se is of tangential importance, it can also be modeled as a stratifiying factor. Stratified models allow the baseline hazard to vary across groups while constraining the effects of covariates to be the same across groups. Why does variation in the baseline hazard function across groups imply nonproportionality of the grouping variable's effect? Consider a parametric model (a Gompertz model, in particular) for the log-hazard, which is linear in *time T*, with an interaction effect of X_j with *time*. Its form is

$$\ln[h_i(t)] = \beta_0 + \gamma_0 T + \beta_1 X_1 + \cdots + \beta_j X_j + \gamma_1 X_j T + \cdots + \beta_K X_K.$$

The effect of time is then $\gamma_0 + \gamma_1 X_j$, which implies that the manner in which the log-hazard varies with time depends on the value of X_j. In the Cox model the parametric form of the relationship with *time*, rather than being linear, can take any form, but the idea is the same. Interactions of covariates with *time* imply either that the effect of the covariate depends on time, as shown in the previous paragraph, or that *the relationship of the hazard with time depends on the covariate or grouping factor*.

In stratified models, each stratum, or value of the grouping factor, has a separate baseline hazard function. Thus, the Cox model for each stratum, *s*, is (Hosmer and Lemeshow, 1999)

$$h_{si}(t) = h_{s0}(t) \exp(\mathbf{x}^{i\prime}\boldsymbol{\beta}),$$

where $s = 1, 2, \ldots, S$. As there is no subscript on $\boldsymbol{\beta}$, the effects of covariates on the hazard are held to be the same for each stratum. The contribution to the partial likelihood for the *s*th stratum is (Hosmer and Lemeshow, 1999)

$$PL_s(\boldsymbol{\beta} \mid \mathbf{t}, \mathbf{x}) = \prod_{i=1}^{k_s} \frac{\exp(\mathbf{x}_s^{i\prime}\boldsymbol{\beta})}{\sum_{o \in R(t_s(i))} \exp(\mathbf{x}_s^{o\prime}\boldsymbol{\beta})},$$

where k_s refers to the total number of events in the sth stratum and $k = k_1 + k_2 + \cdots + k_S$. The full stratified partial likelihood is (Hosmer and Lemeshow, 1999)

$$PL(\boldsymbol{\beta} \mid \mathbf{t}, \mathbf{x}) = \prod_{s=1}^{S} PL_s(\boldsymbol{\beta} \mid \mathbf{t}, \mathbf{x}). \tag{11.15}$$

The PL estimate of $\boldsymbol{\beta}$ is obtained by maximizing function (11.15) with respect to the parameters.

A few comments about stratification are in order. First, the stratifying variable is not in the model, so one loses the ability to estimate the effect of the stratifying variable itself. This is not of particular concern if the stratifying variable is primarily a control rather than a focus variable. Moreover, the effects of all covariates whose values are constant within each stratum are incorporated into the stratum-specific baseline hazard function (Hosmer and Lemeshow, 1999). This implies that one cannot estimate the effects of any other characteristics that only vary across strata rather than within strata. This can be seen as a plus, too, since these factors are all being held constant. That is, any nuisance factors that vary across strata are being automatically controlled in the estimation process. Stratification also offers an advantage over simply incorporating into the model an interaction of the stratifying factor with *time* (Allison, 1995). Including a term in the model of the form $G * time$, where G denotes the stratifying variable, assumes a linear interaction with *time*, which may not be the correct form. With stratification, the functional form of the interaction with *time* is accounted for automatically by allowing separate baseline hazard functions that remain unspecified.

Testing Parameter Invariance. A key assumption of stratified analysis is that the parameter vector, $\boldsymbol{\beta}$, is invariant over strata. If this is violated, it simply means that covariates have different effects, depending on stratum, and the stratified model is not warranted. It is a simple matter to test this assumption, however (Klein and Moeschberger, 1997). One simply estimates the stratified Cox model and obtains $-2\ln L$ for it, denoted $-2\ln L_{st}$. Then one estimates the Cox model for each stratum separately, denoting $-2\ln L$ for each analysis as $-2\ln L_j$, where $j = 1, 2, \ldots, S$. Minus twice the log-likelihood for the model that allows $\boldsymbol{\beta}$ to vary over strata is then

$$-2\ln L_J = \sum_{j=1}^{S} -2\ln L_j.$$

The test statistic for the null hypothesis of parameter invariance is

$$\chi_{st}^2 = -2\ln L_{st} - (-2\ln L_J). \tag{11.16}$$

If $\boldsymbol{\beta}$ is invariant over strata, this statistic has the chi-squared distribution with $K(S-1)$ degrees of freedom. If the test is significant and invariance is rejected, one can always just report the results of estimating the model separately for each stratum.

Union Disruption Example. The last two models in Table 11.3 illustrate the results of stratification. Above, we saw that cohabiting status was seen to interact with time (model 2). Therefore, I estimated a model including only the main effects of covariates (i.e., no interactions with *time*) as both unstratified (model 3) and stratified by cohabiting status (model 4). The unstratified version, model 3, assumes the same baseline hazard across cohabiting statuses, an assumption that has already been rejected. The three strata for model 4 are *marrieds, continuous cohabitors,* and *cohabitors who subsequently married.* Although substantive conclusions are not dramatically altered with stratification, there are noticeable reductions in the effects of *being in a first union, having a union birth,* and *the style of conflict resolution* in the stratified model compared to the unstratified one. Hazard ratios are interpreted in the same manner as in unstratified models, so $\exp(-.606) = .546$ is the ratio of hazards for those with a union birth, versus others, and so on. That is, although the baseline hazards differ across strata, *having a union birth* is posited to magnify the baseline hazard to the same degree in each stratum. Are the effects of covariates invariant across cohabiting-status groups? This was tested by performing the test in equation (11.16). First, I estimated the model (model 3) for each cohabiting-status group separately and summed $-2\ln L$ for the three analyses to arrive at 3660.921 for $-2\ln L_J$. Minus twice the log-likelihood for model 4 is 3678.86. The test statistic is $\chi_{st}^2 = 3678.86 - 3660.921 = 17.939$ and has $7(3-1) = 14$ degrees of freedom. As $p > .2$, the result is nonsignificant, suggesting that the stratified model is valid.

Assessing Model Fit

Discriminatory Power. As always, the degree to which a given model fits the data is of considerable interest. This can be assessed in terms of either the model's discriminatory power or its empirical consistency. As to the first attribute, there is even less agreement on which R^2 analog is best in survival analysis than in the techniques covered in earlier chapters. An investigation by Schemper and Stare (1996) considered several possible measures of discriminatory power for survival models. Nonetheless, the authors admonish that there is no uniformly superior measure, particularly since performance depends heavily on the degree of censoring. The measures most recommended by the authors tend to be computationally cumbersome. However, a measure that appears to perform reasonably well under all conditions and is simple to calculate is the generalized R^2 introduced in Chapter 7, a measure also recommended by other authors (Allison, 1995; Hosmer and Lemeshow, 1999). For the Cox model, the formula is

$$R_G^2 = 1 - \exp\left(\frac{-\mathrm{LR}\chi^2}{n}\right).$$

As an example, discriminatory power for model 2 in Table 11.3 is

$$R_G^2 = 1 - \exp\left(\frac{-329.853}{1230}\right) = .235.$$

As Schemper and Stare (1996) have found R^2 analogs in survival analysis to range typically between .10 and .45, model 2 can be said to exhibit a moderate amount of discriminatory power.

Empirical Consistency. A test for goodness of fit, or empirical consistency, of the Cox model has been recently developed by May and Hosmer (1998). It has been designed in the same spirit as the Hosmer and Lemeshow (2000) goodness-of-fit test for logistic regression. The procedure is as follows. First, fit the Cox model of interest and obtain **b**, the estimate of the parameter vector. Second, calculate the estimated risk score, r_i, for each person as $\hat{r}_i = \mathbf{x}^{i\prime}\mathbf{b}$. (The risk score is nothing more than the linear predictor for each case.) Third, order the risk scores from lowest to highest and then group individuals according to deciles of risk. If there are too few cases to create 10 groups, the authors suggest that a minimum number of groups should be five. Fourth, create dummy variables representing the group each case falls into; for 10 groups, there should be 9 dummies, etc. Fifth, add the block of dummies to the model of interest in a separate step and test whether the block is significant using $\Delta\chi^2$. If the block is not significant, there is evidence that the model has an adequate fit to the data.

As an example, I performed the test on model 4 in Table 11.2 using deciles of risk scores. This is not the final model of interest, as it treats the time-varying covariates as time-invariant. However, the calculation of risk scores with time-varying covariates is problematic since people's risk scores change over time when time-varying covariates are involved. Model 4, on the other hand, facilitates the calculation of a single risk score for each case. I calculated risk scores using the formula $\hat{r}_i = -.028$ *relationship duration* $- .074$ *female's age at union* $- \cdots - 1.457$ *union birth*. Based on a frequency distribution for the scores, I identified the deciles of risk and created nine dummies to represent them in the model. The $LR\chi^2$ values for the models with and without the decile-of-risk dummies are 431.4674 and 417.3661, respectively. The test statistic is therefore $\Delta\chi^2 = 431.4674 - 417.3661 = 14.1013$, which, with nine degrees of freedom, is not significant ($p = .119$). In this case, then, there is no evidence to suggest that the model lacks empirical consistency.

EXERCISES

11.1 The following partial table refers to 1000 domestic violence offenders followed for one year after release from arrest. The event of interest (a "failure") was the first subsequent arrest for domestic violence recorded by police. Subjects were considered censored if they were arrested for a different offense during the follow-up period or if they had not been arrested by the end of the study.

Time (weeks) Interval	Number Failed	Number Censored
[0,10)	25	3
[10,20)	30	8
[20,30)	28	7
[30,40)	46	19
[40,∞)	82	752

Show the estimated survival and hazard functions for these data using the life-table estimators.

11.2 The following data represent times to arrest or censoring for six of the offenders from Exercise 11.1, along with a dummy variable for whether the offender was *employed* at the time of arrest:

Subject	Censoring Indicator[a]	Weeks	Employed
1	1	39	0
2	1	17	1
3	1	46	1
4	1	12	0
5	0	22	0
6	0	52	1

[a] 1 = uncensored; 0 = censored.

If the Cox estimate for the effect of *employment* is -1.073, give $-2\ln L$ for the Cox model for these data with *employment* as the predictor.

11.3 For the data in Exercise 11.2, the standard error of the effect of *employment* is 1.235. Also, $-2\ln L$ for a model with no covariates is 10.386. Use both $LR\chi^2$ and Wald tests to test whether the effect of *employment* is significant. Regardless of significance, interpret the effect of *employment* on the hazard of rearrest for domestic violence.

11.4 Suppose that the number of weeks to rearrest for domestic violence after an initial arrest has an exponential distribution with $\lambda = .0035$. Give:
   (a) The probability of arrest within one year after the initial arrest.
   (b) The probability of surviving one year without being arrested.
   (c) The probability of arrest between weeks 35 and 40.

11.5 The survival function for the Weibull distribution is (Blossfeld et al., 1989) $S(t) = \exp[(-\lambda t)^{\sigma}]$, where λ and σ are positive constants. Show that this distribution implies that $\ln[-\ln S(t)]$ is linear in the log of t.

11.6 Show that if time has an exponential density with $\lambda = 1$, the log of time has a Gumbell density. [*Hint:* Write the expression for F(t). Then let $Y = \ln t$ and let G(y) be the distribution function for Y. Then note that G(y) = P($Y \leq y$) = P($\ln t \leq y$) = P($t \leq e^y$). Now express P($t \leq e^y$) in terms of F(t). This gives you G(y), and taking its first derivative with respect to y gives the density of Y. This is a well-known technique for finding density and distribution functions for a function of some variable, where the variable has some known distribution. See, e.g., Hoel et al. (1971).]

11.7 Based on model 1 in Table 11.2, give the predicted survival time for a couple together for 2 months at wave 1, in which the woman was 19 at the start of the union, in which the man had been married before, in which both partners had drinking problems, whose *open disagreement* score was 30, and whose *conflict resolution* score was 3.

11.8 Suppose that there is a time-varying dummy, M, for marital status such that $M = 1$ if married and 0 if unmarried. Continuing with the substantive example from Exercise 11.1, say that we have two subjects with the following values for M at the times indicated:

Time (weeks)	M for Case 1	M for Case 2
10	0	1
15	0	0
30	1	1
42	1	0

Let the Cox model for the hazard of rearrest be $\ln h_i(t) = \lambda_0(t) + \delta M_i(t)$, where $\delta \neq 0$. Demonstrate, using these data points, that the ratio of hazards for case 1 versus case 2 is nonproportional, due to the time-varying covariate in the model.

Use the following information for Exercises 11.9 to 11.11. Yamaguchi (1991) employed Cox regression to explore the impact of several factors on the hazard of dropping out of college. The sample consisted of 265 students from the High School & Beyond Survey who were high school seniors in 1980 and who subsequently entered four-year colleges. The follow-up period was 1980–1984. The event constituting a "failure" consisted of either dropping out of the initial college of choice or transferring to a different college. Those who graduated or were still in the same school at the end of the study (1984) were considered censored observations. Duration in college was measured in months. The predictors used were:

SEX = *dummy for being female.*
GRD = *self-reported high-school grades (1 = "mostly A's" to 8 = "mostly D's").*
PRT = *dummy for being a part-time student ("full-time student" is the contrast).*
LAG = *time lag (months) between high-school graduation and college entry.*
  MS = *time-varying dummy for being married at time t.*
EMP = *time-varying dummy for being employed at time t.*

11.9   The first model investigated by Yamaguchi resulted in the following equation:

$$\hat{r}_i = .324 \text{ SEX} + .285 \text{ GRD} + 1.462 \text{ PRT} + .127 \text{ LAG} - .086 \text{ PRT} * \text{LAG}.$$

LR $\chi^2_{(5)} = 49.57$. Interpret the results by interpreting all coefficients. (All effects except for SEX are significant at $p < .05$.)

11.10 Suppose that the baseline survivor function for the first five months is 1, .995, .991, .926, .914. Give the estimated survival function for a male full-time student whose high-school grades were mostly B's (GRD = 3), and who waited 12 months before starting college.

11.11 The second model investigated produced the following equation:

$$\hat{r}_i = .340 \text{ SEX} + .289 \text{ GRD} + 1.368 \text{ PRT} + .125 \text{ LAG} - .081 \text{ PRT} * \text{LAG} \\ + 1.255 \text{ MS} + .512 \text{ EMP}.$$
LR $\chi^2_{(7)} = 62.70$.

   (a) Test whether the effects of both time-varying covariates simultaneously equal zero.
   (b) Interpret the effects of marriage and employment.
   (c) Estimate the discriminatory power of the model.

11.12 Teachman (2003) examined the influence on the hazard of first union formation (either marriage or cohabitation) of a number of demographic covariates for a sample of 7477 women taken from the 1995 National Survey of Family Growth. One model produced the following hazard ratios (net of other covariates) for ever having lived before age 16 in various alternative family forms, compared to continuous habitation with both biological parents (the reference group):

Mother-only family: 1.10.
Step-parent family: 1.21.
Parent and cohabiting partner family: 1.26.
Two cohabiting parents family: 1.12.
Residual family type: 1.24.

Interpret the effect of living in a family with *two cohabiting parents*. Give the hazard ratios for living in a *mother-only family* versus living in each of the following family arrangements: *a step-parent family*, a family consisting of *one parent and cohabiting partner*, and *a two cohabiting parents* family.

11.13 A Cox regression for the log-hazard of union disruption based on the 1230 couples in the NSFH used the following predictors: *relationship duration in months* (DURATION), *female's age at union inception* (FAGUNION), *whether either partner experienced a parental divorce* (PARENTAL), *whether the household had children under 5* (CHDN5), and *whether either partner was a minority* (MINORITY). Partial results are:

Variable	b	$\hat{\sigma}_b$
DURATION	$-.015$	.006
FAGUNION	$-.042$	.009
PARENTAL	.220	.133
CHDN5	.103	.120
MINORITY	.200	.129
$-2 \ln L_0$	4599.743	
$-2 \ln L_1$	4556.715	

(a) Is the model significant as a whole?

(b) Compute the Wald χ^2 for each coefficient. Which coefficients are significant?

(c) What is the discriminatory power of the model?

11.14 Added to the model in Exercise 11.13 is a nonproportional effect of PARENTAL, of the form PARENTAL * TIME, producing:

Variable	b
PARTIME	.00314
$-2 \ln L$	4556.388

(The effect of PARENTAL in this model is .079.) Test whether PARENTAL has a significant nonproportional effect. Irrespective of the outcome of the test, show the hazard ratio for those with versus without parental divorce experience at 1 month's, 10 months', 30 months', and 50 months' duration.

11.15 A Cox regression for the log-hazard of union disruption based on the 1230 couples in the NSFH used the following wave 1 predictors: *male partner was violent* (HEHIT), *female partner was violent* (SHEHIT), *male's relationship happiness* (HUSHAP), *female's relationship happiness* (WIFHAP), *male's relationship commitment* (HCOMMIT), *female's relationship commitment* (WCOMMIT). Partial results are as follows:

Variable	b	$\hat{\sigma}_b$
HEHIT	.316	.248
SHEHIT	.132	.242
HUSHAP	−.084	.051
WIFHAP	−.144	.047
HCOMMIT	−.095	.073
WCOMMIT	−.425	.070
$LR\chi^2$	150.153	

(a) Interpret the significant (at $p \le .05$) effects.
(b) By what percent is the hazard of disruption elevated when the male has been violent versus when the female has been violent.
(c) If the estimated covariance of the parameter estimates for HEHIT and SHEHIT is −.05, give a 95% confidence interval for the ratio of the hazard of disruption for couples characterized by male violence versus couples characterized by female violence.
(d) What is the discriminatory power of the model?

11.16 The following data represent age at first sexual intercourse or censoring for a sample of six teenage girls, along with *mother's educational level* and a dummy for *family type* (1 = "two-parent biological," 0 = "other"):

Subject	Censoring Indicator[a]	Age	Family Type	Mother's Education
1	1	16	1	13
2	0	19	1	15
3	1	17	1	20
4	1	15	0	12
5	1	16	0	16
6	0	14	0	17

[a] 1 = uncensored; 0 = censored.

If the estimate of the effect of *mother's education* from a Cox model stratified by *family type* is −.144, give $-2 \ln L$ for the stratified Cox model.

11.17 The following data represent times in months to voluntary job termination or censoring for five employees, along with the cumulative number of *days of sick leave* taken by the start of the study:

Subject	Censoring Indicator[a]	Months	Days of Sick Leave
1	1	14	2
2	1	3	0
3	1	9	12
4	0	15	1
5	0	2	4

[a] 1 = uncensored; 0 = censored.

If the Cox estimate for the effect of *days of sick leave* is .070, give $-2\ln L$ for the Cox model with *days of sick leave* as the predictor.

Use the Union Disruption Data for the following computer exercises involving Cox regression.

11.18 Estimate a Cox model for the hazard of disruption as a function of MEDUC, FEDUC, MINCOME, FINCOME, HE1, and WE1. Adjust for left truncation using the method, outlined by Allison (1995), in which one of the predictors is made into a time-varying covariate as follows. A covariate, say MEDUC, is labeled MEDUCTD. MEDUCTD is then set to missing while SURVIVAL is less than DURATION; otherwise, it is set equal to MEDUC. MEDUCTD is then used in the model in place of MEDUC.

(a) Report the parameter estimates and note which ones are significant.

(b) Interpret the effects of FEDUC and MINCOME.

(c) Test (at $\alpha = .05$) for the equality of the effects of male and female education as well as the equality of effects of male and female *income* using individual Wald tests.

11.19 Test whether the effects of MEDUC versus FEDUC and the effects of MINCOME vs. FINCOME are equal using a global test for both equalities simultaneously.

11.20 Estimate the model in Exercise 11.15 stratified by MINORITY (minority status of the couple), and test whether the parameter vector is invariant over strata.

11.21 Estimate a model of union disruption as a function of DURATION, FAGUNION, FIRSTUNI, ALCDRUG, DISAGMT, COMSTYLE, and the interaction of DISAGMT with COMSTYLE. Estimate the impact of DISAGMT at 1 standard deviation above the mean level of *communication style*, and test whether it is significant at this level of COMSTYLE. (*Hint*: Use targeted centering.)

11.22 Perform May and Hosmer's (1998) goodness-of-fit test for model 3 in Table 11.2.

CHAPTER 12

Multistate, Multiepisode, and Interval-Censored Models in Survival Analysis

CHAPTER OVERVIEW

In Chapter 11 we considered various ways of modeling single, nonrepeatable events, using Cox's semiparametric regression model as our primary analytic vehicle. Often, however, event histories are more complex. For example, instead of duration at risk terminating in only one possible state, there may be multiple destination states that terminate duration at risk. I have already alluded to one such possibility: that marriages may terminate in death as well as in separation or divorce. Or, rather than being a one-time occurrence, the event of interest may be repeatable. Examples of repeatable events are getting married, taking a job, getting a promotion, being arrested, or getting pregnant. In these instances, each person may contribute several survival spells or episodes to the data file. Or, events may take place in discrete rather than continuous time. An example already given is promotion to a higher academic rank in institutions of higher education. Even if events occur in continuous time, however, duration at risk may only be recorded in terms of time intervals rather than exact times. Such discrete or interval-censored data require different modeling techniques than those heretofore discussed.

In this chapter we expand our toolkit for survival analysis by considering the modeling of multiple events, the modeling of repeated events, and the modeling of data that are discrete or interval-censored. I begin with an exploration of models for multiple events, focusing on the competing risks model. An alternative two-step model is also considered for cases in which the competing risks model is not warranted. I then move to a discussion of models for repeated events. Because repeated

Regression with Social Data: Modeling Continuous and Limited Response Variables,
By Alfred DeMaris
ISBN 0-471-22337-9   Copyright © 2004 John Wiley & Sons, Inc.

418

episodes for a given person are likely to be correlated, I discuss several approaches to handling the dependence of survival times across cases. Finally, I detail the approach taken when data are either interval-censored or truly discrete. As this situation necessitates the transformation of one's data, I discuss the data-transformation process first and then consider how survival analysis is accomplished with the transformed data file. Data for illustrating the techniques are, again, drawn from the author's own research.

MULTISTATE MODELS

Survival models in which duration-at-risk can terminate in any of several possible destination states are termed *multistate models* (Blossfeld et al., 1989). In many instances, event histories are characterized by multiple event types rather than a single type. For example, duration in unmarried cohabitation can terminate in either separation or marriage. Duration in a given job can end in either voluntary job termination, involuntary job termination, a transition to a job with a different firm, or a promotion (or demotion) within the same firm. Duration in college can end in dropping out of school, a transfer to a different college, or graduation. The most common approach to analyzing these types of event histories is the *competing risks model*. The model is predicated on two key assumptions. First, it is assumed that the transition to each event type is governed by a separate causal process. An alternative model that assumes the same underlying process for all transitions is discussed below. Second, it is assumed that conditional on model covariates, the multiple event types are independent of each other. That is, the hazard of any given event type is unrelated to the hazard of any other event type. With respect to duration in college, for example, this would mean that those who are at especially high risk of transferring to another school are not, as a result, at any higher or lower risk for dropping out than others. A model that allows for dependence among events is also discussed below.

Modeling Type-Specific Hazard Rates

Suppose that for each person, i, there are a total of $m = 1, 2, \ldots, Q$ different possible destination states that could terminate his or her duration at risk. Further, let T_i represent the variable containing the survival time for the ith person and let M_i represent the variable denoting the event type that the ith person transitioned to at time $T_i = t$. Competing risks models utilize *type-specific hazard rates* of the form

$$h_{im}(t) = \lim_{\Delta t \to 0} \frac{P(t \le T_i < t + \Delta t, M_i = m \mid T_i \ge t)}{\Delta t}. \tag{12.1}$$

The probability in the numerator of equation (12.1) is identical, apart from the i subscript, to that in equation (11.1.) defining the hazard for single events, except for the addition of "$M_i = m$." This probability is now interpreted as the *conditional probability of event occurrence in some small interval of time, given survival up to the beginning of that interval and given that the event is of the mth type.* In words, $h_{im}(t)$ is the instantaneous rate of occurrence of the mth event type to the ith person at time t given survival up to that time. One way of thinking about these hazards is to define T_{im} as the time at which the mth event type occurred to the ith case and to imagine that each case has a set of $T_{i1}, T_{i2}, \ldots, T_{iQ}$ times attached to it. However, in that the occurrence of one event removes one from the risk of occurrence of any other events—and hence the name *competing* risks—we only observe one time, T_i, the one that is the smallest (Allison, 1995). As before, of course, observations who are still at risk at the end of the study or who are lost to follow-up are considered censored. The overall hazard of event occurrence at any given time is simply the sum of hazards for all possible event types. That is,

$$h_i(t) = \sum_{m=1}^{Q} h_{im}(t).$$

A Cox model for competing risks takes the form

$$h_{im}(t) = h_{0m}(t) \exp(\mathbf{x}^{i'}\boldsymbol{\beta}_m). \tag{12.2}$$

Noteworthy is the fact that the baseline hazard function and the coefficient vector are both subscripted with "m." This implies that separate processes are allowed to characterize each type-specific hazard: Each can have a different baseline hazard function and a different set of effect parameters. [More generally, each hazard can even include different explanatory variables or be modeled using a different parameterization of survival time; see Allison (1995).] It turns out that the partial likelihood function for model (12.2) factors into distinct partial likelihoods for each event type if the competing events are considered censored cases (Allison, 1995). What this means is that the parameters of equation (12.2) can be estimated by estimating a Cox model for each separate event type while treating the competing events and censored cases all as censored observations. Hence, estimating a competing risks model involves no new techniques over those that were covered in Chapter 11. The analysis of each different transition produces a $LR\chi^2$ test and a set of estimates that are asymptotically normally distributed, enabling the usual Wald tests and confidence intervals pertaining to the given event type.

One new question arises, however. We might ask whether a separate model is really necessary for each event type, instead of simply treating all events the same and estimating equation (11.11), the Cox model for a single event. A test statistic for the null hypothesis that the same process determines all transitions is formed as follows. First, estimate equation (12.2) in the manner described above. Then sum all M of the minus twice log-likelihoods for the resulting Cox models as

$$-2 \ln L_M = \sum_{m=1}^{Q} -2 \ln L_m.$$

Then code as the same "event" all M of the different event types and estimate a Cox model for the occurrence of an "event." Denote $-2 \ln L$ for this model as $-2 \ln L_C$. Assume there are K predictors in the model, in general. The test statistic for testing model invariance to event type is

$$\Delta\chi^2_{[K(M-1)]} = -2\ln L_c - (-2\ln L_M), \qquad (12.3)$$

which under the null hypothesis of invariance is distributed as chi-squared with $K(M-1)$ degrees of freedom. The term $K(M-1)$ represents the difference in the number of parameters estimated to construct L_C versus L_M. A significant result supports the competing risks approach.

Example: Transitions Out of Cohabitation

In a recent article the author examined survival time in the state of unmarried cohabitation for 411 cohabiting couples in the NSFH (DeMaris, 2001). Couples were all cohabiting in wave 1 of the survey, at which time several characteristics of the couples were measured. In wave 2 of the survey interviewers recorded whether the couple was still together or not. If the couple had separated, the date of separation was recorded. If the couple had married, the date of marriage was recorded. Thus, two transitions were possible: separation and marriage. In all, 85 couples were still cohabiting unmarried, while 173 had married and 153 had separated. Those who were still cohabiting unmarried were considered censored observations.

Demographic predictors of the hazard of a transition included *female's age at union inception, relationship duration* (in months) as of the wave 1 survey, *whether either partner was a minority*, and *whether the couple experienced a birth to the union between waves 1 and 2*, the last being a time-varying covariate. Several other measures tapped the quality of the relationship as measured in wave 1: *frequency of open disagreement, conflict resolution style*, each partner's *relationship happiness* (on a scale from 1 = "very unhappy" to 7 = "very happy"), and each partner's perception of *relationship stability* (on a scale from 1 = "very high probability of separating" to 5 = "very low probability of separating"). The focus variables pertained to intimate violence experienced by the couple. Measures of intimate violence consisted of dummies for whether or not the male or the female *had been violent* with each other in the past year and whether or not the relationship *had been characterized by intense male violence*. The latter was coded 1 if the male had been the only violent partner, or both were violent but the male's violence was more frequent, or the female was the only injured partner. The primary hypothesis was that violence by either partner, and especially *intense male violence*, would elevate the risk of separation and diminish the risk of marriage, net of other regressors.

I estimated a competing risks model by first specifying a Cox model for the hazard of separation, treating both continuous cohabitation and marriage as censored cases. Then I specified a Cox model for the hazard of marriage, treating continuous cohabitation and separation as censored cases. As couples had been cohabiting for anywhere from less than one month to over 18 years when interviewed in wave 1, both models were adjusted for left truncation. The results are shown in Table 12.1.

At first glance, it appears that the transition to marriage is better accounted for than the transition to separation. The discriminatory power, using R^2_G, of models for

Table 12.1   Competing Risks Models for Exits from Unmarried Cohabitation

Predictor	Transition to Separation[a]	Transition to Marriage[b]
Open disagreement	−.013	.073**
Conflict resolution style	−.155*	.138*
Male's violence	−.620	.755
Female's violence	.479	−1.286*
Intense male violence	.731*	−.307
Union birth[c]	−.878*	−2.367**
Female's age at union	−.035**	−.013
Relationship duration (months)	−.006	−.010
Minority couple	−.062	−.765***
Male's relationship happiness	.009	.196*
Female's relationship happiness	−.153	−.179*
Male's relationship stability	−.212	.321**
Female's relationship stability	−.027	.312*
Likelihood ratio χ^2	51.896***	86.126***
Degrees of freedom	13	13

Note: $n = 411$.

[a] Survival time is censored upon transition to marriage.

[b] Survival time is censored upon transition to separation.

[c] Time-varying covariate.

* $p < .05$. ** $p < .01$. *** $p < .001$.

separation and marriage are .118 and .189, respectively. Only four effects are significant for the risk of separation. Predictably, couples with a more positive *conflict resolution style* have lower hazards of separation than others. Also, consistent with the earlier study of union disruption reported in Chapter 11, couples who were older at union inception, as tapped by the *female's age at union inception*, have lower risks of separating. Although neither partner's violence per se has an effect on separation, couples characterized by *intense male violence* have hazards of separation that are higher than others' by a factor of $\exp(.731) = 2.077$; or their hazards of separating are about twice as high as others'. A birth to the union between survey waves reduces the hazard of separation, similar to its effect found in Chapter 11 for the sample that included marrieds.

Several effects are significant when predicting the transition to marriage. A more positive *conflict resolution style* hastens the transition to marriage, as does a greater *frequency of disagreement*. The latter effect may well be due to dissatisfaction with being unmarried and conflicts over whether or when to make the transition to marriage. Surprisingly, the only effect of factors measuring violence is for *violence by the female partner*, which lowers the transition rate to marriage. I have earlier speculated that as it is still customary for the male to propose marriage, her violence may inhibit that step on his part (DeMaris, 2001). *Minority couples* have lower hazards of marriage, net of other factors. Somewhat counterintuitively, experiencing a birth to the union inhibits the transition to marriage. However, this effect is consistent with others' findings. In particular, Wu and Balakrishnan (1995) suggest that those who are comfortable with having children outside of marriage are probably more ideologically committed to long-term cohabitation as an alternative to marriage. The remaining indices of marital quality have predictable effects except for the *female's relationship happiness*: The happier she is with the relationship, the lower the hazard of marriage. Most likely this just reflects the fact that if women are happy with the relationship as is, they are reluctant to formalize the union.

Test for Model Invariance. Although regressor effects appear to be quite different for the event of separation, as opposed to marriage, this could simply be the result of sampling error. As mentioned, the null hypothesis of model invariance is that the same baseline hazard function and the same parameter vector characterize each transition type. This hypothesis was evaluated by estimating a Cox model for any transition out of cohabitation, where marriage and separation are treated as the same event, and employing the test statistic in equation (12.3). Minus twice the log likelihood for the combined-event model was 2907.772, while $-2 \ln L$ for the models for separation and marriage, respectively, were 1314.038 and 1512.889. The test statistic is, therefore, $\Delta \chi^2 = 2907.772 - (1314.038 + 1512.889) = 80.845$. As the model has 13 covariates, the degrees of freedom equal $13(2 - 1) = 13$. At a p-value of less than .00001, the test is quite significant, suggesting that the two events are characterized by different hazard processes.

Alternative Modeling Strategies

Two-Step Approach. As indicated, the competing risks model presumes that transitions to different end states are characterized by different causal processes acting in parallel (Allison, 1995). However, some multistate processes may follow a different pattern. For example, the transition to a romantic union from the single state may be determined primarily by one set of factors, such as subjects' opportunities to meet potential partners or their own attractiveness in the marriage market. However, whether that union is an unmarried cohabitation rather than a marriage may be more strongly determined by other factors, such as subjects' educational level, religiosity, or family background. This raises the possibility of an alternative modeling strategy in which one mechanism is allowed to govern the timing of a transition, in general, and a separate mechanism governs the type of transition, given that a transition occurs. Dubbed the *two-step approach* by Hachen (1988), the model is estimated by first estimating a survival model—say, the Cox model—for the transition to any state. In this case, all outcome states are treated the same (i.e., they are all coded as the same state). In a second step, only the subjects who made a transition are selected, and a multinomial (or binary) logit model is estimated to examine the impact of model covariates on which state is selected. In this second analysis, the log of survival time is included as a covariate (Allison, 1995). Note that this second step is *not* an event history analysis.

The choice of whether to employ the competing risks or two-step approach is strictly a theoretical decision. Allison (1995) suggests that the two-step approach is especially appropriate if the different destination states are alternative means for achieving the same goal, as in married versus unmarried cohabitation. By this criterion, the competing-risks model is clearly more appropriate for the analysis of cohabiting transitions, as separation and marriage fulfill very different goals for cohabitors. Hachen (1988) provides detailed guidelines concerning which model to use, but these are more conceptually complex. Let P(m) represent the conditional probability that the mth state is entered, given transition to some state. Hachen suggests that the two-step model is to be preferred whenever the effect of covariates on a transition, in general, is invariant to changes in the P(m) for $m = 1, 2, \ldots, Q$. One example given by Hachen considers the effect of taking a high school sex-education class on the first type of contraception used during sexual intercourse. Suppose for some reason the availability of, say, IUDs to adolescents were suddenly curtailed. If the effect of high school education on the hazard of first contraceptive use in general is not affected by the lowered probability of IUD use, the two-step model should be employed. Otherwise, the competing risks model would be preferable.

Dependence of Events. The assumption of competing risks models that alternative destination states are independent of each other may often be untenable. Instead, unmeasured factors may link each of the states. Using, again, the formation of the first romantic union as an example, it is likely that unmeasured characteristics of individuals, such as a need for intimate companionship, raise or lower the risk of union formation, in general. Therefore, the hazard of cohabitation and the hazard of marriage would tend to be correlated across cases. Hill et al. (1993, p. 247) maintain that hazard models that ignore this type of dependence among hazards "may provide inaccurate estimates of base hazard rates or parameters." The authors have formulated a shared unmeasured risk factors (SURF) model to adjust for correlated hazards, which can be estimated using conventional software. The technique currently has several limitations, however. It has only been formulated for the case of two competing risks. Moreover, only a positive correlation between risks is allowed for; and the approach assumes that the two-step model is appropriate. If these conditions are satisfactory, the model can be estimated via a two-stage procedure that is detailed in the authors' article (Hill et al., 1993).

MULTIEPISODE MODELS

Models for repeated events are termed *multiepisode* (Blossfeld et al., 1989) or *recurrent event models* (Hosmer and Lemeshow, 1999). Rather than contributing just one spell to the data file, each case now contributes $e = 1, 2, \ldots, E$ potential spells to the file, where e represents a given event number, and E the total number of events experienced by the ith case. Each spell is a survival time in the nonevent state until the event recurs. There are several potential ways of modeling repeated events (see, e.g., Hosmer and Lemeshow, 1999). The model discussed here is that advocated by

Allison (1984, 1995) and is relatively flexible and easy to estimate. Its form is

$$h_{ie}(t) = h_0(t - t_{e-1}) \exp(\mathbf{x}^{ie\prime}\boldsymbol{\beta}). \tag{12.4}$$

The notation needs some clarification. The response, $h_{ie}(t)$, is the hazard of the eth event for the ith person. The baseline hazard function, $h_0(t - t_{e-1})$, is assumed to be the same for each event. However survival time is denoted as "$t - t_{e-1}$," where t is survival time until the eth event or censoring and t_{e-1} is survival time until the *previous* event. The difference is the survival time for the current, or eth, event. In other words, survival time for each spell begins at inception of risk for the *current* event, which is normally the time of occurrence of the *previous* event (although there are exceptions, e.g., the example of unemployment spells below). This follows the assumption that cases are not at risk for a subsequent event until after the occurrence of a prior event. [An alternative is to reckon survival time for all spells from the inception of risk for the first event; see Hosmer and Lemeshow (1999) for a discussion of alternative start times for recurrent event data.] The e superscript on $\mathbf{x}$ indicates that covariate values may change with the risk periods for each event. Covariates may also change over time *within* each risk period, as always. The recurrence of an event terminates a given spell. Spells that are ongoing at the end of the study constitute censored cases. Estimation is straightforward: One simply treats each spell for a given case as a separate observation and then pools all spells over all n cases in the sample to arrive at a total file of nE observations. One then estimates the Cox model for "an event" in the usual fashion.

Example: Unemployment Spells

As an example, Goza and DeMaris (2003) examined transitions out of unemployment for a sample of 283 Brazilian immigrants residing in the United States and Canada in 1990–1991. The authors' study was primarily focused on testing predictions from job search theory regarding the duration of unemployment for immigrants. Each respondent in the study experienced, on average, about 2.2 periods of unemployment, ranging from under one month to as much as five years in duration. Inception of risk for each spell was the date of loss of the previous job, rather than the date on which the previous job began. Spells that ended in reemployment were considered uncensored cases. Unemployment spells that were ongoing when data collection ended were considered censored. When all spells were pooled over all cases, a total of 620 unemployment spells constituted the analytic sample. Of these, 578 were uncensored and 42 were censored. A series of Cox models for the hazard of reemployment was estimated using these data and is shown in Table 12.2.

Model 1 in Table 12.2 contains the eight focus variables in the study: *age at the start of unemployment, duration in months on the previous job, the respondent's number of relatives living in North America, cumulative time in years in North America, the log of monthly income on the last job, education in years of schooling, a dummy for being female*, and a dummy for *English proficiency* being self-rated as at least "good" when entering North America. All effects are significant at the .05

Table 12.2 **Cox Multiepisode Models for Transitions Out of Unemployment**

Predictor	Model 1	Model 2	Model 3[a]	Model 4[b]
Age at unemployment	−.118*	−.124*	−.136*	−.113
Duration of previous job (months)	−.026**	−.028**	−.030**	−.026**
Number of relatives in North America	.020*	.019*	.184	.021*
Cumulative time in North America	−.136***	−.113*	−.152***	−.136**
Log of monthly income last job	.061***	.064***	.068***	.080***
Education	.051***	.049***	.058***	.052***
Female	−.351**	−.349**	−.371**	−.369***
Good English proficiency	−.509**	−.509**	−.549*	−.477*
Cumulative months unemployed		−.018		
Cumulative no. previous jobs		.001		
$\hat{\theta}$				.128**

Note: $n = 620$ unemployment spells.

[a] Shared-frailty model.

[b] Stratified by job-spell number.

$* \, p < .05. \, ** \, p < .01. \, *** \, p < .001.$

level. Longer spells of unemployment are associated with being older at the start of the unemployment spell, spending more time in the previous position, having been in North America longer, being female, and surprisingly, having good English proficiency. The last effect is perhaps the result of those with better English skills having higher expectations for good positions and therefore holding out longer for them (Goza and DeMaris, in 2003). On the other hand, shorter unemployment spells obtain for those with more relatives in North America and those with more human capital in the form of attained educational level and income from the last job.

Nonindependence of Survival Times

Model 1 is the most restrictive possible model in the sense that it makes several assumptions about the data that may or may not be tenable. To begin, it is assumed that the 620 spells constitute a set of independent observations, given model covariates. In all likelihood this assumption is violated since the same people contribute more than one spell to the data set. Therefore any unmeasured factor that might elevate (or diminish) the rate of reemployment for a given person will tend to shorten (or lengthen) successive spells for the same person, resulting in correlated survival times. One solution recommended by Allison (1984) is to add covariates that tap into characteristics of the person's prior event history, such as the number of events prior to the current spell or the length of the previous spell. In that independence of observations is assumed to hold *net of* model covariates, including the proper covariates might render the independence assumption tenable. As a result, model 2 includes two additional factors pertaining to respondents' job histories: *The cumulative number of months of unemployment experienced* and *the cumulative number of jobs taken by the respondent.* Neither additional factor is significant. More important, however, none of the other effects are appreciably altered when these factors are controlled.

A second approach to the dependence problem is to adjust the standard errors of coefficients to account for the dependence of survival times within a person's event history. Lin and Wei (1989) proposed a "sandwich" estimator for the covariance matrix of parameter estimates from a potentially misspecified Cox model. The resulting estimates of variances and covariances of Cox regression coefficients are robust to various types of misspecification, including dependence among observations. (As we will discover below, the problem of dependence can be cast in terms of an omitted variable; hence such misspecification is in the form of omitted-variable bias.) The robust estimator of $V(\mathbf{b})$ is available in mainstream software packages such as SAS and STATA. Model 1 was reexamined employing the robust standard errors of coefficients (results not shown). However, again, results were not altered appreciably. The largest change in standard error estimates was for the coefficient of *good English proficiency*: The unadjusted standard error was .182, while the robust estimate was .192. The coefficient remained significant at $p < .008$ in either case.

A third approach to handling nonindependence of survival times is to employ the *fixed-effects partial likelihood* (FEPL) approach (Allison, 1996). This strategy utilizes a *fixed-effects model* for the hazard, which takes the form

$$h_{ie}(t) = h_0(t - t_{e-1})\alpha_i \exp(\mathbf{x}^{ie\prime}\boldsymbol{\beta}). \tag{12.5}$$

The additional multiplicative term in the model, α_i, represents a fixed constant that characterizes each person. These constants represent individual-level factors affecting the hazard, which are responsible for the correlated survival times for events pertaining to the same person. Normally, these factors are unmeasured and are correspondingly referred to as factors reflecting *unmeasured heterogeneity* in the hazard. A convenient method of estimation absorbs the α_i into the baseline hazard function, thus converting equation (12.5) into

$$h_{ie}(t) = h_{0i}(t - t_{e-1}) \exp(\mathbf{x}^{ie\prime}\boldsymbol{\beta}). \tag{12.6}$$

The reader should notice the resemblance of equation (12.6) to the stratified version of the Cox model presented in Chapter 11. In fact, equation (12.6) is estimated simply by estimating the Cox model while stratifying on individuals. However, a major drawback to this approach, as mentioned in Chapter 11, is that only the effects of covariates whose values vary across or within spells may be estimated. In other words, individual characteristics such as gender or race that remain constant across the spells for a given person cannot be assessed. Nevertheless, all such characteristics are implicitly controlled during model estimation. In the unemployment example the only covariates in model 1 that vary over unemployment spells for a given immigrant are *age at unemployment, duration of previous job, cumulative time in North America*, and *log of monthly income*. A FEPL model containing just these four covariates was therefore estimated (results not shown). The only significant effect was for *log of monthly income*, and its value was .059, which tends to agree with the covariate's effect in model 1.

A fourth approach to the dependence problem is similar in spirit to the unobserved-heterogeneity approach just discussed. However, instead of being fixed, the heterogeneity term is modeled as a random variable and referred to as a *frailty*. Frailty models account for a burgeoning literature within survival analysis (see, e.g., Aalen, 1994; Blossfeld and Hamerle, 1990; Blossfeld and Rohwer, 1995; Galler and Poetter, 1990; Hosmer and Lemeshow, 1999; Klein and Moeschberger, 1997; Land et al., 2001; McGilchrist and Aisbett, 1991) and are becoming more available in mainstream software. When the frailty characterizes a group of observations, the model is referred to as a *shared-frailty model*. In recurrent-event data, for example, the shared-frailty model is applicable since each person contributes a group of observations to the data in the form of multiple episodes. The Cox shared-frailty model for multiepisode data is (Klein and Moeschberger, 1997)

$$h_{ie}(t) = h_0(t - t_{e-1})\upsilon_i \exp(\mathbf{x}^{ie\prime}\boldsymbol{\beta}), \qquad (12.7)$$

where the frailty, υ_i, is assumed to have some density with a mean of 1 and a variance of θ. The gamma density is often chosen for its mathematical tractability. Frailties greater than 1 imply a greater hazard of event occurrence, net of covariates. That is, these people are more "frail," or susceptible to the event. Frailties less than 1 indicate greater resistance to the event, or longer survival times. The difference between equations (12.5) and (12.7) is subtle but important. In equation (12.5), α_i is a fixed effect, whereas in equation (12.7), υ_i is a random variable with some population distribution. By "fixed effect" is meant that there is some finite set of α_i in the population that are constant values over repeated sampling. That is, each sample of people is a sample from the same set of limited α_i values, with the same value of α characterizing potentially many people in the population. On the other hand, υ in equation (12.7) is a random variable that is not fixed over repeated sampling, and may in fact be unique to each person. The set, assumed to be infinite, of all possible υ_i in the population is represented by some distribution function.

Model 3 in Table 12.2 is the Cox shared-frailty model for the unemployment data, assuming a gamma distribution for the frailties. The model was estimated using STATA. The estimate of the frailty variance is .128 and is significantly different from zero according to a likelihood-ratio test. It is therefore important to take account of individual frailties in the model. That said, however, results are, again, not radically altered compared to model 1, except that in model 3 the number of relatives in North America no longer has a significant effect on the hazard.

A word of caution is in order regarding unobserved heterogeneity. In the Cox model, the shape of the hazard function is ignored. However, in parametric models that specify some form for the hazard, care must be exercised in interpreting the effect of a hazard that appears to be declining over time. Unobserved heterogeneity, if unaccounted for, can artificially generate a declining hazard. The reason is that over time, the frailest people experience the event and drop out of the risk set. This leaves a risk set that is composed increasingly of the most "resistant" individuals, making it appear that the risk for the event is declining over time (Blossfeld and Rohwer, 1995). As unobserved heterogeneity is rarely completely accounted for, a declining hazard function should always be regarded with some skepticism.

Model Variation across Spells

An additional assumption of model 1 is that each successive spell of unemployment is characterized by the same hazard function for reemployment. This, of course, may not be the case. Many times the first episode of any event has a different survival trajectory than later episodes. To investigate this possibility, I created a variable representing the sequencing of unemployment spells. This variable, *job-spell number*, was coded 1 for the first unemployment spell, 2 for the second unemployment spell, 3 for spell numbers 3 or 4, and 4 for spell numbers 5 or higher. Later spells were collapsed due to relatively few people having more than four unemployment episodes. I then dummied up *job-spell number* and added the three dummies to model 1 (results not shown), producing a base model for testing interaction. In the next step, I allowed the *job-spell number* dummies to interact with *time* (results not shown) in order to capture changes in baseline hazards across unemployment spells. The likelihood-ratio test for the addition of the interaction terms produced a nonsignificant result ($\Delta\chi^2_{(3)} = 2.2902$, $p = .514$), suggesting that the baseline hazard function did not vary across unemployment spells. Nevertheless, model 4 in Table 12.2 reestimates model 1 but stratifies by *job-spell number*. Again, the results are not appreciably altered, compared to model 1, except that *age at unemployment* in model 4 is not quite significant.

A final simplifying assumption made in model 1 is that the effects of covariates on the hazard of reemployment are the same for each successive unemployment episode. Again, such an assumption may not be tenable. We might expect, for example, that having many relatives to help in the job search would have a stronger effect in later unemployment periods than in earlier ones. During an initial episode of unemployment, people may rely on only their own credentials to find jobs. After repeated unemployment spells, however, people may begin to call on their larger kin network for additional aid in helping to secure a good position. A general test for the interaction of model 1 covariates with *job-spell number* was effected by creating the cross-products of the three *job-spell number* dummies with the eight covariates in model 1 and then adding these to the base model described above (results not shown). Due to small cell sizes, however, the interaction of *job-spell 4* with *good English proficiency* could not be entered into the model. The test for interactions resulted in a $\Delta\chi^2$ of 66.566, which, with 23 *df* is significant at $p < .00001$. This suggests that one or more regressors has different effects, depending on the sequencing of the unemployment spell. Model 1 was therefore estimated separately for observations defined by different values of *job-spell number* (results not shown). As expected, the effect of having relatives in North America only becomes significantly positive in the later unemployment episodes—*job-spell numbers 3* or *4*. At the same time, the effects of *gender* and human capital (i.e., *income* and *education*) are only significant in earlier episodes—*job-spell numbers 1* or *2*—and decline to nonsignificance thereafter.

MODELING INTERVAL-CENSORED DATA

As alluded to previously, survival time may be a discrete rather than a continuous variable. Or, even if theoretically continuous, all that may be recorded is whether or not events have been experienced in some interval of time. In either case, the data are described as *interval-censored* and the appropriate model is a *discrete-time hazard model*. However, a distinction should be made from the outset between a *discrete-time analysis* vs. a *discrete-time model*. If survival time is truly discrete, the model being estimated is a discrete-time model. If survival time, although interval-censored, is truly continuous, on the other hand, then employing the discrete-time approach results only in a discrete-time analysis. That is, the model for a discrete-time process is being used to approximate the underlying continuous-time model.

Discrete-Time Hazard Model and Estimation

Suppose that survival time for the *i*th case, denoted again by T_i, is now a discrete variable representing time periods in which events occur. For example, in the study of time to promotion among college professors by Long et al. (1993), T_i was a discrete variable measured in single years, since promotions take effect at the beginning of the academic year. Survival time was therefore measured in years from the date of hire (for promotion to associate professor) or from the previous promotion (for promotion to full professor) until the current promotion or censoring. If, in the union disruption data, survival time were only measured in six-month intervals, then $T_i = 1, 2, 3, \ldots,$ would capture the six-month interval in which separations or censoring occurred. In either case, the discrete-time hazard, P_{it}, is defined as a probability

$$P_{it} = P(T_i = t \,|\, T_i \geq t). \tag{12.8}$$

That is, P_{it} is the conditional probability that an event occurs at time (or in time interval) t to the *i*th case given that no event occurs before time t (Allison, 1982).

A popular model for the discrete-time hazard as a function of covariates is the logit model, which employs the log-odds of event occurrence at time t as the response

$$\ln \frac{P_{it}}{1 - P_{it}} = \alpha(t) + \mathbf{x}^{i\prime}\boldsymbol{\beta}, \tag{12.9}$$

where $\alpha(t)$ represents some function of time, and $\mathbf{x}^{i\prime}\boldsymbol{\beta}$, as in the Cox model, represents a weighted sum of covariates times parameters, excluding an intercept term. The term $\alpha(t)$ captures the manner in which the log-odds of event occurrence depends on time. As explained below, this term is quite flexible and can take a variety of forms. Also, note that when the probability of an event in any given time interval is small, the log-odds of event occurrence is approximately equal to the *log-hazard* of event occurrence. Why? Notice that when P_{it} is small, say a value of .05 or less, then $1 - P_{it}$ is approximately 1. For example, $1 - .05 = .95 \approx 1$. Therefore, $P_{it}/(1 - P_{it}) \approx P_{it}/1 = P_{it}$. Hence, equation (12.9) can be regarded as a model for the log-hazard of event occurrence given small P_{it}'s in each interval, and interpreted accordingly.

Likelihood Function for Discrete-Time Data. Estimation of equation (12.9) is typically performed using a transformed data set. To provide the rationale for this transformation, I now consider the likelihood function for model (12.9). This explication is, admittedly, algebraically tedious. But it is key to the reader's understanding of the procedure employed to estimate discrete-time models. To continue, let δ_i be a censoring indicator for the ith case, coded 1 if an event occurs at time t_i, and 0 if the case is censored at time t_i. The likelihood function for model (12.9) is

$$L(\beta \mid \mathbf{t}, \mathbf{x}^i) = \prod_{i=1}^{n} [P(T_i = t_i)]^{\delta_i} [P(T_i > t_i)]^{1-\delta_i}. \tag{12.10}$$

Now, for those who experience the event at time t_i, their contribution to the likelihood is $P(T_i = t_i)$. For those who are censored at t_i, the contribution is $P(T_i > t_i)$. Further, if P_{it_i} is the probability that the ith case experiences the event at time t_i, then $1 - P_{it_i}$ is the probability that no event is experienced at t_i. And by probability rules, the probability of surviving to time t_i without experiencing the event is simply the product of $(1 - P_{it_i})$ over all values of T_i up to time t_i. That is,

$$P(T_i > t_i) = \prod_{j=1}^{t_i} (1 - P_{ij}) \tag{12.11}$$

and the probability that the event does not occur before t_i but then occurs *at* time t_i is then P_{it_i} times the probability of surviving to time $t_i - 1$:

$$P(T_i = t_i) = P_{it_i} \prod_{j=1}^{t_i-1} (1 - P_{ij}). \tag{12.12}$$

Substituting expression (12.11) for $P(T_i > t_i)$ and expression (12.12) for $P(T_i = t_i)$ into equation (12.10), the likelihood function can be expressed as

$$L(\beta \mid \mathbf{t}, \mathbf{x}^i) = \prod_{i=1}^{n} \left[P_{it_i} \prod_{j=1}^{t_i-1} (1 - P_{ij}) \right]^{\delta_i} \left[\prod_{j=1}^{t_i} (1 - P_{ij}) \right]^{1-\delta_i}. \tag{12.13}$$

Taking logs of both sides of equation (12.13), we have the log-likelihood for the discrete-time model:

$$\begin{aligned}
\ln L(\beta \mid \mathbf{t}, \mathbf{x}^i) &= \sum_{i=1}^{n} \left\{ \ln \left[P_{it_i} \prod_{j=1}^{t_i-1} (1 - P_{ij}) \right]^{\delta_i} + \ln \left[\prod_{j=1}^{t_i} (1 - P_{ij}) \right]^{1-\delta_i} \right\} \\
&= \sum_{i=1}^{n} \left[\delta_i \ln P_{it_i} + \delta_i \sum_{j=1}^{t_i-1} \ln (1 - P_{ij}) + (1 - \delta_i) \sum_{j=1}^{t_i} \ln (1 - P_{ij}) \right] \\
&= \sum_{i=1}^{n} \left[\delta_i \ln P_{it_i} + \delta_i \sum_{j=1}^{t_i-1} \ln (1 - P_{ij}) + \sum_{j=1}^{t_i} \ln (1 - P_{ij}) - \delta_i \sum_{j=1}^{t_i} \ln (1 - P_{ij}) \right]
\end{aligned}$$

$$= \sum_{i=1}^{n} \left[\delta_i \ln P_{it_i} + \delta_i \sum_{j=1}^{t_j-1} \ln(1 - P_{ij}) + \sum_{j=1}^{t_i} \ln(1 - P_{ij}) - \delta_i \ln(1 - P_{it_i}) \right.$$

$$\left. - \delta_i \sum_{j=1}^{t_j-1} (1 - P_{ij}) \right]$$

$$= \sum_{i=1}^{n} \left[\delta_i \ln \frac{P_{it_i}}{1 - P_{it_i}} + \sum_{j=1}^{t_i} \ln(1 - P_{ij}) \right],$$

or

$$\ln L(\beta \mid \mathbf{t}, \mathbf{x}^i) = \sum_{i=1}^{n} \delta_i \ln \frac{P_{it_i}}{1 - P_{it_i}} + \sum_{i=1}^{n} \sum_{j=1}^{t_i} \ln(1 - P_{ij}). \tag{12.14}$$

Now, suppose that we define a dummy variable, y_{ij}, such that $y_{ij} = 1$ if the ith case experiences an event at time t_i, and 0 otherwise. We can then rewrite equation (12.14) as

$$\ln L(\beta \mid \mathbf{t}, \mathbf{x}^i) = \sum_{i=1}^{n} \sum_{j=1}^{t_i} y_{ij} \ln \frac{P_{ij}}{1 - P_{ij}} + \sum_{i=1}^{n} \sum_{j=1}^{t_i} \ln(1 - P_{ij}). \tag{12.15}$$

Notice that the second sum in the first term on the right-hand side of equation (12.15) will be zero until time t_i, at which point the first term on the right-hand side of equation (12.15) will be identical to the first term in equation (12.14). However, equation (12.15) is now in a recognizable form. Consider the log-likelihood function for the logistic regression model, which I will denote as $\ln L_{LR}$ [see equation (7.13)]. Letting $\pi_i = \exp(\mathbf{x}^{i'}\beta)/[1 + \exp(\mathbf{x}^{i'}\beta)]$ in equation (7.13), we have

$$\ln L_{LR} = \sum_{i=1}^{n} [y_i \ln \pi_i + (1 - y_i) \ln(1 - \pi_i)]$$

$$= \sum_{i=1}^{n} [y_i \ln \pi_i + \ln(1 - \pi_i) - y_i \ln(1 - \pi_i)]$$

$$= \sum_{i=1}^{n} y_i \ln \frac{\pi_i}{1 - \pi_i} + \sum_{i=1}^{n} \ln(1 - \pi_i). \tag{12.16}$$

What the reader should now recognize is that the log-likelihood for the discrete-time hazard model in equation (12.15) is just the log-likelihood for a logistic regression model [e.g., equation (12.16)] for which the units of analysis are *time periods* rather than individuals. This means that model (12.9) is estimated by first converting person-level data, in which people are the analytical units, into *person-period* data, in which people's time periods are the analytical units. One then estimates equation (12.9) as an ordinary logistic regression model using the person-period data (as demonstrated below).

Approximating a Continuous-Time Process. If time is truly discrete, equation (12.9) is an appropriate model. However, if time is truly continuous but the data are interval-censored, the appropriate model is the *interval-censored Cox model* (Allison, 1995; Hosmer and Lemeshow, 1999):

$$\ln[-\ln(1 - P_{it})] = \alpha(t) + \mathbf{x}^{i\prime}\boldsymbol{\beta}, \qquad (12.17)$$

which is simply the complementary log-log model applied to person-period data. If survival times are generated by the Cox model and then grouped into intervals, the corresponding model for P_{it} is equation (12.17) (Allison, 1982, 1995). This implies that the parameter vector, $\boldsymbol{\beta}$, in equation (12.17) is identical to that in the underlying Cox model and can therefore be given the same interpretation. Moreover, the parameter vector in equation (12.17) is invariant to interval length, which is not true of model (12.9). This doesn't mean that sample *estimates* of equation (12.17) are invariant to interval length, but rather only that the underlying theoretical model is (Allison, 1982). At any rate, equation (12.17) is the correct model to use whenever time is theoretically continuous and the Cox model is presumed to underlie the data.

Converting to Person-Period Data

The process of converting from person- to person-period data is straightforward. It merely involves expanding individual observations by the number of time intervals they contribute to the process under study. That is, a given case's data record is replicated (including the original record) as many times as its number of intervals of survival time. Covariate values are either duplicated in each replication if they are time-invariant, or changed appropriately on successive replications if they are time-varying. A binary indicator of event occurrence takes the value 0 in each replication if the case is censored. If the case is uncensored, the event indicator is coded 0 in each replication except the final one, in which it is coded 1. A variable, T, representing the time-interval number, is also recorded with the values $1, 2, 3, \ldots, \tau$, where τ represents the last time interval observed for that case. Once the data are converted, the event indicator can be regressed on a set of covariates, using model (12.9) or (12.17). The term $\alpha(t)$ in the model utilizes some parameterization for T. The most general parameterization is simply to represent T with a series of $\tau - 1$ dummies in the model. This is equivalent to the unspecified function of time implicit in the Cox model.

As an example, let's consider the conversion of the data on union disruption for the 1230 NSFH couples from Chapter 11. In this case, time is theoretically continuous, which is why the Cox model was used for the analyses in Chapter 11. Suppose, however, that we simulate the situation in which survival time is recorded only in six-month intervals. We begin by recoding survival time for each couple according to the number of six-month intervals it represents. For the moment I ignore left truncation and treat inception of risk as beginning at the wave 1 survey. The maximum time recorded from wave 1 until disruption or censoring was 86 months. I therefore created a variable, INTERVAL, ranging from 1 to 15, representing survival time— in six-month increments—for each couple. Those whose survival time was less than or equal to six months were coded 1, those surviving more than six months but no more than a year were coded 2, and so on. I then replicated each couple's data up to a maximum of 15 times, depending on their value of INTERVAL. For example, if INTERVAL was 7, the couple's record was replicated seven times. If INTERVAL was 15, it was replicated 15 times, and so on. In the resulting couple-period data set, INTERVAL was relabeled TIME. The event indicator, named DISRUPTD, was coded as 0 on each replicated record for censored cases. For uncensored cases, DIS-RUPTD was coded 0 for each replicated record except the last, where it was coded 1. In all, 12,480 records were produced in the couple-period data set. Table 12.3 further illustrates the conversion process.

Panel A in the table shows the couple-level data for three couples. The first couple, ID 18, is a continuously cohabiting couple who separated 33 months after wave 1. They had been together nine months before the initial survey. The second couple was cohabiting at wave 1 and had been living together three months before the start of the survey. They then married 13 months after wave 1, and separated 38 months after wave 1. The third couple had been married for 21 months prior to wave 1 and was censored 67 months after wave 1 by the wave 2 survey. This couple had a birth to the union 51 months after the initial survey.

Panel B shows the couple-period version of the data. To account for left-truncation, survival time prior to the start of the study was used to update the TIME variable accordingly for each couple. Hence, as couple 18 was not observed until the second six-month interval of their risk period, TIME begins with the value 2 for this couple. Their total survival time was $9 + 33 = 42$ months, which puts their survival time in the seventh interval. Hence, the maximum TIME value for this couple is 7. Notice that DISRUPTD is coded 0 for times 2 through 6 and then changes to 1 when TIME $= 7$. The second couple, ID 630, had been together for only three months before wave 1. As they are still coming under observation in the first time interval, TIME begins at 1 for them. Couple 630 got married $3 + 13 = 16$ months after inception of the union, which puts their marriage in the third interval. Thus, the time-varying indicator of transition to marriage, COHTOMAR, changes from 0 to 1 at TIME $= 3$. As their survival time is 41 months, they are last observed when TIME $= 7$, at which time DISRUPTD again changes to 1. Finally, the married couple's survival time is $21 + 67 = 88$ months. However, as they are observed for only 67 months, they contribute only 12 (since $67/6 = 11.17$) records to the expanded data set. Nevertheless, they do not come under observation until the fourth time interval, hence TIME begins at 4 for this couple. Their birth took place $21 + 51 = 72$ months after inception of the marriage, which puts their birth in the 12th interval. Thus, the time-varying covariate UNBIRTH is coded 0 until TIME $= 12$, at which point it changes to 1. They were last observed at TIME $= 15$, where DISRUPTD remains at 0. Notice that time-invariant covariates are duplicated as is on all records for all couples.

Discrete-Time Analysis: Examples

Union Disruption. Table 12.4 presents the results of estimating discrete-time models for the log-hazard of union disruption using the couple-period data for our 1230 NSFH couples. The "logit model" is equation (12.9), employing 14 dummies for the

Table 12.3 Union Disruption Data in Couple versus Couple-Period Formats

A. *Couple Data*

ID	FAGUNION	DURATION	DURMOS	APART	UNIBIR	DURTOBIR	CHONLY	COHMAR	DURTOMAR
18	28.75	9	33	1	0		1	0	
630	23.92	3	38	1	0		0	1	13
25	21.67	21	67	0	1	51	0	0	

B. *Couple-Period Data*

ID	FAGUNION	TIME	UNBIRTH	COHONLY	COHTOMAR	DISRUPTD
18	28.75	2	0	1	0	0
18	28.75	3	0	1	0	0
18	28.75	4	0	1	0	0
18	28.75	5	0	1	0	0
18	28.75	6	0	1	0	0
18	28.75	7	0	1	0	1
630	23.92	1	0	0	0	0
630	23.92	2	0	0	0	0
630	23.92	3	0	0	1	0
630	23.92	4	0	0	1	0
630	23.92	5	0	0	1	0
630	23.92	6	0	0	1	0
630	23.92	7	0	0	1	1
25	21.67	4	0	0	0	0
25	21.67	5	0	0	0	0
25	21.67	6	0	0	0	0
25	21.67	7	0	0	0	0
25	21.67	8	0	0	0	0
25	21.67	9	0	0	0	0
25	21.67	10	0	0	0	0
25	21.67	11	0	0	0	0
25	21.67	12	1	0	0	0
25	21.67	13	1	0	0	0
25	21.67	14	1	0	0	0
25	21.67	15	1	0	0	0

(Continued)

Table 12.3   (Continued)

Variable Definitions	
ID	Couple identification number
FAGUNION	Female's age at union inception
DURATION	Number of months from inception of union until beginning of observation period
DURMOS	Number of months from beginning of observation period until either disruption or censoring
APART	Censoring indicator at the couple level
UNIBIR	Time-invariant dummy for a union birth
DURTOBIR	Number of months from beginning of observation period until a union birth
COHONLY	Dummy for continuous cohabitation
COHMAR	Time-invariant dummy for cohabitations that transitioned to marriage
DURTOMAR	Number of months from beginning of observation period until the transition to marriage
TIME	Number of the current time interval
UNBIRTH	Time-varying dummy for occurrence of a union birth
COHTOMAR	Time-varying dummy for cohabitations that transitioned to marriage
DISRUPTD	Censoring indicator at the couple-period level

15 time intervals representing survival time. Actually, after adjusting for left trunca-tion, there were 20 six-month intervals of survival time. But I collapsed intervals 15 to 20 into the value 15 in order to have enough uncensored observations in each interval. The logit model is an approximation to the Cox model shown as model 5 in Table 11.2. However, there are two key differences. First, the Cox model includes prior relationship duration. This is not possible with the discrete-time approach since prior duration is already incorporated into the time-interval dummies. Second, the discrete-time model includes the effect of time on the log-hazard, which is ignored in the Cox model.

The time dummies seem to suggest somewhat of a declining trend in the hazard, in that with the exception of time interval 9, log-hazards are elevated in the first three intervals, compared to other intervals, and significantly so, compared to the last interval. Otherwise, covariate effects in the logit model tend to mirror those in the Cox model except for the regressors *alcohol or drug problem* and *open disagree-ment*, which are not significant in the logit model. The proper interpretations of logit model effects would be in terms of the odds of dissolution. However, given the small probabilities of disruption in each time interval, as seen in Table 11.1, the effects can safely be interpreted in terms of hazard ratios. Hence, each unit improvement in *conflict resolution style* is estimated to reduce the hazard of disruption at any given time by a factor of $\exp(-.205) = .815$, and so on. A more accurate estimate of the "underlying" Cox model is shown as model 1 in Table 12.4. This is equation (12.17), again estimated using time-interval dummies. Although the effects are similar to those in the logit model, the coefficients for *continuously cohabiting, union birth*,

Table 12.4 Discrete-Time Approximations to Continuous-Time Models of Union Disruption

Predictor	Logit Model	Complementary Log-Log Models	
		Model 1	Model 2
Intercept	−1.344*	−1.517**	−1.132*
Time interval 1	1.373***	1.335***	
Time interval 2	.951**	.948**	
Time interval 3	.658*	.649*	
Time interval 4	.455	.416	
Time interval 5	.108	.129	
Time interval 6	.362	.359	
Time interval 7	.338	.337	
Time interval 8	.496	.487	
Time interval 9	.661*	.644*	
Time interval 10	−.043	−.046	
Time interval 11	.324	.322	
Time interval 12	.204	.199	
Time interval 13	.528	.523	
Time interval 14	.533	.518	
Time interval number			−.037*
Female's age at union	−.056***	−.054***	−.054***
Both in a first union	−.628***	−.609***	−.612***
Alcohol or drug problem	.284	.287	.275
Continuously cohabiting	1.643***	1.573***	1.587***
Cohabiting to married [a]	.355	.349	.351
Union birth [a]	−.568***	−.555***	−.547***
Open disagreement	.027	.027	.025
Conflict resolution style	−.205***	−.190***	−.188***

Note: $n = 1230$; number of couple periods is 12,480.

[a] Time-varying covariate.

* $p < .05.$ ** $p < .01.$ *** $p < .001.$

and *conflict resolution style*, in particular, are closer to the Cox results. Again, exponentiating the coefficients in model 1 provides estimates of the hazard ratios for unit increases in the predictors.

Parameterizing the Hazard Function. An advantage of the discrete-time approach is that one can explore various parameterizations of the hazard as a function of *time* to see which best fits the data. If these functions imply nested models, we can use $\Delta\chi^2$ to test whether the effect of *time* can be represented more parsimoniously than in the model that employs the time-interval dummies. I therefore examined a series of nested models, beginning with a model excluding the effect of *time* altogether. That is, I began by simply omitting the time-interval dummies from the model and tested whether a significant loss in fit resulted. If no loss in fit were experienced, a constant-hazard model would be indicated. However, $\Delta\chi^2$ for the constant-hazard model (results not shown), compared to model 1, was 26.826, which, with 14 *df* was significant at $p = .02$. I next fitted a series of polynomial models in *time* (a variable whose values represent time intervals) beginning with a linear term for *time*, then adding a quadratic term, a cubic term, and a quartic term, and compared all to model 1. The linear model had $\Delta\chi^2 = 20.801$, which, with 13 *df*, was not quite significant ($p = .077$). Adding a quadratic term did not improve fit, although quadratic, cubic, and quartic models also resulted in no significant loss in fit, compared to model 1. Due to its greater parsimony, however, I present the linear model (model 2) in Table 12.4. Again, results are approximately the same as for the other two models in the table. The significant and negative effect of *time interval number* in model 2 indicates, as previously suggested, that the hazard of disruption is declining with time. However, recall that unmeasured heterogeneity could also be responsible for such a trend. A discrete-time model that adjusts for unmeasured heterogeneity has been discussed by Land et al., (2001).

Advantages of the Discrete-Time Approach. The discrete-time approach has some clear advantages over the Cox model and over parametric models such as the exponential or Weibull. Therefore, even with continuous-time data, it may at times be advantageous to convert one's data to a discrete-time format in order to benefit from these features. First, there is the issue of tied survival times. For example, for the unemployment data considered in Chapter 11 (as well as below), fully 33.9% of spells were tied at a survival time of one month. About 15% were tied at two months, 10.2% at one-half a month, and 9.2% at three months. When there are many tied survival times in the data, the Cox model becomes unreliable (Yamaguchi, 1991). On the other hand, ties pose no problem for the discrete-time approach. Second, estimation of the Cox model becomes quite time-consuming when there are many time-varying covariates in the model. With the discrete-time approach, the number of such covariates is immaterial, as they are simply incorporated directly into the data set. Third, software for Cox models typically renders the creation of time-varying covariates transparent to the analyst. One just has to trust that they are being created correctly. In the discrete-time method, one can visually inspect the records to ensure correct coding. Fourth, the discrete-time approach allows one to explore the shape of the hazard function and to test various parameterizations of *time* against the unspecified function of *time* implied by time-interval dummies. Finally, as with the parametric models mentioned in Chapter 11, the discrete-time method allows for estimation of the hazard function as well as the survival function.

Estimation of hazard and survival functions employing, say, model 2 in Table 12.4 is straightforward. The estimate of the hazard at time t is recovered from the equation

$$\hat{P}_{it} = 1 - \exp[-\exp(\beta_0 + \beta_1 t + \mathbf{x}^{i\prime}\mathbf{g})],$$

where $\mathbf{x}^{i\prime}\mathbf{g}$ is the linear combination of covariates and parameter estimates, apart from the intercept and the linear effect of time. The survival function, denoted $S_{it} = P(T_i > t_i)$, is estimated using a recursion formula that is initialized at the value 1:

$$\hat{S}_{it} = 1(1 - \hat{P}_{i1})(1 - \hat{P}_{i2})(1 - \hat{P}_{i3}) \cdots (1 - \hat{P}_{it}).$$

[See Singer and Willett (1993) for programming suggestions for estimating P_{it} and S_{it} using SAS.] Figures 12.1 and 12.2 display the survival and hazard functions at mean values of model covariates for those in first unions as opposed to others, based on model 2 in Table 12.4.

Cohabiting Transitions. The technique for estimating competing-risks models when time is discrete or interval-censored parallels the method just articulated for the single-event case. However, the likelihood function no longer factors into separate components for each event type, as it does in continuous time (Allison, 1982). Instead, parameter estimates are obtained by maximizing the joint likelihood involving all event types simultaneously. This is accomplished readily with multinomial logistic regression applied to person-period data. For example, in applying the discrete-time approach to the 411 cohabiting transitions analyzed above using the Cox model, I created couple-period data in a manner similar to that described above for

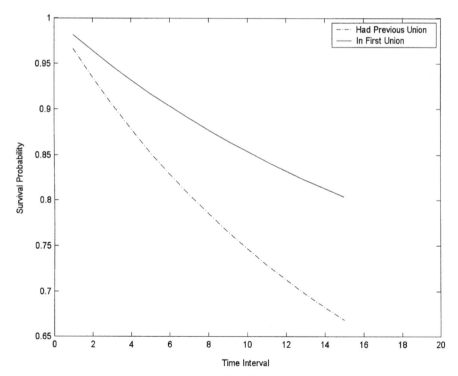

Figure 12.1 Survival curves for event of union disruption by whether in a first union, based on model 2, Table 12.4.

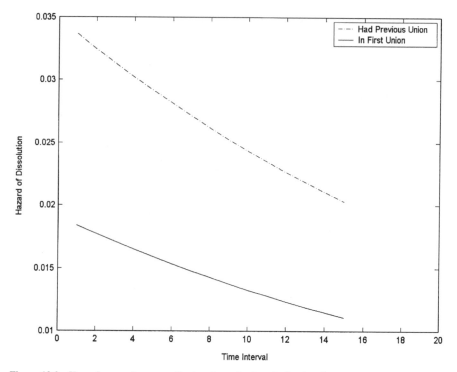

Figure 12.2 Hazard curves for event of union disruption by whether in a first union, based on model 2, Table 12.4.

the union disruption data. Once again, I created 15 six-month time intervals and replicated couples' records up to a maximum of 15 times, resulting in a total of 2265 couple-periods. Due to the longer time these couples had been together prior to wave 1, however, adjusting for left truncation resulted in values for TIME as high as 44, equivalent to 264 months. Nevertheless, in estimating the model, I collapsed the *time* variable down to five categories in order to have ample cell sizes for each type of transition. The event indicator, called COMPRISK, was coded according to SAS's convention for multinomial logistic regression (in PROC CATMOD) in which the highest code is the reference group. In this case, the reference group consisted of the censored cases—the continuous cohabitors—and was coded 3 in all periods. For uncensored cases, the event indicator was coded 3 until the last interval, at which point it was coded 1 for those who separated and 2 for those who married. The results for the multinomial logit model are shown in Table 12.5. Substantively, the findings are quite similar to those from the Cox model in Table 12.1, except that in the discrete-time formulation the effect of *female violence* on the transition to marriage is no longer significant.

Transitions Out of Unemployment. As a final example of discrete-time analyses, I reestimated the risk of reemployment for Brazilian immigrants, earlier examined using

Table 12.5 Discrete-Time (Multinomial Logit) Approximations to Competing-Risks Models for Exits from Unmarried Cohabitation

Predictor	Separation vs. Continuous Cohabitation	Marriage vs. Continuous Cohabitation
Intercept	.754	−6.895***
Time interval 1	1.378***	2.137***
Time interval 2	.609	1.304***
Time interval 3	.471	1.488***
Time interval 4	−.039	.938*
Open disagreement	−.010	.077**
Conflict resolution style	−.153*	.153*
Male's violence	−.573	.677
Female's violence	.397	−1.169
Intense male violence	.766*	−.347
Union birth[a]	−1.045**	−2.467***
Female's age at union	−.040***	−.014
Minority couple	−.091	−.798***
Male's relationship happiness	.046	.202*
Female's relationship happiness	−.168*	−.182*
Male's relationship stability	−.226	.297*
Female's relationship stability	−.037	.277*

Note: $n = 411$; number of couple periods is 2265.

[a] Time-varying covariate.

$* p < .05. ** p < .01. *** p < .001.$

Table 12.6 Discrete-Time Approximations to Multiepisode Models for Transitions Out of Unemployment

Predictor	Logit Model	Complementary Log-Log Models	
		Cox Approximation	Weibull Approximation
Intercept	−1.516***	−1.513***	−1.477***
Time interval 2	1.168***	.916***	
Time interval 3	1.763***	1.336***	
Time interval 4	1.520***	1.180***	
Time interval 5	3.596***	2.356***	
Log of time interval			1.149***
Age at unemployment	−.161*	−.129*	−.124*
Duration of previous job (months)	−.045***	−.029***	−.029**
Number of relatives in North America	.029*	.022*	.024*
Cumulative time in North America	−.184***	−.141***	−.145***
Log of monthly income last job	.104***	.065***	.065***
Education	.061**	.048***	.048**
Female	−.504***	−.381***	−.382***
Good English proficiency	−.676**	−.546**	−.530**

Note: $n = 620$ job spells; number of spell-periods is 1485.

$* p < .05. ** p < .01. *** p < .001.$

the Cox model. In this instance, the 620 unemployment spells represent the original "person-level" data. *Months of unemployment* for each spell were partitioned into five time intervals, each spell's covariates were replicated up to five times, and a total of 1485 spell-periods were created. An event indicator for reemployment was coded 0 in all spell-periods for censored cases, but changed to 1 in the last period for those who were reemployed. Both logit and complementary log-log models were then employed to estimate the binary indicator of reemployment. Table 12.6 shows the results and is to be compared to model 1 in Table 12.2, which employs the Cox model. The logit and Cox approximation models both simulate the Cox model in that time is coded as a series of dummies (omitting the first time interval) and is therefore left unspecified. However, an assessment of various parametric forms for the hazard function as articulated in Chapter 11 pointed to the Weibull distribution as being most appropriate for survival time. The Weibull model is therefore simulated in the third column of Table 12.6, by using *log of time-interval* to represent time periods, rather than the time-interval dummies. In all three models, there is the suggestion of an increasing hazard of reemployment over time. The results of all three models, but especially of the two complementary log-log models, are quite close in spirit and in parameter values to their Cox counterpart in Table 12.2.

EXERCISES

12.1 A competing-risks model for transitions out of the current job for a population of employees has the following form, where $X = IQ$ and transition types are: $1 =$ "quit," $2 =$ "fired," and $3 =$ "promoted":

$$h_{i1}(t) = .02 \exp(-.005X),$$
$$h_{i2}(t) = .005 \exp(-.009X),$$
$$h_{i3}(t) = .015 \exp(.007X).$$

Give the hazard of a job transition of any kind after three years of employment for someone with an IQ of 120.

12.2 Based on the information in Exercise 12.1, by what percent is the hazard changed for each 10-point increase in IQ for (**a**) the hazard of quitting, (**b**) the hazard of being fired, and (**c**) the hazard of being promoted?

12.3 In the employment-transitions problem of Exercise 12.1, if all transitions are treated the same and a Cox model is estimated for the hazard of any transition, the likelihood is .0011. On the other hand, a competing-risks model produces the following likelihoods for quitting, being fired, and getting promoted, respectively: .3, .25, and .655. Test whether the same model applies to each event type.

12.4   Axinn and Thornton (1993) examined the influence of mothers' and children's *attitudes toward marriage and cohabitation*, in 1980, on children's union-formation experience between 1980 and 1985. They estimated equations for (**a**) the cohabitation rate, in which risk sets consisted of those not yet cohabiting or married; (**b**) the marriage rate, in which risk sets consisted of those not yet cohabiting or married; (**c**) the marriage rate, in which risk sets consisted of those not yet married; and (**d**) the union formation rate, in which risk sets consisted of those not yet cohabiting or married. If marriage and cohabitation are regarded as competing risks, which two equations would represent a competing-risks model? Why?

12.5   In the Axinn and Thornton (1993) study mentioned in Exercise 12.4, which equations must be employed to get the appropriate likelihoods for testing model invariance over event types in a competing-risks model of marriage versus cohabitation?

12.6   In the study of Exercise 12.4, the following equation characterizes the interaction of *mother's attitude toward cohabitation* (higher scores indicate more favorable attitudes) with *child's gender*: $\ln \hat{h}_i(t) = \ln h_0(t) + \mathbf{x}'\mathbf{g} + 1.00$ *mother's attitude* $+ .51$ *female* $+ 1.17$ *mother's attitude * female*, where $\mathbf{x}'\mathbf{g}$ represents the linear combination of control variables and parameter estimates. Interpret this interaction effect.

12.7   In the study of the hazard of reemployment of Brazilian immigrants discussed in this chapter, suppose that we have a male immigrant with the following history: first laid off from work in September 1985. Rehired in December 1985. Quit his job in June 1987. Got another job in April 1988. Promoted in October 1988. Fired in October 1990. Got a new job in December 1990. Quit that job in May 1990. Still unemployed when interviewed in October 1991. Translate this job history into a set of unemployment spells. Show duration unemployed in months and censoring status for each spell.

12.8   Suppose that a discrete-time approach is taken for the analysis of the hazard of reemployment, in which survival is recorded in three-month intervals. How many spell-periods would be contributed by the immigrant in Exercise 12.7, and in how many of these would the binary response variable—the event indicator—be coded 1?

12.9   Goza and DeMaris (2003) examined several models for the hazard of reemployment using the unemployment data on Brazilian immigrants discussed in the text. In one discrete-time model employing the log of time interval (LOGTIME), they find a nonproportional effect of CANADA (dummy for residing in CANADA versus the U.S.) of the form $\ln[-\ln(1 - P_{it})] = -1.5974 + \mathbf{x}'\mathbf{g} + 1.7198$ LOGTIME $+ .4064$ CANADA $- .6924$ CANADA *

LOGTIME. Interpret the effect of Canadian residence on the hazard of reemployment (recalling that the coefficients can be interpreted as effects on the log-hazard in the complementary log-log model). At what value of TIME is the coefficient for Canadian residence approximately zero?

12.10 Using the "Cox approximation" model in Table 12.6, give the estimated hazards and survival probabilities for the first two time intervals for a male immigrant who was 40 years old at the beginning of unemployment, who had been in his previous job for six months, who had three relatives in North America (NA), who had been in NA for five years, who was making $25,000.00 per year in his last job, who had a high school education, and whose English proficiency was "good."

12.11 Using model 2 in Table 12.4, give the estimated hazards and survival probabilities for the first two time intervals for a continuously cohabiting couple in which the female was 19 at inception of the union, both partners are in a first union, neither partner has problems with substance abuse, there was no union birth, the open disagreement score is 5.5, and the conflict resolution style score is 2.5.

12.12 Long et al. (1993) used a discrete-time model to examine the process of promotion in rank for 556 male and 450 female professors in the field of biochemistry. All people had held positions as assistant professors in research universities at some point in their careers. Letting $\alpha' t$ represent a particular parameterization of time, their model was expressed as

$$P_{it} = \frac{1}{1 + \exp(-\alpha' t - \mathbf{x}^{it\prime} \boldsymbol{\beta})},$$

where the t superscript on the covariate vector denotes the potential presence of time-varying covariates. What type of model is this? (*Hint*: Rewrite the equation in a more easily recognizable form.)

12.13 In the Long et al. (1993) study, time was parameterized as a fourth-order polynomial in *years in rank*. Letting t denote *years in rank*, the equation for men's hazard of promotion to associate professor, as a function of *years in rank*, and apart from covariates, was

$$\ln \frac{P_{it}}{1 - P_{it}} = -12.221 + 4.507t - .694t^2 + .043t^3 - .001t^4.$$

Give the estimated hazards and the survival probabilities for the first four years in rank for a man whose covariate values were all equal to zero.

12.14 In the Long et al. (1993) study, the equation for the hazard of promotion to associate professor as a function of *years in rank*, for women, was, apart from the covariates,

$$\ln\frac{P_{it}}{1 - P_{it}} = -6.383 + 1.541t - .151t^2 + .004t^3 + .000t^4.$$

Give the equation for the effect of years in rank, apart from covariates, as a function of *gender*, using a dummy, F, for being female, and a set of coefficients for the interaction of *gender* with *time*. That is, the equation should take the form

$$\ln\frac{P_{it}}{1 - P_{it}} = \alpha + \delta F + at + bt^2 + ct^3 + dt^4 + g_1Ft + g_2Ft^2 + g_3Ft^3 + g_4Ft^4.$$

12.15 The Long et al. (1993) study employed the following covariates (among others): (**a**) *years between receiving the PhD and entering the current rank*; (**b**) *prestige of the PhD-granting department*; (**c**) *whether the doctorate was in a medical area*; (**d**) *prestige of the current employing institution*; (**e**) *whether the current job was in the PhD-granting institution*; and (**f**) *the square root of the number of articles published since entering the current rank*. Which of these are time-invariant, and which are time-varying, covariates? Why?

12.16 Using the *cohabiting transitions dataset*, estimate the multistate model in Table 12.1 via the two-step approach discussed in the text, despite this approach being theoretically questionable for this problem. Be sure to include both DURATION and the log of SURVIVAL in the logit step of the model. Show estimates for both equations, with significant effects starred, and provide a general interpretation of the results.

12.17 Using the *cohabiting transitions dataset*, estimate the competing risks model of Table 12.1, allowing nonproportional effects of HEHIT, SHEHIT, and TERROR2. Show estimates for both equations, with significant effects starred, and provide a general interpretation of the results.

12.18 Using the *unemployment dataset*, estimate a Cox multiepisode model for the hazard of reemployment using the predictors PRVJBS, JOBDUR, FEMALES, NEWEDUC, LOGINC, and CANADA. Show estimates for the equation, with significant effects starred, and provide a general interpretation of the results.

12.19 Using the *unemployment dataset*, estimate the model of Exercise 12.18, allowing a nonproportional effect for JOBDUR. Show estimates for the equation, with significant effects starred. Test for a significant nonproportional effect using both LRχ^2 and Wald χ^2 tests. Interpret the nature of the nonproportional effect by showing how the effect of JOBDUR varies with survival time.

12.20 Reestimate the model of Exercise 12.19 using the discrete-time approach. Employ five time intervals, as follows: DURUNEM $<$ 1, 1 $\leq$ DURUNEM $<$ 2, 2 $\leq$ DURUNEM $<$ 4, 4 $\leq$ DURUNEM $<$ 6, and DURUNEM $\geq$ 6. Then use the complementary log-log model, along with the log of time interval, to approximate the Weibull model. Show estimates for the equation, with significant effects starred, and provide a general interpretation of the results, particularly as they differ from those found in Exercise 12.19.

APPENDIX A

Mathematics Tutorials

APPENDIX OVERVIEW

In the event that your mathematics skills are rusty, the following sections contain brief tutorials in basic algebra, summation notation, covariance algebra, derivatives, and matrix algebra. Although you certainly can read and digest all of this material before proceeding with the rest of the book, most readers will probably not want to do that. Therefore, these tutorials are each designed to be discrete sections that can be reviewed separately, according to the reader's need to refresh his or her skills.

I. BASIC ALGEBRA

Algebra involves operations using arbitrary numbers, or variables, as well as specific numbers, or constants. In the following sections, letters such as a, b, c, x, w, y, or z denote any arbitrary real numbers (in this book we will only be concerned with real numbers), whereas 3 or π denote constants. (The symbol π denotes an irrational number representing the ratio of the area of a circle to the square of its radius, and equaling approximately 3.14159.) Further, I follow the mathematical notational convention that exponents only apply to the expression immediately to the left of the exponent unless parentheses group elements for exponentiation. Thus, if $x = 3$, then $2x^2 = 2(3^2) = 18$. But $(2x)^2 = [2(3)]^2 = 36$.

The following algebraic rules apply to any real numbers.

A. Commutative Property of Numbers

(1) $a + b = b + a$.
Example: $3 + 7 = 7 + 3 = 10$.

Regression with Social Data: Modeling Continuous and Limited Response Variables,
By Alfred DeMaris
ISBN 0-471-22337-9   Copyright © 2004 John Wiley & Sons, Inc.

(2) $ab = ba$.

Example: $6(9) = 9(6) = 54$.

B. Associative Property of Numbers

(1) $a + (b + c) = (a + b) + c$.

Example: $7 + (11 + 19) = (7 + 11) + 19 = 37$.

(2) $a(bc) = (ab)c$.

Example: $2[4(8)] = [2(4)]8 = 64$.

C. Distributive Property of Numbers

(1) $a(b + c) = ab + ac$.

Example: $3(4 + 5) = 3(4) + 3(5) = 12 + 15 = 27$.

(2) $(a + b)c = ac + bc$.

Example: $(6 + 2)4 = 6(4) + 2(4) = 24 + 8 = 32$.

D. Repeated Application of Distributive Property

Since $a(b + c) = ab + ac$, if we let $a = m + n$, then $a(b + c) = (m + n)(b + c) = (m + n)b + (m + n)c = mb + nb + mc + nc$.

Note: Subtraction and division are neither commutative nor associative:

$$a - b \neq b - a$$

$$a - (b - c) \neq (a - b) - c$$

$$\frac{b}{a} \neq \frac{a}{b}$$

$$\frac{a/b}{c} \neq \frac{a}{b/c}$$

E. Identities

(1) 0 is the identity element for addition and subtraction. That is,

$$a + 0 = 0 + a = a.$$

$$a - 0 = a \quad \text{(however, } 0 - a = -a\text{)}.$$

(2) 1 is the identity element for multiplication and division. That is,

$$1(a) = (a)1 = a.$$

$$a/1 = a \quad \text{(however, } 1/a \text{ does not equal } a\text{)}.$$

F. Inverses

For every real number a, $-a$ is the additive inverse of a, such that

$$a + (-a) = 0.$$

For every real number a, $1/a$ is the multiplicative inverse, or reciprocal, of a, such that

$$(1/a)a = a(1/a) = (a/a) = 1.$$

G. Other Properties of Addition, Subtraction, Multiplication, and Division

If $a = b$, then for any number c:

(1) $a + c = b + c$,

(2) $a - c = b - c$,

(3) $ac = bc$,

(4) $a/c = b/c$ provided that c is not 0.

H. Factorials

For any integer, N, the symbol $N!$ (pronounced "en factorial") is defined by $N! = N(N-1)(N-2)(N-3) \cdots (2)(1)$. For example, $5! = 5(4)(3)(2)(1) = 120$.

I. Properties Associated with Zero

(1) $0\,(a) = a(0) = 0$; therefore, it is also true that $0(0) = 0$.

(2) $a/0$ is an undefined operation; therefore, $0/0$ is also undefined.

(3) Any number (except 0) raised to the 0 power equals 1.

(4) 0^0 is undefined (however, $0^1 = 0$).

(5) $0! = 1$.

(6) If $a/b = 0$, then $a = 0$.

(7) If $ab = 0$, then either $a = 0$ or $b = 0$ or both a and b equal zero.

J. Rules for Negative Signs

(1) $-(-a) = a$.

(2) $(-a)b = -ab = a(-b)$.

(3) $(-a)(-b) = ab$.

(4) $(-1)a = -a$.

K. Rules for Fractions

(1) $\dfrac{a}{-b} = \dfrac{-a}{b} = -\dfrac{a}{b}.$

Example: $\dfrac{2}{-3} = \dfrac{-2}{3} = -\dfrac{2}{3}.$

(2) $\dfrac{-a}{-b} = -\dfrac{-a}{b} = -\dfrac{a}{-b} = \dfrac{a}{b}.$

Example: $\dfrac{-3}{-7} = -\dfrac{-3}{7} = -\dfrac{3}{-7} = \dfrac{3}{7}.$

(3) $\dfrac{a}{b}\dfrac{c}{d} = \dfrac{ac}{bd}.$

Example: $\dfrac{2}{3}\left(\dfrac{5}{7}\right) = \dfrac{(2)(5)}{(3)(7)} = \dfrac{10}{21}.$

(4) $\dfrac{ad}{bd} = \dfrac{a}{b}.$

Example: $\dfrac{3\pi}{4\pi} = \dfrac{3}{4}.$

(5) $\dfrac{a}{d} + \dfrac{b}{d} = \dfrac{a+b}{d}.$

Example: $\dfrac{3}{17} + \dfrac{9}{17} = \dfrac{12}{17}.$

(6) $\dfrac{a}{b} + \dfrac{c}{d} = \dfrac{ad+bc}{bd}.$ Why? Since

$$\frac{a}{b} = \frac{ad}{bd} \text{ and } \frac{c}{d} = \frac{bc}{bd},$$

then

$$\frac{a}{b} + \frac{c}{d} = \frac{ad}{bd} + \frac{bc}{bd} = \frac{ad+bc}{bd}.$$

Example: $\dfrac{xz}{y} + \dfrac{xw}{y^2} = \dfrac{xzy+xw}{y^2}.$

Elaboration: By rule I.E(2), multiplying xz/y by 1 leaves the term unchanged. However, if we write 1 as y/y, we have, by rule I.K(3), $(xz/y)(y/y) = xyz/y^2$. Now both terms in the example have the same denominator, hence the numerators can be summed, as in rule I.K(5), and the result follows.

(7) $\dfrac{a/b}{c/d} = \dfrac{a}{b}\dfrac{d}{c} = \dfrac{ad}{bc}.$

Example: $\dfrac{2\pi/x}{xyz/3\pi} = \dfrac{6\pi^2}{x^2yz}.$

(8) $\dfrac{a}{b} = \dfrac{c}{d}$ if and only if $ad = bc$.

Example: $\dfrac{2x}{3} = \dfrac{3}{4}$ implies that $8x = 9$ or $x = \dfrac{9}{8}$.

L. Laws of Exponents

Let x and y be real numbers, and suppose that a and b are positive real numbers. Then

(1) $b^x b^y = b^{x+y}$.

Example: $(27^{4/18})(27^{2/18}) = 27^{6/18} = 27^{1/3} = 3$.

(2) $\dfrac{b^x}{b^y} = b^{x-y}$.

Example: $\dfrac{x^3}{x^5} = x^{3-5} = x^{-2} = \dfrac{1}{x^2}$.

(3) $(b^x)^y = b^{xy}$.

Example: $(3^\pi)^2 = 3^{2\pi} = 995.04$.

(4) $(ab)^x = a^x b^x$.

Example: $(9x)^2 = 81x^2$.

(5) $\left(\dfrac{a}{b}\right)^x = \dfrac{a^x}{b^x}$.

Example: $\left(\dfrac{xy}{3z}\right)^2 = \dfrac{x^2 y^2}{9z^2}$.

(6) $b^{-x} = \dfrac{1}{b^x}$.

Example: $15^{-4} = \dfrac{1}{15^4}$.

M. Working with Rational Exponents

Frequently, we encounter expressions such as $\sqrt{a}$ or $\sqrt[3]{a}$ or $(\sqrt[3]{a})^2$. Each expression involves a rational exponent or an exponent that is the ratio of integers. In particular:

$$\sqrt{a} = a^{1/2}.$$

$$\sqrt[3]{a} = a^{1/3}.$$

$$\left(\sqrt[3]{a}\right)^2 = a^{2/3}.$$

In general, if p and q are integers, $a^{p/q}$ is defined to be $(\sqrt[q]{a})^p$, or, equivalently, $\sqrt[q]{a^p}$. The laws of exponents enumerated above apply also to rational exponents.

Some examples are

$$(-32)^{4/3} = (\sqrt[5]{-32})^4 = (-2)^4 = 16$$

$$16^{-5/2} = \frac{1}{16^{5/2}} = \frac{1}{(\sqrt{16})^5} = \frac{1}{4^5} = \frac{1}{1024}$$

$$(-16)^{-2/5} = \frac{1}{(-16)^{2/5}} = \frac{1}{(\sqrt[5]{-16})^2} = \frac{1}{(-1.7411)^2} = \frac{1}{3.0314}.$$

Rational exponents are frequently easier to work with than numbers expressed using root signs. For example, suppose that we wish to evaluate $[\sqrt[5]{(\sqrt{2})^3}][\sqrt[5]{(\sqrt{2})^{17}}]$. Using rational exponents, we write

$$\left[\sqrt[5]{(\sqrt{2})^3}\right] \text{ as } [(2^{1/2})^3]^{1/5} = (2^{3/2})^{1/5} = 2^{3/10},$$

and we write

$$\left[\sqrt[5]{(\sqrt{2})^{17}}\right] \text{ as } [(2^{1/2})^{17}]^{1/5} = (2^{17/2})^{1/5} = 2^{17/10}.$$

Thus,

$$\left[\sqrt[5]{(\sqrt{2})^3}\right]\left[\sqrt[5]{(\sqrt{2})^{17}}\right] = (2^{3/10})(2^{17/10}) = 2^{3/10+17/10} = 2^{20/10} = 2^2 = 4.$$

N. Logarithms

Definition. The *logarithm* to the base q of a number x, symbolized "$\log_q x$," is the power to which one would have to raise the number q in order to get x. For example:

$$\log_2 8 = 3 \quad \text{because} \quad 2^3 = 8.$$
$$\log_2 16 = 4 \quad \text{because} \quad 2^4 = 16.$$
$$\log_4 \frac{1}{16} = -2 \quad \text{because} \quad 4^{-2} = \frac{1}{16}.$$
$$\log_4 2 = \frac{1}{2} \quad \text{because} \quad 4^{1/2} = 2.$$
$$\log_{64} 4 = \frac{1}{3} \quad \text{because} \quad 64^{1/3} = 4.$$

Rules for Logarithms

(1) $\log_q 1 = 0$.
(2) $\log_q ac = \log_q a + \log_q c$.
(3) $\log_q \dfrac{a}{c} = \log_q a - \log_q c$ for $c \neq 0$.

(4) $\log_q a^r = r \log_q a$.

(5) $\log_q \dfrac{1}{c} = -\log_q c$ for $c \neq 0$.

The natural logarithm, denoted "ln" (or "log") is very important in statistics. It is the logarithm to the base e. The number e (after the German mathematician, Euler) is an irrational number defined as follows: $e = \lim_{x \to \infty}(1 + 1/x)^x$. The expression to the right of the equals sign here means (in English) "the value approached by (i.e., the limiting value, or limit, of) the expression $(1 + 1/x)^x$, as x is allowed to increase without bound." This value can be estimated by plugging a very large value of x into the expression $(1 + 1/x)^x$. For example, if $x = 999999999$, we have $(1 + 1/999999999)^{999999999} = 2.71828182438$, or 2.718, to three decimal places. The natural log of x, written $\ln x$ or $\log x$, is therefore the power to which one must raise e to get x. It is defined only for $x > 0$.

Examples:

$$\ln 1 = 0 \quad \text{because} \quad e^0 = 1.$$

$$\ln 10 = 2.3026 \quad \text{because} \quad e^{2.3026} = 10.$$

$$\ln \frac{1}{2} = -.693 \quad \text{because} \quad e^{-.693} = \frac{1}{2}.$$

$$\ln 500 = 6.215 \quad \text{because} \quad e^{6.215} = 500.$$

Antilog Function. The function e^x or $\exp(x)$ is called the *exponential function*. This is also known as the *antilogarithm*, or *antilog*, of x, since it "undoes" the log function. Another way of stating this is that the exponential function is the inverse of the log function, and vice versa. That is, $e^{\ln x} = x$ and $\ln e^x = x$.

For example, $\ln e^2$ is the power to which one must raise e to get e^2, and this is obviously 2, hence $\ln e^2 = 2$. On the other hand, if one has the log of a number, and one raises e to that power, one recovers the number again. Thus, 6.215 is the log of 500, so $e^{6.215}$, or $e^{\log 500}$, gives us 500 again.

O. Absolute Value

The absolute value of a number a, denoted $|a|$, is defined as follows:

(1) $|a| = a$ if $a \geq 0$,

(2) $|a| = -a$ if $a < 0$.

Examples: $|3| = 3$ (since $3 > 0$); and $|-3| = 3$ [since $-3 < 0$, its absolute value is $-(-3)$, or 3].

(3) For any real number, a, $\sqrt{a^2} = |a|$.

Examples: $\sqrt{3^2} = |3| = 3$; $\sqrt{(-3)^2} = |(-3)| = 3$.

If a and b are real numbers and n is an integer:

(4) $|ab| = |a||b|$.

Example: $|3(-4)| = |-12| = 12 = |3||-4|$.

(5) $\left|\dfrac{a}{b}\right| = \dfrac{|a|}{|b|}$.

Example: $\left|\dfrac{-2}{5}\right| = \dfrac{|-2|}{|5|} = \dfrac{2}{5}$.

(6) $|a^n| = |a|^n$.

Example: $|(-3)^3| = |-3|^3 = 3^3 = 27$.

Note: In general, $|a \pm b| \neq |a| \pm |b|$. For example, $|-3 + 4| = 1 \neq |-3| + |4| = 7$.

P. Functions

Functions and functional notation are very important in mathematics and statistics. In the models in this book, for example, an outcome variable is typically modeled as a function of one or more explanatory variables. Let's define this notion more precisely.

Definition. A *function* is a rule of correspondence between the elements of two sets, set x and set y. The elements are the respective values of x and y. It is usually posited that y is a function of x, denoted $y = f(x)$, meaning that the values of y depend on the values of x, often via some mathematical formula involving x. The function itself is the rule that assigns a unique element from the set y to each element in the set x.

The simplest function is a constant: $y = f(x) = 5$ [or $f(x) = 5$; or $y = 5$]. This says that the set y consists of one element, 5, which corresponds to each and every element in the set x. Another example is $f(x) = 4x - 7$. In this case, $4x - 7$ is the rule that assigns y to a given x value. When $x = 0$, $y = -7$. When $x = 1$, $y = -3$, and so on. The notation "$f(x)$" is arbitrary. We could just as well use $g(x)$, $h(x)$, $\theta(x)$, or $\pi(x)$. As another example: $g(x) = x^2 - 3x + 1$.

It is important to understand that in the notation "$f(x)$," x is just a generic label or a dummy variable. That is, it represents whatever is the subject of the operations called for in the function. For example, if we let $f(x) = x^2 + 10$, this notation essentially says: "The outcome—$f(x)$, or y—of this function is arrived at by squaring the input—x—and then adding 10 to that result." Hence, $f(x + 3) = (x + 3)^2 + 10$, $f(\ln x) = (\ln x)^2 + 10$, $f(e^x) = (e^x)^2 + 10$, and so on.

To speak of y as being a function of x also implies dependence of y on x. That is, the value of y is *generated* by applying the rule (or formula) to different values of x. It is in this sense that we will use the term *function* in this book. In social research, y is the outcome variable, whose value is generated (in a causal sense) by values of one or more explanatory variables.

Functions can be expressed in many different ways, with many different variables:

$A = f(r) = \pi r^2$ (the area of a circle is a function of its radius, r).

$V = f(r,h) = \pi r^2 h$ (the volume of a cylinder is a function of its radius, r, and its height, h).

$V = f(r) = \frac{4}{3}\pi r^3$ (the volume of a sphere is a function of its radius, r).

Functions can also be *implicit*; that is, the rule is unspecified:

Teacher salary = f(county of employment).

Job satisfaction = f(salary, autonomy, responsibility, role specificity).

Linear Functions. Especially important in this book, and in statistics generally, are linear functions. *Y is a linear function of one or more x's if it can be expressed as a weighted sum of x-values times constants plus (possibly) other constants.* For example, equations of the form $y = a + bx$ are linear functions of a single x. This is a weighted sum of a constant, b, times x, plus another constant, a. As another example, $y = a + b_1 x_1 + b_2 x_2 + b_3 x_3$ is a linear function of x_1, x_2, and x_3.

The function $y = a + bx$ is linear in the sense that if the set of points (x,y) is plotted on a two-dimensional graph, they will all fall on a straight line. [Correspondingly, when y is a linear function of several x's, the set of points $(x_1, x_2, \ldots, x_k, y)$ falls on a single hyperplane.]

Example. Let $y = 2 + 3x$. Figure A.1 presents a graph of this equation. The equation is defined by two important components: the intercept, a (2, in this case), and the slope, b (3, here). The intercept is the value of y when $x = 0$. It is also the value of y where the line of (x,y) points, implied by the equation, crosses the y-axis. The slope of the equation indicates the number of units y changes as x increases by 1 unit. It is also known as the ratio of the "rise" in y to the "run" in x, or *slope = rise/run*. For example, if x increases 5 units, from 0 to 5, y increases from $2 + 3(0) = 2$ to $2 + 3(5) = 17$. This is an increase of 15 units. The unit increase in y per unit increase in x, however, is $15/5 = 3$ units, which agrees with the slope value of 3 for this equation. The resulting (x,y) points achieved by plugging sample values of x into this equation lie on the line indicated in the figure.

Point–Slope Form of a Line. If we know the slope of a linear equation and any point on the equation, we can easily recover the equation for the line. Hence, if x_0, y_0 is a point on the line and b is the slope, the general equation for the line is $(y - y_0) = b(x - x_0)$. As an example, let's find the equation for the line with slope 5.2 that passes through the point (2,9). Solution: The equation would be $(y - 9) = 5.2(x - 2)$ or $y = 9 + 5.2x - 10.4$. The resulting equation in the form $y = a + bx$ is $y = -1.4 + 5.2x$. It is easily verified that the point (2,9) is on the line, since if $x = 2$, we have $y = -1.4 + 5.2(2) = 9$.

Nonlinear Function. Y can also be a *nonlinear function* of x. Examples are $y = a +$

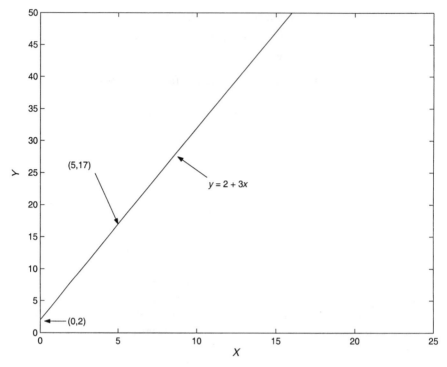

Figure A.1   Linear relationship between Y and X when $Y = 2 + 3X$.

$\ln x$, $y = a + e^x$ or $y = a + bx + cx^2$. In each case, x enters the equation in some form other than simply being multiplied by a constant. The graphs of such equations would be curves rather than straight lines. In the third example here, $y = a + bx + cx^2$, notice that y is a nonlinear function of x but it is *linear in the parameters*, a, b, and c. That is, if we were to fix the value of x (say, by setting $x = 2$) in this equation, and evaluate it for different values of a, b, and c, it becomes a function of these parameters rather than of x. In this event, notice that y would be a linear function of the parameters, since y is a weighted sum of the parameter values times constants. The constants for a, b, and c, respectively, are 1, x, and x^2. The concept of *linearity in the parameters* becomes important when we consider multiple linear regression.

Q. Exercises

(1) Find $(3x + 2)(y + z)$.

(2) Find $(2x + \sqrt{y} + z)(x^2 + \sqrt{y} + 4w)$.

(3) Solve $\frac{1}{2}x + 12 = \sqrt{24}$.

(4) Evaluate $\dfrac{6!}{2!(6-2)!}$.

(5) Find $\dfrac{ab}{c} + \dfrac{3ab\pi}{cd}$.

(6) Find $\dfrac{x}{zw} + \dfrac{y}{2w}$.

(7) Solve $\dfrac{\sqrt{x}}{25} = \dfrac{4}{9}$.

(8) Find $(3x^3)(5x^2)$.

(9) Find $\left(\dfrac{x}{2y}\right)^z$.

(10) Find $\dfrac{(\sqrt{2})^3}{\sqrt[3]{2}}$.

(11) Simplify $\ln\dfrac{x^2}{\sqrt{z}}$.

(12) Find $f(x) = \frac{1}{2} + 9x$ for $x = -\frac{1}{2}, \frac{1}{3}, 2$.

(13) Find $f(x) = x^2 - 3x + 1$ for $x = -\frac{1}{2}, \frac{1}{3}, 2$.

(14) Let $f(x) = \sqrt{x}$ and $g(x) = 2^x$. Find $f(g(x))$ for $x = 3$.

(15) Let $f(x) = e^x$ and $g(x) = -3 + 5x$. Find $f(g(x))$ for $x = \frac{1}{2}$.

(16) Let $f(x) = 2 + \frac{1}{2}x$ and $g(x) = -3 + 5x$. Find $f(g(x))$.

(17) Find the equation for the line with slope 18 that passes through the (x,y) point $(-\frac{1}{2}, \frac{1}{2})$.

II. SUMMATION NOTATION AND RULES OF SUMMATION

A. Summation Notation

Anyone who has ever taken a statistics course is certainly familiar with summation notation. It is a compact means of expressing statistical formulas involving operations on n sample values. For example, if we have five sample values of y: y_1, y_2, y_3, y_4, and y_5, the sum of these five values, in summation notation, is $\sum_{i=1}^{5} y_i$, which indicates the sum $y_1 + y_2 + y_3 + y_4 + y_5$. The i below the summation symbol is called the *index of summation*. It indicates which elements are to be summed and ranges in integer increments from 1 to n. Since y is subscripted with i, "y_i" indicates the y-values that are to be summed and is known as a "variable with respect to the summation."

B. Rules of Summation

As it is often useful to recast sums into a different form for computational (or theoretical) reasons, the following set of rules for working with sums will be helpful:

(1) $\sum_{i=1}^{n} c = nc$. (The sum of a constant, n times, is n times that constant.)

(2) $\sum_{i=1}^{n} cx_i = c\sum_{i=1}^{n} x_i$. (The sum of a constant times a variable with respect to the summation is equal to the constant times the sum of the variable.)

(3) $\sum_{i=1}^{n}(x_i \pm y_i \pm zi) = \sum_{i=1}^{n} x_i \pm \sum_{i=1}^{n} y_i \pm \sum_{i=1}^{n} z_i$. (The sum of a sum or difference of terms is equal to the sum or difference of the separately summed terms.)

(4) $\sum_{i=1}^{n}(x_i y_i) \neq \sum_{i=1}^{n} x_i \sum_{i=1}^{n} y_i$. (It is not generally true that the sum of the product of terms is equal to the product of the separately summed terms.)

(5) $\sum_{i=1}^{n} \frac{x_i}{y_i} \neq \frac{\sum_{i=1}^{n} x_i}{\sum_{i=1}^{n} y_i}$. (It is not generally true that the sum of the ratio of terms is equal to the ratio of the separately summed terms.)

C. Working with Summations

Let's apply rules II.B(1) through II.B(3) to produce computing formulas for two well-known statistical measures.

(1) Sample Variance. The sample variance of x is defined as $s_x^2 = \sum_{i=1}^{n}(x_i - \bar{x})^2/(n-1)$, or, omitting the index of summation for simplicity's sake, we have $s_x^2 = \sum(x - \bar{x})^2/(n-1)$. (We omit the index of summation when it is clear which element is the variable with respect to summation, as it is in this case.) Now, by the rules above,

$$\sum(x - \bar{x})^2 = \sum(x^2 - 2x\bar{x} + \bar{x}^2)$$

(expanding the term inside the sum)

$$= \sum x^2 - \sum 2x\bar{x} + \sum \bar{x}^2$$

[by rule II.B(3)]

$$= \sum x^2 - 2\bar{x} \sum x + n\bar{x}^2$$

[by rules II.B(2) and II.B(1), respectively]

$$= \sum x^2 - 2\bar{x}n\bar{x} + n\bar{x}^2$$

(because $\bar{x} = \sum x/n$ implies that $\sum x = n\bar{x}$)

$$= \sum x^2 - 2n\bar{x}^2 + n\bar{x}^2$$

$$= \sum x^2 - n\bar{x}^2.$$

Thus, a computing formula for the sample variance is

$$s_x^2 = \frac{\sum x^2 - n\bar{x}^2}{n - 1}.$$

(2) Sample Covariance. The sample covariance between x and y is defined as

$$\text{cov}(x,y) = \frac{\sum (x - \bar{x})(y - \bar{y})}{n - 1}.$$

Again, let's work with the sum in the numerator:

$$\sum (x - \bar{x})(y - \bar{y}) = \sum [xy - x\bar{y} - \bar{x}y + (\bar{x})(\bar{y})]$$

(expanding the term inside the sum)

$$= \sum xy - \sum x\bar{y} - \sum \bar{x}y + \sum (\bar{x})(\bar{y})$$

[by rule II.B(3)]

$$= \sum xy - \bar{y} \sum x - \bar{x} \sum y + n(\bar{x})(\bar{y})$$

[by rules II.B(2) and II.B(1)]

$$= \sum xy - n(\bar{x})(\bar{y}) - n(\bar{x})(\bar{y}) + n(\bar{x})(\bar{y})$$

$$= \sum xy - n(\bar{x})(\bar{y}).$$

A computing formula for the covariance is, therefore,

$$\text{cov}(x,y) = \frac{\sum xy - n(\bar{x})(\bar{y})}{n - 1}.$$

The reader will notice, in the derivations above, that both $\bar{x}$ and $\bar{y}$ are treated as constants. This, in fact, is the case with respect to the summation. That is, once the n cases have been employed to compute the means of x and y, these entities are then constant values with respect to any further summations.

D. Weighted Sums

One sum that is particularly important in statistics is the *weighted sum.* A weighted sum of a set of variables, $x_1, x_2, \ldots, x_k$, takes the form $\sum w_k x_k$, where the w_k are the weights and k is the index of summation. In this case, each x_k has an associated weight, w_k, by which it is multiplied. This type of sum is also called a *linear combination* or a *linear composite* of the x's. If the weights, moreover, sum to 1, the result is some type of mean. For example, the sample mean is a weighted sum, where each x is given the same weight, $1/n$. This is easily seen, since the sample mean can be expressed as $\bar{x} = \sum (1/n)\, x$. This clearly has the form $\sum w_k x_k$, where $k = i$, and $w_i = 1/n$. Weighted means are linear combinations of *means* (say, from different subpopulations), where the weights are not the same for each mean. For example, the weights might be the proportion of the overall population that falls into each subpopulation. If the weights sum to 1, the result is the mean, or "average," of the means. Finally, a *linear contrast* is a weighted sum in which the weights sum to zero. Linear contrasts are widely used to test various hypotheses about sample means in analysis of variance (ANOVA).

E. Exercises

(1) Prove that it is always true for any set of x-values that $\sum(x - \bar{x}) = 0$.

(2) Prove that for any set of n sample values $x_1, x_2, \ldots, x_n$, adding a constant, c, to each value changes the mean from $\bar{x}$ to $\bar{x} + c$ but does not affect the sample variance of X. That is, prove that $\text{Mean}(X + c) = \bar{x} + c$ and that $s^2_{x+c} = s^2_x$.

(3) Prove that if any set of X-values is standardized, by converting each X-value to Z via the formula $z = (x - \bar{x})/s_x$, the mean of the standardized scores is always zero and the standard deviation of the standardized scores is always 1. That is, prove that it is always the case that $\bar{z} = 0$ and $s_z = 1$. [*Hint*: Write out formulas for the mean and standard deviation using Z instead of X, then substitute $(x - \bar{x})/s_x$ for Z and use the summation rules to simplify the result.]

III. COVARIANCE ALGEBRA

Many theoretical derivations of importance to statistics depend on making use of covariance algebra. Covariance algebra consists of a set of algebraic rules for finding variances and covariances involving variables and constants. These rules make it possible to find variances of terms, and covariances between terms, which appear at first glance to be quite complicated.

A. Definition

$\text{Cov}(x,y) = E(x - \mu_x)(y - \mu_y)$. The *population* (or *theoretical*) *covariance* between x and y is the expected value, or average, of the cross-product of deviation scores in x with deviation scores in y. [The sample estimator of this quantity is given in Section II.C(2). Notice the difference in notation between "Cov" for the population entity and "cov" for the sample entity.] The covariance is a quantitative measure of how two variables vary together. Positive covariances reflect situations in which large (small) values of x are associated with large (small) values of y. Negative covariances indicate that large (small) values of x are associated with small (large) values of y.

B. Basic Rules of Covariance Algebra

Let W, X, Y, and Z be variables, and let a, b, c, and d be constants, in a given set of data. Then:

(1) $\text{Cov}(X,Y) = \text{Cov}(Y,X)$. Covariance is symmetric with respect to the order of the variables.

(2) $\text{Cov}(X,c) = 0$. The covariance of a variable with a constant is zero. This makes intuitive sense, since one of these "variables" isn't varying at all.

(3) $V(X) = \text{Cov}(X,X)$. The variance of a variable is the covariance of that variable with itself.

(4) $V(cX) = c^2 V(X)$. The variance of a constant times a variable is the square of the constant times the variance of the variable.

(5) $\text{Cov}(aX,bY) = ab \, \text{Cov}(X,Y)$. The covariance of variables multiplied by constants is the product of the constants times the covariance of the variables. Or, constants can be "pulled through" covariance operations.

(6) $\text{Cov}(aX + bY, cW + dZ) = ac \, \text{Cov}(X,W) + ad \, \text{Cov}(X,Z) + bc \, \text{Cov}(Y,W) + bd \, \text{Cov}(Y,Z)$. The covariance of two linear combinations is a linear combination of the individual covariances. This rule shows a simple technique for finding the covariance of any two terms. Let's find that last covariance again, to see how this works:

Step 1. Multiply the terms on each side of the comma together using regular algebra:

$$(aX + bY)(cW + dZ) = aXcW + aXdZ + bYcW + bYdZ.$$

Step 2. In the resulting sum, separate the original components of each term with commas:

$$aX,cW + aX,dZ + bY,cW + bY,dZ.$$

Step 3. Take the sum of the covariances of the terms joined by commas:

$$\text{Cov}(aX,cW) + \text{Cov}(aX,dZ) + \text{Cov}(bY,cW) + \text{Cov}(bY,dZ)$$

Step 4. Apply the appropriate basic rules above to reduce the result to an expression involving the sum of constants times covariances of variables or constants times variances of variables:

$$= ac \, \text{Cov}(X,W) + ad \, \text{Cov}(X,Z) + bc \, \text{Cov}(Y,W) + bd \, \text{Cov}(Y,Z).$$

Notice that I've applied rule III.B(5) to the expression in step 3 to arrive at the result in step 4.

C. Applications

Application 1. In linear regression, we regress Y on a set of X's: $X_1, X_2, \ldots, X_K$ in a particular sample. Each b_k (i.e., each regression coefficient) has an associated sampling variance. Each pair of regression coefficients, such as b_1 and b_2, say, has a sampling covariance. Sampling variances and covariances make sense only in the context of repeated sampling. That is, the current sample is only one of an infinite number of possible samples of size n that could be drawn from the population. Hence, the current (i.e., those from your particular sample) regression coefficients are only one set from an infinite number of sets of regression coefficients that could be obtained by regressing Y on the X's in each of the infinite number of samples. The variance of b_k, denoted $V(b_k)$, is then a quantitative measure of the variation in b_k one would encounter from performing all of these different sample regressions. The covariance of, say, b_1 with b_2, denoted $Cov(b_1, b_2)$, is a measure of the extent to which the values of b_1 and b_2, from all of these regressions, would covary.

Okay, suppose that you have two coefficients in a particular sample, say b_1 and b_2, and you want to test whether the population analogs of these coefficients are equal. That is, you want to test the null hypothesis $H_0: \beta_1 = \beta_2$, or $H_0: \beta_1 - \beta_2 = 0$. The test is a t test of the form

$$t_{(n-K-1)} = \frac{b_1 - b_2}{SE(b_1 - b_2)},$$

where $SE(b_1 - b_2)$ is the estimated standard error of $b_1 - b_2$, the difference in the sample coefficients. This standard error is the estimate of the square root of $V(b_1 - b_2)$, the variance of $b_1 - b_2$. How do we find this variance?

Realize first that b_1 and b_2 are variables over repeated sampling, and their difference is therefore also a variable, so we can use covariance algebra to find the difference between two variables:

$$V(b_1 - b_2) = Cov(b_1 - b_2, b_1 - b_2)$$

[by rule III.B(3); notice that the constant multiplier of b_2 here is "-1"]

$$= Cov(b_1, b_1) - Cov(b_1, b_2) - Cov(b_2, b_1) + Cov(b_2, b_2)$$

[by rule III.B(6)]

$$= V(b_1) + V(b_2) - 2\,Cov(b_1, b_2)$$

[by rules III.B(1) and III.B(3)]. The sampling variance of $b_1 - b_2$ can therefore be estimated by plugging sample estimates of $V(b_1)$, $V(b_2)$, and $Cov(b_1, b_2)$ into this last expression and then taking its square root. The required sample estimates can be found in the *variance–covariance matrix of parameter estimates*, which is an optional part of standard regression output.

Application 2. Continuing our linear regression example, suppose that we estimate an interaction model with two explanatory variables, X and Z. Our sample equation is $\hat{y} = a + bX + cZ + dXZ$, where XZ is the cross-product of the variables X and Z. We are interested in whether the impact of X on Y is significant at a particular level of Z, say at z. As explained in Chapter 3, the partial slope for the effect of X on Y at a particular level of Z is a function of Z. To see this, we factor all multipliers of X

in the regression equation $\hat{y} = a + bX + cZ + dXZ$, and we get $\hat{y} = a + cZ + (b + dZ)X$. Hence, the partial slope for X is the coefficient of X in this rewritten equation, or $b + dZ$. That is, the partial slope depends on the particular value of Z at which it is evaluated. When $Z = z$ (a particular value of Z), the partial slope is $b + dz$. The test statistic for the significance of this partial slope is, like any test statistic for a partial slope, the partial slope estimate divided by its estimated standard error. This is a t-test statistic with $n - K - 1$ degrees of freedom under the null hypothesis, where $K =$ the total number of regressors in the model (in this simple case, $K = 3$).

The test is

$$t_{(n-K-1)} = \frac{b+dz}{SE(b-dz)}.$$

How do we find $SE(b + dz)$? You guessed it—we use covariance algebra. We have to find an estimate of $V(b + dz)$ and then take its square root. Now, remember that if Z is fixed over repeated sampling (or if our results are conditional on the particular values of Z in our sample), z is a constant throughout this process (of repeated sampling, that is). That is, it doesn't change over repeated sampling; only b and d vary. Don't confuse the b and d in this example with the b and d in the covariance algebra rules. In that case they were constants. Now they're estimated regression coefficients, and therefore variables! So

$$V(b + dz) = \text{Cov}(b + dz, b + dz)$$

[applying rule III.B(3)]

$$= \text{Cov}(b,b) + \text{Cov}(b,dz) + \text{Cov}(dz,b) + \text{Cov}(dz,dz)$$

[by rule III.B(6)]

$$= V(b) + 2z\,\text{Cov}(b,d) + z^2 V(d)$$

[using rules III.B(3), III.B(4), and III.B(5)].

Again, estimates of the required variances and covariances can be obtained from the variance–covariance matrix of parameter estimates.

Application 3. As a third example of applying covariance algebra, I prove that the correlation between x and y is 1 in absolute value whenever y is a perfect linear function of x. First, consider the formula for the correlation coefficient: $r_{xy} = \text{cov}(x,y)/s_x s_y$. That is, the correlation between x and y is the covariance of x with y, divided by the product of their standard deviations. Since the product of the standard deviations can also be written as the square root of the product of the variances of x and y, we also have that $r_{xy} = \text{cov}(x,y)/\sqrt{v(x)v(y)}$.

Now, suppose that y is a perfect linear function of x. That is, suppose that $y = a + bx$. Then $\text{cov}(x,y) = \text{cov}(x,\ a + bx) = b\ \text{cov}(x,x) = b\ v(x)$ [by rules III.B(2),

III.B(3), and III.B(5)]. Also, $v(y) = v(a + bx) = \text{cov}(a + bx, a + bx) = b^2 v(x)$ [also by rules III.B(2), III.B(3), and III.B(5)]. Hence,

$$r_{xy} = \frac{\text{cov}(x,y)}{\sqrt{v(x)v(y)}} = \frac{bv(x)}{\sqrt{v(x)b^2v(x)}} = \frac{bv(x)}{v(x)\sqrt{b^2}}$$

(the variance of x is always positive, but b may not be) $= b/|b| = b/b = 1$ if $b > 0$, or $= b/-b = -1$ if $b < 0$.

D. Exercises

For variables X, Y, and ε and constants α, β, a, b, c, and d:

(1) Find the covariance of $\sqrt{5}\,X$ with $\sqrt{5}\,Y$.
(2) Find the covariance of $a + bX$ with $c + dY$.
(3) Find the covariance of $\frac{1}{2}X + 3Y$ with $2X - 9Y$ if $\text{Cov}(X,Y) = 1.2$, $V(X) = .5$, and $V(Y) = 2.5$.
(4) Find the variance of $2X - 3Y$ if $\text{Cov}(X,Y) = 1.2$, $V(X) = .5$, and $V(Y) = 2.5$.
(5) Suppose that $Y = \alpha + \beta X + \varepsilon$. Find the variance of Y in terms of the variance of X and the variance of ε if $\text{Cov}(X,\varepsilon) = 0$.
(6) Show that the correlation between X and Y is unchanged if X is multiplied by a positive constant, c, and Y is multiplied by a positive constant, d, where the correlation between any two variables U and V is $\rho_{UV} = \text{Cov}(U,V)/\sqrt{V(U)V(V)}$. [*Note*: The implication is, of course, that the correlation between two variables is unchanged by a rescaling of the variables (e.g., the correlation between height in feet and weight in pounds is the same as the correlation between height in meters and weight in grams).]
(7) Show that the covariance between two standardized variables, Z_x and Z_y, is the same as their correlation. [*Hint*: Use the definition of correlation in problem (6) and define $Z_x = (X - \mu_x)/\sigma_x$ and $Z_y = (Y - \mu_y)/\sigma_y$.]

IV. DERIVATIVES

A. Introduction

The derivative is a very important tool for the quantification of effects of explanatory variables in regression models. (It is also central to the estimation of parameters for these models.) In this tutorial, I define the derivative and partial derivative and give a series of algebraic rules for finding derivatives in various situations.

The derivative is the solution to one of the fundamental problems of calculus, the tangent problem (Anton, 1984): Given a function $Y = f(x)$ and a point $P(x_0, y_0)$ on its graph, find the equation of the tangent line, T, to the graph at P. Figure A.2 illustrates this problem. The tangent line, T, is the line that touches $f(x)$ only at point P. Finding the equation of this line requires determining its slope, which I denote as b_{tan}. Once

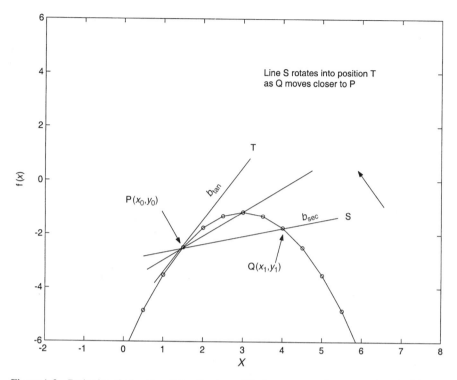

Figure A.2 Derivative of a function of X as the slope of the line tangent to the curve at the point $P(x_0, y_0)$.

this slope is known, given that we know a point on the line, $P(x_0, y_0)$, the equation for the line is simply $(y - y_0) = b_{tan}(x - x_0)$ (see Section (I.P).

As we saw earlier (i.e., in Section I.P), the slope of a straight line indicates the number of units that Y changes as X increases by 1 unit, *along the line*. However, for a nonlinear function of X, as shown in the figure, the slope of the line passing through point P on the curve is continually changing as P moves along the curve. Similarly, along the curve, f(x), Y does not change at a uniform rate as X increases. Rather, at some points along the curve Y is changing very rapidly with X (the slope at P is very steep), while at other points Y is changing more slowly (the slope at P is very shallow). Only along a straight line does Y change at a uniform rate with X. Hence, for any function in general, the slope, b_{tan}, indicates the *instantaneous rate of change* of Y with increase in X at a particular point, x.

Generally, the change in a function f(x) as X increases from x_0 to x_1 is $f(x_1) - f(x_0)$. The ratio of this change to the increase in X is then $[f(x_1) - f(x_0)]/(x_1 - x_0)$. This represents the *average rate of change* in the function with change in X over the interval from x_0 to x_1. This ratio is also the slope of the line joining point $P(x_0, y_0)$ and point $Q(x_1, y_1)$ in Figure A.2, called a *secant line* for the curve and denoted "S" in the figure. We denote the slope of this line as b_{sec}. This slope is a very crude approximation to b_{tan}. However, if we move point Q closer to point P and then recalculate

$b_{sec} = [f(x_1) - f(x_0)]/(x_1 - x_0)$, b_{sec} will become a better approximation to b_{tan}. If we continue in this fashion, moving Q increasingly closer to P and calculating b_{sec} each time, the line S will rotate toward a "limiting" position as Q converges to the point P. That position will be T, the tangent line of interest. This notion motivates the following definition.

Definition. The slope, b_{tan}, of the line tangent to the curve at the point (x_0, y_0) is

$$b_{tan} = \lim_{x_1 \to x_0} \frac{f(x_1) - f(x_0)}{x_1 - x_0}.$$

In English, this says that the slope of the line tangent to a function at a point is *the value approached by the ratio of the change in the function to the change in x at that point as the change in x diminishes to zero* (i.e., as x_1 converges to x_0). This expression is so important that it is given some special notation (that is more consistent with the "English interpretation" above). First, let $\Delta x = x_1 - x_0$, so that $x_1 = x_0 + \Delta x$. Also let $b_{tan} = f'(x_0)$. This slope is also called the *first derivative of* f *with respect to* X *at the point* x_0. In this altered notation, the first derivative is defined as

$$f'(x_0) = \lim_{\Delta x \to 0} \frac{f(x_0 + \Delta x) - f(x_0)}{\Delta x}.$$

Thus, the first derivative is the slope of the line tangent to the curve, $f(x)$, at x_0 [or at the point $P(x_0, y_0)$]. More intuitively, as noted previously, it is *the instantaneous rate of change in Y along the function* $f(x)$, *with increase in X, at the point* x_0. Because of this interpretation, the first derivative is ideal for quantifying the causal impact of X on Y, at least at a particular value of X.

It is typical to present the formula for the first derivative in terms of x, rather than x_0, making the notation more general. Thus, the first derivative of any function $f(x)$, at the point x, is

$$f'(x) = \lim_{\Delta x \to 0} \frac{f(x + \Delta x) - f(x)}{\Delta x}$$

Example 1. Let's see how this works. Suppose that we calculate the first derivative of $f(x)$ at the point x, where $f(x) = \alpha + \beta x$. (Notice that this is the simple linear regression function, as for the mean of Y at a given x.) We have

$$f'(x) = \lim_{\Delta x \to 0} \frac{\alpha + \beta(x + \Delta x) - (\alpha + \beta x)}{\Delta x}$$

$$= \lim_{\Delta x \to 0} \frac{\alpha + \beta x + \beta \Delta x - \alpha - \beta x}{\Delta x}$$

$$= \lim_{\Delta x \to 0} \frac{\beta \Delta x}{\Delta x} = \lim_{\Delta x \to 0} \beta = \beta.$$

The last equality results because the expression we are computing the limiting value for (i.e., β) is no longer a function of Δx. In fact, it is a constant value regardless of what Δx converges to. Hence its limit, as Δx converges to zero, remains at that value.

This example demonstrates that the slope of f(x) in the simple linear regression model is β, a constant. That is, E(Y) changes at a constant rate with increase in X, at any point x, along the regression line.

Example 2. Now suppose that $f(x) = x^2$. Again, the derivative is

$$f'(x) = \lim_{\Delta x \to 0} \frac{f(x + \Delta x) - f(x)}{\Delta x}$$

$$= \lim_{\Delta x \to 0} \frac{(x + \Delta x)^2 - x^2}{\Delta x}$$

$$= \lim_{\Delta x \to 0} \frac{x^2 + 2x\,\Delta x + (\Delta x)^2 - x^2}{\Delta x}$$

$$= \lim_{\Delta x \to 0} \frac{\Delta x (2x + \Delta x)}{\Delta x}$$

$$= \lim_{\Delta x \to 0} 2x + \Delta x = 2x + 0 = 2x.$$

In this case, the first derivative, $f'(x) = 2x$, is a function of x. That is, the rate of change of y with increase in x depends on the value of x. Thus, unlike a linear function of x, this nonlinear function of x (i.e., x^2) has a nonconstant slope. For example, when $x = 1$, the slope is $2(1) = 2$, so y is increasing at twice the rate of x. When $x = 2$, on the other hand, the slope is $2(2) = 4$, so y is increasing at four times the rate of x, and so on.

B. Rules for Finding Derivatives

Although we have defined the first derivative in terms of a limiting value above, it is not necessary to compute this limit every time we want to find the first derivative. Fortunately, there is a set of algebraic rules that provide derivatives for the functions that we will encounter. First, let us introduce some additional notation. The first derivative of f(x) with respect to x is also denoted $\frac{d}{dx} f(x)$. This notation will be especially useful when more complicated functions are involved. The rules are:

(1) $\frac{d}{dx}(c) = 0$, where c is a constant (i.e., the derivative of a constant with respect to x equals zero).

Example: $\frac{d}{dx}(5) = 0$.

(2) $\frac{d}{dx} e^x = e^x$.

(3) $\frac{d}{dx} \ln x = \frac{1}{x}$.

(4) $\frac{d}{dx} cf(x) = c\frac{d}{dx}f(x)$, where c is a constant.

Example: $\dfrac{d}{dx}(3.75\ln x) = 3.75\dfrac{d}{dx}\ln x = \dfrac{3.75}{x}$.

(5) $\dfrac{d}{dx}x^r = rx^{r-1}$, where r is any real-valued exponent.

Examples: $\dfrac{d}{dx}x^2 = 2x; \ \dfrac{d}{dx}x^5 = 5x^4; \ \dfrac{d}{dx}x^{2/3} = \dfrac{2}{3}x^{-1/3}; \ \dfrac{d}{dx}4x^3 = 12x^2$.

(6) $\dfrac{d}{dx}[f(x) \pm g(x)] = \dfrac{d}{dx}f(x) \pm \dfrac{d}{dx}g(x)$.

Example: $\dfrac{d}{dx}(\ln x + x^2) = \dfrac{d}{dx}(\ln x) + \dfrac{d}{dx}(x^2) = \dfrac{1}{x} + 2x$.

(7) $\dfrac{d}{dx}[f(x)g(x)] = f(x)g'(x) + g(x)f'(x)$.

Example: Find $\dfrac{d}{dx}(\sqrt{x}\ln x)$.

Solution: Let $f(x) = \sqrt{x} = x^{1/2}$ and $g(x) = \ln\ x$. Then $f'(x) = \frac{1}{2}x^{-1/2} = 1/2\sqrt{x}$, and $g'(x) = 1/x$. Hence,

$$f(x)g'(x) + g(x)f'(x) = \sqrt{x}\dfrac{1}{x} + (\ln x)\dfrac{1}{2\sqrt{x}} = \dfrac{1}{\sqrt{x}} + \dfrac{\ln x}{2\sqrt{x}} = \dfrac{2 + \ln x}{2\sqrt{x}}.$$

(8) $\dfrac{d}{dx}\left[\dfrac{f(x)}{g(x)}\right] = \dfrac{g(x)f'(x) - f(x)g'(x)}{[g(x)]^2}$.

Example: Find $\dfrac{d}{dx}\left(\dfrac{e^x}{\sqrt{x}}\right)$.

Solution: Let $f(x) = e^x$ and $g(x) = \sqrt{x}$. Then $f'(x) = e^x$ and $g'(x) = \frac{1}{2}x^{-1/2} = 1/2\sqrt{x}$. Hence,

$$\dfrac{g(x)f'(x) - f(x)g'(x)}{[g(x)]^2} = \dfrac{\sqrt{x}e^x - e^x(1/2\sqrt{x})}{x} = e^x\dfrac{2x - 1}{2x^{3/2}}.$$

One final rule that is particularly useful is based on the composition of two functions, $f(x)$ and $g(x)$.

Definition. Given two functions, $f(x)$ and $g(x)$, the *composition of* $f(x)$ *with* $g(x)$ is the function defined by $f(g(x))$.

This needs some explanation. The composition of f with g essentially takes $g(x)$ as the input to $f(x)$. That is, "x" in $f(x)$ becomes "$g(x)$." We then perform on $g(x)$ the operations specified by the function $f(x)$. Another way to view this operation is to call $f(x)$ the "outside function" and $g(x)$ the "inside function." Then $f(g(x))$ is the outside function evaluated at the inside function of x. [Exercises (14) to (16) in Section I.Q already involved compositions of functions.]

Example. Suppose that $f(x) = x^2$ and $g(x) = \ln\ x$. Then the composition $f(g(x))$ is $(\ln x)^2$. Notice that $g(x) = \ln\ x$ becomes "x," or the input, to the function $f(x) = x^2$.

As another example, suppose that $f(x) = e^x$ and $g(x) = \alpha + \beta x$. Then $f(g(x)) = e^{\alpha + \beta x}$. As a third example, suppose that $f(x) = x^2$ and $g(x) = y - a - bx$. Then $f(g(x)) = (y - a - bx)^2$.

Chain Rule. *If* $g(x)$ *is differentiable at the point* x *and* $f(x)$ *is differentiable at the point* $g(x)$, *the composition of* f *with* g, $f(g(x))$, *is differentiable at the point* x. Moreover,

$$(9) \quad \frac{d}{dx}[f(g(x))] = f'[g(x)]g'(x) = \left[\frac{d}{dg(x)}f(g(x))\right]\left[\frac{d}{dx}g(x)\right].$$

That is, the derivative of $f(g(x))$ is the derivative of the outside function evaluated at the inside function times the derivative of the inside function with respect to x.

Example 1. Find $\dfrac{d}{dx}\ln x^2$.

Solution: Let $f(x) = \ln x$ and $g(x) = x^2$. That is, $\ln x$ is the outside function and x^2 is the inside function, so that $f(g(x))$ is $\ln(x^2) = \ln x^2$. Then

$$\frac{d}{dx}\ln x^2 = \left(\frac{d}{dx^2}\ln x^2\right)\left(\frac{d}{dx}x^2\right) = \frac{1}{x^2}(2x) = \frac{2}{x}.$$

Example 2. Find $\dfrac{d}{dx}(5 - x^2)^3$.

Solution: One way to solve this, of course, is to find the cube of $(5 - x^2)$ and then take the first derivative with respect to x of the resulting expression. This is the "long way around." Using rule IV.B(9), however, we have

$$\frac{d}{dx}(5 - x^2)^3 = \frac{d}{d(5 - x^2)}(5 - x^2)^3\frac{d}{dx}(5 - x^2) = 3(5 - x^2)^2(-2x) = -6x(5 - x^2)^2.$$

The reader can verify that the same answer is arrived at going the long way around.

Example 3. Find $\dfrac{d}{dx}e^{5\sqrt{x}}$.

Solution: Let $f(x) = e^x$ and $g(x) = 5\sqrt{x}$. Then

$$\frac{d}{dx}e^{5\sqrt{x}} = \frac{d}{d5\sqrt{x}}e^{5\sqrt{x}}\frac{d}{dx}5\sqrt{x} = e^{5\sqrt{x}}\left(\frac{5}{2\sqrt{x}}\right) = \frac{5e^{5\sqrt{x}}}{2\sqrt{x}}.$$

Higher-Order Derivatives. The second-, third-, and higher-order derivatives of $f(x)$ with respect to x are just the derivatives of the next-lower-order derivative. That is,

$$f''(x) = \frac{d}{dx}f'(x) \text{ is the second derivative of } f(x) \text{ with respect to } x,$$

$$f'''(x) = \frac{d}{dx}f''(x) \text{ is the third derivative of } f(x) \text{ with respect to } x, \text{ and so on.}$$

Example. Find the second derivative of $f(x) = x^3$ with respect to x.

Solution: $f'(x) = 3x^2$, $f''(x) = \dfrac{d}{dx}(3x^2) = 6x$.

In this book we are not concerned with derivatives any higher than second order.

C. Partial Derivatives

Partial derivatives are derivatives with respect to a particular variable, say x, for functions in more than one variable. Suppose that $y = f(x,w,z)$. That is, f is now a function of three independent variables, x, w, and z. Suppose that we fix w and z at the respective values of w_0 and z_0. Then $y = f(x,w_0,z_0)$ is only a function of x. In other words, holding w and z constant at the values of w_0 and z_0, respectively, the values of y change only as we plug different values of x into the function. These resulting values of y (as x varies) then represent the "curve" of the function as x takes on different values. This brings us to the following.

Definition. The first partial derivative of $f(x,w_0,z_0)$ with respect to x, at x_0, is the slope of the tangent line to the curve $f(x,w_0,z_0)$ at the point (x_0,w_0,z_0).

The first partial derivative of $f(x,w,z)$ with respect to x is denoted:

$$\frac{\partial}{\partial x} f(x,w,z) \quad \text{or} \quad f_x(x,w,z).$$

The partial derivative (unless otherwise stated, we assume that this refers to the *first* partial derivative) is calculated by treating all other independent variables in the function as though they are constants. In that we are fixing them at specific values (e.g., w_0,z_0), this makes intuitive sense.

Example 1. Find $\dfrac{\partial}{\partial x}(.5 + 1.2x + 3w - \sqrt{2}\,z)$.

Solution: In that w and z are being treated as constants, the only term involving x is "1.2x." Hence, $\dfrac{\partial}{\partial x}(.5 + 1.2x + 3w - \sqrt{2}\,z) = 1.2$.

Example 2. Find $\dfrac{\partial}{\partial x}(xwz^2 + w\sqrt{x} + z^3 \ln x)$.

Solution: $\dfrac{\partial}{\partial x}(xwz^2 + w\sqrt{x} + z^3 \ln x) = wz^2 + \dfrac{1}{2}wx^{-1/2} + \dfrac{z^3}{x} = wz^2 + \dfrac{w}{2\sqrt{x}} + \dfrac{z^3}{x}.$

Example 3. Find $\dfrac{\partial}{\partial X_3}(\alpha + \beta_1 X_1 + \beta_2 X_2 + \beta_3 X_3 + \beta_4 X_2 X_3)$.

Solution: $\dfrac{\partial}{\partial X_3}(\alpha + \beta_1 X_1 + \beta_2 X_2 + \beta_3 X_3 + \beta_4 X_2 X_3) = \beta_3 + \beta_4 X_2.$

Second Partial Derivatives. We can compute second- and higher-order partial derivatives. These are just derivatives of the next-lower-order partial derivative.

Again, we will only be concerned with first- and second-order partial derivatives in this work. Because the function in question consists of more than one independent variable, second-order derivatives can be constructed either with respect to the same, or a different, variable than is true for the first partial derivative. Hence, for $y = f(x,z)$ we have

$$\frac{\partial^2 y}{\partial x^2} = \frac{\partial}{\partial x}\left(\frac{\partial y}{\partial x}\right) : \text{the second partial derivative of } y \text{ with respect to } x,$$
also denoted $f_{xx}(x,z)$.

$$\frac{\partial^2 y}{\partial z^2} = \frac{\partial}{\partial z}\left(\frac{\partial y}{\partial z}\right) : \text{the second partial derivative of } y \text{ with respect to } z,$$
also denoted $f_{zz}(x,z)$.

$$\frac{\partial^2 y}{\partial z \partial x} = \frac{\partial}{\partial z}\left(\frac{\partial y}{\partial x}\right) : \text{the mixed second partial with respect to } x, \text{ then } z,$$
also denoted $f_{xz}(x,z)$.

$$\frac{\partial^2 y}{\partial x \partial z} = \frac{\partial}{\partial x}\left(\frac{\partial y}{\partial z}\right) : \text{the mixed second partial with respect to } z, \text{ then } x,$$
also denoted $f_{zx}(x,z)$.

Second partial derivatives with respect to each variable or each pair of variables can be constructed for functions of several variables as well as for functions of just two variables. As with first partial derivatives, the second partial derivative with respect to a given variable is constructed by treating all other variables as constants.

Example 1. Find $\dfrac{\partial^2}{\partial x^2}(xwz^2 + w\sqrt{x} + z^3 \ln x)$.

Solution: From Example 2 in Section IV.C we found that $\dfrac{\partial}{\partial x}(xwz^2 + w\sqrt{x} + z^3 \ln x) = wz^2 + \dfrac{1}{2}wx^{-1/2} + \dfrac{z^3}{x}$. Hence,

$$\frac{\partial^2}{\partial x^2}(xwz^2 + w\sqrt{x} + z^3 \ln x) = \frac{\partial}{\partial x}\left(wz^2 + \frac{1}{2}wx^{-1/2} + z^3 x^{-1}\right) = -\frac{1}{4}wx^{-3/2} - z^3 x^{-2}$$

$$= \frac{-w}{4x^{3/2}} - \frac{z^3}{x^2}.$$

Example 2. Find $\dfrac{\partial^2}{\partial z^2}(xwz^2 + w\sqrt{x} + z^3 \ln x)$.

Solution: First, we note that $\dfrac{\partial}{\partial z}(xwz^2 + w\sqrt{x} + z^3 \ln x) = 2xwz + 3z^2 \ln x$. Then we have that

$$\frac{\partial^2}{\partial z^2}(xwz^2 + w\sqrt{x} + z^3 \ln x) = \frac{\partial}{\partial z}(2xwz + 3z^2 \ln x) = 2xw + 6z \ln x.$$

Example 3. Find $\dfrac{\partial^2}{\partial z \partial x}(xwz^2 + w\sqrt{x} + z^3 \ln x)$.

Solution: We saw that $\dfrac{\partial}{\partial x}(xwz^2 + w\sqrt{x} + z^3 \ln x) = wz^2 + \dfrac{1}{2}wx^{-1/2} + \dfrac{z^3}{x}$, Hence,

$$\frac{\partial^2}{\partial z\,\partial x}(xwz^2 + w\sqrt{x} + z^3 \ln x) = \frac{\partial}{\partial z}\left(wz^2 + \frac{1}{2}wx^{-1/2} + z^3 x^{-1}\right) = 2wz + 3z^2 x^{-1}.$$

Example 4. Find $\dfrac{\partial^2}{\partial x\,\partial z}(xwz^2 + w\sqrt{x} + z^3 \ln x)$.

Solution:

$$\frac{\partial^2}{\partial x\,\partial z}(xwz^2 + w\sqrt{x} + z^3 \ln x) = \frac{\partial}{\partial x}\left[\frac{\partial}{\partial z}(xwz^2 + w\sqrt{x} + z^3 \ln x)\right]$$

$$= \frac{\partial}{\partial x}(2xwz + 3z^2 \ln x) = 2wz + 3z^2 x^{-1}.$$

It is not just coincidence that the mixed second partial derivatives in Examples 3 and 4 are equal. By theorem, the mixed second partial derivatives are equal for a wide class of functions (Munem and Foulis, 1984).

D. Exercises

(1) Find $\dfrac{d}{dx}(x^3)$.

(2) Find $\dfrac{d}{dx}(x^{4/5})$.

(3) Find $\dfrac{d}{dx}(3x^4 + 2x^3 - x^2 + 1)$.

(4) Find f'(x) for f(x) = ln $x + x^3$ and evaluate f'(3). [*Hint:* First find f'(x) in general, then plug in "3" for x and evaluate the result.]

(5) Find f'(−2) for f(x) = $e^x - x^2$.

(6) Find f'(x) if f(x) = $x^3 e^x$.

(7) Find f'(x) if f(x) = x ln x.

(8) Find f'(x) if f(x) = $\sqrt{x}\ln x^2$.

(9) Find f'(x) if f(x) = $\dfrac{\sqrt{x}}{\ln x}$.

(10) Find f''(x) if f(x) = x^4.

(11) Find f''(x) if f(x) = ln x.

(12) Find f''(x) if f(x) = e^{5x}.

(13) Find f'(x) if f(x) = $(\ln x)^2$.

(14) Find f''(x) if f(x) = e^{x^2}.

(15) Find $\dfrac{d}{dx}\left(\dfrac{e^x}{1 + e^x}\right)$.

(16) Find $\dfrac{d}{dx}\left(\dfrac{e^{\alpha + \beta x}}{1 + e^{\alpha + \beta x}}\right)$.

(17) Find $f'(x)$ if $f(x) = \ln x^2$.

(18) Find $f'(x)$ if $f(x) = (x + 2)^4$.

(19) Find $f'(x)$ if $f(x) = \left(\dfrac{1}{2}x + \ln x\right)^3$.

(20) Find $\dfrac{\partial}{\partial x}(x^2 + 3xy + y^2)$.

(21) Find $\dfrac{\partial}{\partial x}\left(\dfrac{e^{\alpha+\beta x+\gamma w+\lambda z}}{1 + e^{\alpha+\beta x+\gamma w+\lambda z}}\right)$.

V. MATRIX ALGEBRA

Matrix algebra is a compact and powerful form of algebra that makes possible computations that would be extremely cumbersome to perform in ordinary, called *scalar*, algebra. In this tutorial, I define the major elements and tools of matrix algebra and demonstrate various applications of its use in statistics. Many students find matrix notation and matrix manipulations to be especially arcane. In this tutorial I will try, as much as possible, to demystify these topics.

A. Why Matrix Algebra?

Sometimes you just can't "get there from here" very well without matrix algebra. Let's consider an example. Suppose that we wish to find the expected value and variance of the ordinary least squares (OLS) estimator of β_k in the linear regression model. That is, we need to find $E(b_k)$ and $V(b_k)$, applying rules for finding the expectation and variance of sample statistics. With just one regressor in the model, this isn't too daunting in scalar algebra. The formula for b is just

$$b = \frac{\sum (X - \bar{X})(Y - \bar{Y})}{\sum (X - \bar{X})^2}.$$

However, with two regressors in the model, the formulas for b_1 and b_2 are

$$b_1 = \frac{\sum (X_2 - \bar{X}_2)^2 \sum (X_1 - \bar{X}_1)(Y - \bar{Y}) - \sum (X_1 - \bar{X}_1)(X_2 - \bar{X}_2)\sum (X_2 - \bar{X}_2)(Y - \bar{Y})}{\sum (X_1 - \bar{X}_1)^2 \sum (X_2 - \bar{X}_2)^2 - \left[\sum (X_1 - \bar{X}_1)(X_2 - \bar{X}_2)\right]^2},$$

$$b_2 = \frac{\sum (X_1 - \bar{X}_1)^2 \sum (X_2 - \bar{X}_2)(Y - \bar{Y}) - \sum (X_1 - \bar{X}_1)(X_2 - \bar{X}_2)\sum (X_1 - \bar{X}_1)(Y - \bar{Y})}{\sum (X_1 - \bar{X}_1)^2 \sum (X_2 - \bar{X}_2)^2 - \left[\sum (X_1 - \bar{X}_1)(X_2 - \bar{X}_2)\right]^2}.$$

The reader can well imagine how cumbersome these expressions get with even more regressors in the model. In contrast, the formula in matrix notation for the vector, or set, of regression estimates, **b**, for any number of regressors in the model, is $\mathbf{b} = (\mathbf{X'X})^{-1}\mathbf{X'y}$, which is considerably more parsimonious. Using this expression, finding the expectation and variance of **b** is quite straightforward, as is demonstrated in Section V.J below.

B. Notation and Definitions

Notation. The following notational conventions will be used throughout this section: Capital letters (e.g., A, B, X, Z) denote matrices. Lowercase letters in boldface (e.g., **a**, **b**, **x**, **z**) represent vectors. Lowercase letters at the end of the alphabet (e.g., u, v, w, x, y, z) represent variables. Lowercase letters at the beginning of the alphabet (e.g., a, b, c, d) represent constants.

Definitions. A *vector* is a collection of values, or *scalars*, arranged in a *column*. For example,

$$\mathbf{x} = \begin{bmatrix} 1 \\ 2 \\ 3 \end{bmatrix}$$

is a vector consisting of the three scalars (numbers) 1, 2, and 3. A vector can also be written as a row of numbers. When it is, it is referred to as a *row vector* and given a transpose (**x'**) superscript. In fact, the *transpose* of a column vector is the same vector written as a row. Thus,

$$\mathbf{x'} = [1 \quad 2 \quad 3].$$

In contrast to some works, this book will always use a transpose sign to indicate a row vector.

A *matrix* is simply a collection of column (row) vectors:

$$X = \begin{bmatrix} 1 & 4 & 7 \\ 2 & 5 & 8 \\ 3 & 6 & 9 \end{bmatrix}.$$

Matrices are distinguished according to the number of rows and columns in the matrix. These constitute the dimensions, or *order*, of a matrix and are often shown as subscripts to the matrix symbol. In that the matrix, X, above, has three rows and three columns, it would be denoted X_{33}. In general, X_{rc} is a matrix with r rows and c columns. Individual elements inside a matrix are denoted, in the abstract, with i and j subscripts to indicate their row and column location within the matrix. Thus, x_{ij} is the element in the ith row and jth column of the matrix X. For example, in the matrix immediately above, $x_{32} = 6$.

A vector is a matrix with either r or c equal to 1. Vectors can also be distinguished by dimensional subscripts. For example,

$$\mathbf{x}_{31} = \begin{bmatrix} 1 \\ 2 \\ 3 \end{bmatrix},$$

whereas

$$\mathbf{x}'_{13} = [1 \quad 2 \quad 3].$$

However, most of the time, the dimensions of a matrix or a vector will be omitted from the symbol, for expressive economy.

A matrix can be shown in more compact notation by showing only its columns, written as vectors. For example, if the columns of X are denoted

$$\mathbf{x}_1 = \begin{bmatrix} 1 \\ 2 \\ 3 \end{bmatrix}, \quad \mathbf{x}_2 = \begin{bmatrix} 4 \\ 5 \\ 6 \end{bmatrix}, \quad \text{and} \quad \mathbf{x}_3 = \begin{bmatrix} 7 \\ 8 \\ 9 \end{bmatrix},$$

we can write X as

$$X = [\mathbf{x}_1 \quad \mathbf{x}_2 \quad \mathbf{x}_3],$$

where in this case, the subscripts on the columns indicate the column numbers rather than their dimensions. Notice that when we write X as $[\mathbf{x}_1 \quad \mathbf{x}_2 \quad \mathbf{x}_3]$, it has the form of a row vector. In fact, a matrix could be described as a vector whose individual elements are themselves vectors. Writing X as $[\mathbf{x}_1 \quad \mathbf{x}_2 \quad \mathbf{x}_3]$ is also referred to as *partitioning* X by its columns. This is not the only way to partition a matrix. It can also be partitioned by submatrices within the overall matrix. However, we will frequently find it useful to partition X by its columns (or its rows) and then to treat it in subsequent operations as though it is a vector, which, in fact, it is. Examples will be seen shortly.

Similarly, if the rows of X are denoted

$$\mathbf{x}^{1\prime} = [1 \quad 4 \quad 7], \quad \mathbf{x}^{2\prime} = [2 \quad 5 \quad 8], \quad \text{and} \quad \mathbf{x}^{3\prime} = [3 \quad 6 \quad 9],$$

we can also write X as

$$X = \begin{bmatrix} \mathbf{x}^{1\prime} \\ \mathbf{x}^{2\prime} \\ \mathbf{x}^{3\prime} \end{bmatrix}.$$

In this case, X is shown as partitioned by its rows and has the appearance of a column vector, which, again, it is. It is a column vector whose individual elements are themselves row vectors. In general, I use $\mathbf{x}_i$ to denote the *i*th column of a matrix, X, and $\mathbf{x}^{i\prime}$ to denote the *i*th row of that matrix. This notation allows an unequivocal language for representing rows and columns of a matrix, as well as their transposes, without ambiguity. As noted above, the transpose of a column vector is that column written as a row. Hence,

$$\mathbf{x}'_1 = [1 \quad 2 \quad 3].$$

The transpose of a row vector is, conversely, that vector written as a column. In that the row vector already has a transpose sign, transposing it again eliminates the transpose sign:

$$\left(\mathbf{x}^{1\prime}\right)' = \mathbf{x}^1 = \begin{bmatrix} 1 \\ 4 \\ 7 \end{bmatrix}.$$

Note that for any matrix X, $\mathbf{x}_i$, the ith column of the matrix, and $\mathbf{x}^i$, the transpose of the ith row of the matrix, are not generally equal. This is easily seen in the example matrix X above, since

$$\mathbf{x}_1 = \begin{bmatrix} 1 \\ 2 \\ 3 \end{bmatrix} \neq \begin{bmatrix} 1 \\ 4 \\ 7 \end{bmatrix} = \mathbf{x}^1.$$

Two vectors are *equal* only if they consist of exactly the same elements. Two matrices are equal only when they have the same elements and these elements are in the *same positions* in the matrix. For example, $X = \begin{bmatrix} 1 & 4 & 7 \\ 2 & 5 & 8 \\ 3 & 6 & 9 \end{bmatrix}$ is *not* equal to $Y = \begin{bmatrix} 1 & 2 & 3 \\ 4 & 5 & 6 \\ 7 & 8 & 9 \end{bmatrix}$, even though both matrices consist of the same elements.

The transpose of a matrix X, denoted X', is a matrix whose rows are the columns of X, or equivalently, whose columns are the rows of X. For example, Y immediately above is the transpose of X. That is, $Y = X'$, since the first column of X, $\begin{bmatrix} 1 \\ 2 \\ 3 \end{bmatrix}$, forms the first row of Y, and so on.

The transpose of a partitioned matrix is the transposed matrix of transposed submatrices. This sounds complicated. Let's see what it means, by finding the transpose of X again, but this time writing X partitioned by its columns. That is, $X = [\mathbf{x}_1 \quad \mathbf{x}_2 \quad \mathbf{x}_3]$, so

$$X' = \begin{bmatrix} \mathbf{x}'_1 \\ \mathbf{x}'_2 \\ \mathbf{x}'_3 \end{bmatrix} = \begin{bmatrix} 1 & 2 & 3 \\ 4 & 5 & 6 \\ 7 & 8 & 9 \end{bmatrix} = Y.$$

Notice that X' is created by transposing the matrix X from a row vector to a column vector, while at the same time transposing all of the submatrices, which are column vectors, into row vectors. We could achieve the same result by partitioning X by its rows and then transposing it:

$$X = \begin{bmatrix} \mathbf{x}^{1\prime} \\ \mathbf{x}^{2\prime} \\ \mathbf{x}^{3\prime} \end{bmatrix} \quad \text{implies that} \quad X' = \lfloor \mathbf{x}^1 \quad \mathbf{x}^2 \quad \mathbf{x}^3 \rfloor = \begin{bmatrix} 1 & 2 & 3 \\ 4 & 5 & 6 \\ 7 & 8 & 9 \end{bmatrix}.$$

Here we see that X' has been created by transposing X from a column to a row vector while transposing the rows into columns.

A *square matrix* is one with the same number of rows as columns. X and X′ above are square matrices since they each have three rows and three columns. In general, a square matrix is of the form A_{nn}, where n is the number of rows and columns in the matrix. The *diagonal* of a square matrix consists of the elements in the ith row and ith column for $i = 1, 2, \ldots, n$. For example, the diagonal of X is $\begin{bmatrix} 1 \\ 5 \\ 9 \end{bmatrix}$. (We follow the convention of denoting the diagonal of a matrix as a column vector.) Notice that these are the elements in, respectively, the first row, first column; the second row, second column; and the third row, third column, of X. The *trace* of a square matrix, or tr(A), for A square, is the sum of the diagonal elements. Hence tr(X) $= 1 + 5 + 9 = 15$.

C. Working with Matrices

Vectors. Adding or subtracting two vectors consists of the elementwise addition or subtraction of the components of two vectors, provided that they are of the same order. For example, if $\mathbf{a} = \begin{bmatrix} 3 \\ 5 \end{bmatrix}$ and $\mathbf{b} = \begin{bmatrix} 2 \\ 4 \end{bmatrix}$, then $\mathbf{a} + \mathbf{b} = \begin{bmatrix} 3+2 \\ 5+4 \end{bmatrix} = \begin{bmatrix} 5 \\ 9 \end{bmatrix}$ and $\mathbf{a} - \mathbf{b} = \begin{bmatrix} 3-2 \\ 5-4 \end{bmatrix} = \begin{bmatrix} 1 \\ 1 \end{bmatrix}$. However, addition and subtraction cannot be performed for vectors of different orders, or between vectors and matrices, or between scalars and either vectors or matrices.

Multiplication of vectors takes place in either of two ways. The *inner product* of two vectors is the product of a row vector with a column vector of the same order. It is formed by summing the elementwise products of the corresponding components of each vector. The result is a scalar. For example, $\mathbf{a'b} = \begin{bmatrix} 3 & 5 \end{bmatrix}\begin{bmatrix} 2 \\ 4 \end{bmatrix} = (3)(2) + (5)(4) = 26$.

The *outer product*, on the other hand, is the product of a column vector with a row vector of the same order. The result is a matrix whose elements are all possible products of the elements of the column vector with the elements of the row vector. Hence, $\mathbf{ab'} = \begin{bmatrix} 3 \\ 5 \end{bmatrix}\begin{bmatrix} 2 & 4 \end{bmatrix} = \begin{bmatrix} (3)(2) & (3)(4) \\ (5)(2) & (5)(4) \end{bmatrix} = \begin{bmatrix} 6 & 12 \\ 10 & 20 \end{bmatrix}$. Finally, any vector can be multiplied or divided by a scalar. The result is a vector whose elements are all multiplied or divided by that scalar. For example, $.5\mathbf{a} = \begin{bmatrix} (.5)(3) \\ (.5)(5) \end{bmatrix} = \begin{bmatrix} 1.5 \\ 2.5 \end{bmatrix}$.

Matrices. Adding or subtracting two matrices consists of the elementwise addition or subtraction of the components of two matrices, provided that they are of the same order. Hence if

$$A = \begin{bmatrix} 2 & 5 \\ 7 & 8 \end{bmatrix} \quad \text{and} \quad B = \begin{bmatrix} 3 & 9 \\ 12 & 4 \end{bmatrix},$$

then

$$A + B = \begin{bmatrix} 2+3 & 5+9 \\ 7+12 & 8+4 \end{bmatrix} = \begin{bmatrix} 5 & 14 \\ 19 & 12 \end{bmatrix},$$

and

$$A - B = \begin{bmatrix} 2-3 & 5-9 \\ 7-12 & 8-4 \end{bmatrix} = \begin{bmatrix} -1 & -4 \\ -5 & 4 \end{bmatrix}.$$

Matrix multiplication is *not* simply the elementwise product of corresponding components in two matrices (although one type of matrix product, the *Hadamard product, is* formed in this fashion, but is not used very much). To begin, matrix multiplication can only take place between two matrices that are *conformable* for multiplication. A and B are conformable if B has the same number of rows as A has columns. That is, A_{rc} and B_{cq} are conformable for multiplication since A has c columns and B has c rows. The result of the product AB is a matrix with r rows and q columns. Hence, two matrices are conformable if the column dimension of the first is equal to the row dimension of the second, and the order of their product is then the row dimension of the first by the column dimension of the second. As an example,

$$A_{32} = \begin{bmatrix} 1 & 2 \\ 3 & 4 \\ 5 & 6 \end{bmatrix} \text{ and } B_{24} = \begin{bmatrix} 2 & 4 & 6 & 8 \\ 1 & 3 & 5 & 9 \end{bmatrix} \text{ are conformable, since A has two columns}$$

and B has two rows. Their product, AB, will be a 3×4 matrix.

The product is constructed by taking the inner products of each row vector in A with each column vector in B. That is, *the (i, j)th element of AB is the inner product of the ith row of A with the jth column of B.* Let's see how this works by computing AB:

The $(1,1)$th element of AB is the inner product of the first row of A with the first column of B: $[1 \quad 2]\begin{bmatrix} 2 \\ 1 \end{bmatrix} = 4$. The $(2,1)$th element of AB is the inner product of the second row of A with the first column of B: $[3 \quad 4]\begin{bmatrix} 2 \\ 1 \end{bmatrix} = 10$. The $(3,1)$th element of AB is the inner product of the third row of A with the first column of B: $[5 \quad 6]\begin{bmatrix} 2 \\ 1 \end{bmatrix} = 16$. The $(1,2)$th element of AB is the inner product of the first row of A with the second column of B: $[1 \quad 2]\begin{bmatrix} 4 \\ 3 \end{bmatrix} = 10$. We continue in this fashion until we get to the last element of AB, the $(3,4)$th: $[5 \quad 6]\begin{bmatrix} 8 \\ 9 \end{bmatrix} = 94$. The complete product is

$$AB = \begin{bmatrix} 1 & 2 \\ 3 & 4 \\ 5 & 6 \end{bmatrix} \begin{bmatrix} 2 & 4 & 6 & 8 \\ 1 & 3 & 5 & 9 \end{bmatrix} = \begin{bmatrix} 4 & 10 & 16 & 26 \\ 10 & 24 & 38 & 60 \\ 16 & 38 & 60 & 94 \end{bmatrix}.$$

Another way to understand this operation is to partition A by its rows and B by its columns. This treats A like a column vector and B like a row vector. AB is then an outer product whose individual elements are inner products of rows of A with columns of B. In particular, if $A = \begin{bmatrix} \mathbf{a}^{1\prime} \\ \mathbf{a}^{2\prime} \\ \mathbf{a}^{3\prime} \end{bmatrix}$ and $B = [\mathbf{b}_1 \quad \mathbf{b}_2 \quad \mathbf{b}_3 \quad \mathbf{b}_4]$ then

$$AB = \begin{bmatrix} \mathbf{a}^{1\prime} \\ \mathbf{a}^{2\prime} \\ \mathbf{a}^{3\prime} \end{bmatrix} [\mathbf{b}_1 \quad \mathbf{b}_2 \quad \mathbf{b}_3 \quad \mathbf{b}_4] = \begin{bmatrix} \mathbf{a}^{1\prime}\mathbf{b}_1 & \mathbf{a}^{1\prime}\mathbf{b}_2 & \mathbf{a}^{1\prime}\mathbf{b}_3 & \mathbf{a}^{1\prime}\mathbf{b}_4 \\ \mathbf{a}^{2\prime}\mathbf{b}_1 & \mathbf{a}^{2\prime}\mathbf{b}_2 & \mathbf{a}^{2\prime}\mathbf{b}_3 & \mathbf{a}^{2\prime}\mathbf{b}_4 \\ \mathbf{a}^{3\prime}\mathbf{b}_1 & \mathbf{a}^{3\prime}\mathbf{b}_2 & \mathbf{a}^{3\prime}\mathbf{b}_3 & \mathbf{a}^{3\prime}\mathbf{b}_4 \end{bmatrix}$$

This result demonstrates the method of matrix multiplication we just articulated.

The product of a matrix, A, with a vector, $\mathbf{c}$, is formed in a similar fashion, except that a vector consists of only one column, so the process is somewhat simpler. Again, the process is easy to see if A is partitioned by its rows. In this case, A takes the form of a column vector and $\mathbf{c}$ takes the form of a constant. The result is treated like the product of a vector with a constant. Hence,

$$A\mathbf{c} = \begin{bmatrix} \mathbf{a}^{1\prime} \\ \mathbf{a}^{2\prime} \\ \mathbf{a}^{3\prime} \end{bmatrix} \mathbf{c} = \begin{bmatrix} \mathbf{a}^{1\prime}\mathbf{c} \\ \mathbf{a}^{2\prime}\mathbf{c} \\ \mathbf{a}^{3\prime}\mathbf{c} \end{bmatrix}.$$

That is, the result is a *vector* whose elements are inner products of the rows of A with the vector $\mathbf{c}$. For example, if $c = \begin{bmatrix} 5 \\ 7 \end{bmatrix}$, then

$$A\mathbf{c} = \begin{bmatrix} 1 & 2 \\ 3 & 4 \\ 5 & 6 \end{bmatrix} \begin{bmatrix} 5 \\ 7 \end{bmatrix} = \begin{bmatrix} 19 \\ 43 \\ 67 \end{bmatrix}.$$

Notice that A_{32} and $\mathbf{c}_{21}$ are conformable for multiplication, and the result, $A\mathbf{c}$, is a 3×1 vector. Finally, the product or the quotient of a matrix with a scalar is formed by multiplying (dividing) each element of the matrix by that scalar.

D. Special Matrices

A variety of special matrices are used in matrix algebra, and these will be defined here.

(1) Diagonal Matrix. A diagonal matrix is a square matrix having zeros for all of its off-diagonal elements. For example:

$$D = \begin{bmatrix} 3 & 0 & 0 \\ 0 & 5 & 0 \\ 0 & 0 & 9 \end{bmatrix} \text{ is a diagonal matrix.}$$

Premultiplying a matrix A by a diagonal matrix D results in a matrix whose *rows* are multiplied by the corresponding diagonal elements in D.

Example:
$$\begin{bmatrix} \frac{1}{3} & 0 \\ 0 & \frac{1}{2} \end{bmatrix}\begin{bmatrix} 2 & 4 \\ 6 & 8 \end{bmatrix} = \begin{bmatrix} (\frac{1}{3})2 + (0)6 & (\frac{1}{3})4 + (0)8 \\ (0)2 + (\frac{1}{2})6 & (0)4 + (\frac{1}{2})8 \end{bmatrix} = \begin{bmatrix} (\frac{1}{3})2 & (\frac{1}{3})4 \\ (\frac{1}{2})6 & (\frac{1}{2})8 \end{bmatrix}.$$

Postmultiplying a matrix A by a diagonal matrix D results in a matrix whose *columns* are multiplied by the corresponding diagonal elements in D.

Example:
$$\begin{bmatrix} 2 & 4 \\ 6 & 8 \end{bmatrix}\begin{bmatrix} \frac{1}{3} & 0 \\ 0 & \frac{1}{2} \end{bmatrix} = \begin{bmatrix} 2(\frac{1}{3}) + 4(0) & 2(0) + 4(\frac{1}{2}) \\ 6(\frac{1}{3}) + 8(0) & 6(0) + 8(\frac{1}{2}) \end{bmatrix} = \begin{bmatrix} 2(\frac{1}{3}) & 4(\frac{1}{2}) \\ 6(\frac{1}{3}) & 8(\frac{1}{2}) \end{bmatrix}.$$

(2) Null Matrix. A null matrix is a matrix with all elements zero. It will be distinguished from the scalar zero by placing it in boldface type.

Example: $\mathbf{0} = \begin{bmatrix} 0 & 0 \\ 0 & 0 \end{bmatrix}$ is a null matrix.

Analogous to the rule in scalar algebra, $A \pm \mathbf{0} = A$ and $A\mathbf{0} = \mathbf{0}A = \mathbf{0}$ for any matrix A, provided that A and $\mathbf{0}$ are conformable for these respective operations.

(3) Identity Matrix. An identity matrix, denoted by I, is a square matrix with 1's on the diagonal and zeros everywhere else.

Example: $I = \begin{bmatrix} 1 & 0 & 0 \\ 0 & 1 & 0 \\ 0 & 0 & 1 \end{bmatrix}$ is an identity matrix of order 3.

Identity matrices are analogous to the scalar 1: $AI = IA = A$ for any matrix A.

(4) Idempotent Matrix. An idempotent matrix A is one with the property that $A^2 = A$.

Example. Identity matrices are all idempotent, since $I^2 = II = I$. The second equality in this last expression follows from the property that any matrix is unchanged when multiplied by an identity matrix.

(5) Symmetric Matrix. A symmetric matrix A has the property that $A = A'$.

Example: $\begin{bmatrix} 1 & 2 & 3 & 4 \\ 2 & 5 & 6 & 9 \\ 3 & 6 & 7 & 8 \\ 4 & 9 & 8 & 1 \end{bmatrix}$ is symmetric, since if you transpose it, you get the same

matrix again. Covariance and correlation matrices for a set of K variables are always symmetric matrices.

(6) Orthogonal Matrix. An orthogonal matrix A has the property that $A'A = AA' = I$. To elaborate further, we must first define normality and orthogonality with respect to vectors. A vector **x** is normal if $\mathbf{x'x} = 1$. First, note that $\mathbf{x'x}$ is simply

the sum of squared elements of the vector **x** (as the reader can easily verify for himself or herself, using any vector **x**). Thus, a normal vector is such that the sums of squares of its elements equals 1. Two vectors, **x** and **y**, are orthogonal if $\mathbf{x'y} = \mathbf{y'x} = 0$. A set of vectors is orthonormal if each vector is normal and they are all pairwise orthogonal. A square matrix whose columns constitute an orthonormal set of vectors is an orthogonal matrix.

Example:

$$P = \begin{bmatrix} \dfrac{1}{\sqrt{3}} & \dfrac{1}{\sqrt{3}} & \dfrac{1}{\sqrt{3}} \\ \dfrac{1}{\sqrt{2}} & \dfrac{-1}{\sqrt{2}} & 0 \\ \dfrac{1}{\sqrt{6}} & \dfrac{1}{\sqrt{6}} & \dfrac{-2}{\sqrt{6}} \end{bmatrix}$$

is orthogonal, as is readily verified. The reader can also verify that $P'P = PP' = I$.

(7) Centering Matrix. Centering matrices are used to represent quantities such as $\sum (X - \overline{X})^2$ for a set of variable scores $x_1, \ldots, x_n$. To describe this matrix, we must first define the summing vector. A vector whose elements are all 1's is represented as **1** and is referred to as a *summing vector*. This label arises from the fact that $\mathbf{1'x} = \mathbf{x'1} = \sum x$, as the reader can easily verify with any vector **x**. For a **1**-vector of order n (i.e., having n elements), the square matrix $J_{nn} = \mathbf{11'}$ is a matrix of order n, all of whose elements are 1's. Further, the matrix $\overline{J} = (1/n)J_{nn}$ is a square matrix all of whose elements are $1/n$. A centering matrix, C, is then defined as $C = (I - \overline{J})$, *where* I *is of the same order as* $\overline{J}$. Let's see where the centering matrix gets its appellation.

For a vector **x** of variable scores, C**x** is such that $C\mathbf{x} = (I - \overline{J})\mathbf{x} = I\mathbf{x} - \overline{J}\mathbf{x}$. Now $I\mathbf{x} = \mathbf{x}$ [see V.D(3)]. But what is $\overline{J}\mathbf{x}$? $\overline{J}\mathbf{x} = (1/n)\mathbf{11'x} = \mathbf{1}[(1/n)\mathbf{1'x}] = \mathbf{1}\overline{x}$, which equals a vector all of whose elements are $\overline{x}$. Let's denote this vector of means by $\overline{\mathbf{x}}$. Thus, $C\mathbf{x} = (I - \overline{J})\mathbf{x} = \mathbf{x} - \overline{\mathbf{x}}$, a vector consisting of x scores that have been centered or deviated from their means. This formulation is shown below to provide the basis of matrix formulas for sample variances and covariances.

E. Rules for Matrix Expressions

Rules for matrix expressions are analogous to those for algebraic expressions, with some key differences. They are:

(1) *Commutative property*: $A + B = B + A$.

(2) *Associative property 1*: $A + (B + C) = (A + B) + C$.

(3) *Associative property 2*: $(AB) C = A(BC) = ABC$.

(4) *Scalar property*: $cA\mathbf{x} = Ac\mathbf{x} = A\mathbf{x}c$ for any scalar c. That is, the order of multiplication is invariant with respect to scalars.

Rules V.E(1) to V.E(4) are similar to rules in scalar algebra. The following properties, however, are unique to matrix algebra:

(5) *Distributive property 1*: $A(B + C) = AB + AC$. Note that A is on the *left* in both terms here. This is not just coincidental. The result is not generally equal to $AB + CA$ or $BA + AC$ or $BA + CA$. In fact, given conformability requirements, BA and CA may be undefined.

(6) *Distributive property 2*: $(B + C)A = BA + CA$, which is not generally equal to $AB + AC$.

(7) $AB + CA$ cannot be factored.

(8) In scalar algebra $a + ca$ is factored as $(1 + c)a$. Similarly, in matrix algebra, $A + CA$, for A and C being square matrices, is factored as $(I + C)A$, where I is an identity matrix of the same order as A and C. That is, I is analogous to the scalar 1 in factoring operations. Also, the expression $\mathbf{x}'\mathbf{x} + \mathbf{x}'C\mathbf{x}$ is factored as $\mathbf{x}'(I + C)\mathbf{x}$. In the same vein, $BA - 2B$ is factored as $B(A - 2I)$.

(9) Unlike the situation in scalar algebra, $AB = \mathbf{0}$ does not imply that either A or B is a null matrix.

Example: $AB = \mathbf{0}$ for $A = \begin{bmatrix} 1 & 1 \\ 1 & 1 \end{bmatrix}$ and $B = \begin{bmatrix} 1 & 1 \\ -1 & -1 \end{bmatrix}$. Moreover, for

$$X = \begin{bmatrix} 1 & 2 & 5 \\ 2 & 4 & 10 \\ -1 & -2 & -5 \end{bmatrix}, X^2 = \mathbf{0} \text{ even though X is not } \mathbf{0}.$$

(10) $(ABC)' = C'B'A'$. That is, the transpose of a product of matrices is the product of the transposed matrices, in reverse order.

(11) $(A \pm B)' = A' \pm B'$.

(12) $\text{tr}(A \pm B) = \text{tr}(A) \pm \text{tr}(B)$.

(13) $\text{tr}(A) = \text{tr}(A')$.

(14) $\text{tr}(ABC) = \text{tr}(CAB) = \text{tr}(BCA)$. That is, the trace of a matrix product is invariant to elementwise cycling of individual matrices in the product. Note that elementwise cycling means the last matrix in the product is "cycled" to the front of the product, and this process is repeated to produce all possible nonredundant orderings of the product. The product BAC, for example, would *not* constitute an elementwise cycling of ABC.

(15) A scalar equals its own trace [i.e., $\text{tr}(c) = c$].

(16) A scalar equals its own transpose [i.e., $c' = c$].

F. Matrix Equations and Their Solutions

One of the most valuable uses of matrix algebra is in solving systems of linear equations. As an example, the following is a system of linear equations in two unknowns, x_{11} and x_{21}:

$$x_{11} + x_{21} = 20$$
$$x_{11} - x_{21} = -12.$$

This system of equations can be written as a matrix equation of the form $Ax = y$, where A is a matrix of constants, x is the vector of unknowns, x_{11} and x_{21}, and y is the vector of constants on the right-hand side of the equation. For this system we

have $A = \begin{bmatrix} 1 & 1 \\ 1 & -1 \end{bmatrix}$, $x = \begin{bmatrix} x_{11} \\ x_{21} \end{bmatrix}$, and $y = \begin{bmatrix} 20 \\ -12 \end{bmatrix}$. The reader can verify that the

left-hand side of this system can be expressed as the vector Ax.

The Inverse. Solving this system amounts to finding the inverse matrix of A, denoted A^{-1}, if it exists. (Below we consider conditions necessary for the inverse of a matrix to exist.) The *inverse* of a *square matrix* A (the inverse exists only for square matrices) is the matrix A^{-1}, with the following property: $AA^{-1} = A^{-1}A = I$. This is analogous to the inverse, a^{-1}, of a scalar, a, since $a\,a^{-1} = a^{-1}a = 1$. Multiplication by the inverse matrix is the scalar analog of division. (Note that dividing by a matrix is an undefined operation.) The equation $ax = y$ has solution $x = a^{-1}y$, in scalar algebra, because we can multiply both sides of the equation by a^{-1} to isolate x on the left-hand side. That is, $ax = y$ implies that $a^{-1}ax = a^{-1}y$ or $x = a^{-1}y$. similarly, the matrix equation $Ax = y$ is solved by premultiplying both sides of the equation by A^{-1}: $Ax = y$ implies that $A^{-1}Ax = A^{-1}y$, or $Ix = A^{-1}y$, or $x = A^{-1}y$. In the example,

$A^{-1} = \begin{bmatrix} \frac{1}{2} & \frac{1}{2} \\ \frac{1}{2} & -\frac{1}{2} \end{bmatrix}$ (the reader can verify that $AA^{-1} = A^{-1}A = I$ in this case). The

solution is therefore $x = \begin{bmatrix} \frac{1}{2} & \frac{1}{2} \\ \frac{1}{2} & -\frac{1}{2} \end{bmatrix}\begin{bmatrix} 20 \\ -12 \end{bmatrix} = \begin{bmatrix} 4 \\ 16 \end{bmatrix}$.

The Determinant. Finding the inverse matrix can be complex and time consuming. Fortunately, we can let the computer do the actual work for us. Nevertheless, to further our understanding, we'll discuss how to find the inverse of a second-order matrix, as that is quite simple. First, we need to find the *determinant* of the matrix. The determinant of a square matrix A, denoted $|A|$, is a scalar value that is related to the degree of linear independence among the columns of the matrix. (We will take up the issue of linear independence below.) For any 2×2 matrix A, where

$A = \begin{bmatrix} a_{11} & a_{12} \\ a_{21} & a_{22} \end{bmatrix}$, $|A|$ is equal to $(a_{11})(a_{22}) - (a_{21})(a_{12})$. For higher-order matrices, the

idea is the same, except that the determinant becomes a weighted sum of determinants of lower-order submatrices within the larger matrix.

Example. The determinant of $A = \begin{bmatrix} 1 & 1 \\ 1 & -1 \end{bmatrix}$ is $|A| = (1)(-1) - (1)(1) = -2$.

Finding the Inverse. The inverse of a second-order matrix A, having the form $\begin{bmatrix} a_{11} & a_{12} \\ a_{21} & a_{22} \end{bmatrix}$, is

$$A^{-1} = \frac{1}{|A|}\begin{bmatrix} a_{22} & -a_{12} \\ -a_{21} & a_{11} \end{bmatrix} = \frac{1}{a_{11}a_{22} - a_{21}a_{12}}\begin{bmatrix} a_{22} & -a_{12} \\ -a_{21} & a_{11} \end{bmatrix}.$$

Here we see that the inverse is the reciprocal of the determinant times a matrix in which the positions of a_{11} and a_{22} are interchanged, while the signs of a_{12} and a_{21} are reversed.

Example. The determinant of $A = \begin{bmatrix} 1 & 1 \\ 1 & -1 \end{bmatrix}$ was already found to be -2; hence, A^{-1} is

$$-\frac{1}{2}\begin{bmatrix} -1 & -1 \\ -1 & 1 \end{bmatrix} = \begin{bmatrix} \frac{1}{2} & \frac{1}{2} \\ \frac{1}{2} & -\frac{1}{2} \end{bmatrix},$$

which agrees with our previous result.

Multiplication Rule. The inverse of a product of matrices is the product of their individual inverses, in reverse order, provided that the individual inverses exist. That is, $(ABC)^{-1} = C^{-1}B^{-1}A^{-1}$, provided that A^{-1}, B^{-1}, and C^{-1} all exist. It is easy to see why this is true, since $(ABC)^{-1}(ABC) = C^{-1}B^{-1}A^{-1}ABC = C^{-1}B^{-1}IBC = C^{-1}B^{-1}BC = C^{-1}IC = C^{-1}C = I$.

Transpose Rule. The inverse of the transpose is the transpose of the inverse. That is, $(A')^{-1} = (A^{-1})'$.

G. Linear Dependence and Rank

Let's consider again the product Ax. The result is a vector. For a 3×3 matrix A and a 3×1 vector x, Ax can be represented as

$$[a_1 \quad a_2 \quad a_3]\begin{bmatrix} x_{11} \\ x_{21} \\ x_{31} \end{bmatrix} = [a_1 x_{11} + a_2 x_{21} + a_3 x_{31}].$$

This formulation exploits the fact that A can be partitioned by its columns and then treated like a row vector (which it is). Hence the resulting product, Ax, can be treated like an inner product of two vectors, giving us the result on the right-hand side of the equation. The resulting vector, $[a_1 x_{11} + a_2 x_{21} + a_3 x_{31}]$, is therefore seen to be a linear combination, or weighted sum, of the columns of A, with the weights being the components of x. Now if there is a nonnull (i.e., nonzero) vector x such that $Ax = 0$, then provided that no column of A is null, the columns of A are said to be *linearly dependent*. What this means is that one column of A is a linear combination of the other columns, since $Ax = 0$ implies that $a_1 x_{11} + a_2 x_{21} + a_3 x_{31} = 0$, or

$$x_{11}a_1 = -x_{21}a_2 - x_{31}a_3, \text{ or } a_1 = \left(-\frac{x_{21}}{x_{11}}\right)a_2 + \left(-\frac{x_{31}}{x_{11}}\right)a_3.$$

If there is no nonnull x such that $Ax = 0$, the columns of A are linearly *independent*. Linear dependence is especially important in linear regression, where the columns of the "design matrix," X, represent one's independent variables (as we will shortly show). It is assumed in linear regression that the columns of X are linearly independent. If they are, instead, linearly dependent, the least-squares estimates are undefined. Unless one predictor is a perfect linear combination of other predictors, linear dependence is rare. However, near-perfect linear dependence, a not-so-rare phenomenon, is the problem known as *multicollinearity*.

Rank. The *rank* of a matrix A, denoted r(A), is the *number of linearly independent columns in the matrix*. (The number of linearly independent columns of a matrix is the same as the number of linearly independent rows.) The rank of a matrix has a direct bearing on whether the inverse of a matrix exists: *If the rank of an $n \times n$ matrix equals n, the matrix has an inverse*. Also: *A has an inverse if and only if $|A|$ is not zero*. This principle is apparent in the definition of the inverse for a 2×2 matrix above, since if $|A|$ is zero, the inverse is undefined. Hence, a matrix A has an inverse only if its columns are linearly independent, in which case its determinant is nonzero.

Example. The 2×2 matrix $A = \begin{bmatrix} 1 & 2 \\ 2 & 4 \end{bmatrix}$ has only one linearly independent column,

since the second column is twice the first. We would therefore expect that the inverse of this matrix doesn't exist. This is clear from the fact that $|A| = (1)(4) - (2)(2) = 0$; hence, the inverse is undefined.

H. Eigenvalues and Eigenvectors

Eigenvalues and eigenvectors are very important tools in the diagnosis of and remedy for multicollinearity. Therefore, I introduce them briefly here.

Definitions. Given a square matrix A_{nn}, we ask whether there is a scalar λ and a vector $\mathbf{u}$ such that $A\mathbf{u} = \lambda\mathbf{u}$. (In other words, the scalar acts like the matrix in products with the vector.) If so, then λ is called an *eigenvalue* of A and $\mathbf{u}$ is called an *eigenvector* of A. It turns out that, for any A_{nn}, there are n such λ's and at least n such $\mathbf{u}$'s. It should, however, be noted that the eigenvalues of A are roots of a polynomial equation of degree n and are not necessarily real numbers unless A is symmetric. The eigenvalues of a symmetric matrix are all real.

Example. $\begin{bmatrix} 1 & 4 \\ 9 & 1 \end{bmatrix}$ has two eigenvalues, -5 and 7, with associated eigenvectors

$$\mathbf{u}_1' = [2 \quad -3],$$
$$\mathbf{u}_2' = [2 \quad 3].$$

The reader can easily verify that

$$\begin{bmatrix} 1 & 4 \\ 9 & 1 \end{bmatrix} \begin{bmatrix} 2 \\ -3 \end{bmatrix} = -5 \begin{bmatrix} 2 \\ -3 \end{bmatrix},$$

$$\begin{bmatrix} 1 & 4 \\ 9 & 1 \end{bmatrix} \begin{bmatrix} 2 \\ 3 \end{bmatrix} = 7 \begin{bmatrix} 2 \\ 3 \end{bmatrix}.$$

Two principles connect eigenvalues to other matrix properties:

(1) *The trace of a matrix is equal to the sum of its eigenvalues.*
(2) *The determinant of a matrix is equal to the product of its eigenvalues.*

Rule 2 helps to convey somewhat of an intuitive feeling for eigenvalues. They are associated with the degree of linear dependence in a matrix, since if the determinant is zero, meaning that one or more columns of a matrix is a linear combination of the other columns, one or more of the eigenvalues must therefore also be zero.

Spectral Decomposition of a Symmetric Matrix. A special application of eigenvalues and eigenvectors is the *spectral decomposition* of a symmetric matrix. It shows how a symmetric matrix (a correlation matrix, for example) can be shown to be a weighted sum of its eigenvalues times other matrices. These other matrices consist of the outer products of the matrix's eigenvectors with themselves. This formulation is especially important for understanding the remedy for multicollinearity known as *principal components regression*.

Suppose that we have a symmetric matrix, A. Suppose, further, that λ_j and $\mathbf{u}_j$ are the jth eigenvalue and associated eigenvector, respectively, of A. If A is $n \times n$, there are n eigenvalues and eigenvectors associated with this matrix. If A is symmetric, it turns out that its eigenvectors are all pairwise orthogonal. Without any loss of generality, we can normalize the eigenvectors so that each is also normal. The jth eigenvector, $\mathbf{u}_j$, is normalized by multiplying it by $1/\sqrt{\mathbf{u}_j'\mathbf{u}_j}$. The resulting normalized eigenvectors will be both normal and orthogonal. Hence, if we collect them in a matrix U, it will be an orthogonal matrix with the property that $UU' = I$. We then note that

$$A = AI = AUU'.$$

Now suppose, for argument's sake, that A is 2×2, so that U is also 2×2. Then

$$UU' = [\mathbf{u}_1 \quad \mathbf{u}_2]\begin{bmatrix} \mathbf{u}_1' \\ \mathbf{u}_2' \end{bmatrix} = \mathbf{u}_1\mathbf{u}_1' + \mathbf{u}_2\mathbf{u}_2' = \sum \mathbf{u}_j\mathbf{u}_j'.$$

This illustrates that, in general,

$$A = AUU' = A\sum \mathbf{u}_j\mathbf{u}_j' = \sum A\mathbf{u}_j\mathbf{u}_j' = \sum \lambda_j\mathbf{u}_j\mathbf{u}_j'.$$

(Recall that by the definition of eigenvalues and eigenvectors, $A\mathbf{u}_j = \lambda_j\mathbf{u}_j$.) The last sum on the right is called the *spectral decomposition* of a symmetric matrix. It shows that any symmetric matrix A can be shown to be a *sum* of matrices, each of which is the product of an eigenvalue of A times the outer product of its associated eigenvector with itself.

I. Expectation and Variance of Vectors

The expectation of a random variable, X, is denoted by $E(X)$. It is the average, or mean, value of X in the population and is usually given the symbol μ. The variance of a random variable, $V(X)$, is the average squared deviation of X from its mean. That is, $V(X) = E(X - \mu)^2$. We can define similar properties for vectors of random variables.

Suppose that $\mathbf{x}$ is a vector of, say, three random variables: $\mathbf{x} = \begin{bmatrix} X_1 \\ X_2 \\ X_3 \end{bmatrix}$. Notice now

that the elements of this vector are not specific values. Rather, they are variables. That is, instead of talking about these variables individually, we have collected them in a vector, $\mathbf{x}$. The expected value of the vector $\mathbf{x}$, denoted $E(\mathbf{x})$, is simply the vector of expected values of the individual variables. That is,

$$E(\mathbf{x}) = \begin{bmatrix} E(X_1) \\ E(X_2) \\ E(X_3) \end{bmatrix} = \begin{bmatrix} \mu_1 \\ \mu_2 \\ \mu_3 \end{bmatrix} = \mathbf{\mu}.$$

Notice that $\mathbf{\mu}$ is the symbol for the vector of variable means.

The variance of the vector $\mathbf{x}$ is defined as $V(\mathbf{x}) = E[(\mathbf{x} - \mathbf{\mu})(\mathbf{x} - \mathbf{\mu})']$. Notice here that $(\mathbf{x} - \mathbf{\mu})$ is a column vector of the variables deviated from their means. Its transpose $(\mathbf{x} - \mathbf{\mu})'$ is therefore a row vector of the variables deviated from their means. Thus, $(\mathbf{x} - \mathbf{\mu})(\mathbf{x} - \mathbf{\mu})'$ is an outer product of two vectors, or a matrix. In other words, the variance of a vector is a matrix of expected values. For our three-variable vector, $\mathbf{x}$, here's what $(\mathbf{x} - \mathbf{\mu})(\mathbf{x} - \mathbf{\mu})'$ looks like:

$$(\mathbf{x} - \mathbf{\mu})(\mathbf{x} - \mathbf{\mu})' = \begin{bmatrix} X_1 - \mu_1 \\ X_2 - \mu_2 \\ X_3 - \mu_3 \end{bmatrix} [X_1 - \mu_1 \quad X_2 - \mu_2 \quad X_3 - \mu_3]$$

$$= \begin{bmatrix} (X_1 - \mu_1)^2 & (X_1 - \mu_1)(X_2 - \mu_2) & (X_1 - \mu_1)(X_3 - \mu_3) \\ (X_2 - \mu_2)(X_1 - \mu_1) & (X_2 - \mu_2)^2 & (X_2 - \mu_2)(X_3 - \mu_3) \\ (X_3 - \mu_3)(X_1 - \mu_1) & (X_3 - \mu_3)(X_2 - \mu_2) & (X_3 - \mu_3)^2 \end{bmatrix}.$$

Hence,

$$V(\mathbf{x}) = E[(\mathbf{x} - \mathbf{\mu})(\mathbf{x} - \mathbf{\mu})']$$

$$= \begin{bmatrix} E(X_1 - \mu_1)^2 & E(X_1 - \mu_1)(X_2 - \mu_2) & E(X_1 - \mu_1)(X_3 - \mu_3) \\ E(X_2 - \mu_2)(X_1 - \mu_1) & E(X_2 - \mu_2)^2 & E(X_2 - \mu_2)(X_3 - \mu_3) \\ E(X_3 - \mu_3)(X_1 - \mu_1) & E(X_3 - \mu_3)(X_2 - \mu_2) & E(X_3 - \mu_3)^2 \end{bmatrix} = V.$$

Since $E(X_i - \mu_i)^2$ is the variance of X_i, and $E(X_i - \mu_i)(X_j - \mu_j)$ is the covariance of X_i with X_j, V is referred to as the *variance–covariance matrix* for $\mathbf{x}$.

We have seen above that the product of, say, a 3×3 matrix A with a 3×1 vector $\mathbf{x}$ has the form

$$A\mathbf{x} = \begin{bmatrix} \mathbf{a}^{1\prime} \\ \mathbf{a}^{2\prime} \\ \mathbf{a}^{3\prime} \end{bmatrix} \mathbf{x} = \begin{bmatrix} \mathbf{a}^{1\prime}\mathbf{x} \\ \mathbf{a}^{2\prime}\mathbf{x} \\ \mathbf{a}^{3\prime}\mathbf{x} \end{bmatrix}.$$

That is, each element of the resulting matrix is a linear combination of the elements of the vector $\mathbf{x}$. In general, if A is a matrix of constants and $\mathbf{x}$ is a vector of random variables with $E(\mathbf{x}) = \mu$ and $V(\mathbf{x}) = V$, then $\mathbf{y} = A\mathbf{x}$ is called a *linear transformation* of the vector $\mathbf{x}$, and

$$E(\mathbf{y}) = E(A\mathbf{x}) = AE(\mathbf{x}) = A\mu,$$
$$V(\mathbf{y}) = AVA'.$$

J. Applications

In this section we examine some common statistical applications of matrix algebra.

Application 1. Expressing a sample variance–covariance matrix of K variables in matrix notation. Suppose that we have a sample of measurements on K variables, X_1, $X_2, \ldots, X_K$, for n persons. Let's construct a matrix expression for the variance–covariance matrix of these X's, in matrix algebra. First, for the sake of simplicity, let's assume that $K = 3$, so that our variables are X_1, X_2, and X_3. Further, let's assume that $n = 5$. Then, our data matrix can be expressed as

$$X = \begin{bmatrix} x_{11} & x_{12} & x_{13} \\ x_{21} & x_{22} & x_{23} \\ x_{31} & x_{32} & x_{33} \\ x_{41} & x_{42} & x_{43} \\ x_{51} & x_{52} & x_{53} \end{bmatrix}.$$

Each row consists of the variable scores for one observation, and each column represents all of the scores for one variable. In general, such a data matrix, X, can be constructed for any value of K or n. Now consider a given column of X, say the first. Denote this as $\mathbf{x}_1$. We saw above that if C is the centering matrix, $C\mathbf{x}_1 = \mathbf{x}_1 - \bar{\mathbf{x}}_1$ is the vector of $\mathbf{x}_1$ scores deviated from their means. Now, the inner product of this vector with itself is the sum of squares of the elements of the vector. That is,

$$(C\mathbf{x}_1)'(C\mathbf{x}_1) = (\mathbf{x}_1 - \bar{\mathbf{x}}_1)'(\mathbf{x}_1 - \bar{\mathbf{x}}_1) = \sum_{i=1}^{n}(x_{i1} - \bar{x}_1)^2.$$

(The reader can verify that this would be true, using the first column of the matrix X above.) Now $(C\mathbf{x}_1)'(C\mathbf{x}_1) = \mathbf{x}_1'C'C\mathbf{x}_1$ [by rule V.E.(10)] $= \mathbf{x}_1'C\mathbf{x}_1$, since C is both symmetric and idempotent. At this point, let's pause and show that C is symmetric and idempotent. Recall that $C = (I - \bar{J}) = I - (1/n)\mathbf{11}'$. Thus, $C' = [I - (1/n)\mathbf{11}']' = I' - \mathbf{11}'(1/n) = I - (1/n)\mathbf{11}' = C$ [by rules V.E (10), V.E (11), and the fact that I is symmetric, as the reader can easily verify]. Hence, C is symmetric. Now $C^2 = (I - \bar{J})(I - \bar{J}) = I^2 - I\bar{J} - \bar{J}I + \bar{J}^2 = I - \bar{J} - \bar{J} + \bar{J}^2 = I - \bar{J} = C$ [since I, as noted previously, is idempotent, and $\bar{J}^2 = (1/n)\mathbf{11}'(1/n)\mathbf{11}' = (1/n^2)\mathbf{11}'\mathbf{11}' = (1/n^2)\mathbf{1}n\mathbf{1}' = (1/n)\mathbf{11}' = \bar{J}$]. Therefore, C is also idempotent.

Since $\mathbf{x}_1'C\mathbf{x}_1 = \sum_{i=1}^{n}(x_{1i} - \bar{x}_1)^2$, we have that

$$\frac{1}{n-1}\mathbf{x}_1'C\mathbf{x}_1 = \frac{\sum(x_1 - \bar{x}_1)^2}{n-1} = s_1^2,$$

the sample variance of X_1. Now what is $[1/(n-1)]\mathbf{x}_1'C\mathbf{x}_2$? Recall that $C\mathbf{x}_1 = \mathbf{x}_1 - \bar{\mathbf{x}}_1$. Similarly, $C\mathbf{x}_2 = \mathbf{x}_2 - \bar{\mathbf{x}}_2$. Hence,

$$(\mathbf{x}_1 - \bar{\mathbf{x}}_1)'(\mathbf{x}_2 - \bar{\mathbf{x}}_2) = (C\mathbf{x}_1)'C\mathbf{x}_2 = \mathbf{x}_1'C'C\mathbf{x}_2 = \mathbf{x}_1'C\mathbf{x}_2 = \sum_{i=1}^{n}(x_{1i} - \bar{x}_1)(x_{2i} - \bar{x}_2).$$

That is, $\mathbf{x}_1'C\mathbf{x}_2$ is the corrected (by the mean) sum of cross-products of X_1 with X_2. Therefore,

$$\frac{1}{n-1}\mathbf{x}_1'C\mathbf{x}_2 = \frac{1}{n-1}\sum_{i=1}^{n}(x_{1i} - \bar{x}_1)(x_{2i} - \bar{x}_2) = \mathrm{cov}(X_1, X_2).$$

With these results established, we now consider the matrix product $[1/(n-1)]X'CX$ for the X matrix above. Partitioning X by its columns gives us $X = [\mathbf{x}_1 \ \ \mathbf{x}_2 \ \ \mathbf{x}_3]$. Hence $X'CX$ is

$$\begin{bmatrix} \mathbf{x}_1' \\ \mathbf{x}_2' \\ \mathbf{x}_3' \end{bmatrix} C[\mathbf{x}_1 \ \ \mathbf{x}_2 \ \ \mathbf{x}_3].$$

Treating C as a constant (which is legitimate as long as conformability for multiplication is satisfied), this equals

$$\begin{bmatrix} \mathbf{x}_1'C\mathbf{x}_1 & \mathbf{x}_1'C\mathbf{x}_2 & \mathbf{x}_1'C\mathbf{x}_3 \\ \mathbf{x}_2'C\mathbf{x}_1 & \mathbf{x}_2'C\mathbf{x}_2 & \mathbf{x}_2'C\mathbf{x}_3 \\ \mathbf{x}_3'C\mathbf{x}_1 & \mathbf{x}_3'C\mathbf{x}_2 & \mathbf{x}_3'C\mathbf{x}_3 \end{bmatrix}.$$

Thus,

$$\frac{1}{n-1}X'CX = \begin{bmatrix} \dfrac{1}{n-1}\mathbf{x}_1'C\mathbf{x}_1 & \dfrac{1}{n-1}\mathbf{x}_1'C\mathbf{x}_2 & \dfrac{1}{n-1}\mathbf{x}_1'C\mathbf{x}_3 \\ \dfrac{1}{n-1}\mathbf{x}_2'C\mathbf{x}_1 & \dfrac{1}{n-1}\mathbf{x}_2'C\mathbf{x}_2 & \dfrac{1}{n-1}\mathbf{x}_2'C\mathbf{x}_3 \\ \dfrac{1}{n-1}\mathbf{x}_3'C\mathbf{x}_1 & \dfrac{1}{n-1}\mathbf{x}_3'C\mathbf{x}_2 & \dfrac{1}{n-1}\mathbf{x}_3'C\mathbf{x}_3 \end{bmatrix},$$

which is the sample variance–covariance matrix of the variables in X. Notice that the sample variances of X_1, X_2, and X_3 are on the diagonal, while the covariances are everywhere else.

Application 2. Multiple regression in matrix algebra. Let's examine the matrix formulation for the multiple regression model. First, we consider the multiple regression equation for the ith observation, with, say, two regressors in the model—again, for simplicity. The model is $Y_i = \beta_0 + \beta_1 X_{i1} + \beta_2 X_{i2} + \varepsilon_i$. To estimate the model using ordinary least squares it is assumed, among other things, that the ε_i are independent and identically distributed random variables with a mean of zero and a variance of σ^2 for all i. Let's again suppose that $n = 5$, to make things manageable. With these specifications, the data can easily be represented as follows. The set of five Y scores can be represented by the vector

$$\mathbf{y} = \begin{bmatrix} y_{11} \\ y_{21} \\ y_{31} \\ y_{41} \\ y_{51} \end{bmatrix}.$$

The scores on the independent variables for the five cases can be represented with the regressor matrix

$$\mathbf{X}_{52} = \begin{bmatrix} x_{11} & x_{12} \\ x_{21} & x_{22} \\ x_{31} & x_{32} \\ x_{41} & x_{42} \\ x_{51} & x_{52} \end{bmatrix}.$$

The error terms can be represented by the vector

$$\boldsymbol{\varepsilon} = \begin{bmatrix} \varepsilon_{11} \\ \varepsilon_{21} \\ \varepsilon_{31} \\ \varepsilon_{41} \\ \varepsilon_{51} \end{bmatrix}.$$

Finally, the vector of parameters can be represented by the vector

$$\boldsymbol{\beta} = \begin{bmatrix} \beta_0 \\ \beta_1 \\ \beta_2 \end{bmatrix}.$$

Now we make the following modifications in notation. First, we drop the second subscript on the terms in $\mathbf{y}$ and $\boldsymbol{\varepsilon}$, for simplicity. The remaining subscript simply indexes the ith case, where $i = 1, \ldots, 5$. We have

$$\mathbf{y} = \begin{bmatrix} y_1 \\ y_2 \\ y_3 \\ y_4 \\ y_5 \end{bmatrix}, \boldsymbol{\varepsilon} = \begin{bmatrix} \varepsilon_1 \\ \varepsilon_2 \\ \varepsilon_3 \\ \varepsilon_4 \\ \varepsilon_5 \end{bmatrix}.$$

Next, we add a column of ones to the matrix X to accommodate the equation intercept (as the reader will see shortly). The resulting matrix, called the *design* matrix, is

$$\mathbf{X} = \begin{bmatrix} 1 & x_{11} & x_{12} \\ 1 & x_{21} & x_{22} \\ 1 & x_{31} & x_{32} \\ 1 & x_{41} & x_{42} \\ 1 & x_{51} & x_{52} \end{bmatrix}.$$

The regression equation for the first observation is $y_1 = \beta_0(1) + \beta_1 x_{11} + \beta_2 x_{12} + \varepsilon_1$. (Notice how "1" is the "variable" whose coefficient is β_0 here.) Another way to write this is $y_1 = \mathbf{x}^{1\prime} \boldsymbol{\beta} + \varepsilon_1$, where $\mathbf{x}^{1\prime}$ is the first row of the X matrix and $\boldsymbol{\beta}$ is the parameter vector. (The reader can verify that this is the equation for y_1 by performing the operations on the right-hand side of this equation, using the first row of X and the parameter vector, plus the error term.) In general, we write $y_i = \mathbf{x}^{i\prime} \boldsymbol{\beta} + \varepsilon_i$ as the model for the ith observation on Y. To write all five equations at once, we employ matrix notation: $\mathbf{y} = X\boldsymbol{\beta} + \boldsymbol{\varepsilon}$. In general, $\mathbf{y} = X\boldsymbol{\beta} + \boldsymbol{\varepsilon}$ is the matrix expression for the regression of the observed y values on the predictor set, regardless of sample size. To understand why this expression works, we partition X according to its rows. Then we have

$$
\mathbf{y} = \begin{bmatrix} y_1 \\ y_2 \\ y_3 \\ y_4 \\ y_5 \end{bmatrix} = \begin{bmatrix} \mathbf{x}^{1\prime} \\ \mathbf{x}^{2\prime} \\ \mathbf{x}^{3\prime} \\ \mathbf{x}^{4\prime} \\ \mathbf{x}^{5\prime} \end{bmatrix} \boldsymbol{\beta} + \begin{bmatrix} \varepsilon_1 \\ \varepsilon_2 \\ \varepsilon_3 \\ \varepsilon_4 \\ \varepsilon_5 \end{bmatrix}.
$$

Notice that in this formulation, I am depicting $\boldsymbol{\beta}$ as though it is a constant, since to show it as a column vector makes it appear as though it is not conformable for multiplication with X. However, its conformability for multiplication with X is clear, since X is 5×3 and $\boldsymbol{\beta}$ is 3×1. The result, however, is a 5×1 vector, of which the ith element is $\mathbf{x}^{i\prime} \boldsymbol{\beta}$.

Recall, from above, the assumption that the ε_i are independent and identically distributed random variables with a mean of zero and a variance of σ^2 for all i. These assumptions can also be expressed in matrix form. They are $E(\boldsymbol{\varepsilon}) = \mathbf{0}$ and $V(\boldsymbol{\varepsilon}) = \sigma^2 I$. That is, if the expected value of each error is zero, the expected value of the vector $\boldsymbol{\varepsilon}$ is a vector of zeros. Hence, $E(\mathbf{y}) = E(X\boldsymbol{\beta} + \boldsymbol{\varepsilon}) = E(X\boldsymbol{\beta}) + E(\boldsymbol{\varepsilon}) = X\boldsymbol{\beta} + \mathbf{0} = X\boldsymbol{\beta}$. The term $\sigma^2 I$ needs some further explanation. Remember that the variance of a vector of variables (and the ε's *are*, in fact, theoretical random variables) is a variance–covariance matrix for the variables. What this second assumption is saying is that the variance–covariance matrix for the error terms is of the form $\sigma^2 I$, where I is $n \times n$. That is, in our simple example, the variance–covariance matrix for the five error terms is of the form

$$
\sigma^2 \begin{bmatrix} 1 & 0 & 0 & 0 & 0 \\ 0 & 1 & 0 & 0 & 0 \\ 0 & 0 & 1 & 0 & 0 \\ 0 & 0 & 0 & 1 & 0 \\ 0 & 0 & 0 & 0 & 1 \end{bmatrix} = \begin{bmatrix} \sigma^2 & 0 & 0 & 0 & 0 \\ 0 & \sigma^2 & 0 & 0 & 0 \\ 0 & 0 & \sigma^2 & 0 & 0 \\ 0 & 0 & 0 & \sigma^2 & 0 \\ 0 & 0 & 0 & 0 & \sigma^2 \end{bmatrix} = V(\boldsymbol{\varepsilon}).
$$

As we can see, this means that the variances of the error terms are a constant value of σ^2 for each observation, and the covariances among the error terms are all zero.

The vector of parameter *estimates* is

$$
\mathbf{b} = \begin{bmatrix} b_0 \\ b_1 \\ b_2 \end{bmatrix}.
$$

These are found by minimizing the sum of squared residuals with respect to the parameter values. The least-squares solution vector, **b**, is found by solving the *normal equations*, which in matrix form are

$$X'Xb = X'y.$$

The least-squares solution vector is therefore

$$b = (X'X)^{-1}X'y.$$

Now we come full circle and answer the original question posed at the beginning of this particular tutorial (see Section V.A): How do we find the expected value and variance of a coefficient estimate in the linear regression model? In fact, let's find the expected value and variance of the entire vector of linear regression estimates. First, we will assume that the X-values are fixed over repeated sampling. This fixed-X assumption is a standard assumption in linear regression, although it is routinely violated. Nevertheless, the results we present hold asymptotically regardless of the nature of the X's (see, e.g., Greene, 2003). Moreover, we assume that we have a sample of n observations and $p = K + 1$ regressors, including the equation intercept, so that **y** and **ε** have dimension $n \times 1$, X has dimension $n \times p$, and **β** has dimension $p \times 1$.

If X is fixed, the $p \times n$ matrix $(X'X)^{-1}X'$ is a matrix of constants. Call this matrix A. Recall, in general, that if A is a matrix of constants, then $y = Ax$ is called a *linear transformation* of the vector **x**, and $E(y) = AE(x)$, $V(y) = AVA'$. Now let **b** be **y** here, and let **y** be **x**. Then $b = Ay$ and we have that $E(b) = AE(y) = (X'X)^{-1}X'E(y) = (X'X)^{-1}X' X\beta = I\beta = \beta$. This shows that the vector of estimates, **b**, is *unbiased* for the parameter vector, **β**. Now what about $V(b)$? First, we need to observe that if $y = X\beta + \varepsilon$, then $V(y) = V(X\beta + \varepsilon) = V(\varepsilon) = \sigma^2 I = V$. (The term $X\beta$ has no variance over repeated sampling, since X is fixed and **β** is also a collection of constants.) We then have

$$V(b) = AVA' = (X'X)^{-1}X'\sigma^2 IX(X'X)^{-1}$$

$$= \sigma^2(X'X)^{-1}X'X(X'X)^{-1}$$

$$= \sigma^2(X'X)^{-1}I = \sigma^2(X'X)^{-1}.$$

Substituting the estimate of σ^2 [which is $SSE/(n - K - 1)$] into this last expression gives us an estimate of the variance–covariance matrix of the regression parameter estimates.

Application 3. Using matrix calculations to find the estimates of b_0 and b_1 in a simple linear regression. Just for practice, let's use the matrix expression for **b**, $b = (X'X)^{-1}X'y$, to calculate **b** for a simple linear regression model of four observations. It is then left as an exercise for the reader to verify that the same estimates are obtained using the traditional SLR formulas for the intercept and slope (see Chapter 2). The four X-values are 2, 3.3, 3.9, and 7. The four Y-values are, respectively, 5, 2,

3, and 9. The relevant vectors and matrices are, therefore,

$$y = \begin{bmatrix} 5 \\ 2 \\ 3 \\ 9 \end{bmatrix},$$

$$X = \begin{bmatrix} 1 & 2 \\ 1 & 3.3 \\ 1 & 3.9 \\ 1 & 7 \end{bmatrix}.$$

To find the least-squares solution vector, we calculate

$$X'X = \begin{bmatrix} 1 & 1 & 1 & 1 \\ 2 & 3.3 & 3.9 & 7 \end{bmatrix} \begin{bmatrix} 1 & 2 \\ 1 & 3.3 \\ 1 & 3.9 \\ 1 & 7 \end{bmatrix} = \begin{bmatrix} 4 & 16.2 \\ 16.2 & 79.1 \end{bmatrix}$$

(as the reader can verify). To find the inverse of this 2×2 matrix, we first find the determinant, which is $(4)(79.1) - (16.2)(16.2) = 53.96$. The inverse matrix is then

$$\frac{1}{53.96} \begin{bmatrix} 79.1 & -16.2 \\ -16.2 & 4 \end{bmatrix} = \begin{bmatrix} 1.4659 & -.3002 \\ -.3002 & .0741 \end{bmatrix}.$$

(as the reader can verify). In a similar vein, we have

$$X'y = \begin{bmatrix} 1 & 1 & 1 & 1 \\ 2 & 3.3 & 3.9 & 7 \end{bmatrix} \begin{bmatrix} 5 \\ 2 \\ 3 \\ 9 \end{bmatrix} = \begin{bmatrix} 19 \\ 91.3 \end{bmatrix}$$

(as, again, the reader should verify). Finally, we have

$$\mathbf{b} = (X'X)^{-1}X'y = \begin{bmatrix} 1.4659 & -3.002 \\ -3.002 & .0741 \end{bmatrix} \begin{bmatrix} 19 \\ 91.3 \end{bmatrix} = \begin{bmatrix} .4418 \\ 1.0638 \end{bmatrix}$$

The reader should also verify this last calculation. The sample regression equation is, therefore,

$$Y = .4418 + 1.0638X + e.$$

K. Exercises

(1) Evaluate $\mathbf{x}'\mathbf{y}$ for $\mathbf{x} = \begin{bmatrix} \frac{1}{2} \\ -\frac{1}{2} \end{bmatrix}$, $\mathbf{y} = \begin{bmatrix} 4 \\ 6 \end{bmatrix}$.

(2) Evaluate $\mathbf{x'z}$ for $\mathbf{x} = \begin{bmatrix} 2 \\ 4 \\ 6 \end{bmatrix}$ and $\mathbf{z} = \begin{bmatrix} 3 \\ -3 \\ 12 \end{bmatrix}$.

(3) Evaluate $\begin{bmatrix} 1 & 2 \\ 3 & 4 \end{bmatrix} \begin{bmatrix} 3 \\ 9 \end{bmatrix}$.

(4) Evaluate $\begin{bmatrix} 1 & 2 \\ 3 & 4 \end{bmatrix} \begin{bmatrix} -2 & 1 \\ 1.5 & -.5 \end{bmatrix}$.

(5) Evaluate $\begin{bmatrix} 1 & 1 & 2 \\ 2 & 1 & 2 \\ 2 & 2 & 1 \end{bmatrix} \begin{bmatrix} 1 & -1 & 0 \\ 2 & 0 & -2 \\ 1 & 1 & 3 \end{bmatrix}$.

(6) Evaluate $\mathbf{xy'}$ for $\mathbf{x} = \begin{bmatrix} \frac{1}{2} \\ -\frac{1}{2} \end{bmatrix}$ and $\mathbf{y} = \begin{bmatrix} 4 \\ 6 \end{bmatrix}$.

(7) Evaluate $\begin{bmatrix} x_{11} & x_{12} \\ x_{21} & x_{22} \end{bmatrix} \begin{bmatrix} d_1 & 0 \\ 0 & d_2 \end{bmatrix}$.

(8) Evaluate $\begin{bmatrix} d_1 & 0 \\ 0 & d_2 \end{bmatrix} \begin{bmatrix} x_{11} & x_{12} \\ x_{21} & x_{22} \end{bmatrix}$.

(9) Find A^{-1} if $A = \begin{bmatrix} 2 & 4 \\ 6 & 8 \end{bmatrix}$ and verify that $A^{-1}A = AA^{-1} = I$.

(10) Find A^{-1} if $A = \begin{bmatrix} \frac{1}{2} & -\frac{1}{2} \\ 2 & 9 \end{bmatrix}$ and verify that $A^{-1}A = AA^{-1} = I$.

(11) Verify that the inverse of $A = \begin{bmatrix} 1 & 2 & 3 \\ 4 & 5 & 6 \\ 7 & 8 & 10 \end{bmatrix}$ is $-\frac{1}{3}B$, where $B = \begin{bmatrix} 2 & 4 & -3 \\ 2 & -11 & 6 \\ -3 & 6 & -3 \end{bmatrix}$.

(12) Verify that the inverse of $P = \dfrac{1}{15} \begin{bmatrix} 5 & -14 & 2 \\ -10 & -5 & -10 \\ 10 & 2 & -11 \end{bmatrix}$ is P'. Matrices with

this property are orthogonal, that is, if $\mathbf{u}_j$ is the jth column of the matrix, then $\mathbf{u}_j'\mathbf{u}_j = 1$ for all j, and $\mathbf{u}_i'\mathbf{u}_j = 0$ whenever $i \neq j$. Verify that the columns of P have these properties.

(13) Solve the following systems $A\mathbf{x} = \mathbf{y}$ by finding $\mathbf{x} = A^{-1}\mathbf{y}$:

System a: $2x_1 - 3x_2 = 5$; $4x_1 + x_2 = 9$.

System b: $3x_1 + 5x_2 = 0$; $2x_1 - 4x_2 = 7$.

(14) For the data $y = 6, 2, 3, 0$ and $x = 1, 2, 3, 4$, find the vector **b** of least-squares estimates for the regression of y on x using matrix algebra. Also, if $SSE = \mathbf{y}'\ \mathbf{y} - \mathbf{b}'X'\mathbf{y}$, give the estimated variance–covariance matrix of regression parameter estimates.

(15) For the data $y = 1, 1.5, 1, 3$ and $x = -.5, 0, .5, 10$, find the vector **b** of least squares estimates for the regression of y on x using matrix algebra. Also, give the estimated variance–covariance matrix of regression parameter estimates.

APPENDIX B

Answers to Selected Exercises

This appendix contains short answers to the odd-numbered chapter exercises, plus short answers to all exercises in Appendix A. Note that the term *self-correcting* means that it is obvious when the correct solution has been reached.

CHAPTER 2

2.1. (a) b_0: the estimated average score on the first exam for those with a *college GPA* of zero is 25.092; b_1: each unit increase in *college GPA* is expected to result in a 16.783-point increase in the first exam score. (b) $\hat{\sigma}_{b_0} = 6.347$; $\hat{\sigma}_{b_1} = 2.027$. (c) $F = 68.579$; $p < .0001$. (d) $r^2 = .244$; $r^2_{adj} = .241$. (e) $\hat{\sigma}^2 = 219.717$. (f) 75.441. (g) (12.810, 20.756).

2.3. The orthogonality assumption is not very reasonable. For example, health positively affects coital activity and is negatively correlated with age. Therefore, the true model for coital frequency is more likely $Y = \beta_0 + \beta_1$ *male age* $+ \beta_2$ *male health* $+ \varepsilon'$. When we estimate $Y = \beta_0 + \beta_1$ *male age* $+ \varepsilon$, $\varepsilon = \beta_2$ *male health* $+ \varepsilon'$. So even if Cov(*male age*, ε') = 0, Cov(*male age*, ε) = Cov(*male age*, β_2 *male health* $+ \varepsilon'$) = β_2 Cov(*male age, male health*) $\neq 0$.

2.5. $b_0 = \bar{y} - b_1\bar{x}$ implies that $\bar{y} = b_0 + b_1\bar{x}$, which means that the point $(\bar{x}, \bar{y})$ is on the regression line.

2.7. Self-correcting.

2.9. (a) corr(SAT,GPA) = .834. (b) $t_{(8)} = 4.28$; $p < .01$. Yes. (c) $\hat{E}(GPA) = 1.049 + .337SAT$. (d) 2.67. (e) $-.27$.

2.11. (a) $b_0 = 13.052$; $b_1 = -.7681$. (b) r = $-.364$. (c) b_0: the estimated average *homicide rate* for cities with a zero *reading quotient* is 13.052; b_1: being a unit higher on the *reading quotient* is associated, on average, with a *homicide rate*

Regression with Social Data: Modeling Continuous and Limited Response Variables,
By Alfred DeMaris
ISBN 0-471-22337-9   Copyright © 2004 John Wiley & Sons, Inc.

that is lower by .7681. **(d)** $t_{(52)} = -2.818$; $p < .01$. Yes, there is a significant linear relationship. **(e)** .1325. **(f)** 7.58.

2.13. Via SAS: The Shapiro–Wilk test produces a p-value of .0003, suggesting that the null hypothesis of normally distributed errors should be rejected.

2.15. **(a)** $\hat{y} = 2.7575 + .1542x$. **(b)** $\hat{\sigma}^2 = 3.0811$. **(c)** $r^2 = .1039$. It has modest discriminatory power, at best. **(d)** $F_{(2,6)} = .9703/3.7847 = .2564$; $p > .7$. We cannot reject the null hypothesis that the model is empirically consistent with the data. **(e)** $r^2_{adj} = -.0081$.

2.17. Self-correcting.

2.19. $r^2_{adj} = 1 - \dfrac{n-1}{n-2}\dfrac{SSE}{TSS} = 1 - \dfrac{SSE}{TSS} + \dfrac{SSE}{TSS} - \dfrac{n-1}{n-2}\dfrac{SSE}{TSS} = r^2 +$

$\dfrac{SSE}{TSS} - \dfrac{n-1}{n-2}\dfrac{SSE}{TSS} = r^2 + \dfrac{SSE}{TSS}\left(1 - \dfrac{n-1}{n-2}\right)$, and this is $< r^2$, since

$\dfrac{SSE}{TSS}\left(1 - \dfrac{n-1}{n-2}\right) < 0$ for $n \geq 3$.

2.21. Self-correcting.

2.23. corr(error,STATMOOD) = 0. This is an artifact of OLS estimates; corr(error,COLGPA) = .4603. This suggests that there is some other systematic factor (namely, COLGPA) in the error term that should be entered as a regressor in the model; corr(error,SCORE) = .4738. This suggests that there is some other systematic factor (namely, SCORE) in the error term that should be entered as a regressor in the model; corr(error,EXAM1) = .95908. This suggests that error accounts for a substantial part of the variation in EXAM1; corr($\hat{y}$, STATMOOD) = 1.0. This is due to the fact that $\hat{y}$ is an exact linear function of STATMOOD; corr($\hat{y}$, error) = 0. This follows from the fact that the error term is uncorrelated with STATMOOD; corr($\hat{y}$, EXAM1) = .2832. This is equal to the correlation between STATMOOD and EXAM1, since $\hat{y}$ is just a linear "translation" of STATMOOD.

2.25. $s_y^2 = b_1^2 s_x^2 + s_e^2$.

2.27. Self-correcting.

CHAPTER 3

3.1. $X_2 = 3 \Rightarrow \hat{y} = (b_0 + 3b_2) + b_1 X_1$; $\qquad X_2 = 6 \Rightarrow \hat{y} = (b_0 + 6b_2) + b_1 X_1$;
$X_2 = 9 \Rightarrow \hat{y} = (b_0 + 9b_2) + b_1 X_1$.

3.3. **(a)** .454. **(b)** .454.

3.5. $F_{(5,408)} = 3.802$, $p = .0022$. The addition of these variables appears to produce a significant improvement in the model.

3.7. **(a)** $F_{(4,209)} = 13.899$, $p < .00001$. Reject H_0. At least one coefficient is not zero. **(b)** .2101. **(c)** 24.586. **(d)** COLGPA: $t = 1.597/.604 = 2.644$; significant at $p = .009$; SCORE: $t = .296/.110 = 2.691$; significant at $p = .008$; HOURS:

$t = -.490/.145 = -3.379$; significant at $p = .0009$; PREVMATH: $t = 1.516/.344 = 4.407$; significant at $p = .000017$. **(e)** 7.275. **(f)** Not really, since zero is outside of the observed, and in some cases, logical range of most of the regressor values. No one has zero *college GPA, math diagnostic score,* or *hours in the current semester.*

3.9. The plot of standardized residuals against $\hat{y}$ from the regression reveals no major irregularities, and all z_e are under 4 in absolute value. Similarly, the partial regression leverage plots reveal no noticeable nonlinear trends. In conclusion, the model appears, at least, to demonstrate empirical consistency.

3.11. **(a)** .2052. **(b)** $F_{(6,228)} = 11.07$, $p < .00001$. Yes, the model appears to have some utility for predicting STATMOOD. **(c)** .475. **(d)** 23.9679.

3.13. t for MCHATT versus FCHATT effects is .2599 and is not significant; t for MALEAGE versus FEMAGE effects is $-.6164$ and is not significant; t for MEDUC versus FEDUC effects is .2934 and is not significant.

3.15. The test of equality of male versus female parent effects is $F_{(4,347)} = .0949$, $p > .98$. There is not enough evidence to conclude that there is a difference in male versus female parent effects on *offspring's sexual adventurism.*

3.17. **(a)** If $r_{x_1 x_2} = 0$ then $b_1 = r_{y x_1}(s_y / s_{x_1})$, which is the b for the SLR of Y on X_1. **(b)** As $r^2_{x_1 x_2}$ approaches 1.0, the denominator of b_1 becomes ever smaller, meaning that the magnitude of b_1 grows ever larger.

3.19. $E(Y | x_k + 1, \mathbf{x}_{-k}) - E(Y | x_k, \mathbf{x}_{-k})$ (note that X_j is one of the other $X_1, X_2, \ldots, X_K$)

$$= \beta_0 + \beta_1 X_1 + \beta_2 X_2 + \cdots + \beta_k(x_k + 1) + \cdots + \beta_K X_K + \gamma(x_k + 1)X_j$$
$$- \beta_0 - \beta_1 X_1 - \beta_2 X_2 - \cdots - \beta_k x_k - \cdots - \beta_K X_K - \gamma x_k X_j$$
$$= \beta_k(x_k + 1) - \beta_k x_k + \gamma(x_k + 1)X_j - \gamma x_k X_j$$
$$= \beta_k(x_k + 1 - x_k) + \gamma[(x_k + 1)X_j - x_k X_j]$$
$$= \beta_k(1) + \gamma(x_k X_j + X_j - x_k X_j) = \beta_k + \gamma X_j.$$

3.21. At 2 years below mean *years in rank* the estimated partial slope is $1008.267 - 15.481(-2) = 10039.229$; at 2 years above mean *years in rank* it is $1008.267 - 15.481(2) = 977.305$; at 9 years above mean *years in rank* it is $1008.267 - 15.481(9) = 868.938$. Estimated mean salaries: 57444.749, 57410.961, 57351.832, respectively.

3.23. **(a)** Net of other factors, for kids with an average level of *parental monitoring,* a unit increase in their *sexual permissiveness* is worth, on average, a .1625-unit increase in their *sexual adventurism.* **(b)** Partial slope: $.1625 + .024$ MONITOR. Partial slope at 1 standard deviation above mean *parental monitoring*: .2441. **(c)** At $+1$ *SD* above mean *parental monitoring*: $t_{(352)} = 4.253$, $p < .001$; significant. At 2 *SD* below mean *parental monitoring*: $t_{(352)} = -.0079$; not significant.

3.25. $F_{(7,228)} = 3.8805$, $p < .001$. The models for STATMOOD do appear to be significantly different for males and females.

3.27. Nested $F = 2.627$, which implies a t of 1.6208, $p = .1058$. t for difference $= 1.624$, $p = .1051$. t for $\theta = 1.624$, $p = .1051$.

CHAPTER 4

4.1. **(a)** $\hat{y} = 1.5658 + .4101$ PRESCHDN. **(b)** $r = .341$, $r^2 = .116$. **(c)** Two-sample t test: $t = 7.36, p < .001$; test for r: $t = 7.38, p < .001$; test for d: $t = 7.35, p < .001$.

4.3. **(a)**

	SOPH	JUN	POST
Senior	-1	-1	-1
Sophomore	1	0	0
Junior	0	1	0
Postgrad/grad	0	0	1

(b) $\hat{y} = 78.806 + 2.789$ SOPH-4.125 JUN$+5.258$ POST.

4.5. **(a)** $F_{(4,1255)} = 33.135$, $p < .00001$. Net of *gender, marital status* has a significant effect on *cohabitation attitude*. **(b)** $F_{(4,1251)} = 1.045, p > .38$. There is not enough evidence to conclude that there is an interaction effect in the population. **(c)** Cell means: married female $= 5.603$; married male $= 5.688$; widowed female $= 4.802$; widowed male $= 5.474$; divorced female $= 6.522$; divorced male $= 6.662$; separated female $= 6.909$; separated male $= 5.95$; never married female $= 7.013$; never married male $= 7.396$.

4.7. **(a)** Married couples: 8.171; cohabiting couples: 11.485. **(b)** Married couples: 8.23. cohabiting couples: 10.55. The adjusted means are closer together, suggesting that part of the gap in *sexual frequency* between marrieds and cohabitors is accounted for by differences in *age* and *relationship duration* between these types of couples.

4.9. **(a)** Interpretations:

- *Intercept:* Males in couples where both are working full time do, on average, 23.766% of the housework, when MINCOME and FINCOME are both zero (this is, of course, nonsensical, but is nevertheless the interpretation!).

- *MPARTIME:* Controlling for partners' *incomes* and females' *employment status*, males' average percent contribution to housework is 2.104 more when males work part time, compared to when they work full time.

- *MUNEMP:* Controlling for partners' *incomes* and females' *employment status*, unemployed males do, on average, .177% less housework, compared to those who work full time.

- *FPARTIME:* Controlling for partners' *incomes* and females' *employment status*, the average percent contribution to housework of male partners of women who work part time is 6.043 less, compared to males whose partners work full time.

- *FUNEMP:* Controlling for partners' *incomes* and females' *employment status*, the average percent contribution to housework of male partners of unemployed women is 4.738 less, compared to males whose partners work full time.

(b) Adjusted mean percent of housework done by males when: both work full time: 25.881; male works part time, female full time: 27.985; male unemployed, female full time: 25.704; male works full time, female part time: 19.838; male works full time, female unemployed: 21.143; both unemployed: 20.966; both work part time: 21.942; male works part time, female unemployed: 23.247; male unemployed, female part time: 19.661.

4.11. **(a)** $F_{(2,226)} = 3.176$, $p < .05$. Yes, it is. **(b)** Controlling for *couple conflict*, women with *natural children* are, on average, 5.0846 units lower on *depressive symptomatology*, compared to those without children. **(c)** Since both types of kids are associated with greater conflict, but less depression with conflict controlled, and conflict is strongly positively related to depression, omitting conflict from the model reverses the impact of kids on depression.

4.13. Seniors: 75.892; sophomores: 79.522; juniors: 76.366; postgrad/grads: 79.398. The means are the same, within rounding error, as in Exercise 4.12(b).

4.15. **(a)** Interpretation of rank effects: Full professor: net of the covariates, full professors' average salary is $17,500 higher than for assistant professors. Associate professor: net of the covariates, associate professors' average salary is $6213.53 higher than for assistant professors. Instructor/lecturer: net of the covariates, instructor/lecturers make, on average, $7870.20 less than assistant professors. **(b)** Adjusted means: full professors = 56243; associate professors = 44956.526; assistant professors = 38743; instructor/lecturers = 30872.797.

4.17. **(a)** $F_{(171,215)} = 1.2459$, $p = .064$. There is not enough evidence to reject homogeneity of error variance. **(b)** Chow test: $F_{(15,386)} = .895, p = .57$. The models do not appear to differ for males versus females. **(c)** Effect of COITFREQ: $-.0183 + 1.2076$ MALEHIT $- .2841$ FEMAHIT. Interpretation: *coital frequency* has virtually no effect on the female's *depressive symptomatology* if neither partner has been violent; it increases *depressive symptomatology* if the male has been violent; and it lowers *depressive symptomatology* if the female has been violent.

4.19. Intercept constrained: $F_{(10,1853)} = 2.381$. Intercept unconstrained: $F_{(8,1853)} = 2.336$.

4.21. Constrained intercept: $F_{(12,701)} = 1.422$. Unconstrained intercept: $F_{(11,701)} = 1.361$.

4.23. The interaction is disordinal in Z, but ordinal in X.

4.25. Self-correcting.

CHAPTER 5

5.1. **(a)** *ps*, husbands: $-12.343 + 2(1.944)$ *power. ps*; wives: $-11.567 + 2(2.537)$ *power*. **(b)** Husbands: *ps* at *power* $= 1$ is -8.455. *ps* at *power* $= 4$ is 3.209; wives: *ps* at *power* $= 1$ is -6.493. *ps* at *power* $= 4$ is 8.729. **(c)** Husband's minimum occurs at *power* $= 3.175$; wife's minimum occurs at *power* $= 2.28$.

5.3. With $\gamma_0 = 1.5$, $\gamma_1 = .25$, $\varepsilon_1 = .5$, $\varepsilon_2 = -.5$, we have:

X	1	2	3	4	5
Model (5.13):					
Y_1	3.176	4.077	5.236	6.723	8.632
Y_2	1.168	1.500	1.926	2.473	3.176
Model (5.12):					
Y_1	2.426	2.973	3.676	4.577	5.736
Y_2	1.426	1.973	2.676	3.577	4.736
Variances of Y:					
$S^2_{5.13}$	2.016	3.320	5.478	9.031	14.884
$S^2_{5.12}$	.5	.5	.5	.5	.5

5.5. (a) $F_{(2,7)} = .151$, n.s. A linear model appears adequate for the data. (b) A quiz score of 60 implies $\hat{y} = 49.152$; a quiz score of 92 implies: $\hat{y} = 91.903$. (c) ps at mean -1 SD(Quiz): .929. ps at mean(Quiz): 1.376. ps at mean $+1$ SD(Quiz): 1.823.

5.7. (a) $\log \hat{y} = 2.773 -.024$ MALEAGE. (b) Each additional year older the male partner is results in a $100(e^{-.024} - 1) = 2.37\%$ reduction in coital frequency. For a 10-year increase in age, the reduction is 21.34%. (c) Male age $= 25$: $\hat{y} = 12.043$. Male age $= 35$: $\hat{y} = 9.474$. Male age $= 55$: $\hat{y} = 5.862$. (d) .127. (e) When the residuals are plotted against $\log \hat{y}$, they appear to have zero mean, but their variance appears to increase with $\log \hat{y}$. According to a formal test, the normality of the errors would be rejected. So the assumptions on the errors may not be warranted.

5.9. (a) $\hat{y} = 1.693 - .508x^{1/3} - .896z + .365zx^{1/3}$. (b) $\dfrac{\partial \hat{y}}{\partial x} = \dfrac{-.508+.365z}{3x^{2/3}}$; $\dfrac{\partial \hat{y}}{\partial z} = -.896 + .365x^{1/3}$. (c) Effect of X at $\bar{z} - 1s_z$: $\dfrac{-.204}{3x^{2/3}}$; at $\bar{z}$: $\dfrac{1.252}{3x^{2/3}}$; at $\bar{z} + 1s_z$: $\dfrac{2.707}{3x^{2/3}}$.
(d) Effect of Z at $\bar{x} - 1s_x$: $-.471$; at $\bar{x}$: $-.293$; at $\bar{x} + 1s_x$: $-.184$. (e) See *Instructor's Solutions Manual*.

5.11. (a) See *Instructor's Solutions Manual*. The trend resembles a segment I curve. (b) $F_{(8,406)} = 2.444$, $p = .014$. It appears that the data trend is significantly nonlinear. (c) $F_{(7,406)} = .521$, n.s. It appears that a quadratic model adequately captures the nonlinearity in the relationship. (d) $\hat{y} = 1.824 -.013$ DURYRS $- .0002$ DURYRS^2. Slope at mean $- 1$ SD: $-.007$; at mean: $-.013$; at mean $+ 1$ SD: $-.019$. A segment I curve is indicated.

5.13. (a) At $x = 0$: $\dfrac{2}{5 + 2(0)} = .4$. At $x = 2.5$: $\dfrac{2}{5 + 2(2.5)} = .2$. At $x = 5$: $\dfrac{2}{5 + 2(5)} = .133$. (b) At $x = 0$: 1.609. At $x = 2.5$: 2.303. At $x = 5$: 2.708.
(c) See *Instructor's Solutions Manual*.

5.15. (a) U-shaped curve. (b) U-shaped curve. (c) Segment I curve. (d) Inverted U-shaped curve.

5.17. At $\bar{z} - 1s_z$: $-3.263 + .5x$. At $\bar{z}$: $-3 + .5x$. At $\bar{z} + 1s_z$: $-2.738 + .5x$.

5.19. At $\bar{z} - 1s_z$: $-2.378 - .375x$. At $\bar{z}$: $-3 + .5x$. At $\bar{z} + 1s_z$: $-3.263 + 1.375x$.

5.21. (a) Linear model. (c) Nonlinear interaction in X. Linear interaction in Z.

5.23. (a) Nonlinear model. (c) Nonlinear interaction in X and in Z.

5.25. (a) Nonlinear model. (c) Nonlinear interaction in X and in Z.

CHAPTER 6

6.1. (a) Self-correcting. (b) $R_{xx} = \begin{bmatrix} 1 & .876 \\ .876 & 1 \end{bmatrix}$.

6.3. Self-correcting.

6.5.

$$R_{xx} = \begin{bmatrix} 1 & .675 \\ .675 & 1 \end{bmatrix} \Rightarrow R_{xx}^{-1} = \frac{1}{1 - .675^2} \begin{bmatrix} 1 & -.675 \\ -.675 & 1 \end{bmatrix} = \begin{bmatrix} \dfrac{1}{1 - .675^2} & \dfrac{-.675}{1 - .675^2} \\ \dfrac{-.675}{1 - .675^2} & \dfrac{1}{1 - .675^2} \end{bmatrix}$$

6.7. $\lambda \mathbf{uu}' = \lambda \begin{bmatrix} u_1 \\ u_2 \\ . \\ . \\ . \\ u_n \end{bmatrix} \mathbf{u}' = \begin{bmatrix} \lambda u_1 \mathbf{u}' \\ \lambda u_2 \mathbf{u}' \\ . \\ . \\ . \\ \lambda u_n \mathbf{u}' \end{bmatrix}$, and each $\lambda u_j \mathbf{u}'$ is a scalar multiple of the same

vector, $\mathbf{u}'$, so each $\lambda u_j \mathbf{u}'$ is an exact linear function of every other vector (e.g., $\lambda u_2 \mathbf{u}' = \frac{u_2}{u_1} \lambda u_1 \mathbf{u}'$, and so on). Therefore there is only one linearly independent row in the matrix, hence the matrix has rank 1.

6.9. (a) h_{ii} for cases 1 through 9, respectively: .475, .1646, .1646, .1112, .1112, .1112, .1572, .1572, .5479. (b) $t_9 = -2.172$; $dffits_9 = -2.392$; $dfbetas_9 = -1.521$; $D_9 = 1.867$.

6.11. (a) $E(\mathbf{b}_1^s) = \begin{bmatrix} .25 \\ .15 \\ .00 \end{bmatrix}$. (b) The cross-product term, and therefore the interaction effect, are completely suppressed when the intervening variables are omitted.

6.13. $\mathbf{b}_{pc}^s = \begin{bmatrix} .35 \\ .35 \end{bmatrix}$.

6.15. (a) $V(b_4^s) = \sigma_*^2 \left(\dfrac{.0137^2}{2.9305} + \dfrac{.9999^2}{.9996} + \dfrac{.0017^2}{.0689} + \dfrac{.0014^2}{.00091806} \right) = \sigma_*^2 (1.00248)$.
   (b) $p_{13} = .000408$; $p_{23} = .00000024$; $p_{33} = .02198$; $p_{43} = .9776$.

6.17. Self-correcting.

6.19. WLS estimate of the equation is: $10.8721 + .1993$ ADVENTRE $- .0852$ FSTYLE1 $+ .2041$ MSEXATT $+ .1988$ FSEXATT. $R_{WLS}^2 = .1049$.

6.21. Gender difference in exam scores at: *college GPA* $= 2.5$: $d = 1.429, p > .68$, n.s.; at *college GPA* $= 3.0$: $d = -2.149$, $p > .34$, n.s.; at *college GPA* $= 3.5$: $d = -5.727, p > .07$, n.s.

6.23. One serious near linear dependency involving YRDG, YRBG, and PRIOREXP. The *VIF*'s for their coefficients (52.836, 49.587, and 14.719, respectively) suggest that the coefficient variances are substantially affected.

6.25. Partial principal component results in the form of VARIABLE (b_{pc}) are: INTERCEPT (7873.500), YRDG (−13.897), YRBG (−80.556), PRIOREXP (179.726). (See *Instructor's Solutions Manual* for the full set of estimates.) The ridge estimates are the most intuitively appealing, since one would expect that, all else equal, YRDG, YRBG, and PRIOREXP would each have a *positive* effect on salary.

CHAPTER 7

7.1. −3.0089.

7.3. $R_L^2 = .1058$. $R_G^2 = .1328$. $R_{GSC}^2 = .1795$.

7.5. .0176.

7.7. .718.

7.9. Logit: −.0093; probit: −.0082.

7.11. Baseline $\hat{\pi} = .1062$. Change in probability: $\Delta\hat{\pi} = -.0042$.

7.13. (a) The odds of *violence* for those with no *alcohol or drug problems* at average *economic disadvantage* is .143. Having *alcohol or drug problems* magnifies the odds by a factor of 2.876, net of *economic disadvantage*. Each unit increase in *economic disadvantage* magnifies the odds by a factor of 1.031, controlling for *substance abuse problems*. (b) For those without *alcohol/drug problems* with mean *economic disadvantage*, the change in the probability of *violence* for a unit increase in ECNDISAD is $\Delta\hat{\pi} = .0034$. For those *with alcohol/drug problems*, on the other hand, the change is $\Delta\hat{\pi} = .0064$.

7.15. The model χ^2's are: logit model $1 = 128.634$; logit model $2 = 275.859$; probit model $1 = 126.631$; probit model $2 = 273.353$. The test for the additional conflict block is: logit: $\chi_{(3)}^2 = 147.225$; probit: $\chi_{(3)}^2 = 146.722$. Both tests are significant at $p < .0001$.

7.17. (a) .638. (b) .886. (c) .725.

7.19. Logit: $\Delta\hat{\pi} = .044$; probit: $\Delta\hat{\pi} = .04$.

7.21. Partial results are as follows. Logit equation: $\ln \hat{O} = -1.0825 + 1.1050$ *lived apart* $-.2936$ *number of children* $+ \cdots$; model $\chi_{(5)}^2 = 62.4470$. Probit equation: $\Phi^{-1}(\hat{\pi}) = -.8325 + .6055$ *lived apart* $-.1550$ *number of children* $+ \cdots$; model $\chi_{(5)}^2 = 63.1250$. (See *Instructor's Solutions Manual* for the full set of estimates.) Interpretations: Logit: Each additional year the couple has been together lowers the odds of *female aggression* by $100[\exp(-.1219) - 1] = 11.5\%$. Probit: Each additional year the couple has been together reduces the latent *female-aggression scale* by .0614, on average.

7.23. Probit: $\Delta\hat{\pi} = .0512$; logit: $\Delta\hat{\pi} = .0362$.

7.25. Probit: $R_L^2 = .258$; $R_{MZ}^2 = .53$; logit: $R_L^2 = .255$; $R_{MZ}^2 = .564$.

7.27. Interpretations: Each unit increase in GPA adds .464 to the probability of an A. Those taught with the PSI method have .379 greater probability of getting an A. Each unit increase in TUCE adds .01 to the probability of an A. Estimated probability: $\hat{\pi}_{A,11} = -.078$.

7.29. Interpretations: Each unit increase in GPA raises the odds of getting an A by a factor of 16.88. Those taught with the PSI method have 10.791 times higher odds of getting an A. Each unit increase in TUCE raises the odds of getting an A by a factor of 1.1. Estimated probability: $\hat{\pi}_{A,11} = .024$. $R_L^2 = .374$.

CHAPTER 8

8.1. Self-correcting.

8.3. At -1 *SD economic disadvantage*: $\psi = 1.848$. At $+.5$ *SD economic disadvantage*: $\psi = 3.142$. At $+1.5$ *SD economic disadvantage*: $\psi = 4.476$. At -1 *SD economic disadvantage*, the impact of *alcohol/drug problem* is reduced by about 30%. At $+.5$ *SD*, it's increased by 19.3%, and at $+1.5$ *SD*, it's increased by 70%.

8.5. At $+.5$ *SD*: $-.041$. At $+1.5$ *SD*: $-.016$. At 2 *SD*: $-.003$.

8.7. Self-correcting.

8.9. Self-correcting.

8.11. With 0 = "intense male violence," 1 = "physical aggression," and 2 = "nonviolence," the estimated probabilities are: $P(Y=0) = .085$; $P(Y=1) = .225$; $P(Y=2) = .69$.

8.13. (a) $\chi^2_{(12)} = 11.5317$, $p = .484$. There is not enough evidence to reject the proportional odds hypothesis. (b) Interpretations: Controlling for other regressors, each unit increase in COLGPA magnifies the odds of a higher grade by a factor of 8.929. Controlling for other regressors, each unit increase in *math diagnostic score* magnifies the odds of a higher grade by a factor of 1.305. Controlling for other regressors, each unit increase in STATMOOD magnifies the odds of a higher grade by 1.064. Controlling for other regressors, each unit increase in the ratio of *study hours* to *TV hours* diminishes the odds of a higher grade by about 2%.

8.15. Model $\chi^2_{(10)} = 182.2603$, $p < .00001$. The model is significant as a whole. $R_L^2 = .126$.

8.17. Sample interpretations (see *Instructor's Solutions Manual* for the complete set of interpretations): AGE: Controlling for other regressors, each additional year of age increases the odds of "no prohibition" by a factor of 1.022, and increases the odds of "prohibition for all" by a factor of 1.029. MALE: Controlling for other regressors, men's odds of "no prohibition" are 26.125% lower than women's, and their odds of "prohibition for all" are 38.097% lower than women's. On treating PORN18 as ordinal: no. (See explanation in *Instructor's Solutions Manual*.)

8.19. P(no prohibition) = .026. P(prohibition for all) = .178. P(prohibition if <18) = .796.

8.21. Sample interpretations (see *Instructor's Solutions Manual* for the complete set of interpretations): EDUCAT: Controlling for other regressors, each additional year of schooling reduces the odds of "easier" by 14.23%, and the odds of "more difficult" by 10.631%. MALE: Controlling for other regressors, men's odds of "easier" are 28.794% lower than women's, and their odds of "more difficult" are 5.578% lower than women's. On treating DIVLAW as ordinal: no. (See explanation in *Instructor's Solutions Manual*.)

8.23. P(easier) = .248. P(more difficult) = .699. P(keep as is) = .053.

8.25. (a) Sample interpretations (see *Instructor's Solutions Manual* for the complete set of interpretations): SCORE: Controlling for other regressors, a unit increase in SCORE magnifies the odds of an A by 1.451, and the odds of a B by 1.304. MALE: Controlling for other regressors, males' odds of an A are .573 times females' odds, whereas their odds of a B are .998 times those of females. (b) Model $\chi^2_{(12)} = 97.96$, $p < .00001$. (c) College GPA is the only discriminator of an A versus a B grade. (d) $\chi^2_{(7)} = 31.287$, $p < .0001$. The grades of A and B are not collapsible with respect to the predictors. (e) Test for the proportional odds assumption: $\chi^2_{(7)} = 9.7946$, $p = .2005$. There is insufficient evidence to reject the proportional odds assumption. Sample interpretations (see *Instructor's Solutions Manual* for the complete set of interpretations): SCORE: Controlling for other regressors, a unit increase in SCORE increases the odds of a better grade by 34%, an effect that is significant at $p < .0001$. MALE: Controlling for other regressors, men's odds of a better grade are 38.4% lower than women's, a nonsignificant effect.

CHAPTER 9

9.1. $E(y|y>8) = 13.152$. $SD(y|y>8) = 3.172$.

9.3. $E(y|z>5) = 12.278$. $SD(y|z>5) = 3.884$.

9.5. 87.365.

9.7. 22.289.

9.9. Self-correcting.

9.11. $R_p^2 = .254$.

9.13. The proportion due to change in the mean is .377. The proportion due to change in P(uncensored) = .623.

9.15. $[corr(y,\hat{y})]^2 = .638$.

9.17. (a) $\hat{y} = 43.819$. (b) 83.9% of the effect of *child abuse* is due to its effect on elevating average PTSD, while 16.1% of its effect is due to raising the risk for PTSD.

9.19. $\rho_{eu} = \beta_3\gamma_3 V(x_3)$, which is > 0.

9.21. Bias(b_1) = .697, therefore b_1 will overestimate β_1.

9.23. There is little evidence of selection bias, since neither LAMBDA (Heckit) nor $\hat{\rho}$ (ML) is significant, suggesting that the association between remaining in the class and score on *exam 2* is pretty much accounted for by the explanatory variables. Also, the ML and OLS estimates tend to be quite similar. Interestingly, students with a stronger *attitude toward statistics* appear to be *less* likely to have stayed in the class!

9.25. The ML procedure did not converge in this case. But judging from Heckit, selection bias does not appear to be a problem. The only significant predictor of inclusion is *paying one's own tuition*. Its effect suggests that those who pay their own way are more likely to be nonmissing on the response. Otherwise, none of the focus variables are significant predictors of inclusion, and LAMBDA is also nonsignificant. As a result, there is relatively little difference between OLS and Heckit estimates of the substantive equation.

CHAPTER 10

10.1. $\displaystyle\sum_{y=1}^{\infty} \frac{e^{-\mu}\mu^y}{y!(1-e^{-\mu})} = \frac{1}{1-e^{-\mu}}\sum_{y=1}^{\infty}\frac{e^{-\mu}\mu^y}{y!} = \frac{1}{1-e^{-\mu}}(1-e^{-\mu}) = 1.$

10.3. (a) $f(0) = .041$. (b) $f(3) = .223$. (c) $f(10) = .0013$.

10.5. (a) $f(0) = .0001$. (b) $f(3) = .0015$. (c) $f(10) = .00997$.

10.7. (a) $p = .238$. (b) $f(0) = .238$. (c) $f(3) = .105$. (d) $f(10) = .016$.

10.9. $\displaystyle L(\beta|y,x) = \sum_{i=1}^{n}\left[y_i\left(\sum_{k=0}^{K}\beta_k X_{ik}\right) - \exp\left(\sum_{k=0}^{K}\beta_k X_{ik}\right) - \ln y_i! - \right.$
$$\left. \ln\left(1 - \exp\left(-\exp\left(\sum_{k=0}^{K}\beta_k X_{ik}\right)\right)\right)\right].$$

10.11. $f(0) = e^{-\mu} \equiv \dfrac{e^{-\mu}\mu^0}{0!}$. $f(1) = \dfrac{\mu}{1}e^{-\mu} \equiv \dfrac{e^{-\mu}\mu^1}{1!}$. $f(2) = \dfrac{\mu}{2}\mu e^{-\mu} = \dfrac{\mu^2 e^{-\mu}}{2} \equiv \dfrac{e^{-\mu}\mu^2}{2!}$.

$f(3) = \dfrac{\mu}{3}\dfrac{\mu^2 e^{-\mu}}{2} = \dfrac{\mu^3 e^{-\mu}}{6} \equiv \dfrac{e^{-\mu}\mu^3}{3!}$. $f(4) = \dfrac{\mu}{4}\dfrac{\mu^3 e^{-\mu}}{6} = \dfrac{\mu^4 e^{-\mu}}{24} \equiv \dfrac{e^{-\mu}\mu^4}{4!}$.

10.13. $\hat{f}(0|x,\hat{\beta}) = .575$. $\hat{f}(1|x,\hat{\beta}) = .138$. $\hat{f}(2|x,\hat{\beta}) = .076$. $\hat{f}(3|x,\hat{\beta}) = .050$. $\hat{f}(4|x,\hat{\beta}) = .035$.

10.15. (a) Proportionate change is $(1 + 1/N)^d - 1$. (b) $.014$.

10.17. (a) Cohabitors' expected number of children in the household is estimated to be about 45% lower than that of marrieds. (b) $\hat{f}(0|x,\hat{\beta}) = .388$. $\hat{f}(1|x,\hat{\beta}) = .367$. $\hat{f}(2|x,\hat{\beta}) = .174$. (c) $\chi^2_{(4)} = 129.8468$, $p < .00001$. Reject H_0. At least one coefficient is nonzero. $R_L^2 = .105$.

10.19. (a) The likelihood of being in the structural-zero group is higher if a couple is cohabiting, was older at inception of the union, has been together longer, and has a higher household income. The expected number of children, given that a couple is at risk for having children, is, correspondingly, lower for couples with the same characteristics. (b) Since $V = 10 > 1.96$, the ZIP is to be preferred over the PRM. (c) $R_L^2 = .214$.

10.21. **(a)** Selected estimates (significance levels) are: LOGTIME .0472 (.0029); ETHNIC .3633 (.0000); EDUCCL $-.0684$ (.0005). (See *Instructor's Solutions Manual* for the full set of estimates.) **(b)** Test 1 [H_1 is $V(Y) = \mu + \alpha\mu$]: $\hat{\alpha} = 2.798$, $p = .0000$. Test 2 [H_1 is $V(Y) = \mu + \alpha\mu^2$]: $\hat{\alpha} = 2.607$, $p = .0000$. Conclusion: the data are overdispersed. **(c)** $r^2 = .1139$; $R_D^2 = .1827$; $R_L^2 = .1437$.

10.23. **(a)** Selected estimates for the log odds of being a structural zero (μ) are: LOGTIME .0321 (.0453); ETHNIC $-.5456$ (.0677); EDUCCL .0148 ($-.0546$). (See *Instructor's Solutions Manual* for the full set of estimates.) **(b)** $V = 27.2561$ for ZIP/PRM, so ZIP is preferred. **(c)** $R_L^2 = .3516$; $r^2 = .0604$.

10.25. See *Instructor's Solutions Manual* for the graph.

CHAPTER 11

11.1. Hazard estimates are: $\hat{h}_1 = .003$; $\hat{h}_2 = .003$; $\hat{h}_3 = .003$; $\hat{h}_4 = .005$. Survival estimates are: $\hat{S}_1 = 1.000$; $\hat{S}_2 = .975$; $\hat{S}_3 = .945$; $\hat{S}_4 = .916$; $\hat{S}_5 = .869$.

11.3. Wald: $\chi^2 = .755$, n.s. LR$\chi^2 = .811$, n.s. Interpretation: the hazard of rearrest for domestic violence at any given time is 65.8% lower for those who were employed when first arrested.

11.5. $S(t) = \exp(-\lambda t^\sigma) \Rightarrow \ln S(t) = -\lambda t^\sigma \Rightarrow -\ln S(t) = \lambda t^\sigma \Rightarrow \ln[-\ln S(t)] = \ln\lambda + \sigma\ln t$.

11.7. 11.29 months.

11.9. Sample interpretations (see *Instructor's Solutions Manual* for the complete set of interpretations): Net of other covariates, women's hazard of dropping out is 38.265% higher than men's at any given time, and the hazard of dropping out is increased by 32.976% for each unit decrement in *high-school grades* at any given time.

11.11. **(a)** $\Delta\chi^2_{(2)} = 13.13$, $p = .0014$. There is evidence that at least one of the effects is nonzero. **(b)** *Marriage*: net of other covariates, at any given time being married raises the hazard of dropping out by a factor of 3.508. *Employment*: net of other covariates, at any given time being employed raises the hazard of dropping out by a factor of 1.669. **(c)** $R_G^2 = .211$.

11.13. **(a)** LR$\chi^2_{(5)} = 43.028$, $p < .0001$. The model is significant. **(b)** Wald chi-squareds: DURATION: $\chi^2 = 6.25$; FAGUNION: $\chi^2 = 21.778$; PARENTAL: $\chi^2 = 2.736$; CHDN5: $X^2 = .737$; MINORITY: $\chi^2 = 2.404$. DURATION and FAGUNION have significant effects. **(c)** $R_G^2 = .034$.

11.15. **(a)** Interpretations: each unit increase in the *female partner's relationship happiness* reduces the hazard of disruption by 13.4%; each unit increase in the *female partner's commitment* reduces the hazard of disruption by 34.6%. **(b)** 20.2%. **(c)** (.480, 3.013). **(d)** $R_G^2 = .115$.

11.17. $-2\ln L = 6.086$.

11.19. Global test: $\Delta\chi^2_{(2)} = 3.7116$, $p = .156$, n.s. There is insufficient evidence to suggest that there are any gender differences in the effects of *income* or *education* on the hazard of disruption.

11.21. The effect of DISAGMT at 1 standard deviation above the mean of COM-STYLE is .02283, a value that is not significant.

CHAPTER 12

12.1. .048.

12.3. $\Delta\chi^2_{(2)} = 7.598$, $p = .022$. Reject H_0; conclude that the three events are characterized by different models.

12.5. Equations (a), (b), and (d).

12.7.

Unemployment Spell	Duration Unemployed	Censoring Indicator
1	3	1
2	10	1
3	2	1
4	17	0

12.9. When TIME = 1 (i.e., in the first interval), the hazard of reemployment is 50% higher for immigrants residing in Canada. However, each unit increase in LOGTIME brings a 50% reduction in the *effect* of Canadian residence. The coefficient is approximately zero when TIME = 1.8.

12.11. The hazards for the first two intervals are: $\hat{P}_{i1} = .191$; $\hat{P}_{i2} = .184$. The survival function for the first two intervals is: $\hat{S}_{i1} = .809$; $\hat{S}_{i2} = .660$.

12.13. The hazards for the first four years are: $\hat{P}_{i1} = .00023$; $\hat{P}_{i2} = .0035$; $\hat{P}_{i3} = .0205$; $\hat{P}_{i4} = .0573$. The survival function for the first four years is: $\hat{S}_{i1} = .9998$; $\hat{S}_{i2} = .996$; $\hat{S}_{i3} = .976$; $\hat{S}_{i4} = .92$.

12.15. (a) and (c) are time invariant, since their values are fixed at the time the employee enters the given rank (either assistant or associate professor). (b), although treated as time-invariant in the study, could easily be regarded as time varying, since the prestige of an institution can change over time. (d), (e), and (f) are time varying because their values change with years in rank.

12.17. Selected estimates for the log-hazard of separation (marriage) are DIS-AGMT −.009 (.073); COMSTYLE −.151 (.142); HEHIT −1.467 (2.712). (See *Instructor's Solutions Manual* for the full set of estimates.) Interpretation: the main effects of covariates not involved in an interaction with survival time are similar to those in Table 12.1. The block of crossproduct terms is not significant when added to either model, according to $\Delta\chi^2$ tests. However, there is a significant nonproportional effect for *male violence* in the model for marriage, which suggests that *male violence* accelerates the transition to marriage at first, but over time this effect is diminished.

12.19. Selected estimates for the log-hazard of reemployment are PRVJBS −.045*; JOBDUR −.052***; FEMALES −.314**. (See *Instructor's Solutions*

Manual for the full set of estimates.) Tests: $\Delta\chi^2 = 7.9135$, $p = .005$. Wald $\chi^2 = 8.2846$, $p = .004$. By either test, the interaction is significant. Interpretation: At the time of unemployment, each additional month on the previous job reduces the hazard of reemployment by about 5%. After being unemployed for 17.33 months, however, the *duration of the previous job* no longer has any impact on the hazard of reemployment.

APPENDIX A

I.Q **(1)** $3xy + 2y + 3xz + 2z$.

(2) $2x^3 + x^2\sqrt{y} + x^2z + 2x\sqrt{y} + y + \sqrt{y}z + 8xw + 4\sqrt{y}w + 4wz$.

(3) $x = 2\sqrt{24} - 24$.

(4) 15.

(5) $\dfrac{ab(d + 3\pi)}{cd}$.

(6) $\dfrac{2x + yz}{2zw}$.

(7) $x = 123.4568$.

(8) $15x^5$.

(9) $\dfrac{x^z}{2^z y^z}$.

(10) $2^{7/6}$.

(11) $2\ln x - \frac{1}{2}\ln z$.

(12) $f(-\frac{1}{2}) - 4$; $f(\frac{1}{3}) = 3.5$; $f(2) = 18.5$.

(13) $f(-\frac{1}{2}) = 2.75$; $f(\frac{1}{3}) = \frac{1}{9}$; $f(2) = -1$.

(14) $f(g(3)) = \sqrt{8}$.

(15) $f(g(\frac{1}{2})) = .6065$.

(16) $f(g(x)) = x$.

(17) $y = 9.5 + 18x$.

II.E **(1)** Self-correcting.

(2) Self-correcting.

(3) Self-correcting.

III.D **(1)** $5\,\mathrm{Cov}(X,Y)$.

(2) $bd\,\mathrm{Cov}(X,Y)$.

(3) -65.2.

(4) 10.1.

(5) $\beta^2\,V(X) + V(\varepsilon)$.

(6) Self-correcting.

(7) Self-correcting.

IV.D **(1)** $3x^2$.

(2) $\dfrac{4}{5x^{1/5}}$.

(3) $12x^3 + 6x^2 - 2x$.

(4) $f'(x) = \dfrac{1}{x} + 3x^2$; $f'(3) = 27\frac{1}{3}$.

(5) $f'(-2) = 4.1353$.

(6) $e^x(x^3 + 3x^2)$.

(7) $1 + \ln x$.

(8) $\dfrac{4 + 2\ln x}{2\sqrt{x}}$.

(9) $\dfrac{\ln x - 2}{2\sqrt{x}(\ln x)^2}$.

(10) $12x^2$.

(11) $-\dfrac{1}{x^2}$.

(12) $25e^{5x}$.

(13) $\dfrac{2\ln x}{x}$.

(14) $2e^{x^2}(2x^2 + 1)$.

(15) $\dfrac{e^x}{(1 + e^x)^2}$.

(16) $\beta\dfrac{e^{\alpha + \beta x}}{(1 + e^{\alpha + \beta x})^2}$.

(17) $\dfrac{2}{x}$.

(18) $4(x + 2)^3$.

(19) $3\left(\dfrac{1}{2}x + \ln x\right)^2\left(\dfrac{1}{2} + \dfrac{1}{x}\right)$.

(20) $2x + 3y$.

(21) $\beta\dfrac{e^{\alpha + \beta x + \gamma w + \lambda z}}{(1 + e^{\alpha + \beta x + \gamma w + \lambda z})^2}$.

V.K (1) -1.

(2) 66.

(3) $\begin{bmatrix} 21 \\ 45 \end{bmatrix}$.

(4) $\begin{bmatrix} 1 & 0 \\ 0 & 1 \end{bmatrix}$.

(5) $\begin{bmatrix} 5 & 1 & 4 \\ 6 & 0 & 4 \\ 7 & -1 & -1 \end{bmatrix}$.

(6) $\begin{bmatrix} 2 & 3 \\ -2 & -3 \end{bmatrix}$.

(7) $\begin{bmatrix} x_{11}d_1 & x_{12}d_2 \\ x_{21}d_2 & x_{22}d_2 \end{bmatrix}$.

(8) $\begin{bmatrix} d_1x_{11} & d_1x_{12} \\ d_2x_{21} & d_2x_{22} \end{bmatrix}$.

(9) $-\dfrac{1}{8}\begin{bmatrix} 8 & -4 \\ -6 & 2 \end{bmatrix}$.

(10) $\dfrac{2}{11}\begin{bmatrix} 9 & \frac{1}{2} \\ -2 & \frac{1}{2} \end{bmatrix}$.

(11) Self-correcting.

(12) Self-correcting.

(13) System a: $\begin{bmatrix} x_1 \\ x_2 \end{bmatrix} = \begin{bmatrix} 2.2857 \\ -.1429 \end{bmatrix}$, within rounding error. System b: $\begin{bmatrix} x_1 \\ x_2 \end{bmatrix}$

$$= \begin{bmatrix} 1.5909 \\ -.9545 \end{bmatrix}, \text{ within rounding error.}$$

(14) $\mathbf{b} = \begin{bmatrix} 7 \\ -1.7 \end{bmatrix}, \hat{V}(\mathbf{b}) = \begin{bmatrix} 3.225 & -1.075 \\ -1.075 & .43 \end{bmatrix}$.

(15) $\mathbf{b} = \begin{bmatrix} 1.1702 \\ .1809 \end{bmatrix}, \hat{V}(\mathbf{b}) = \begin{bmatrix} .0360 & -.0036 \\ -.0036 & .0014 \end{bmatrix}$.

References

Aalen, O. O. (1994). Effects of frailty in survival analysis. *Statistical Methods in Medical Research, 3*, 227–243.

Agresti, A. (1984). *Analysis of ordinal categorical data.* New York: Wiley.

Agresti, A. (1989). Tutorial on modeling ordered categorical response data. *Psychological Bulletin, 105*, 290–301.

Agresti, A. (2002). *Categorical data analysis* (2nd ed.). Hoboken, NJ: Wiley.

Agresti, A., and Finlay, B. (1997). *Statistical methods for the social sciences* (3rd ed.). Upper Saddle River, NJ: Prentice Hall.

Aiken, L. S., and West, S. G. (1991). *Multiple regression: Testing and interpreting interactions.* Thousand Oaks, CA: Sage.

Albert, A., and Anderson, J. A. (1984). On the existence of maximum likelihood estimates in logistic regression models. *Biometrika, 71*, 1–10.

Aldrich, J. H., and Nelson, F. D. (1984). *Linear probability, logit, and probit models.* Beverly Hills, CA: Sage.

Allison, P. D. (1982). Discrete-time methods for the analysis of event histories. In S. Leinhardt (Ed.), *Sociological Methodology 1982* (pp. 61–98). San Francisco, CA: Jossey-Bass.

Allison, P. D. (1984). *Event history analysis: Regression for longitudinal event data.* Beverly Hills, CA: Sage.

Allison, P. D. (1995). *Survival analysis using the SAS system: A practical guide.* Cary, NC: SAS Institute.

Allison, P. D. (1996). Fixed-effects partial likelihood for repeated events. *Sociological Methods & Research, 25*, 207–222.

Allison, P. D. (1999). Comparing logit and probit coefficients across groups. *Sociological Methods & Research, 28*, 186–208.

Anton, H. (1984). *Calculus with analytic geometry* (2nd ed.). New York: Wiley.

Axinn, W. G., and Thornton, A. (1993). Mothers, children, and cohabitation: The intergenerational effects of attitudes and behavior. *American Sociological Review, 58*, 233–246.

Regression with Social Data: Modeling Continuous and Limited Response Variables,
By Alfred DeMaris
ISBN 0-471-22337-9   Copyright © 2004 John Wiley & Sons, Inc.

Balzer, W., Boudreau, N. S., Hutchinson, P., Ryan, A. M., Thorsteinson, T., Sullivan, J., Yonker, R., and Snavely, D. (1996). Critical modeling principles when testing for gender equity in faculty salary. *Research in Higher Education, 37*, 633–658.

Barker, L., and Brown, C. (2001). Logistic regression when binary predictor variables are highly correlated. *Statistics in Medicine, 20*, 1431–1442.

Begg, C. B., and Gray, R. (1984). Calculation of polychotomous logistic regression parameters using individualized regressions. *Biometrika, 71*, 11–18.

Belsley, D. A., Kuh, E., and Welsch, R. E. (1980). *Regression diagnostics: Identifying influential data and sources of collinearity.* New York: Wiley.

Bennett, N. G., Blanc, A. K., and Bloom, D. E. (1988). Commitment and the modern union: Assessing the link between premarital cohabitation and subsequent marital stability. *American Sociological Review, 53*, 127–138.

Bentler, P. M., and Chou, C. (1988). Practical issues in structural modeling. In J. S. Long (Ed.), *Common problems/proper solutions: Avoiding error in quantitative research* (pp. 161–192). Newbury Park, CA: Sage.

Berk, R. A. (1983). An introduction to sample selection bias in sociological data. *American Sociological Review, 48*, 386–398.

Bickel, P. J., and Doksum, K. A. (1977). *Mathematical statistics: Basic ideas and selected topics.* Oakland, CA: Holden-Day.

Blossfeld, H., and Hamerle, A. (1990). Unobserved heterogeneity in hazard rate models: A test and an illustration from a study of career mobility. In K. U. Mayer and N. B. Tuma (Eds.), *Event history analysis in life course research* (pp. 241–252). Madison, WI: University of Wisconsin Press.

Blossfeld, H., and Rohwer, G. (1995). *Techniques of event history modeling.* Mahwah, NJ: Erlbaum.

Blossfeld, H., Hamerle, A., and Mayer, K. U. (1989). *Event history analysis.* Hillsdale, NJ: Erlbaum.

Bollen, K. A. (1989). *Structural equations with latent variables.* New York: Wiley.

Booth, A., Johnson, D. R., White, L., and Edwards, J. N. (1984). Women, outside employment, and marital instability. *American Journal of Sociology, 90*, 567–583.

Boudreau, N., Sullivan, J., Balzer, W., Ryan, A. M., Yonker, R., Thorsteinson, T., and Hutchinson, P. (1997). Should faculty rank be included as a predictor variable in studies of gender equity in university faculty salaries? *Research in Higher Education, 38*, 297–312.

Bradley, C. S. (2000). Does self-esteem contribute to good grades at the college level? A sociological examination of the factors that impact academic performance in college. Unpublished master's thesis, Bowling Green State University.

Breen, R. (1996). *Regression models: Censored, sample selected, or truncated data.* Thousand Oaks, CA: Sage.

Breslow, N. E. (1974). Covariance analysis of censored survival data. *Biometrics, 30*, 89–99.

Breusch, T., and Pagan, A. (1979). A simple test for heteroscedasticity and random coefficient variation. *Econometrica, 47*, 1287–1294.

Cameron, A. C., and Trivedi, P. K. (1990). Regression-based tests for overdispersion in the Poisson model. *Journal of Econometrics, 46*, 347–364.

Cameron, A. C., and Trivedi, P. K. (1998). *Regression analysis of count data.* Cambridge: Cambridge University Press.

Cameron, A. C., and Windmeijer, F. A. G. (1997). An R-squared measure of goodness of fit for some common nonlinear regression models. *Journal of Econometrics, 77,* 329–342.

Caudill, S. B. (1988). An advantage of the linear probability model over probit or logit. *Oxford Bulletin of Economics and Statistics, 50,* 425–427.

Chow, G. (1960). Tests of equality between sets of coefficients in two linear regressions. *Econometrica, 28,* 591–605.

Chung, K. L. (1974). *A course in probability theory* (2nd ed.). San Diego, CA: Academic Press.

Clark, M. L. (2001). Factors associated with rule violations in correctional facilities. Unpublished master's thesis, Bowling Green State University.

Cleary, P. D., and Angel, R. (1984). The analysis of relationships involving dichotomous dependent variables. *Journal of Health and Social Behavior, 25,* 334–348.

Clogg, C. C., and Haritou, A. (1997). The regression method of causal inference and a dilemma confronting this method. In V. R. McKim and S. P. Turner (Eds.), *Causality in crisis? Statistical methods and the search for causal knowledge in the social sciences* (pp. 83–112). Notre Dame, IN: University of Notre Dame Press.

Clogg, C. C., Petkova, E., and Haritou, A. (1995). Statistical methods for comparing regression coefficients between models. *American Journal of Sociology, 100,* 1261–1293.

Collett, D. (1994). *Modelling survival data in medical research.* London: Chapman & Hall.

Cox, D. R. (1972). Regression models and life tables. *Journal of the Royal Statistical Society, 34,* 187–220.

Cox, D. R. (1975). Partial likelihood. *Biometrika, 62,* 269–276.

Cragg, J. G. (1971). Some statistical models for limited dependent variables with application to the demand for durable goods. *Econometrica, 39,* 829–844.

Cragg, J. G., and Uhler, R. S. (1970). The demand for automobiles. *Canadian Journal of Economics, 3,* 386–406.

DeMaris, A. (1991). A framework for the interpretation of first-order interaction in logit modeling. *Psychological Bulletin, 110,* 557–570.

DeMaris, A. (1992). *Logit modeling: Practical applications.* Newbury Park, CA: Sage.

DeMaris, A. (1993). Odds versus probabilities in logit equations: A reply to Roncek. *Social Forces, 71,* 1057–1065.

DeMaris, A. (1995). A tutorial in logistic regression. *Journal of Marriage and the Family, 57,* 956–968.

DeMaris, A. (1997). Elevated sexual activity in violent marriages: Hypersexuality or sexual extortion? *Journal of Sex Research, 34,* 361–373.

DeMaris, A. (2000). 'Till discord do us part: The role of physical and verbal conflict in union disruption. *Journal of Marriage and the Family, 62,* 683–692.

DeMaris, A. (2001). The influence of intimate violence on transitions out of cohabitation. *Journal of Marriage and Family, 63,* 235–246.

DeMaris, A. (2002a). Covariance structure models. In M. Wiederman and B. E. Whitley (Eds.), *Handbook for conducting research on human sexuality* (pp. 289–325). Hillsdale, NJ: Erlbaum.

DeMaris, A. (2002b). Regression models. In M. Wiederman and B. E. Whitley (Eds.), *Handbook for conducting research on human sexuality* (pp. 255–287). Hillsdale, NJ: Erlbaum.

DeMaris, A. (2002c) Explained variance in logistic regression: A Monte Carlo study of proposed measures. *Sociological Methods & Research, 31,* 27–74.

DeMaris, A. (2003). Logistic regression. In I. B. Weiner (Series Ed.), W. Velicer and J. Schinka (Vol. Eds.), *Handbook of psychology:* Vol. 2, *Research methods in psychology* (pp. 509–532). New York: Wiley.

DeMaris, A., and Kaukinen, C. (2003). Violent victimization and women's mental and physical health: Evidence from a national sample. Unpublished manuscript, Bowling Green State University.

DeMaris, A., Benson, M. L., Fox, G. L., Hill, T., and Van Wyk, J. (2003). Distal and proximal factors in domestic violence: A test of an integrated model. *Journal of Marriage and Family, 65,* 652–667.

Draper, N. R., and Smith, H. (1998). *Applied regression analysis* (3rd ed.). New York: Wiley.

DuMouchel, W. H., and Duncan, G. J. (1983). Using sample survey weights in multiple regression analyses of stratified samples. *Journal of the American Statistical Association, 78,* 535–543.

Dunteman, G. H. (1989). *Principal components analysis.* Newbury Park, CA: Sage.

Efron, B. (1977). The efficiency of Cox's likelihood function for censored data. *Journal of the American Statistical Association, 76,* 312–319.

Fair, R. C. (1978). A theory of extramarital affairs. *Journal of Political Economy, 86,* 45–61.

Fox, J. (1997). *Applied regression analysis, linear models, and related methods.* Thousand Oaks, CA: Sage.

Freedman, D. A. (1997a). From association to causation via regression. In V. R. McKim and S. P. Turner (Eds.), *Causality in crisis? Statistical methods and the search for causal knowledge in the social sciences* (pp. 113–161). Notre Dame, IN: University of Notre Dame Press.

Freedman, D. A. (1997b). Rejoinder to Spirtes and Scheines. In V. R. McKim and S. P. Turner (Eds.), *Causality in crisis? Statistical methods and the search for causal knowledge in the social sciences* (pp. 177–182). Notre Dame, IN: University of Notre Dame Press.

Galler, H. P., and Poetter, U. (1990). Unobserved heterogeneity in models of unemployment duration. In K. U. Mayer and N. B. Tuma (Eds.), *Event history analysis in life course research* (pp. 226–240). Madison, WI: University of Wisconsin Press.

Ganzach, Y. (1997). Misleading interaction and curvilinear terms. *Psychological Methods, 2,* 235–247.

Gill, J. (2001). *Generalized linear models: A unified approach.* Thousand Oaks, CA: Sage.

Goza, F., and DeMaris, A. (2003). Unemployment transitions among Brazilians in the United States and Canada. *International Migration, 41,* 127–152.

Graybill, F. A. (1976). *Theory and application of the linear model.* Boston: Duxbury Press.

Greene, W. H. (1994). Accounting for excess zeros and sample selection in Poisson and negative binomial regression models. Working Paper EC-94-10, Department of Economics, New York University.

Greene, W. H. (1998). *LIMDEP version 7.0 user's manual* (rev. ed.). Plainview, NY: Econometric Software.

Greene, W. H. (2003). *Econometric analysis* (5th ed.). Upper Saddle River, NJ: Prentice Hall.

Grogger, J. T., and Carson, R. T. (1991). Models for truncated counts. *Journal of Applied Econometrics, 6,* 225–238.

Guo, G. (1993). Event-history analysis for left-truncated data. In P. V. Marsden (Ed.), *Sociological methodology 1993* (pp. 217–243). Cambridge, MA: Blackwell.

Hachen, D. S., Jr. (1988). The competing risks model: A method for analyzing processes with multiple types of events. *Sociological Methods & Research, 17,* 21–54.

Hadi, A. S., and Ling, R. F. (1998). Some cautionary notes on the use of principal components regression. *American Statistician, 52,* 15–19.

Hamerle, A. (1991). On the treatment of interrupted spells and initial conditions in event history analysis. *Sociological Methods & Research, 19,* 388–414.

Hanuschek, E. A., and Jackson, J. E. (1977). *Statistical methods for social scientists.* New York: Academic Press.

Hardy, M. A. (1993). *Regression with dummy variables.* Newbury Park, CA: Sage.

Hauck, W. W., and Donner, A. (1977). Wald's test as applied to hypotheses in logit analysis. *Journal of the American Statistical Association, 72,* 851–853.

Heckman, J. J. (1979). Sample selection bias as a specification error. *Econometrica, 47,* 153–161.

Hill, D. H., Axinn, W. G., and Thornton, A. (1993). Competing hazards with shared unmeasured risk factors. In P. V. Marsden (Ed.), *Sociological methodology 1993* (pp. 245–277). Cambridge, MA: Blackwell.

Hoel, P. G., Port, S. C., and Stone, C. J. (1971). *Introduction to probability theory.* Boston: Houghton Mifflin.

Holland, P. W. (1986). Statistics and causal inference. *Journal of the American Statistical Association, 81,* 945–960.

Holland, B. S., and Copenhaver, M. D. (1988). Improved Bonferroni-type multiple testing procedures. *Psychological Bulletin, 104,* 145–149.

Holm, S. (1979). A simple sequentially rejective multiple test procedure. *Scandinavian Journal of Statistics, 6,* 65–70.

Hosmer, D. W., and Lemeshow, S. (1999). *Applied survival analysis: Regression modeling of time to event data.* New York: Wiley.

Hosmer, D. W., and Lemeshow, S. (2000). *Applied logistic regression* (2nd ed.). New York: Wiley.

Johnson, M. P. (1995). Patriarchal terrorism and common couple violence: Two forms of violence against women. *Journal of Marriage and the Family, 57,* 283–294.

Jolliffe, I. T. (1986). *Principal component analysis.* New York: Springer-Verlag.

Kalbfleisch, J. D., and Prentice, R. L. (1980). *The statistical analysis of failure time data.* New York: Wiley.

Kalmuss, D. (1984). The intergenerational transmission of marital aggression. *Journal of Marriage and the Family, 46,* 11–19.

Kerlinger, F. N. (1986). *Foundations of behavioral research* (3rd ed.). New York: Holt, Rinehart and Winston.

King, G. (1988). Statistical models for political science event counts: Bias in conventional procedures and evidence for the exponential Poisson regression model. *American Journal of Political Science, 32,* 838–863.

King, G. (1989). Event count models for international relations: Generalizations and applications. *International Studies Quarterly, 33,* 123–147.

Klein, J. P., and Moeschberger, M. L. (1997). *Survival analysis: Techniques for censored and truncated data.* New York: Springer-Verlag.

Knoke, D., Bohrnstedt, G. W., and Mee, A. P. (2002). *Statistics for social data analysis* (4th ed.). Itasca, IL: Peacock.

Korn, E. L., and Simon, R. (1991). Explained residual variation, explained risk, and goodness of fit. *American Statistician*, *45*, 201–206.

Kramar, A., Faraggi, D., Fortune, A., and Reiser, B. (2001). mROC: A computer program for combining tumour markers in predicting disease states. *Computer Methods and Programs in Biomedicine*, *66*, 199–207.

Laitila, T. (1993). A pseudo-R^2 measure for limited and qualitative dependent variable models. *Journal of Econometrics*, *56*, 341–356.

Lamb, K. A., Lee, G. R., and DeMaris, A. (2003). Union formation and depression: Selection and relationship effects. *Journal of Marriage and Family*, *65*, 953–962.

Lambert, D. (1992). Zero-inflated Poisson regression, with an application to defects in manufacturing. *Technometrics*, *34*, 1–14.

Land, K. C., McCall, P. L., and Nagin, D. S. (1996). A comparison of Poisson, negative binomial, and semiparametric mixed Poisson regression models with empirical applications to criminal careers data. *Sociological Methods & Research*, *24*, 387–442.

Land, K. C., Nagin, D. S., and McCall, P. L. (2001). Discrete-time hazard regression models with hidden heterogeneity: The semiparametric mixed Poisson regression approach. *Sociological Methods & Research*, *29*, 342–373.

Lennon, M. C., and Rosenfield, S. (1994). Relative fairness and the division of housework: The importance of options. *American Journal of Sociology*, *100*, 506–531.

Leonard, K. E., and Senchak, M. (1996). Prospective prediction of husband marital aggression within newlywed couples. *Journal of Abnormal Psychology*, *105*, 369–380.

Lieberson, S. (1997). The big broad issues in society and social history. In V. R. McKim and S. P. Turner (Eds.), *Causality in crisis? Statistical methods and the search for causal knowledge in the social sciences* (pp. 359–385). Notre Dame, IN: University of Notre Dame Press.

Lin, D. Y., and Wei, L. J. (1989). The robust inference for the Cox proportional hazards model. *Journal of the American Statistical Association*, *84*, 1074–1078.

Lipsitz, S. R., Fitzmaurice, G. M., and Molenberghs, G. (1996). Goodness-of-fit tests for ordinal response regression models. *Applied Statistics*, *45*, 175–190.

Long, J. S. (1983). *Covariance structure models: An introduction to LISREL.* Beverly Hills, CA: Sage.

Long, J. S. (1990). The origins of sex differences in science. *Social Forces*, *68*, 1297–1315.

Long, J. S. (1997). *Regression models for categorical and limited dependent variables.* Thousand Oaks, CA: Sage.

Long, J. S., Allison, P. D., and McGinnis, R. (1993). Rank advancement in academic careers: Sex differences and the effects of productivity. *American Sociological Review*, *58*, 703–722.

Longmore, M. A., and DeMaris, A. (1997). Perceived inequity and depression in intimate relationships: The moderating effect of self-esteem. *Social Psychology Quarterly*, *60*, 172–184.

MacDonald, W. L., and DeMaris, A. (2002). Stepfather–stepchild relationship quality: The stepfather's demand for conformity and the biological father's involvement. *Journal of Family Issues*, *23*, 121–137.

Maddala, G. S. (1983). *Limited-dependent and qualitative variables in econometrics.* Cambridge: Cambridge University Press.

May, S., and Hosmer, D. W. (1998). A simplified method of calculating an overall goodness-of-fit test for the Cox proportional hazards model. *Lifetime Data Analysis, 4,* 109–120.

McCallum, R. C., and Mar, C. M. (1995). Distinguishing between moderator and quadratic effects in multiple regression. *Psychological Bulletin, 118,* 405–421.

McClelland, G. H., and Judd, C. M. (1993). Statistical difficulties of detecting interactions and moderator effects. *Psychological Bulletin, 114,* 376–390.

McClendon, M. J. (1994). *Multiple regression and causal analysis.* Itasca, IL: Peacock.

McCullagh, P., and Nelder, J. A. (1989). *Generalized linear models* (2nd ed.). New York: Chapman & Hall.

McDonald, J. F., and Moffitt, R. A. (1980). The uses of tobit analysis. *Review of Economics and Statistics, 62,* 318–321.

McGilchrist, C. A., and Aisbett, C. W. (1991). Regression with frailty in survival analysis. *Biometrics, 47,* 461–466.

McKelvey, R. D., and Zavoina, W. (1975). A statistical model for the analysis of ordinal dependent variables. *Journal of Mathematical Sociology, 4,* 103–120.

McKim, V. R., and Turner, S. P. (1997). *Causality in crisis? Statistical methods and the search for causal knowledge in the social sciences.* Notre Dame, IN: University of Notre Dame Press.

Miles-Doan, R. (1998). Violence between spouses and intimates: Does neighborhood context matter? *Social Forces, 77,* 623–645.

Miller, A. (1954). *The crucible.* New York: Bantam Books.

Mirowsky, J. (1985). Depression and marital power: An equity model. *American Journal of Sociology, 91,* 557–592.

Mirowsky, J., and Hu, P. (1996). Physical impairment and the diminishing effects of income. *Social Forces, 74,* 1073–1096.

Mirowsky, J., and Ross, C. E. (1984). Components of depressed mood in married men and women: The Center for Epidemiological Studies depression scale. *American Journal of Epidemiology, 119,* 997–1004.

Mittlboeck, M., and Schemper, M. (1996). Explained variation in logistic regression. *Statistics in Medicine, 15,* 1987–1997.

Morse, B. J. (1995). Beyond the conflict tactics scale: Assessing gender differences in partner violence. *Violence and Victims, 10,* 251–272.

Mullahy, J. (1986). Specification and testing of some modified count data models. *Journal of Econometrics, 33,* 341–365.

Munem, M. A., and Foulis, D. J. (1984). *Calculus with analytic geometry* (2nd ed.). New York: Worth.

Myers, R. H. (1986). *Classical and modern regression with applications.* Boston: Duxbury.

Nagelkerke, N. J. D. (1991). A note on a general definition of the coefficient of determination. *Biometrika, 78,* 691–692.

Nagler, J. (1994). Scobit: An alternative estimator to logit and probit. *American Journal of Political Science, 38,* 230–255.

Nelder, J. A., and Wedderburn, R. W. M. (1972). Generalized linear models. *Journal of the Royal Statistical Society, 135,* 370–385.

Neter, J., Wasserman, W., and Kutner, M. H. (1985). *Applied linear statistical models* (2nd ed.). Homewood, IL: Irwin.

Overton, R. C. (2001). Moderated multiple regression for interactions involving categorical variables: A statistical control for heterogeneous variance across two groups. *Psychological Methods*, *6*, 218–233.

Pearl, J. (1998). Graphs, causality, and structural equation models. *Sociological Methods & Research*, *27*, 226–284.

Petersen, T. (1985). A comment on presenting results from logit and probit models. *American Sociological Review*, *50*, 130–131.

Pindyck, R. S., and Rubinfeld, D. L. (1981). *Econometric models and economic forecasts* (2nd ed.). New York: McGraw-Hill.

Pregibon, B. (1981). Logistic regression diagnostics. *Annals of Statistics*, *9*, 705–724.

Rao, K. V., and DeMaris, A. (1995). Coital frequency among married and cohabiting couples in the United States. *Journal of Biosocial Science*, *27*, 135–150.

Ratkowsky, D. A. (1990). *Handbook of nonlinear regression models.* New York: Marcel Dekker.

Rubin, D. B. (1974). Estimating causal effects of treatments in randomized and nonrandomized studies. *Journal of Educational Psychology*, *66*, 688–701.

Sampson, R. J., Raudenbusch, S. W., and Earls, F. (1997). Neighborhoods and violent crime: A multilevel study of collective efficacy. *Science*, *277*, 918–924.

Sanchez, L., Manning, W. D., and Smock, P. J. (1998). Sex-specialized or collaborative mate selection? Union transitions among cohabitors. *Social Science Research*, *27*, 280–304.

Santner, T. J., and Duffy, D. E. (1986). A note on A. Albert and J. A. Anderson's conditions for the existence of maximum likelihood estimates in logistic regression models. *Biometrika*, *73*, 755–758.

Schaefer, R. L. (1986). Alternative estimators in logistic regression when the data are collinear. *Journal of Statistical Computing and Simulation*, *25*, 75–91.

Schemper, M., and Stare, J. (1996). Explained variation in survival analysis. *Statistics in Medicine*, *15*, 1999–2012.

Searle, S. R. (1982). *Matrix algebra useful for statistics.* New York: Wiley.

Shapiro, S. S., and Wilk, M. B. (1965). An analysis of variance test for normality (complete samples). *Biometrika*, *52*, 591–611.

Sherman, L. W., Smith, D. A., Schmidt, J. D., and Rogan, D. P. (1992). Crime, punishment, and stake in conformity: Legal and informal control of domestic violence. *American Sociological Review*, *57*, 680–690.

Singer, J. D., and Willett, J. B. (1993). It's about time: Using discrete-time survival analysis to study duration and the timing of events. *Journal of Educational Statistics*, *18*, 155–195.

Smith, D. A., and Brame, R. (2003). Tobit models in social science research: Some limitations and a more general alternative. *Sociological Methods & Research*, *31*, 364–388.

Sobel, M. E. (1998). Causal inference in statistical models of the process of socioeconomic achievement: A case study. *Sociological Methods & Research*, *27*, 318–348.

Spector, L., and Mazzeo, M. (1980). Probit analysis and economic education. *Journal of Economic Education*, *11*, 37–44.

Stets, J. E. (1991). Cohabiting and marital aggression: The role of social isolation. *Journal of Marriage and the Family*, *53*, 669–680.

Stevens, J. (1986). *Applied multivariate statistics for the social sciences.* Hillsdale, NJ: Erlbaum.

Stolzenberg, R. M., and Relles, D. A. (1990). Theory testing in a world of constrained research design: The significance of Heckman's censored sampling bias correction for nonexperimental research. *Sociological Methods & Research, 18,* 395–415.

Stolzenberg, R. M., and Relles, D. A. (1997). Tools for intuition about sample selection bias and its correction. *American Sociological Review, 62,* 494–507.

Straus, M. A. (1979). Measuring intrafamily conflict and violence: The conflict tactics (CT) scales. *Journal of Marriage and the Family, 41,* 75–88.

Sweet, J. A., Bumpass, L. L., and Call, V. (1988). *The design and content of the National Survey of Families and Households.* Madison, WI: University of Wisconsin, Center for Demography and Ecology.

Teachman, J. (2003). Childhood living arrangements and the formation of coresidential unions. *Journal of Marriage and Family, 65,* 507–524.

Theil, H. (1970). On the estimation of relationships involving qualitative variables. *American Journal of Sociology, 76,* 103–154.

Tjaden, P., and Thoennes, N. (1999). *Violence and threats of violence against women and men in the United States, 1994–1996* [computer file]. ICPSR Version, 2566. Denver CO: Center for Policy Research [producer]. Ann Arbor, MI: Inter-University Consortium for Political and Social Research [distributor].

Vuong, Q. H. (1989). Likelihood ratio tests for model selection and non-nested hypotheses. *Econometrica, 57,* 307–333.

Walster, E., Walster, G. W., and Berscheid, E. (1978). *Equity: Theory and research.* Boston: Allyn & Bacon.

Walton, J., and Ragin, C. (1990). Global and national sources of political protest: Third world responses to the debt crisis. *American Sociological Review, 55,* 876–890.

White, H. (1980). A heteroskedasticity-consistent covariance matrix estimator and a direct test for heteroskedasticity. *Econometrica, 48,* 817–838.

Winship, C., and Radbill, L. (1994). Sampling weights and regression analysis. *Sociological Methods & Research, 23,* 230–257.

Wooldridge, J. M. (2000). *Introductory econometrics: A modern approach.* Belmont, CA: South-Western.

Wu, Z., and Balakrishnan, T. R. (1995). Dissolution of premarital cohabitation in Canada. *Demography, 32,* 521–532.

Yamaguchi, K. (1991). *Event history analysis.* Newbury Park, CA: Sage.

Index

Regression with Social Data: Modeling Continuous and Limited Response Variables,
By Alfred DeMaris
ISBN 0-471-22337-9   Copyright © 2004 John Wiley & Sons, Inc.

WILEY SERIES IN PROBABILITY AND STATISTICS

ESTABLISHED BY WALTER A. SHEWHART AND SAMUEL S. WILKS

The *Wiley Series in Probability and Statistics* is well established and authoritative. It covers many topics of current research interest in both pure and applied statistics and probability theory. Written by leading statisticians and institutions, the titles span both state-of-the-art developments in the field and classical methods.

Reflecting the wide range of current research in statistics, the series encompasses applied, methodological and theoretical statistics, ranging from applications and new techniques made possible by advances in computerized practice to rigorous treatment of theoretical approaches.

This series provides essential and invaluable reading for all statisticians, whether in academia, industry, government, or research.

*Now available in a lower priced paperback edition in the Wiley Classics Library.

*Now available in a lower priced paperback edition in the Wiley Classics Library.

*Now available in a lower priced paperback edition in the Wiley Classics Library.

*Now available in a lower priced paperback edition in the Wiley Classics Library.

MASON, GUNST, and HESS · Statistical Design and Analysis of Experiments with Applications to Engineering and Science, *Second Edition*

McCULLOCH and SEARLE · Generalized, Linear, and Mixed Models

McFADDEN · Management of Data in Clinical Trials

McLACHLAN · Discriminant Analysis and Statistical Pattern Recognition

McLACHLAN and KRISHNAN · The EM Algorithm and Extensions

McLACHLAN and PEEL · Finite Mixture Models

McNEIL · Epidemiological Research Methods

MEEKER and ESCOBAR · Statistical Methods for Reliability Data

MEERSCHAERT and SCHEFFLER · Limit Distributions for Sums of Independent Random Vectors: Heavy Tails in Theory and Practice

MICKEY, DUNN, and CLARK · Applied Statistics: Analysis of Variance and Regression, *Third Edition*

*MILLER · Survival Analysis, *Second Edition*

MONTGOMERY, PECK, and VINING · Introduction to Linear Regression Analysis, *Third Edition*

MORGENTHALER and TUKEY · Configural Polysampling: A Route to Practical Robustness

MUIRHEAD · Aspects of Multivariate Statistical Theory

MULLER and STOYAN · Comparison Methods for Stochastic Models and Risks

MURRAY · X-STAT 2.0 Statistical Experimentation, Design Data Analysis, and Nonlinear Optimization

MURTHY, XIE, and JIANG · Weibull Models

MYERS and MONTGOMERY · Response Surface Methodology: Process and Product Optimization Using Designed Experiments, *Second Edition*

MYERS, MONTGOMERY, and VINING · Generalized Linear Models. With Applications in Engineering and the Sciences

NELSON · Accelerated Testing, Statistical Models, Test Plans, and Data Analyses

NELSON · Applied Life Data Analysis

NEWMAN · Biostatistical Methods in Epidemiology

OCHI · Applied Probability and Stochastic Processes in Engineering and Physical Sciences

OKABE, BOOTS, SUGIHARA, and CHIU · Spatial Tesselations: Concepts and Applications of Voronoi Diagrams, *Second Edition*

OLIVER and SMITH · Influence Diagrams, Belief Nets and Decision Analysis

PALTA · Quantitative Methods in Population Health: Extensions of Ordinary Regressions

PANKRATZ · Forecasting with Dynamic Regression Models

PANKRATZ · Forecasting with Univariate Box-Jenkins Models: Concepts and Cases

*PARZEN · Modern Probability Theory and Its Applications

PEÑA, TIAO, and TSAY · A Course in Time Series Analysis

PIANTADOSI · Clinical Trials: A Methodologic Perspective

PORT · Theoretical Probability for Applications

POURAHMADI · Foundations of Time Series Analysis and Prediction Theory

PRESS · Bayesian Statistics: Principles, Models, and Applications

PRESS · Subjective and Objective Bayesian Statistics, *Second Edition*

PRESS and TANUR · The Subjectivity of Scientists and the Bayesian Approach

PUKELSHEIM · Optimal Experimental Design

PURI, VILAPLANA, and WERTZ · New Perspectives in Theoretical and Applied Statistics

PUTERMAN · Markov Decision Processes: Discrete Stochastic Dynamic Programming

*RAO · Linear Statistical Inference and Its Applications, *Second Edition*

RAUSAND and HØYLAND · System Reliability Theory: Models, Statistical Methods, and Applications, *Second Edition*

RENCHER · Linear Models in Statistics

RENCHER · Methods of Multivariate Analysis, *Second Edition*
RENCHER · Multivariate Statistical Inference with Applications
RIPLEY · Spatial Statistics
RIPLEY · Stochastic Simulation
ROBINSON · Practical Strategies for Experimenting
ROHATGI and SALEH · An Introduction to Probability and Statistics, *Second Edition*
ROLSKI, SCHMIDLI, SCHMIDT, and TEUGELS · Stochastic Processes for Insurance and Finance
ROSENBERGER and LACHIN · Randomization in Clinical Trials: Theory and Practice
ROSS · Introduction to Probability and Statistics for Engineers and Scientists
ROUSSEEUW and LEROY · Robust Regression and Outlier Detection
RUBIN · Multiple Imputation for Nonresponse in Surveys
RUBINSTEIN · Simulation and the Monte Carlo Method
RUBINSTEIN and MELAMED · Modern Simulation and Modeling
RYAN · Modern Regression Methods
RYAN · Statistical Methods for Quality Improvement, *Second Edition*
SALTELLI, CHAN, and SCOTT (editors) · Sensitivity Analysis
* SCHEFFE · The Analysis of Variance
SCHIMEK · Smoothing and Regression: Approaches, Computation, and Application
SCHOTT · Matrix Analysis for Statistics
SCHOUTENS · Levy Processes in Finance: Pricing Financial Derivatives
SCHUSS · Theory and Applications of Stochastic Differential Equations
SCOTT · Multivariate Density Estimation: Theory, Practice, and Visualization
* SEARLE · Linear Models
SEARLE · Linear Models for Unbalanced Data
SEARLE · Matrix Algebra Useful for Statistics
SEARLE, CASELLA, and McCULLOCH · Variance Components
SEARLE and WILLETT · Matrix Algebra for Applied Economics
SEBER and LEE · Linear Regression Analysis, *Second Edition*
SEBER · Multivariate Observations
SEBER and WILD · Nonlinear Regression
SENNOTT · Stochastic Dynamic Programming and the Control of Queueing Systems
* SERFLING · Approximation Theorems of Mathematical Statistics
SHAFER and VOVK · Probability and Finance: It's Only a Game!
SMALL and McLEISH · Hilbert Space Methods in Probability and Statistical Inference
SRIVASTAVA · Methods of Multivariate Statistics
STAPLETON · Linear Statistical Models
STAUDTE and SHEATHER · Robust Estimation and Testing
STOYAN, KENDALL, and MECKE · Stochastic Geometry and Its Applications, *Second Edition*
STOYAN and STOYAN · Fractals, Random Shapes and Point Fields: Methods of Geometrical Statistics
STYAN · The Collected Papers of T. W. Anderson: 1943–1985
SUTTON, ABRAMS, JONES, SHELDON, and SONG · Methods for Meta-Analysis in Medical Research
TANAKA · Time Series Analysis: Nonstationary and Noninvertible Distribution Theory
THOMPSON · Empirical Model Building
THOMPSON · Sampling, *Second Edition*
THOMPSON · Simulation: A Modeler's Approach
THOMPSON and SEBER · Adaptive Sampling
THOMPSON, WILLIAMS, and FINDLAY · Models for Investors in Real World Markets
TIAO, BISGAARD, HILL, PEÑA, and STIGLER (editors) · Box on Quality and Discovery: with Design, Control, and Robustness
TIERNEY · LISP-STAT: An Object-Oriented Environment for Statistical Computing and Dynamic Graphics

*Now available in a lower priced paperback edition in the Wiley Classics Library.